Cosmology and Controversy

Cosmology and Controversy

THE HISTORICAL DEVELOPMENT OF

TWO THEORIES OF THE UNIVERSE

HELGE KRAGH

PRINCETON UNIVERSITY PRESS

PRINCETON, NEW JERSEY

Library of Congress Cataloging-in-Publication Data

Kragh, Helge, 1944–
Cosmology and controversy : the historical development of
two theories of the universe / Helge Kragh.
p.   cm.
Includes bibliographical references and index.
ISBN 0-691-02623-8 (cloth : alk. paper)
1. Cosmology—History.   I.Title.
QB981.K73   1996   523.1—dc20   96-5612   CIP

This book has been composed in Times Roman

Princeton University Press books are printed on acid-free paper and meet the guidelines for
permanence and durability of the Committee on Production Guidelines for Book
Longevity of the Council on Library Resources

Printed in the United States of America by Princeton Academic Press

1   2   3   4   5   6   7   8   9   10

TO MIKKEL AND LINE, AND THE MEMORY OF PJEVS

Cosmologists are often in error, but never in doubt.

(L. Landau)

The less one knows about the universe, the easier it is to explain.

(L. Brunschvicg)

CONTENTS

As AN OBJECT of speculation and philosophical thought, often integrated with religious and mythical ideas, cosmology is as old as humankind. The practice of this kind of cosmology—dealing with the view of the world in the widest possible sense—may even be said to be so thoroughly integrated wih the basic mental characteristics of the human race that it defines us as being human. However, it is not this kind of cosmology with which the present work is concerned. Although the scientific study of the universe cannot perhaps be entirely separated from the other sense of "cosmology"—with the vague meaning of world view—it is important to be aware of the dual meaning of the term "cosmology" and the differences between the two meanings. The present work is concerned solely with cosmology in the sense of the scientific study of the universe at large. This scientific cosmology, which deals not only with what happens to be the observed part of the universe, but with the hypothetical world as a whole, is distinctly a modern branch of knowledge. In the relatively brief period from about 1920 to 1970, cosmology changed dramatically, first of all as a result of the insight gained from Einstein's general theory of relativity. The subject became established as a science, not only removing itself from religion and philosophy (although not completely), but also developing from a predominantly mathematical to a physical science.

In spite of cosmology's amazing development in this century and the strong scientific and public interest in the new science of the universe, only very little is known of how this development took place. Since the days of Eddington and Jeans, cosmology has been a favorite area of popularization, a branch of science literature that has exploded during the last ten to twenty years. Most of this literature, usually written by astronomers, physicists, and science journalists, concentrates on the most recent developments and has very little historical perspective. To the extent it includes a historical perspective, it is often distorted and unreliable. The perspective of many astronomers and physicists is encapsulated by the two Russian astrophysicists Yakov Zel'dovich and Igor Novikov, who in the introduction to their *Structure and Evolution of the Universe* (English edition of 1983) wrote that "the history of the Universe is infinitely more interesting than the history of the study of the Universe." Needless to say, perhaps, this is not the view of the present work. It is undoubtedly a view shared by many scientists, but it is also a thoroughly unhistorical statement which fails to recognize the fundamental difference between the physical "history" of the universe and the human history of the ways in which the knowledge of the universe has developed. I find the latter kind of history at least as interesting as the first kind. Furthermore, I find it difficult to separate the scientific quasi history of the universe from the real kind of history, which deals with the ideas, theories, and observations that changed our world view.

It is up to the historians of science to provide a more detailed and critical account of the development of cosmology, but this rich area has been subjected to historical scrutiny in only a few cases, and then mostly for the period before 1940. For reasons that are unclear, postwar cosmology has been almost ignored by modern historians of science (as well as by historians of modern science, should there be any difference). This neglect is all the more puzzling because of the fact that cosmology—in the older, more restricted sense—has traditionally been a central part of the history of the physical sciences. Hundreds or thousands of scholarly works have been devoted to ancient cosmology, to Dante's poetic vision, to Copernicus's revolution, and to Kepler's cosmological work. The result is that we know far more of how the heliocentric system of the "world" came into existence than of the emergence of the big-bang idea in modern cosmology. And yet it would be difficult to argue that the twentieth-century picture of the evolution of the universe is less of an intellectual achievement, or is less revolutionary, than the pictures constructed by Ptolemy, Copernicus, and Kepler. It serves no purpose, and has little meaning, to compare the importance of Copernicus with that of Lemaître, or to compare Tycho Brahe with Fred Hoyle, but it seems to me that the difference in the amount of historical scholarship devoted to the two periods in no way reflects their relative importance.

The present work does not pretend to be a complete history of the development of modern cosmology. To write such a work would be a formidable task, possibly beyond the power of a single author and certainly beyond mine. The reason is that the development of modern cosmology is exceedingly complex and difficult, both technically and conceptually. A proper understanding would require constant attention to areas outside scientific cosmology, such as its philosophical and religious contexts (but also contexts of politics, ideology, and technology); but even within the more narrow scientific limits, the development has been characterized by a confusing variety of approaches and competing theories. Many of these were based on the theory of general relativity, or modifications thereof, but there have also been important developments unrelated or opposed to relativistic cosmology. Since this book is not an attempt to give a comprehensive review of the history of cosmology, many of the less important theories are either ignored or only briefly mentioned. On the other hand, I refer to a substantial part of the papers published in cosmology in the period 1940–65, and it is my hope that the present work, in spite of its obvious weaknesses and lack of completeness, may serve as a starting point for further, more detailed, and more scholarly works in modern cosmology. Should this catalyzing aim be fulfilled, the book has served an important purpose.

I have written the book with a diverse audience in mind. One group of readers I hope to attract are astronomers, physicists, and other scientists working with, or teaching, topics related to cosmology. Maybe I will be able to convince them that the study of the universe has a history which is as rich and interesting as that of the universe. However, this is not a history mainly aimed at scientists. I believe I have something new to tell and that the book will interest also historians of science and ideas. Last but not least I have endeavored to shape the book in a way that makes it attractive to the general reader interested in how the world picture of modern science has emerged. Although not particularly popular, and at times quite demanding, it is my hope that the book will not circulate in academic circles only. I can do no better than quote Dennis Sciama, who in an interview of 1978 said: "None of us can understand why there is a Universe at all, why anything should exist; that's the ultimate question. But while we cannot answer that question, we can at least make progress with the next simpler one, of what the Universe as a whole is like. Everybody must care about that one way or another, more or less. A few of us devote our time to find out, supported financially and spiritually by the whole community. Therefore it's a responsibility to report back to the community the results of our findings or our musings." Much the same can be said about the historians and philosophers who are able to devote their time to finding out what the scientists do.

The plan of the book is built around two grand and persistent themes in post-1920 cosmology, namely, the stationary and the evolutionary universe; or rather, a universe of infinite age and a universe with a beginning in time. These two themes can be followed far back in time, but my concern is only with the modern development where they have been discussed as scientific hypotheses. That this was at all possible was the result of, on the one hand, Einstein's theory of general relativity, and, on the other, progress in observational astronomy. This background is introduced in chapter 1. After the expansion of the universe had been recognized in 1930, most astronomers and physicists accepted that the entire universe is evolving in time, and at the end of that decade the evolution was often interpreted as having started from a superdense state of matter that somehow "exploded." Chapter 2 examines how this idea of a big-bang universe came into existence, first of all through the work of Georges Lemaître. His version of relativistic big-bang cosmology did not attract much interest, however, and it was only after George Gamow had developed it further, along his own lines, that the foundation of modern big-bang cosmology was laid. The development in the United States in the period 1940–53, which is the subject of chapter 3, differed in many respects from the kind of cosmology traditionally cultivated; in particular, Gamow and his coworkers considered the early universe a huge nuclear-physical laboratory and in this way provided cosmology with a valuable content of physics. This turn of cosmology from a mathematical to a physical science is a leading theme in the book. "Physical cosmology" is often seen as a characteristic feature of post-1965 development, but in fact both the program and its partial realization go much farther back in time.

The attempt to present cosmology in terms of physics and not merely mathematics (in conjunction with astronomical data) was not restricted to the big bang evolution program of Gamow and coworkers. It also played a leading role in an entirely different tradition in prewar cosmology, where the universe was considered to be stationary and with a perpetual exchange of energy between matter and radiation. This kind of nonmathematical cosmology, cultivated in particular by William MacMillan and Walther Nernst, is not well known. It is detailed in section 4.1 and there presented as an early version of steady-state cosmology. The emergence of the postwar steady-state theory of Fred Hoyle, Hermann Bondi, and Tommy Gold is the subject of the remainder of chapter 4. A considerable part of the book is concerned with this theory, which is followed from its birth to its death some thirty years later. The reason for devoting so much space to this wrong theory is twofold: first, it is an interesting theory, which is not well known and is often misrepresented in later literature; second, it was of great importance in advancing cosmological knowledge in general and in the process which led to the later, standard big-bang theory. Indeed, the historical development of our present world view cannot be understood without understanding the steady-state theory and the role this theory played in the controversy in the 1950s and 1960s.

This controversy is the book's central topic. The general aspects of the controversy are discussed in considerable detail in chapter 5, which deals mostly with the extrascientific parts of it (philosophical and religious aspects), and not least with the heated debate about continual creation of matter. The attempts to settle the matter by means of observations are discussed in chapter 6. In spite of the very different views of scientists connected with the steady-state program and those favoring an evolutionary big bang world, all involved scientists agreed that the controversy would have to be decided by ordinary scientific methods, namely, comparison of theory with observation. Until about 1960 no clear verdict resulted from the observational tests, but in the following years the picture changed drastically and five years later the relativistic big-bang theory emerged as the victor in the controversy, if not undisputably. At that time the steady-state theory was dying, but not dead. It continued to be developed by Hoyle and a few other astronomers, but in versions that differed from the earlier theory and that failed to attract much interest. The decline of the steady-state theory is the subject of chapter 7, which also contains a brief outline of some of the later developments. However, I have made no attempt to cover the development after the late 1960s. Much information about the last three decades of cosmology can be found in the popular literature and in scientific review articles. To include this development also would require a new and rather different book, as well as an author with a better knowledge of modern cosmology than I have.

The present work has little to say about observational techniques and astronomical instruments, although it can be argued that it was in fact advances in technology that led to the termination of the cosmological controversy. Like all historical works, mine is selective, and I have given high priority to theory rather than experiment. My excuse is that most of the debate took place within a theoretical or conceptual context. The importance of observations was always admitted, but the role played by the technical details of instruments was subordinate to that of theory. I am fully aware that I might have put more emphasis on the observational and instrumental aspects, and that such a change in emphasis might have led to a picture of the cosmological controversy somewhat different from the one presented here. Future scholarship will show how different.

I began to think of this work several years ago, when I was asked by Norriss Hetherington to write a couple of historical review articles for the *Encyclopedia of Cosmology*. It was only then that I realized the sad state of affairs in the historiography of modern cosmology, a state that compares most unfavorably with the situation in, for example, quantum theory and theoretical physics in general. During the early phase of the work I was supported by the Danish Research Council of Humanities in a project dealing with a very different subject (namely, the history of Danish technology). Part of the later work was done at the Dibner Institute for the History of Science and Technology, where I spent the fall term of 1994 and was offered excellent working conditions. I acknowledge a traveling grant from the American Institute of Physics, which allowed me to study taped interviews and other sources at the Center for History of Physics. I am grateful also to a number of people who have commented on parts of the manuscript, provided me with materials, or otherwise discussed the development of modern cosmology with me. They include Sylvan S. Schweber, Christian Klixbüll Jørgensen, Robert Corby Hovis, Benjamin Martin, Manuel Doncel, Jes Madsen, Norriss Hetherington, and Ernan McMullin. Trevor Lipscombe of Princeton University Press has been not only a source of constant encouragement, but also of great help in the production of the final version of the manuscript. I appreciate his kindness and competence. I am particularly grateful to the scientists who have been actively involved in the development I analyze and who have responded to my letters with information I could not otherwise have obtained. My thanks to Ralph Alpher, Hermann Bondi, Tommy Gold, Carl Friedrich von Weizsäcker, Felix Pirani, William Davidson, Robert Herman, William Bonnor, Martin Harwit, Jayant Narlikar, Roger Tayler, Wolfgang Rindler, Igor Novikov, Jim Peebles, Fred Hoyle, and the late Karl Popper for their kind cooperation.

Some of the sections rely upon, or are extended versions of, earlier published articles. I am grateful to the publishers of the relevant journals (*Centaurus* and the *Journal for the History of Astronomy*) for permission to use the material. For permission to quote from archival material I thank the American Institute of Physics (Sources for History of Modern Astrophysics), the Jewish National and University Library (Albert Einstein Archives), the Harvard University Archives (Shapley correspondence), and the Niels Bohr Archive, Copenhagen (Archive for History of Quantum Physics). Other permissions to quote material or reprint illustrations have kindly been granted by F. Hoyle, J. Peebles, T. Gold, R. A. Alpher, and M. Harwit.

Helge Kragh
Oslo, Norway, 1995

Cosmology and Controversy

Background: From Einstein to Hubble

Before Einstein

Cosmology is not, of course, a child of the twentieth century. Concern with the structure and evolution of the world as a whole goes back to time immemorial in the form of mythical and religious conceptions. With the rise of modern science in the seventeenth century the field became the object of natural philosophers, who tried to understand the universe in terms of the new mechanical world picture associated with the great Newton. But the "universe" was at that time a much more limited concept than it became later on, and so-called cosmology (or cosmogony) often dealt with the solar system alone. There were exceptions, however, one of them being the philosopher Immanuel Kant, who in 1755 offered a prophetic and brilliantly argued model of the entire world. In his *Universal Natural History and Theory of the Heavens*, Kant conjectured the existence of other subuniverses outside the Milky Way. Kant's universe was hierarchic and infinite in size as well as time, a conclusion he characteristically obtained from philosophical and theological arguments. A good example of his reasoning is this:

> We come no nearer the infinitude of the creative power of God, if we enclose the space of its revelation within a sphere described with the radius of the Milky Way, than if we were to limit it to a ball an inch in diameter. All that is finite, whatever has limits and a definite relation to unity, is equally far removed from the infinite. Now, it would be absurd to represent the Deity as passing into action with an infinitely small part of His potency, and to think of His Infinite Power—the storehouse of a true immensity of natures and worlds—as inactive, and as shut up eternally in a state not being exercised. . . . For this reason the field of the revelation of the Divine attributes is as infinite as these attributes themselves. Eternity is not sufficient to embrace the manifestations of the Supreme Being, if it is not combined with the infinitude of space.[1]

The argument may sound terribly old fashioned and unscientific, but, as we shall see in later chapters, arguments of an essentially similar kind could also be found in cosmology two hundred years later. Kant believed that the universe, although stable ("which is the mark of the choice of God"), was also in continual evolution on a large scale. Worlds of the size of our Milky Way would decay and disappear, but elsewhere in the infinite universe creation would go on and restore a grand equilibrium. "Nature," wrote Kant, "even in the region where it decays and grows old, advances unexhausted through new scenes, and, at the other boundary of creation in the space of the unformed crude matter, moves on with steady steps, carrying out the plan of the Divine revelation, in order to fill eternity, as well as all the regions of space, with her wonders."[2]

During the nineteenth century the static clockwork universe of Newtonian mechanics was replaced with an evolutionary worldview. It now became accepted that the world has not always been the same, but is the result of a natural evolution from some previous state probably very different from the present one. Because of the evolution of the world, the future is different from the past—the universe acquired a history. The evolutionary worldview was greatly stimulated by Darwinism, but even before Darwin it was suggested as a consequence of the nebular hypothesis of Laplace and William Herschel. According to this hypothesis some of the observed nebulae were protostellar clouds that would eventually condense and form stars and planets. Although the credibility of the nebular hypothesis diminished with the resolution into separate stars of some of the nebulae thought to be clouds of hot gas, the hypothesis continued to enjoy general respect. The Victorian conception of the universe was, in a sense, evolutionary, but the evolution was restricted to the constituents of the universe and did not, as in the world models of the twentieth century, cover the universe in its entirety.[3]

There was, in fact, much uncertainty about the meaning of the term "universe" and the possibility of obtaining scientific knowledge of its constituent bodies. To many scientists, cosmology continued to be confined to the solar system. They received philosophical support from Auguste Comte, the French founder of positivism, who about 1840 concluded that "As for those innumerable stars scattered in the sky, they have scarcely any interest for astronomy other than as markers in our observations." Comte believed that, whereas positive, empirical knowledge could be attained with regard to the solar system, the nature of the stars would forever be beyond scientific insight. Two decades before Bunsen and Kirchhoff started the spectroscopic revolution, Comte claimed confidently that "we can never by any means investigate their [the stars'] chemical composition or mineralogical structure."[4] Fortunately the astronomers, chemists, and physicists declined to let their scientific imagination be limited by Comte's argument.

Another important stimulus for nineteenth-century cosmological thought was the new science of thermodynamics, which from its very beginning was discussed in a cosmological context. William Thomson, later Lord Kelvin, was one of the pioneers of thermodynamics and in many ways representative of the widespread tendency to extrapolate the new laws of physics far beyond the realm of the laboratory. At the Liverpool meeting of the British Association of the Advancement of Science in 1854 he gave a sweeping survey of his cosmological ideas. Inviting his listeners to trace backward in time the actions of the laws of physics, he reasoned that, since the dissipation of energy means a universal change from potential to kinetic and thermal energy, "we find that a time must have been when the earth, with no sun to illuminate it, the other bodies known to us as planets, and the other countless smaller planetary

masses at present seen as the zodiacal light, must have been indefinitely remote from one another and from all other solids in space." Thomson further
speculated that the source of the mechanical energy in the universe might be
sought in "some finite epoch [with] a state of matter derivable from no antecedent by natural laws." However, such an origin of matter and motion, mechan-
ically unexplainable and different from any known process, contradicted
Thomson's sense of both causality and uniformitarianism. "Although we
can conceive of such a state of all matter," he wrote, "yet we have no indications whatever of natural instances of it, and in the present state of science we
may look for mechanical antecedents to every natural state of matter which
we either know or can conceive at any past epoch however remote."[5] A century later the same kind of question of the legitimacy of assuming a "state of
matter derivable from no antecedent by natural laws" was to be heatedly discussed in the controversy between the big-bang and the steady-state theories
of the universe.

In spite of spirited attempts such as those of Kant and Thomson, a scientific
study of the universe became a possibility only with the advances in observational astronomy in the nineteenth century. The great telescopes revealed the
existence of numerous nebulous stellar systems scattered around and provided, together with advances in spectroscopic studies, a new picture of the
universe. It was a grand and exciting picture, but one which was not easily
understood in terms of physical theory. Although thermodynamics occasionally entered cosmological discussions, the theoretical framework of the astronomers built on Newtonian gravitation theory, if sometimes modified in
order to make better sense of the astronomical data. At the time of the First
World War there existed a mathematically sophisticated Newtonian theory of
the universe, and fundamental problems—such as whether the universe is finite or infinite—were discussed among some astronomers and physicists.
However, not only were the observations hopelessly inadequate for deciding
such problems, so was the theoretical framework. Key terms such as "cosmology" and "universe" still had a narrow meaning, in most cases referring to the
objects making up the Milky Way. Whether the other nebulae were located
inside or outside the Milky Way was a matter of some debate, but the general
view was that everything visible in the heavens belonged to our galaxy. As far
as the material content of the world was concerned, it was assumed to be
limited to the area occupied by the Milky Way system. Writing in 1890, the
English astronomer Agnes Clerke summed up the prevailing view as follows:

No competent thinker, with the whole of the available evidence before him, can now,
it is safe to say, maintain any single nebula to be a star system of coordinate rank
with the Milky Way. A practical certainty has been attained that the entire contents,
stellar and nebular, of the sphere belong to one mighty aggregation, and stand in
ordered mutual relations within the limits of one all-embracing scheme—all-embracing, that is to say, so far as our capacities of knowledge extend. With the infinite
possibilities beyond, science has no concern.[6]

The concern of twentieth-century cosmology was, of course, exactly with "the infinite possibilities beyond." We cannot in the present context do justice to pre-Einsteinian cosmology, and it must suffice to note that although scientific cosmology existed well before 1917, it was a kind of cosmology substantially different from the one that emerged in the 1920s.[7]

Albert Einstein did not invent cosmology, but he put it on an entirely new and, as it turned out, extremely fruitful basis. It is generally agreed that the seeds of a revolution in theoretical cosmology were planted when Einstein completed his general theory of relativity in the fall of 1915. On 25 November he read to the Prussian Academy of Sciences the final communication, which contained a consistent set of gravitational equations. One and a half years later, in a paper announced on 8 February 1917, Einstein took the revolutionary step of exploring the consequences of his new theory for no less than the entire universe.[8] The new, relativistic theory of the universe had conceptual roots far back in time, especially in problems discussed by Newton in a famous correspondence with the Reverend Richard Bentley in 1692–93.[9] Newton considered the universe as an infinite container with an infinite number of stars, but in that case it seemed impossible to define the gravitational force acting upon a body in a definite way. Later scientists sought to resolve the dilemma by keeping to Newton's idea of an infinite space, but including a modification of his law of gravitation. In the mid-1890s two German theoreticians, Carl von Neumann and Hugo Seeliger, suggested independently that the amount of matter in the spatially infinite universe was finite. Although this led to a well-defined gravitational force, it also led to a universe which would seem to collapse under the influence of gravitation (as realized by Newton). To avoid this consequence Neumann and Seeliger proposed to change Newton's law of gravitation, which states that the attractive force between two mass points is proportional to their masses and inversely proportional to the distance between them. Instead of the familiar $F = Gmm'/r^2$, where G is Newton's constant of gravitation, they suggested the modified law $F = (Gmm'/r^2)$ $\exp(-\Lambda r)$, where the extra factor is close to 1 ($\Lambda r \approx 0$) for distances not extremely large. Incidentally, the suggested law of force is of the same kind as the law of nuclear forces proposed in Hideki Yukawa's meson theory in 1935. The Neumann-Seeliger proposal amounted to changing Poisson's equation for the gravitational potential φ from $\Delta\varphi = 4\pi G\rho$ to $\Delta\varphi - \Lambda\varphi = 4\pi G\rho$. Here ρ is the mass density and the symbol Δ denotes the Laplace differential operator. If the mass density is known as a function of the space coordinates, the gravitational potential can be calculated. Seeliger, an accomplished mathematician, studied the statistics of star counts and concluded in 1911 that the density of stars decreased rapidly to zero at a distance of about 8000 light years from the earth. This kind of limited sidereal universe received strong support from the Dutch astronomer Jacobus C. Kapteyn, who from a series of works in the 1910s was led to believe that the visible universe was essentially identical with the Milky Way. The Kapteyn universe was ellipsoidal, with the density of stars decreasing gradually with the distance from the center and with the major axis measuring about 16 kpc or 50 000 light years. A universe of the kind pictured by Kapteyn enjoyed widespread acceptance in the 1910s. Although Einstein did not endorse it explicitly, the model formed the background for his work of 1917.

The Einstein World

When Einstein attacked the cosmological problem, he was much aware of the Newtonian anomaly and earlier attempts to solve it, such as that of Neumann and Seeliger. He wrote: "I shall conduct the reader over the road that I have myself travelled, rather a rough and winding road, because otherwise I cannot hope that he will take much interest in the result at the end of the journey. The conclusion I shall arrive at is that the field equations of gravitation which I have championed hitherto still need a slight modification, so that on the basis of the general theory of relativity those fundamental difficulties may be avoided . . . as confronting the Newtonian theory."[10]

That the road to the cosmological theory had been rough and winding, an intellectual tour de force, was also what Einstein wrote to his friend, the Dutch physicist Paul Ehrenfest. In early February 1917 Einstein told him that the work had exposed him "to the danger of being confined in a madhouse."[11] The conceptual problem which Einstein faced was essentially the same as that Newton had struggled with, namely, to formulate boundary conditions for an infinite space. In December 1916 he argued in a letter to his friend Michele Besso that a homogeneous, symmetrical distribution of matter throughout all of infinite space would not be sufficient to produce the stable universe that both he and Besso presupposed. "Only the closedness of the universe can get rid of this dilemma," he wrote, and added that his new idea was "one of great scientific significance [and] not a product of my imagination."[12] Einstein's solution was to circumvent the problem, which he could do by conceiving the universe as a spatially closed continuum in accordance with his general theory of relativity: "If it were possible to regard the universe as a continuum which is *finite (closed) with respect to its spatial dimension*, we should have no need at all of any such boundary conditions. We shall proceed to show that both the general postulate of relativity and the fact of the small stellar velocities are compatible with the hypothesis of a spatially finite universe; though certainly, in order to carry through this idea, we need a generalizing modification of the field equations of gravitation."[13] Einstein thus assumed the universe to be a spatially closed continuum, "spherical" in four dimensions. This model is also referred to as Einstein's "cylinder" world: with two of the spatial dimensions suppressed, the model universe can be pictured as a cylinder where the radius represents the space and the axis the time coordinate. Einstein was also, and naturally so, guided by the available empirical evidence. This suggested that the universe was indeed spatially finite, that it was static, and that it contained a finite amount of matter. In order to keep to what he and most astronomers considered convincing observational evidence, Einstein was led to the following picture: "The curvature of space is variable in time and place, according to the distribution of matter, but we may roughly approximate to it by means of a spherical space. At any rate, this view is logically consistent, and from the standpoint of the general theory of relativity lies nearest at hand; whether, from the standpoint of present astronomical knowledge, it is tenable, will not here be discussed."[14]

Apart from being influenced by the existing discussion of Newtonian cosmology, Einstein was also motivated by the ideas of the famous Austrian physicist and philosopher Ernst Mach. According to Mach's principle (proposed in the 1880s), the laws of mechanics, including the law of inertia, should be seen as purely relational, namely, relative to the universe as a whole. Einstein's version of the principle was rather different; he tended to understand it in the sense that the space-time metric is determined by the masses of the universe, and thus that the local dynamics is conditioned by the universe at large.[15] In general, Mach's principle is interpreted as the assumption that local inertial frames are determined by some average of the motion of the distant celestial objects. Originally Einstein believed that his relativistic theory of cosmology embodied Mach's principle, but in his later years he concluded that the principle could not be harmonized with the general theory of relativity.

In his 1917 paper, "Kosmologische Betrachtungen zur allgemeinen Relativitätstheorie" (cosmological considerations concerning the general theory of relativity), Einstein dressed the ideas mentioned above in mathematical formulas. He began with the gravitational field equations that he had derived in 1915, namely,

$$R_{mn} - \frac{1}{2} g_{mn} R = - \kappa T_{mn}. \tag{1.1}$$

Mathematically, the quantities with double indices are tensors, and, since the indices refer to the four coordinates of space-time, the tensor equation comprises ten second-order differential equations ($R_{mn} = R_{nm}$, etc.; $n, m = 0,1,2,3$). The physical meaning of this equation is that it relates the geometry of space-time (left side) to its physical content (right side). The quantity R_{mn} denotes the Ricci curvature tensor, and R is a curvature invariant derived from R_{mn}. The components of the metric fundamental tensor g_{mn} are functions of the coordinates in the sense that they specify the geometry, the invariant distance ds between two neighboring points in space-time being $ds^2 = g_{mn} dx^m dx^n$, where there is a summation over the values of the indices m and n. In general, the expression for ds^2 thus consists of ten quadratic terms. With a particular set of values for the g coefficients it reduces to the Minkowski line element of special relativity, $ds^2 = c^2 dt^2 - (dx^2 + dy^2 + dz^2)$ where the space part is the ordinary Euclidean space. The constant κ is a quantity that is related to the Newtonian constant of gravitation G by $c^2 \kappa = 8\pi G/c^2$, where c is the velocity of light.[16] Finally, T_{mn} is the energy-momentum (or energy-stress) tensor, which represents various sources of energy and momentum, including pressure and electrical charges. Conservation of energy and momentum is guaranteed by the zero divergence of the left-hand side of equation (1.1).

The field equations (1.1) relate different ontological entities, respectively geometrical and physical. Originally Einstein found this very satisfactory, but he soon changed his mind and came to consider the structure of the field equations a hindrance for the geometrization of physics that he dreamed of. Referring to equation (1.1), he wrote in 1936: "It is similar to a building, one wing of which is made of fine marble (left part of the equation), but the other wing of which is built of low grade wood (right side of equation). The phenomenological representation of matter is, in fact, only a crude substitute for a representation which would correspond to all known properties of matter."[17]

In order to secure a universe static in time, Einstein was led to an important change in his original equations, the "generalizing modification" referred to in the quotation above. His change consisted in adding a term proportional to the metrical tensor. The factor of proportionality soon became known as the cosmological constant, and is always denoted by a Greek lambda, either λ or Λ. With this change, the fundamental equations read

$$ R_{mn} - \frac{1}{2} g_{mn} R - \Lambda g_{mn} = -\kappa T_{mn}. \tag{1.2} $$

Einstein admitted in his 1917 paper that the introduction of the cosmological constant "is not justified by our actual knowledge of gravitation," i.e., that it had an ad hoc character, but he found it "necessary for the purpose of making a quasi-static distribution of matter." The value of the cosmological constant was (and still is) unknown, but in order for the equations to agree with planetary motions it had to be very small: the equations $R_{mn} = 0$, obtained from equations (1.1) for empty space, were known to agree with observations within the solar system and thus the Λ term had to be exceedingly small not to spoil the agreement. Furthermore, the dimension of the cosmological constant is that of the inverse square of a distance. It is helpful to think of the constant as a term which introduces a cosmic repulsion proportional to the distance, negligible at small distances but increasingly important at very large distances. In this picture the evolution of the universe is determined by the competition between the repulsive Λ force and the attractive force of Newtonian gravitation. In Einstein's static universe the two forces are in balance. The cosmological constant in Einstein's theory was essentially the same quantity as that appearing in the earlier Neumann-Seeliger theory, an analogy between classical and relativistic theory admitted by Einstein.

With regard to the cosmological constant, it is important to be aware that the logical structure and principal physical meaning of the field equations remain unchanged. Equation (1.2) is covariant and satisfies conservation of energy and momentum, just as does equation (1.1). Although the cosmological constant later came to be seen as suspect and ad hoc, it played an important and natural role in Einstein's cosmological theory.[18] He saw it as justified in particular by its connection to the mean density of matter in the closed universe. In a letter to Besso of August 1918, Einstein explained:

> Either the universe has a centre, has a vanishing density everywhere, empty at infinity where all the thermal energy is gradually lost as radiation; or, all the points are equivalent on the average, and the mean density is everywhere the same. In either case, one needs a hypothetical constant Λ, which specifies the particular mean density of matter consistent with equilibrium. One perceives at once that the second possibility is more satisfactory, especially since it implies a finite size for the universe. Since the universe is unique, there is no essential difference between considering Λ as a constant which is peculiar to a law of nature or as a constant of integration.[19]

The cosmological constant was introduced in order to maintain a static universe in accordance with observationally based belief, but this does not necessarily mean that it blocked Einstein from predicting the dynamic universe inherent in his theory. The lack of recognition of dynamic solutions, i.e., solutions where the radius of curvature varies with time, was not caused by the presence of the cosmological constant. This is illustrated by the fact that both the first models of the expanding universe, due to Friedmann and Lemaître, arose from considerations operating with a nonzero cosmological constant. Although not necessarily ad hoc, the Λ term certainly makes the field equations a bit more complicated and a bit less appealing. It was such aesthetic considerations that at an early stage made Einstein doubt if the cosmological constant could be justified. In 1919 he described the introduction of the constant as "gravely detrimental to the formal beauty of the theory."[20] However, at that time he could see no alternative, and it was twelve years before he decided that the introduction of the cosmological constant had been a mistake.

The model of the universe derived by Einstein, and, he believed, the only one consonant with his equations, was homogeneously filled with dilute matter and could thus be ascribed a definite mass. He found the cosmological constant to be related to the density (ρ), volume, and mass of his closed universe by the relations

$$\left.\begin{array}{rcl} \Lambda &=& \frac{1}{2}\kappa\rho = R^{-2} \\[4pt] V &=& 2\pi^2 R^3 \\[4pt] M &=& 2\pi^2 \rho R^3 = \pi^2 \left(\frac{32}{\kappa^3 \rho}\right)^{1/2} \end{array}\right\} . \tag{1.3}$$

The first of these relations reflects the key message of relativistic cosmology, that the density of matter determines the radius of curvature of the universe. We also notice that although the field equations do not specify the sign of the cosmological constant, in the Einstein world it is necessarily positive. As to the numerical values of these quantities, claimed to follow from fundamental physical theory, Einstein was understandably cautious. In his letter to Besso of December 1916 he erroneously suggested that $R \approx 10^7$ light years, based on the much too high estimate of $\rho \approx 10^{-22}$ g·cm^{-3}, and he believed that the most distant visible stars were some 10^4 light years away from the earth.[21] In a letter to de Sitter of 12 March 1917 he repeated the suggestion, but he wisely decided not to publish it.[22] What mattered was that the universe, according to Einstein, had a constant positive curvature and thus was spatially closed. Temporally it was infinite, the radius R having the same value at all time.

Non-Static Universes

Einstein's cosmological theory appeared in the midst of the Great War and was therefore unknown to most scientists outside Germany. But Einstein was in contact with the eminent Dutch astronomer Willem de Sitter, a scientist equally at home with astronomical observations and advanced mathematical analysis (the Netherlands remained neutral during the war). Forty-five years old, de Sitter was at the time professor of astronomy at the University of Leiden and best known for his work in celestial mechanics. As a foreign member of the Royal Astronomical Society, de Sitter undertook to give an account of the new theory which, via Eddington (who was secretary of the society), thus became known to the English-speaking world. In fact, de Sitter did more than that, for he extended Einstein's analysis by showing that, contrary to Einstein's contention, the static, matter-filled model was not the only solution to the cosmological field equations.[23] In his third report to the Royal Astronomical Society of 1917, he drew attention to what subsequently became known as the de Sitter solution. (De Sitter modestly termed it solution B, to distinguish it from Einstein's solution A.) De Sitter's model was an empty universe, with $\rho = 0$ and $\Lambda = 3/R^2$, spatially closed in spite of its lack of matter. As in the Einstein model, the pressure was taken to be zero. De Sitter showed that his universe had a peculiar property: if a particle was introduced at a distance r from the origin of a system of coordinates, it would appear as moving away from the observer. It would acquire an outward acceleration $\Lambda c^2 r/3$, corresponding to the repulsive Λ force in the Newtonian analogy. De Sitter summarized the difference betwen the two models as follows: "In A there is a world-matter, with which the whole world is filled, and this can be in a state of equilibrium without any internal stresses or pressures if it is entirely homogeneous and at rest. In B there may, or may not, be matter, but if there is more than one material particle these cannot be at rest, and if the whole world were filled homogeneously with matter this could not be at rest without internal pressure or stress."[24] Most interestingly, de Sitter's model indicated that, as a result of the metric, clocks would appear to run more slowly the farther away they were from the observer. Since frequencies are inverse time-intervals, light would therefore be expected to be received with a smaller frequency, being more redshifted the larger the distance between source and observer. As de Sitter wrote: "The lines in the spectra of very distant stars or nebulae must therefore be systematically displaced towards the red, giving rise to a spurious positive radial velocity." Notice that de Sitter described the velocity as "spurious": it was not a real velocity caused by the expansion of space, but an effect of the particular space-time metric describing this kind of universe. In spite of the redshift built into de Sitter's model, it was, like Einstein's, a static model.

By 1917 there then existed two general-relativistic models of the universe, Einstein's and de Sitter's. Einstein accepted de Sitter's mathematics, but found the new model objectionable from a physical point of view, among other reasons because it contained a world horizon, a distance from beyond which light signals cannot reach the observer.[25] In a letter to de Sitter, Einstein argued that the new solution "does not correspond to any physical possibility."[26] According to Einstein's conception of Mach's principle, the curved space-time (the g_{mn} field) was determined or generated by matter and so de Sitter's empty universe at first made no sense to him. Although de Sitter's model, being devoid of matter, may seem very artificial, it soon became a popular foundation for further theoretical work. It was seen as particularly interesting because of its connection with the observations of apparently systematic redshifts which were reported at the time. Its lack of material content certainly did not prevent researchers from investigating it as a possible model of the real universe. As de Sitter suggested, although the universe is indeed filled with matter, it is known to be of very low density, perhaps so low that his model would apply as a zero-density approximation.

Whatever the credibility of the Einstein and de Sitter models as candidates for the real structure of the universe, from the early 1920s there developed a minor industry based on these two models. It was predominantly a mathematical industry, with mathematically minded physicists and astronomers analyzing the properties of the two solutions and proposing their own modifications. Among the more important participants in this tradition of mathematical cosmology were the great British astronomer Arthur Eddington; the German-Swiss mathematician Herman Weyl; a Hungarian-German physicist, Cornelius Lanczos; the Belgian astrophysicist Georges Lemaître; and the Americans Howard Robertson and Richard Tolman. The basic aim of these investigations was to determine which of the two relativistic models of the static universe was the most satisfactory. This was done primarily by examining them mathematically, whereas comparison with the meager observational data played a subordinate role. The discussion concerning the models of Einstein and de Sitter has been described as a "controversy," but this is hardly an appropriate name for a restricted scientific debate in which few of the participants were metaphysically committed to either of the views.[27] It was, at any rate, a debate of an entirely different kind from the cosmological controversy that raged in the 1950s.

Whether in Einstein's or de Sitter's version, the idea of treating the entire world by means of the relativistic field equations constituted a revolution in the age-old conception of the universe. It is not much of an exaggeration to claim that Einstein invented a new concept of the universe with his theory of general relativity. The world or universe had traditionally been thought of as those parts within the limits of observation (dependent on telescope technology as that limit was). It now became everything, the totality of events in space and time—and all this governed by a single tensor equation. In spite of (or because of?) its revolutionary nature, the change was accepted by only a small number of physicists, mathematicians, and astronomers. The significance of the new relativistic cosmology is evident in retrospect, but it is not reflected in the astronomical literature of the 1920s.[28] To the majority of astronomers, and of course to most laypersons, Einstein's reconceptualization of the universe was unknown, irrelevant, unintelligible, or objectionable. The eminent French mathematician Emile Borel described a common objection in these terms: "It may seem rather rash indeed to draw conclusions valid for the whole universe from what we can see from the small corner to which we are confined. Who knows that the whole visible universe is not like a drop of water at the surface of the earth? Inhabitants of that drop of water, as small relative to it as we are relative to the Milky Way, could not possibly imagine that beside the drop of water there might be a piece of iron or a living tissue, in which the properties of matter are entirely different."[29] This objection (to which Borel did not subscribe) was far from childish and continued to play a role in the cosmological discussion throughout this century. But Einstein, and those who followed him, decided that if cosmology were to progress—become a science—the objection had to be ignored.

During the course of their work to understand and elaborate the two relativistic world models, some scientists proposed solutions that combined features of Einstein's and de Sitter's models. However, with two notable exceptions, the framework of the tradition was essentially confined to static world models. Even disregarding these exceptions—due to Friedmann in 1922 and Lemaître in 1927—the tendency toward the end of the 1920s was to conclude that neither of the two classical solutions could represent the actual universe. In a formal sense, a nonstatic world model was discussed by Lanczos in 1922. By an ingenious change of coordinates Lanczos found a model in which the radius varies hyperbolically with time [namely, as $R \sim \cosh(ct/R_0)$, where R_0 is a length].[30] Later contributions were independently made by Lemaître in 1925 and Robertson in 1928. This work did not amount to giving up the static universe in a physical sense, however. What these scientists did was to transform de Sitter's line element in such a way that it became nonstatic, i.e., so that one or more of the components of g_{mn} depend on the time coordinate. In this case, the metric could be formally written in the form $ds^2 = c^2 dt^2 - F(t)(dx^2 + dy^2 + dz^2)$, where $F(t)$ is some function of the time parameter.

For example, in his work of 1925 Lemaître introduced another division of space and time than that used by de Sitter and was in this way able to derive a model in which "the radius of space is constant at any place, but it is variable with time."[31] The nonstatic feature thus introduced was expressed by the line element

$$ds^2 = c^2dt^2 - \exp(2ct\sqrt{\Lambda/3})(dx^2 + dy^2 + dz^2). \tag{1.4}$$

To a modern reader this looks very much like a universe expanding exponentially in time, which is the way in which the de Sitter model is understood today. However, everything depends on the way the geometrical symbols are interpreted physically. The fact is that during the 1920s such transformations were not seen as implying any change in the physical interpretation, i.e., as indicating a world in evolution. As Gerald Whitrow later expressed it: "What de Sitter had in fact discovered in 1917 was one of the simplest models of an expanding universe. With a more physically appropriate choice of co-ordinates, as was first shown by G. Lemaître in 1925, and independently by H. P. Robertson in 1928, the metric of the de Sitter universe can be expressed [in the form of] the limiting case of an expanding universe as the mean density everywhere tends to zero. We therefore no longer consider de Sitter's as a static universe, its apparent changelessness being a mathematical fiction."[32]

As early as 1922, Friedmann proved that there are no more static solutions to the field equations than those associated with the names of Einstein and de Sitter (apart from the useless one of special relativity—useless because it does not include gravitation). The proof had no impact, but seven years later Tolman repeated it independently.[33] This, together with the ingrained belief in the static nature of the world, led to a state of crisis in mathematical cosmology. If both the Einstein model and the de Sitter model were inadequate, and if these were the only ones, how could cosmology still be based on general relativity? The alternative of abandoning general relativity and returning to some classical framework was not seriously considered within the relativistic tradition. The "obvious" solution, to search for evolutionary models, had already been published at the time, but was as unknown to most cosmologists as it was unwelcome. The conceptual climate that governed mathematical cosmology was that of a physically static universe, and the scientists engaged in the field tried hard to avoid breaking with the paradigm.

Early Observational Cosmology

Redshifted light from stars or galaxies was not a new discovery, but the phenomenon attained cosmological significance only in the light of de Sitter's work. As early as 1912, Vesto Slipher at the Lowell Observatory had found the first Doppler shift for a spiral nebula—a blueshift for the Andromeda galaxy, indicating a motion toward the sun with the amazing velocity of 300 km·s^{-1}, the highest velocity for a celestial body known at the time.[34] His program soon revealed that Andromeda was probably an exception. There turned out to be a marked preponderance of large redshifts, which Slipher interpreted as the result of recessional velocities, possibly indicating some kind of expansion of the system of nebulae. In 1917 he reported measurements of the radial velocities of twenty-five nebulae of which four were receding with a velocity of more than 1000 km·s^{-1}. Slipher inferred at first that the nebulae receded on the north side of our galaxy and approached on the south side. He therefore suggested a hypothesis of galactic drift according to which the observed radial velocities reflected the motion of the Milky Way relative to the nebulae. Slipher kept to this hypothesis until the early 1920s and did not think of connecting his observations with de Sitter's cosmological hypothesis, of which he may have been unaware.[35]

Eight years later, in 1925, Slipher had found Doppler shifts for forty-five nebulae, forty-one of which were redshifts. At that time de Sitter's prediction of a velocity-distance correlation had disseminated to the community of observational astronomers also. It inspired several astronomers to look for such a correlation, and in general increased interest in de Sitter's model at the expense of Einstein's. In the absence of reliable knowledge of the distances of the nebulae, no clear correlation of a universal nature was found at the time, but with the discovery of Cepheid variables in the Andromeda nebula, and later in other spiral nebulae, indications of a redshift-distance relationship gradually became stronger.

Meanwhile, and largely unrelated to the observational results, a few theoreticians had predicted from de Sitter's theory, or modifications of it, that linear relationships between the redshift and the distance would hold either precisely or as a first approximation. This was first shown, if only indirectly, by Weyl in 1923.[36] The following year Ludwik Silberstein, a Polish-born physicist then staying in England, argued explicitly for a relation of the form $\Delta\lambda/\lambda = \pm\ r/R$, where r is the distance of the source of light and the symbol λ denotes the wavelength.[37] As indicated by the double sign, the formula was supposed to be valid for blue- as well as redshifts. Silberstein claimed that his formula was supported by observations of globular clusters, but his claim relied on exclusion of data that did not agree with his prediction. Most astronomers denied that the available data supported a linear relationship of the type suggested by Silberstein. As a consequence his theory was strongly criticized by several leading astronomers, who ridiculed what they called the "Silberstein effect." Partly as a result of the negative reaction to Silberstein's prediction, redshift-distance relations were for a period regarded with skepticism in the astronomical community.

Another theoretical derivation was offered by Lemaître in his work of 1925 where he discussed a nonstatic de Sitter world and—remarkably at the time—linked it to current observations. "Our treatment evidences this non-statical character of de Sitter's world which gives a possible interpretation of the main receding motion of spiral nebulae," he wrote.[38] The Belgian astrophysicist found the redshift formula $\Delta\lambda/\lambda = r/cT$, where r is the distance between the light source and the observer and T the time measured by the observer. Contrary to Silberstein's result only redshifts followed from his theory. However, in spite of this pleasing result, Lemaître concluded that de Sitter's solution had to be abandoned. He seems at the time vaguely to have recognized the possibility of a third alternative, the expanding universe, but only discussed this explicitly two years later. A nonstatic line element similar to Lemaître's was also suggested by Howard Percy Robertson of Princeton University, who in 1928 showed that it yielded a recession of the galaxies in agreement with observed data. Robertson noticed that "t [in equation (1.4)] appears explicitly in the line element, and consequently natural processes are not reversible." He summarized his investigation as follows: "Although space is unlimited the observable world is not, and objects at an appreciable fraction of its radius should show a residual motion of recession; assuming that the known excess of recessional velocity of spiral nebulae is due to this cause, the radius of the observable universe is found to be 2×10^{27} cm [about 650 Mpc]."[39]

The establishment of an empirical relation between redshift and distance became a reality in 1929 through the work of Edwin Powell Hubble, who greatly extended and reinterpreted the research program of Slipher. Born in 1889, Hubble received his Bachelor of Science degree from the University of Chicago in 1910.[40] After three years in Oxford as a Rhodes scholar, where he studied law and Spanish, he returned to the United States with the intention of practicing law. However, although he passed the bar examination, apparently he never turned his intention into reality. Instead he switched to astronomy, in which field he obtained his Ph.D. in 1917. His astronomical career was interrupted when the United States entered the World War. Three days after having passed his final examination at Yerkes Observatory, Hubble joined the army. With the rank of major he was sent to France in September 1918, but the war ended before he went into combat.

After a stay in England, Hubble joined the staff of the Mount Wilson Observatory and began the series of observations that would make him the best known astronomer of the twentieth century. Determinations of distances to very distant nebulae were crucial to Hubble's success, and in this respect the so-called Cepheid method played a decisive role. Astronomers' raw data is the apparent luminosity L, which is a measure of the light energy received on earth per unit area and unit time (and for all wavelengths, in which case L is called the bolometric luminosity). The apparent luminosity is logarithmically related to the apparent magnitude m by $m = -2.5 \log L + C$, where C is a constant that defines zero magnitude and the base of the logarithm is 10. The absolute luminosity is the total power radiated from the source and is similarly related to the absolute magnitude M, namely, as the apparent magnitude of the object if it were at a distance of 10 pc. From this follows the standard relationship

$$m = M + 5 \log r, \tag{1.5}$$

where the distance r is measured in parsecs. If the absolute magnitude can be determined, the distance is known.

Cepheids are stars whose brightness varies regularly, a phenomenon known since 1786 when the variation of δ Cephei was first noticed. One hundred and twenty-two years later, Henrietta Leavitt, then working in Cambridge, England, observed in the Magellanic Clouds sources with a similar periodic change in brightness. More importantly, she found the more luminous stars to have longer periods, and in 1912 she concluded that there was a linear relationship between the period and the logarithm of the apparent magnitudes.[41] This was a most important discovery because it could be used as a distance indicator: the variable stars in the Magellanic Clouds have approximately the same distance to the earth, which means that it is possible to convert the apparent magnitude to absolute magnitude. Leavitt's discovery allowed her to determine the distances of Cepheid variable stars relative to the Magellanic Clouds, but not their absolute distances. This requires a calibration of the period-luminosity function, that is, a distance to the Magellanic Clouds. Such a calibration was first attempted by the Danish astronomer Ejnar Hertzsprung in 1913 and three years later Harlow Shapley completed Hertzsprung's method. By 1918 a single Cepheid associated with a given object was enough to determine the distance of that object.[42]

In October 1923 Hubble found in the Andromeda nebula sources that exhibited the same variation in brigthness as Cepheid stars. He initially believed that he had observed a nova, but soon decided that it was really a Cepheid. With this insight he immediately used the period-luminosity correlation to determine the distance to about 300 000 pc.[43] A few years earlier the Dutch astronomer Adriaan van Maanen had derived the distance to be only 290 pc! (And recall that the Kapteyn universe was only of the order of magnitude 10 pc.) The universe was growing, and radically so. The very large distance strongly indicated that Andomeda was located outside the Milky Way and for this reason Hubble's result was important ammunition for the still controversial idea of "island universes," i.e., the extragalactic nature of the spiral nebulae already advocated by Kant.[44] After Hubble had obtained more variables and checked his results, the discovery was announced at a meeting of the American Astronomical Society on 1 January 1925. Hubble was not present and his paper was read by the Princeton astrophysicist Henry Norris Russell. At the time of the official discovery of Cepheids in spiral nebulae Hubble's results had been circulating among astronomers for nearly a year.[45] The establishment of the island-universe view was an important milestone in the history of astronomy. It was soon followed by an even more important application of the Cepheid method, this time in a cosmological context.

Hubble first referred to the world models of general relativity in a paper of 1926 where he estimated that there were 9×10^{-18} nebulae per cubic parsec, which he translated to a lower limit of the mean density of the universe of $\rho = 1.5 \times 10^{-31}$ g·cm^{-3}. Applying this result to the Einstein universe by means of eq. (1.3), he arrived at a radius of 27 billion (27×10^9) parsecs, about six hundred times the range of the 100-inch reflector at Mount Wilson.[46] In his work of 1926 he seems to have been aware of only the Einstein world model.

By 1928 Hubble was aware of the theoretical discussions concerning the Einstein and de Sitter models, and he now decided to investigate the form of the redshift-distance correlation or "de Sitter effect," if it existed at all. Whether Hubble was also aware of, and perhaps inspired by, the predictions of Weyl, Lanczos, Lemaître, and Robertson, is less certain. He had good contacts with theoretical astronomers such as Robertson and Tolman, and it is

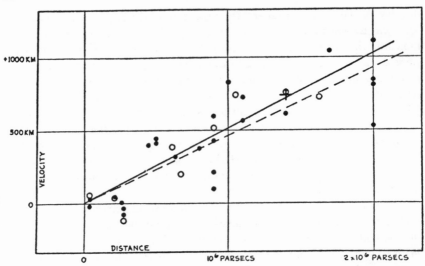

Figure 1.1. Hubble's 1929 plot of the apparent velocities of galaxies as a function of their distances. The two lines, drawn in full and broken, represent two different ways of reducing the data according to the motion of the sun. *Source:* Hubble (1929), p. 172.

possible that he knew about the predictions before his seminal paper of 1929 was published.[47] However, he preferred to present his work as an empirical investigation, and if he was aware of the predictions he did not refer to them or, for that matter, to any other theoretical work except de Sitter's.

Using the new 100-inch telescope at Mount Wilson, Hubble published in 1929 data for forty-six galaxies, of which he believed the distances to twenty-four to be reliably determined.[48] Most of the redshifts were taken from the work of Slipher. Other were new observations made by Milton Humason, who since 1927 had studied the spectra of distant galaxies. Humason had an un-usual career—he has been described as the last highly successful research astronomer to base his career on an eighth-grade education.[49] After having worked as a mule driver on the Sierra Madre trail and been foreman on a Californian ranch, he became janitor at the Mount Wilson Observatory and then, in 1921, advanced to become a spectroscopic observer. In 1929 Huma-son reported the galaxy NGC7619 to have a radial velocity of 3779 km·s^{-1}— twice as large as the previous record—and Hubble used the new value to check the linearity in his sample ranging from −12.7 to −17.7 in absolute magnitude. The result was a confirmation of the linear relationship between redshift and distance that Hubble more or less expected. In a now famous diagram Hubble plotted the velocity of the receding galaxies against their distances, obtaining what he claimed was a reasonably linear correlation (fig-ure 1.1). In fact there was considerable scatter, and the data could just as well have been fitted to a quadratic relation. As to the theoretical significance of his work, Hubble wrote: "The outstanding feature . . . is the possibility that the velocity-distance relation may represent the de Sitter effect, and hence that numerical data may be introduced into discussions of the general curvature of space." In an accompanying paper Humason also suggested the link to de Sitter's theory. The observations would then be explained "both by the apparent slowing-down of light vibrations with distance and by a real tendency of material bodies to scatter in space."[50]

At Harvard, Shapley was not impressed by Hubble's data. He could see no observational evidence in favor of a proportionality between the redshifts and distances of the galaxies and found it premature to conclude either for or against a linear law.[51] A few other astronomers also criticized Hubble's conclusion, but in general it was accepted. With more extended data appearing in 1931,[52] based upon measurements made by Humason, the famous Hubble law was seen as a reality: Out to the unprecedented distance of 32 million parsecs (Mpc), the light from the galaxies arrived redshifted in direct proportion to their magnitude, interpreted as distance (figure 1.2).

The Hubble law is, and was, traditionally expressed as a relation between the distances and the velocities of the galaxies, the latter obtained by interpreting the redshifts as Doppler shifts. In that case,

$$v = Hr, \tag{1.6}$$

where H is a constant of dimension $(\text{time})^{-1}$, usually expressed in the awkward unit $\text{km·s}^{-1}\text{·Mpc}^{-1}$ (Hubble originally used the symbol K for the constant). With this unit, the value of the Hubble constant (as it soon became known) was found to be about 500 in 1929, and was corrected to 558 in 1931. Five years later, in his book *The Realm of the Nebulae*, Hubble used an extended data base to conclude that $H = 526 \text{ km·s}^{-1}\text{·Mpc}^{-1}$. The fastest receding galactic cluster, measured by Humason, was found to have a radial velocity of $42\ 000 \text{ km·s}^{-1}$; this corresponded to a distance of about 80 Mpc, which marked the limit of the Mount Wilson telescope. Whatever the precise value of the recession constant, what Hubble and Humason actually measured were not the velocities of receding galaxies, of course, but their redshifts. These are often expressed in terms of the quantity $z = (\lambda' - \lambda)/\lambda$, where λ' is the received redshifted wavelength and λ is the wavelength as observed in the laboratory. In this case the Hubble law reads $z = Hr/c$. This version has the advantage that it does not refer to the velocity. In fact, the cosmological redshift z is not a Doppler redshift: the cosmological expansion is not a motion of galaxies *through* space, but an expansion *of* space, carrying the galaxies with it. It is only for relatively small velocities that the cosmological redshift coincides with the Doppler redshift and equation (1.6) is valid.

For later purposes it will be useful to rewrite the Hubble law in a form involving magnitudes. From the logarithmic form $\log r = \log(cz) - \log H$ and equation (1.5) it follows that $m = M - 5 \log H + 5 \log(cz) + \text{const}$: a semilogarithmic plot of z against m will give a straight line of the form $\log z = 0.2\ m + \text{const}$. This version of the Hubble law was consistent with the early data, but

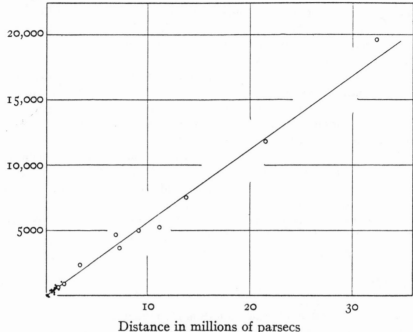

Distance in millions of parsecs

Figure 1.2. The improved Hubble diagram obtained by Hubble and Humason in 1931. It extended to far larger distances and yielded a clearer linear correlation between redshift and distance than in 1929. The slope of the line, the recession constant, was determined to be 558 km·s⁻¹Mpc⁻¹. *Source:* Hubble and Humason (1931), p. 77.

needed to be extended with various empirical correction terms in order to account also for observations of galaxies very far away, where space cannot be assumed to be flat. It proved especially useful in connection with nonstatic cosmologies, and it was in such a context that Hubble introduced it in 1934.

It is sometimes claimed that Hubble discovered the expansion of the universe with his work of 1929. This is a claim that has little merit, although not as little as the strange claim that Hubble also discovered the big bang.[53] The fact is that Hubble tended to ascribe the linear law to a "de Sitter effect," which does not mean that the redshifts of the nebulae are simply caused by their radial velocities. He did present the law in terms of velocities, but it is evident that he referred to "apparent velocities," i.e., redshifts that could conveniently be transformed to velocities by means of the Doppler formula but for which other interpretations might be equally valid. "We [Hubble and Humason] use the term 'apparent' velocities in order to emphasize the empirical features of the correlation," he wrote to de Sitter in 1931. "The interpretation, we feel, should be left to you and the very few others who are competent to discuss the matter with authority."[54] Nowhere in his work of 1929 did he conclude that the galaxies are actually receding from us or otherwise suggest that the universe is expanding. Hubble provided the observational data that led to the proposal and acceptance of the expanding universe, but it was other scientists who first made the proposal.

Furthermore, even after most astronomers had accepted the expanding universe, Hubble chose to emphasize the uncertain observational foundation of the hypothesis. For example, in the mid-1930s he was led to concede that observations might probably be accounted for in terms of nonvelocity redshifts within a static Einstein universe. If observations had to fit an evolving universe, he concluded in 1936, this would be "a curiously small-scale universe" with a "disturbing[ly]" high density. Consequently, he found that "the expanding models are a forced interpretation of the observational results."[55] And in 1953, shortly before his death, he repeated his cautious conclusion, that "it is important that the [Hubble] law be formulated as an empirical relation between observed data."[56] However, in spite of Hubble's cautious attitude, there is no doubt that he personally preferred relativistic expansion models over the nonrelativistic alternatives. The first kind of model agreed with his philosophical outlook and without manipulating his data he did his best to interpret them as favorably as possible with respect to such models.[57]

Although the expanding universe was not yet a reality in 1929, it was in the air. In a popular account of Hubble's work, written in 1929, H. N. Russell asked, "Are the nebulae really flying out in all directions—away from us and therefore from one another—so that the universe of nebulae is expanding without limits into the depths of space?" He admitted that he did not like the scenario: "The best answer that has yet been suggested comes from a peculiar form of the theory of relativity suggested a few years ago by the great Dutch astronomer de Sitter. . . . It would be premature, however, to adopt de Sitter's theory without reservation. The notion that all the galaxies were originally close together is philosophically rather unsatisfactory."[58] Yet the very fact that a question such as Russell's could be asked shows a change in climate. Both theory and observation now indicated that there was something seriously wrong with the static models. Tolman observed that "our assumption of a static line element takes no explicit recognition of any universal evolutionary process which may be going on," and he concluded that "the investigation of non-static line-elements would be very interesting."[59] Similar conclusions were obtained by Robertson, Eddington, and de Sitter. At a meeting of the Royal Astronomical Society on 10 January 1930, Eddington pointed out that, since both the Einstein and the de Sitter solutions had proved inadequate, interest should focus on nonstatic solutions. De Sitter, who was present at the meeting, agreed.[60] Clearly, the expanding universe was ready to be born.

CHAPTER TWO

Lemaître's Fireworks Universe

THE EXPANDING UNIVERSE became a reality during the first half of 1930, when Lemaître's earlier prediction became generally known and Hubble's discovery was interpreted within the new theoretical framework. Although barely noticed at the time, the possibility of an expanding universe had been proposed even before Lemaître's contribution, in work by Friedmann published in 1922. The road to the big bang presupposed, as a matter of logic as well as of historical fact, the notion of an expanding universe. All the same, it is important to keep in mind that the two notions are distinct. Whereas the big bang presupposes an expanding universe (although not necessarily in perpetual expansion), obviously the opposite is not the case. The recognition that the universe is expanding does not logically entail a universe of a finite age, a point shown also by the sequence of historical events: there was considerable resistance to the idea of a finite-age universe among astronomers who otherwise accepted the expansion of the universe.

The Friedmann-Lemaître theory was the first step toward not only the expanding universe but also the idea of a big-bang origin. As will be argued, although the big-bang concept can be read into the works of Friedmann and Lemaître of the 1920s, it was not really there. It was only in 1931 that the first proposal of what can reasonably be called a big-bang theory was made, namely, in Lemaître's hypothesis of what he called the primeval atom. This proposal was not generally accepted during the 1930s, when its originator was largely alone in defending it and attempting to develop it. Yet the notion of an expanding universe won general acceptance, and with it the notion of a world with a finite past. However, world models with a finite age, such as were discussed in that decade, do not necessarily qualify as big-bang models in a physical sense. In this connection we discuss the theories of Milne and Dirac, not only because they are interesting examples of finite-age models, but also because they gave rise to problems and debates that resurfaced after the war, in the controversy between big-bang and steady-state models.

2.1 THE DISCOVERY OF THE EXPANDING UNIVERSE

Friedmann: Mathematics without Physics

In a paper completed in the fall of 1929, Robertson referred in a footnote to some earlier work of A. Friedmann. This was one of the few contemporary references in Western astronomical literature to the Soviet physicist who became so famous later on.[1] Robertson had evidently studied Friedmann, but without recognizing the originality of the work (and Robertson was not the only one to misread Friedmann). Less than a year later, relativists and theoretical astronomers would read the articles in an entirely different light, as the brilliant prediction of the expanding universe and the discovery of a new class of relativistic world models. What were Friedmann's contributions and why was their true nature only recognized so late?

Alexander Alexandrovich Friedmann was born in St. Petersburg (Petrograd, later Leningrad) on 17 June 1888. His father, with the same name, was a ballet dancer and composer. In the local gymnasium Friedmann showed great interest in and talent for mathematics and in 1906 he entered St. Petersburg University to study pure and applied mathematics. The same year he published his first scientific paper. He was influenced in particular by the distinguished mathematician Vladimir Steklov and the Austrian-born theoretical physicist Paul Ehrenfest, who, being married to a Russian, worked in St. Petersburg from 1907 until he went to Leiden in the Netherlands in 1912. After having completed his studies for the master's degree, Friedmann was drawn into meteorology and spent most of a year in Leipzig as a research student under the Norwegian theoretical meteorologist Vilhelm Bjerknes, who had founded an important school there. During the First World War, Friedmann served at the beginning as a soldier at the Austrian front in western Galicia. On 15 February 1915, he wrote in a letter: "My life is fairly quiet apart from such happenings as a shrapnel exploding at a distance of twenty steps, explosion of a detonator of an Austrian bomb at a distance of half a step, when I got off nearly scot-free, and a fall on my face and head resulting in nicking my upper lip and suffering some headaches."[2] Friedmann survived the fighting and in 1916 he was transferred to Kiev's Central Aeronautical Station and the following year to the University of Perm.

After the end of the civil war, Friedmann returned to St. Petersburg, where he worked on hydrodynamics and other subjects in the theoretical department of the Main Geophysical Observatory. During his few years there, he also gave lectures and seminars at the university, the Polytechnical Institute, and the Institute of Railway Engineering, and established himself as the city's leading theoretical physicist. In 1925 he was appointed director of the Main Geophysical Observatory. Although a mathematical physicist, he was also much concerned with observations and practical aspects of geophysics and aeronautics. In the summer of 1925 he went on a daring balloon flight to study the upper atmosphere, reaching the Soviet record altitude of 7400 m. Two months later he died of a disease, probably typhus of the stomach. During his last hours he was in a delirium, which was recorded as follows: "He spoke of students, lectures, recalled his balloon flights, and tried to carry out some calculations. Occasionally it seemed he was giving a lecture."[3]

Friedmann began a serious study of relativity about 1920 and soon mastered the subject. In 1923 he wrote a popular book entitled *The World as Space and Time*, and the following year there appeared a scientific textbook, written together with Vsevelod Frederiks. Only the first volume, dealing with the mathematical foundation, was published, and that with two years' delay because of the shortage of paper. In the theoretical physics seminar at the Physical Institute of Leningrad University, Friedmann and Frederiks reported on the general theory of relativity. Among the members of the seminar was Vladimir Fock, later an outstanding theoretician. In 1963 he reminisced about the reports of Friedmann and Frederiks: "They were in different styles: Frederiks had a deep understanding of the physical side of the theory, but did not like mathematical calculations, while Fridman put the emphasis not on the physics, but on the mathematics. He strove for mathematical rigor and gave great importance to the complete and precise formulation of the initial hypotheses."[4]

In his paper of 1922, Friedmann offered a complete analysis of the solutions of Einstein's cosmological field equations that went beyond the earlier solutions of Einstein and de Sitter as it also included nonstatic solutions.[5] Friedmann did so clearly and explicitly: "The purpose of this note," he wrote, "is firstly to show that the cylindrical [Einstein] and spherical [de Sitter] worlds are special cases of more general assumptions, and secondly to demonstrate the possibility of a world in which the curvature of space is independent of the three spatial coordinates but does depend on time." For closed models he showed that the basic equations governing the time variation of the curvature R (the "size of the universe") can be written

$$\left(\frac{R'}{R}\right)^2 + 2\frac{R''}{R} + \frac{c^2}{R^2} - \Lambda c^2 = 0$$

and

$$3\left(\frac{R'}{R}\right)^2 + 3\frac{c^2}{R^2} - \Lambda c^2 = \kappa\rho c^4,$$

where R' denotes dR/dt and R'' is d^2R/dt^2. By integrating the first equation, he found a class of solutions that, depending on the value of the cosmological constant Λ, includes cyclical universes as well as a family of homogeneously expanding ones. The expanding universe was clearly in Friedmann's work, which also explicitly referred to the finite age of one type of such a universe, originating from a singularity. Referring to what he called "monotonic worlds of the first class"—expanding models—he gave an expression for "the time since [the] beginning of the world" and even wrote about "the creation of the world." He thus introduced into cosmology two concepts of revolutionary importance, the age of the world and the creation of the world. "The time since the creation of the world," he wrote, "is the time which has passed from the moment at which space was [concentrated at] a point ($R = 0$) to the present state ($R = R_0$)." Here we have, for the first time, the idea of an expanding universe originating in a singularity—a big-bang universe. But how much physical significance did Friedmann attach to his words and equations?

It is tempting to read Friedmann's visionary work with modern eyes, but one should be careful not to be misled by later knowledge and put too much emphasis on the words used. The paper was primarily of a mathematical rather than a physical nature, and there was no attempt to connect its findings with astronomical observations. Thus Friedmann did not refer to the redshift measurements at all. This is not surprising in a work of 1922, but he was silent about the problem in his paper of 1924 also. In connection with his discussion of cyclical or oscillatory worlds, he remarked in 1922 that "our knowledge is completely insufficient for a numerical comparison to decide which world is ours." For the sake of illustration he used $\Lambda = 0$ and a mass of the universe equal to 5×10^{21} sun masses, which gives a period of the cyclical world model of about ten billion years. This happens to be in nice accordance with our present estimate, but to Friedmann it was just a numerical example without any deeper meaning. It is impossible to be sure what significance Friedmann attached to his calculations, but the general impression is that he considered them nothing but calculations. The age of the universe appears as a mathematical curiosity, not a possible physical reality. The mathematical character of Friedmann's work is further illustrated by the fact that he indicated a solution with negative density as one of the possible solutions to his equations. That such a solution is physically inadmissible (or at least problematical) does not seem to have interested him.

In his popular book of 1923, published only in Russian, Friedmann elaborated in plain words on his new discovery of world models of variable radius of curvature, among which he seems to have had a preference for the cyclical models. He wrote:

> The variable type of universe represents a great variety of cases; there can be cases of this type when the world's radius of curvature . . . is constantly increasing in time; cases are also possible when the radius of curvature changes periodically: The universe contracts into a point (into nothing) and then again increases its radius from a point up to a certain value, then again, diminishing its radius of curvature, transform itself into a point, etc. This brings to mind what Hindu mythology has to say about cycles of existence, and it also becomes possible to speak about "the creation of the world from nothing," but all this should at present be considered as curious facts which cannot be reliably supported by the inadequate astronomical experimental material.[6]

Whereas in his 1922 paper in *Zeitschrift für Physik* Friedmann only explicitly related the age of the world to cyclical models, in his book he also considered the concept in the case of ever-expanding models. After having repeated the lack of empirical support for such "useless" calculations, he nonetheless felt the tempation "to calculate, out of curiosity, the time which has passed since the moment when the universe was created out of a point up to its present stage." Without revealing the basis for his result, he stated it as "tens of billions of our ordinary years."[7]

If Friedmann seriously believed that his theory allowed for the possibility of our universe being created in a singularity, he did not emphasize or develop the point beyond what has been mentioned. Such a notion, at least formally appearing in Friedmann's work, was at the time not only unwelcome, it was almost unthinkable. For example, in the contemporary work of Lanczos there appeared formally a space-time singularity. Lanczos certainly did not think of it as anything real, but the mere appearance of an initial singularity was widely seen as a blemish. Thus in 1924 Weyl argued that his own version of cosmology "has the great advantage [over Lanczos's] of not introducing a singular initial moment, of conserving the homogeneousness of time."[8] He did not mention Friedmann.

The puzzling thing about Friedmann's work is that it had virtually no impact at all on the development of cosmology.[9] The paper was published in the world's leading journal of physics, was studied by several physicists and astronomers, and even received two public responses from Einstein, at that time at the height of his fame; and yet it was not appreciated or properly understood, meaning that the scientists did not pay attention to the points that were seen later as the very message of the paper. Friedmann was very clear about the expansion of the universe, but to no avail. Neither did it make any difference that the 1922 paper was published in Russian two years later. When Einstein read the paper he first thought that the unknown Russian had committed an error, and he claimed that it actually demonstrated the static nature of relativistic world models. Friedmann read Einstein's comment with surprise, rechecked his calculations, and wrote a letter to Berlin explaining that there was no error on his part. "Bearing in mind the definite interest in the possibility of existence of a non-steady-state universe," Friedmann wrote, "I venture to submit to you my calculations." Because of a heavy travel schedule, Einstein did not read the letter until much later. He was first informed about Friedmann's objections by Yuri Krutkov, a friend and colleague of Friedmann, who happened to be in Leiden at the same time as Einstein, in May 1923. Having discussed the matter with Krutkov, Einstein realized that Friedmann was right. "Petrograd's honour is saved!" Krutkov wrote triumphantly to Friedmann. Einstein publicly retracted his objection, admitting that he had made an error of calculation. "I am convinced that Mr. Friedmann's results are both correct and clarifying," he now wrote. "They show that in addition to the static solutions to the field equations there are time varying solutions with a spatially symmetric structure."[10] So Einstein apparently accepted the solutions corresponding to an evolutionary universe. But he did so in a mathematical sense, a sense according well with the spirit of Friedmann's paper. In his manuscript Einstein originally remarked that of course the evolutionary solutions were of no physical significance—a remark which did not enter the published version.[11]

The fact that Friedmann was satisfied with having saved Petrograd's honor and did not return to his theory of an expanding universe suggests that he may himself have thought about the theory in a similar way. In the summer of 1923 Friedmann visited Berlin. He failed to see Einstein, but visited the two

German astronomers Emmanuel von Pahlen and Erwin Freundlich with whom he discussed the structure of the universe. From Friedmann's account of the meeting one gets the impression that he saw his 1922 theory more as an intellectual game than as a description of the real universe: "Everybody was very impressed by my struggle with Einstein and my eventual victory, it is pleasant for me because of my papers, I shall be able to get them published more easily."[12] That Friedmann did not quite recognize the physical importance of his theory was also the impression of Fock, who recalled: "Fridman more than once said that his task was to indicate the possible solutions of Einstein's equations, and that then the physicists could do what they wished with these solutions."[13]

Most astronomers must have felt the work to be irrelevant because it lacked information about observational consequences. The 1922 paper was listed under "Relativity Theory" in the *Astronomischer Jahresbericht*, the leading abstract journal in astronomy, but only with its title. The following year the journal decided to exclude articles on the theory of relativity judged not to be of astronomical interest. As a result, Friedmann's 1924 paper did not appear in the *Astronomischer Jahresbericht*. To most contemporaries, his theory may have looked as just another exercise in mathematical cosmology and its novelty been unappreciated because of other works on nonstatic line elements that apparently arrived at similar conclusions. In 1928 Frederiks and Anna Schechter, a former research student of Friedmann's, examined optical phenomena in different kinds of cosmological models, including Friedmann's nonstationary system.[14] However, the two Leningrad physicists did not so much as mention that this was an expanding system, and few readers of *Zeitschrift für Physik* would have guessed that it was. As far as Einstein is concerned, his awareness of Friedmann's work certainly did not change his view. In an article for the 1929 edition of the *Encyclopedia Britannica*, he wrote that "nothing certain is known of what the properties of the space-time continuum may be as a whole. Through the general theory of relativity, however, the view that the continuum is infinite in its time-like extent but finite in its space-like extent has gained in probability."[15] There is no trace of an evolutionary universe here.

Friedmann did indeed predict an expanding universe, but not *the* expanding universe. That is, he demonstrated that among the solutions to the cosmological field equations there are some which must be interpreted as a universe in expansion or, generally, evolution. He did not predict or argue that the one and only real universe is of this type. Insofar as Friedmann explicitly referred to the finite age of expanding-universe models, and even to the creation of such a universe, we may see in his work the germ of the big bang. But to credit him with the "discovery" of the big-bang universe would be to go too far. He mentioned it as a possibility among many, without suggesting that this was the possibility realized in nature; and his entire discussion was limited to mathematics, with no attempt to incorporate physics or astronomy. This was the essential difference between Friedmann and Lemaître.

Lemaître, Cosmologist and Priest

The Belgian physicist and astronomer Georges Édouard Lemaître, to whom
we have already referred, was born in Charleroi on 17 July 1894.[16] He started
as a student of engineering at the Catholic University of Louvain, not because
he was particularly interested in becoming an engineer, but because the education might allow him to support the family's frail economy. However, he had
to leave his studies unfinished when the German army invaded Belgium on 4
August 1914. Five days later young Georges volunteered as a soldier in the
Belgian army, where he spent the next fifty-three months. He was engaged in
heavy house-to-house fighting and a witness to the first attack with poison gas
(chlorine) in the history of warfare. In quieter periods, Lemaître found time
to relax by reading classics of physics, including Poincaré's *Lecons sur les
Hypothèses Cosmogonique*. After the war he was awarded a high distinction
(Croix de Guerres avec palmes) for his service. Still in uniform, he started
studies in physics and mathematics at the University of Louvain in early 1919,
and, in 1920, also in theology. Lemaître was ordained a priest in 1923 and in
his later life he continued a career within the clerical establishment parallel
with his scientific career. He was elected a member of the Pontifical Academy
of Science at its establishment in 1936 (when it replaced the Academia dei
Novi Lincei), and from 1960 until his death in 1966 he served as president
for the academy. In his scientific studies, young Lemaître was attracted by
complicated mathematical problems, for which he showed a remarkable talent. His "discovery" of general relativity and Einstein's methods of physics in
the early 1920s made a strong and lasting impression upon him. From this
time Lemaître's ideal of physics was the general theory of relativity, and he
remained throughout his life a devoted, orthodox relativist. In accordance
with Einstein, or what Lemaître read into Einstein, he became a believer in
logical beauty, simplicity, and unity. "Scientific progress," he wrote in private
notes from 1922, "is the discovery of a more and more comprehensive simplicity." And with regard to the possibility of a scientific cosmology: "We
become more and more conscious of the fact that the universe is cogniz-
able. . . . Behind objects which can be touched or looked upon, should be
something hidden."[17]

Lemaître got an opportunity to cultivate his interest in relativity when in
1923–24 he spent a year in Cambridge as a student of Eddington. The British
authority in relativity was impressed by his Belgian postgraduate student
whom he described in a letter as "a very brilliant student, wonderfully quick
and clear-sighted, and of great mathematical ability."[18] Most of the following
year Lemaître spent in the United States, where he worked under Shapley at
the Harvard College Observatory and prepared for a Ph.D. at the Massachu-
setts Institute of Technology. In his application for a fellowship he stated that
his aim was "to get practical knowledge of modern astrophysics works especially in matters connected with the astronomical consequences of the Princi-

ple of Relativity."[19] During this stay Lemaître partially transformed himself from a mathematical physicist into a theoretical astronomer and was exposed to the latest developments in extragalactic astronomy. For example, he recognized at once the cosmological significance of Hubble's discovery of Cepheids in spiral galaxies and visited both Slipher (at the Lowell Observatory) and Hubble (at Mount Wilson) to learn about the most recent measurements of redshifts. This led to his previously mentioned paper on a nonstatic universe with a redshift-distance relation, which in some respects can be seen as anticipating the expanding universe.

Back in Louvain, Lemaître subjected the relativistic world models to a systematic investigation, with the result that he, by and large, reproduced the earlier work of Friedmann, of which he was unaware at the time. However, not only was his work independent of Friedmann's, it also differed from it in some important respects. With a time-dependent space curvature $R(t)$ he found differential equations of the same type as Friedmann's, but including a radiation pressure (he considered the matter pressure to be negligible). In what was the first introduction of thermodynamics in relativistic evolutionary cosmology, he derived the equation for energy conservation, which he wrote as

$$\frac{d\rho}{dt} + 3\frac{R'}{R}\left(\rho + \frac{p}{c^2}\right) = 0$$

with p the radiation pressure. He showed explicitly that the cosmological equations could be satisfied by an expanding universe in which "the radius of the universe increases without limit from an asymptotic value R_0 for $t = -\infty$."[20] As to the value of R_0, which he related to the cosmological constant by Einstein's formula $R_0 = \Lambda^{-1/2}$, Lemaître calculated it from a discussion of data given by Hubble and Gustav Strömberg to be about 2.7×10^8 pc. The present radius of the world was estimated to be about twenty times as large.

The difference between the works of Friedmann and Lemaître lies more in their approach and spirit than in their formal content. Lemaître's was a serious attempt to develop a physically realistic cosmology, not only a mathematical exercise in general relativity. It was clearly a mathematician trained in astronomy who spoke. Whereas words such as galaxy, radiation, and energy did not appear in Friedmann's work, these physical terms were central in Lemaître's discussion. He did not present the expanding universe as just one possibility out of many, but suggested a definite evolutionary model for the real universe. His expanding universe was within the realm of physics and hence required a cause for its explanation, a notion foreign to Friedmann and other researchers. Lemaître suggested tentatively that "the expansion has been set up by the radiation itself," but at the time he was unable to develop this rather obscure suggestion. Another difference was that Lemaître connected his theory with the redshifts of the galaxies, which was a natural continuation of his 1925 work. He explicitly described his model as an expanding one in which the recession of the galaxies cause their received light to be redshifted. "The receding velocities of extragalactic nebulae are a cosmical effect of the expansion of the universe," he wrote. For the redshift he derived the expression $\Delta\lambda/\lambda = v/c = (R'/R)\, r$ with

$$\frac{R'}{R} = \frac{1}{R\sqrt{3}}(1 - 3/y^2 + 2/y^3)^{1/2}$$

and $y = R/R_0$. This gives to a good approximation the linear velocity-distance relation $v = (c/R_0\sqrt{3})\, r$. For the proportionality factor—later the Hubble constant—Lemaître suggested a value of about 625 km·s^{-1}·Mpc^{-1}, which is not very different from the observational result obtained by Hubble two years later. Lemaître based his result on an analysis of data for forty-two galaxies and stressed that the observational material, although subject to great uncertainty, gave support to the hypothesis of a linear velocity-distance relation. The famous Hubble law is clearly in Lemaître's paper. It could as well have been named Lemaitre's law.

It should be emphasized that Lemaître's 1927 model of the world did not have a definite age and that it was far from being a big-bang universe. It evolved gradually from an Einstein universe, the equilibrium state of which was somehow disturbed; in the future it approached asymptotically a de Sitter universe (in its modern interpretation, i.e., exponentially expanding). Since Lemaitre's equations were the same as Friedmann's, the big-bang solution was implicit in his work; but Lemaître did not discuss this possibility which at the time he may have found unphysical. Lemaître was not yet ready to take the crucial step from the expanding to the big-bang universe as implied noncommittally in Friedmann's work. When he became aware of this work later in 1927, in a conversation with Einstein, it does not seem to have made any impact on him. It is of some interest to note that Einstein, as one of the very few relativists, knew about both Friedmann's and Lemaître's work and had actually studied them. His response to Lemaître indicates the same unwillingness to change his position that characterized Einstein's response to Friedmann. He was willing to accept the mathematics, but would have nothing to do with a physically expanding universe, an idea he found "abominable" according to Lemaître's recollection of their encounter.[21]

Lemaître's prediction of an expanding universe suffered the same fate as Friedmann's, but for different reasons. Friedmann was unknown to Western relativists and astronomers, yet his papers were known and, to some extent, discussed. Lemaître, on the other hand, was an insider, a member of the community of astronomers and cosmologists. His paper, contrary to Friedmann's, was addressed to this community and would, one might expect, have been appreciated by astronomers. Lemaître had visited the important centers of astronomy in England and the United States, and was personally acquainted with or had met key persons such as Eddington, Shapley, Slipher, Russell, and

Hubble. All the same, his work of 1927 seems to have been almost completely unknown.[22] There is little doubt that part of the reason was Lemaître's choice of journal, the relatively obscure *Annales Scientifique Bruxelles*. The paper was reprinted later in 1927, but the site of publication, volume 4 of *Publications du Laboratoire d'Astronomie et de Géodésie de l'Université de Louvain*, was not exactly suited for widespread dissemination. One may wonder why he did not think of a more central journal or take other steps to disseminate the news of his theory. The fact is that he did not, and that he did his best, apparently, to let his theory remain unknown.

It would have been natural if he had discussed his new ideas with people in the United States when he stayed there in the spring of 1927 to complete his Ph.D. degree, but apparently he did not. Then in July 1928 Lemaître attended the third General Conference of the International Astronomical Union in Leiden, organized by de Sitter. Surely he would contact de Sitter concerning the expanding universe? As far as we can tell, he did not. It is possible, but not likely, that he was discouraged by Einstein's dismissive response in 1927 and disappointed to discover that he had been preceded by Friedmann. Another, and in my view more likely, explanation is that Lemaître did not care very much for international reputation and that he may have had second thoughts about the soundness of the expanding universe and for that reason did not press the point. When he surveyed the science of cosmology in January 1929 he mentioned Friedmann's work, but he only referred briefly and cautiously to the idea of the expanding universe. In a footnote he pointed out that Friedmann's theory included "most of the notions and results" of his own work of 1927, but this was the only reference to his theory of two years earlier. The vehicle of publication was again a Belgian journal not well known outside the country.[23]

Rediscovery of the Expanding Models, 1930

In any case, things changed dramatically in the early part of 1930. At that time it was recognized that some kind of dynamic universe would have to replace the existing static solutions of Einstein and de Sitter. The climate of the time was nicely described by Eddington a few years later, when he formulated the question as follows: "Shall we put a little motion into Einstein's world of inert matter, or shall we put a little matter into de Sitter's Primum Mobile?"[24] When the news about the meeting of the Royal Astronomical Society of 10 January, where Eddington and de Sitter had advocated some kind of nonstatic universe, reached Louvain, Lemaître finally reacted. Accompanying his letter by one or more copies of his 1927 paper, he wrote to Eddington and reminded him about his theory of 1927, which offered a solution to the dilemma between the Einstein and de Sitter universes. "I had occasion to speak of the matter with Einstein two years ago," Lemaître wrote. "He told me that the theory was right and is all which [needs] to be done, that it was not new but had be[en] considered by Friedman, he made critic[ism]s against which he was obliged to withdraw, but that from the physical point of view it was 'tout à fait abominable.'"[25]

Probably Eddington had earlier received a copy of Lemaître's paper, but either he had not read it or he had not, in 1927, fully understood its importance. Now, three years later, the situation was different and the stage set for an acceptance of the expanding universe. According to George McVittie, at the time a research student of Eddington's and working with him on the stability of the Einstein world: "[I remember] the day when Eddington, rather shamefacedly, showed me a letter from Lemaître which reminded Eddington of the solution to the problem which Lemaître had already given. Eddington confessed that although he had seen Lemaître's paper in 1927 he had forgotten completely about it until that moment. The oversight was quickly remedied by Eddington's letter to *Nature* of 1930 June 7, in which he drew attention to Lemaître's brilliant work of three years before."[26] Eddington also sent, on 19 March, a copy of Lemaître's paper to de Sitter in Leiden, adding on the front page that "This seems a complete answer to the problem we were discussing." He furthermore incorporated the expanding universe in a paper which appeared in May.[27] With Eddington's enthusiastic endorsement, which included an English translation of the 1927 paper,[28] Lemaître's theory was discovered by the international community of astronomers.

Eddington praised Lemaître's "brilliant solution," an evaluation which was shared by de Sitter. In April, de Sitter told Shapley about the new theory which, in the opinion of the Dutch astronomer, was "the true solution, or at least a possible solution, which must be somewhere near the truth."[29] He even found, one year later, that it was "a solution of such simplicity as to make it appear self-evident . . . there cannot be the slightest doubt that Lemaître's theory is essentially true."[30] De Sitter was as quick as Eddington to seize on Lemaître's discovery. In a paper appearing in late May 1930, he gave a brief summary of the new theory, introducing it with the remark: "A non-static solution is contained in a paper by Dr. G. Lemaître, published in 1927, which had failed to attract my notice, but to which my attention was called by Professor Eddington only a few weeks ago."[31] The import of Hubble's data was now clear to de Sitter, who expressed them as the law of expansion $v = (c/2000)r$, where the 2000 is a distance measured in billions of light years. This corresponds to a Hubble constant of 490 $km \cdot s^{-1} \cdot Mpc^{-1}$. One month later de Sitter presented a full investigation of Lemaître's theory which he extended to cover solutions that were not considered by Lemaître in 1927.[32] With the support of Eddington and de Sitter, the Belgian priest suddenly rose to become a celebrated innovator of science. Although Friedmann's work was also rediscovered in the wake of this development, his contributions became to a large extent overshadowed by those of Lemaître and were typically mentioned in footnotes only.

With the "discovery" of the works of Lemaître and Friedmann in 1930, cosmology experienced a paradigmatic shift. It was only now that Hubble's discovery was transformed to become, i.e., interpreted to be, a discovery of the expanding universe. In a formal hypothetical sense, the expansion of the universe can be traced back to Friedmann's work of 1922, and in an observational sense it was supported by Hubble's measurements of 1929. It may even be argued that the expansion had been there all the time since 1917, hidden in Einstein's equations. In any case, it was only with the fusion of theory and observation, as it took place in 1930, that the expanding universe became a reality in a social sense, as a notion widely accepted by the scientific community. With the paradigm shift, works unable to incorporate the idea of universal expansion became obsolete almost overnight. For example, Ludwik Silberstein, the controversial Polish-Italian physicist, published a book on cosmology which appeared in early 1930. Eddington reviewed it shortly after he had become aware of Lemaître's paper and pointed out that the newest development "renders obsolete the contest between Einstein's and de Sitter's cosmogonies" on which Silberstein naturally built.[33]

The interpretation of the observed redshifts in terms of radial velocities was accepted by a majority of scientists, but there were also those who denied it and argued for alternative interpretations which did not involve a universe in expansion. During the 1930s suggestions of such nonvelocity alternatives were common; however, although many astronomers were willing to consider the expanding universe no more than a hypothesis, the alternative schemes failed to gain general support. The first alternative, due to the Bulgarian-born Swiss-American astronomer Fritz Zwicky, came in 1929, shortly after Hubble's paper had appeared.[34] Zwicky proposed a mechanism of gravitational "drag" according to which photons would transfer energy to the intergalactic matter. In that case the frequency ($v = E/h$) would decrease with the distance traversed and then be received redshifted. Zwicky, who developed his hypothesis in several later publications, obtained a relation similar in form to Hubble's and argued that it was in qualitative agreement with the data.

John Stewart, a Princeton physicist, suggested in 1931 another variant of the "tired-light" hypothesis by assuming the frequency to decrease exponentially with the distance.[35] This kind of explanation attracted some interest and was developed in many different versions. Other scientists again, often inspired by Eddington's numerological arguments, suggested that the fundamental constants of nature were not true constants over long periods of time, but might exhibit a secular variation. For example, postulating Planck's constant to decrease in time as $\exp(-Ht)$ allowed Samuel Sambursky to explain the redshifts without recourse to universal expansion.[36] However, these and many other alternatives were generally considered ad hoc and unnecessary.

Although the expanding universe was not without its opponents, many scientists eagerly began to explore the consequences of the new world picture. In England, William McCrea and George McVittie pointed out that Lemaître's theory did not distinguish between contracting and expanding solutions. Why should the equilibrium Einstein world start to expand rather than to contract? Following up on Eddington's idea of instability they investigated the effect of a single condensation and showed that it would produce continual contraction, i.e., cause the universe to collapse.[37] A few months later, McVittie was led to an empirically more satisfactory result by considering an initial state with a large number of randomly located condensations. His analysis now showed that the Einstein world would turn into an indefinitely expanding Lemaître world and that the formation of condensations thus might be the cause of the initial expansion.[38]

It was not only in England that Lemaître's theory caused a reorientation in cosmological research. As soon as Tolman was informed of Lemaître's work (by de Sitter and Eddington) he started a new line of research, systematically investigating evolutionary world models. This work led in 1934 to his seminal *Relativity, Thermodynamics and Cosmology*, possibly the most authoritative and detailed account of theoretical cosmology of the decade. Likewise, Robertson systematized the new relativistic cosmology in an influential article of 1933 and so did the German astronomer Otto Heckmann in 1932.[39] Heckmann, who was born in 1901, was a member of the faculty of the University of Göttingen and was later, in 1942, appointed Director of Hamburg Observatory. Another thorough treatment appeared in a remarkable book by Ernest Williams Barnes, the bishop of Birmingham.[40] Barnes, who was as well versed in physics as in theology, gave a detailed and masterly survey of the latest developments in relativity and cosmology, and paid particular attention to the theory of his fellow physicist-theologian. Lemaître's theory was also subjected to critical analysis by two Bulgarian physicists, G. Maneff and Raschko Zaykoff, who proposed various modifications.[41]

In addition to these scientific developments, the new picture of the world got wide public notice through a number of popular works of which those of Jeans, Eddington, and de Sitter were particularly important.[42] The relativistic, expanding universe entered the pages of the *Times* in May 1932, when several readers expressed the layperson's discomfort with the new and strange cosmology. A curved universe expanding into what? They received replies and lessons from Jeans, who enthusiastically defended the new picture of the world.[43]

Whereas Einstein, the father of relativistic cosmology, had dismissed Lemaître's theory a couple of years earlier, he now accepted the new paradigm. He did so publically in early April 1931, during a visit to Pasadena and the Mount Wilson Observatory. The same year he said a definite goodbye to the cosmological constant, in a paper in which he advocated an oscillatory model (this type of universe is consequently sometimes referred to as a Friedmann-Einstein model).[44] Einstein had been dissatisfied with the cosmological constant for a long time, and now, when he realized that it was unnecessary in evolutionary cosmology, he was pleased to get rid of the term. However, although most cosmologists followed Einstein in putting $\Lambda = 0$, the cosmological constant continued to stay alive, favored by a minority of astronomers, including Eddington and Lemaître. Einstein's 1931 model was of the big bang-type, but he did not comment on this feature (i.e., $R \to 0$ for $t \to t_0$), which he considered troubling and wanted to avoid. By assuming the distribution of matter for very small R to be inhomogeneous he believed that the singularity could be explained away as a mathematical artifact.[45]

The following year, 1932, Einstein collaborated with his friend and former cosmological rival, de Sitter, in suggesting a model of the universe that came to be of considerable importance. Given that the only "directly observed data" were the expansion of the universe and its present mean density, the two scientists noted that observations did not allow derivation of either the sign or the value of the curvature. Therefore, "the question arises whether it is possible to represent the observed facts without introducing a curvature at all."[46] The simple model considered by Einstein and de Sitter had zero curvature and zero pressure, and assumed the cosmological term to be zero as well. It follows that $(R'/R)^2 = \frac{1}{3} \kappa\rho$, and from this that the density is given by

$$\rho = \frac{3H^2}{8\pi G}. \tag{2.1}$$

With $H = 500$ km·s^{-1}·Mpc^{-1}, Einstein and de Sitter found what they considered the reasonable value $\rho = 4 \times 10^{-28}$ g·cm^{-3} and concluded that "at the present time it is possible to represent the facts without assuming a curvature of three-dimensional space." Remarkably from a later perspective, Einstein and de Sitter confined their interest to the relation between density and expansion rate and did not write out the "Einstein–de Sitter solution," that is, the time dependency of the scale factor R. This has the form

$$R(t) = \alpha t^{2/3} + \beta, \tag{2.2}$$

where α and β are constants. From this it follows again that the age of the Einstein–de Sitter universe is $t = 2/3H$, which would imply an age of the world of about 1.2 billion years. But none of this was mentioned by the authors, who seem to have wanted to avoid discussing the big-bang features inherent in the model: whether the constant β is taken to be zero or not, the model clearly indicates an abrupt beginning of the expansion. It is noteworthy that Einstein and de Sitter avoided discussing this feature. In fact, at the time de Sitter was deeply worried about the time scale problem and denied that a definite age of a few billion years could be ascribed to the universe.

The Einstein–de Sitter model played a considerable role in the later development of cosmology, and we shall often meet it in later chapters. Curiously, at the time the model was proposed neither Einstein nor de Sitter seems to have considered it important.[47] The historical importance of the model does not lie so much in its being a possible candidate for the real structure of the world as in its simplicity: the model has often been considered as representative of relativistic evolution models and therefore been used in discussions of the validity of this kind of cosmology. For example, it was in this capacity that the Einstein–de Sitter model was often discussed by steady-state theoreticians.

Lemaître's theory was also taken up in the Soviet Union, where the brilliant twenty-five-year-old Leningrad physicist Matvei Bronstein examined it from the point of view of time asymmetry. As he pointed out, since Lemaître's (and, of course, Friedmann's) theory was derived from general relativity, and since this theory does not distinguish between past and future, then Lemaître's theory is unable to explain why the universe expands rather than contracts. This was the same problem as earlier addressed by McVittie, but Bronstein did not believe that the solution should be found in the initial state of the world. Inspired by Niels Bohr's contemporary ideas of energy nonconservation (see section 3.1) Bronstein instead suggested a modification of the Friedmann-Lemaître equations. For, as he wrote, "A physical theory upon which the solution of the cosmological problem can be based cannot be symmetrical with respect to the interchange of the past and the future."[48] Bronstein's modification was to include a cosmological constant varying in time, $\Lambda = \Lambda(t)$. This change implies that the expansion of the universe will no longer be adiabatic, corresponding to a violation of the energy law. Bronstein considered this justified because the new cosmological equations then became asymmetric in time: although the general theory of relativity would no longer apply to the universe as a whole, he believed that he had found a cosmological "arrow of time" and in it a physical cause of the expansion of the universe. Bronstein's theory received no attention, but the problem he examined, the cosmological direction of time, would later become a central topic in cosmology. Bronstein did not live to experience the renewed interest in the subject. He was executed by a military firing squad on 18 February 1938, falsely accused of being a counterrevolutionary and a spy.

FRIEDMANN-LEMAÎTRE MODELS

By 1933 the formal structure of what became known as Friedmann-Lemaître cosmology had been clarified and systematized in the reviews of Heckmann, de Sitter, Robertson, and Tolman. It will be useful to give a brief summary of some of the knowledge known at that time. The basic equations are the Friedmann-Lemaître equations (or just Friedmann equations), which in the most general case can be written as

$$3\left(\frac{R'}{R}\right)^2 + 3\frac{kc^2}{R^2} = (\Lambda + \kappa\rho c^2)c^2 \tag{2.3}$$

and

$$\left(\frac{R'}{R}\right)^2 + 2\frac{R''}{R} + \frac{kc^2}{R^2} = (\Lambda - p\kappa)c^2, \tag{2.4}$$

where k is the space curvature parameter. The value of k distinguishes between closed space ($k = +1$), flat space ($k = 0$), and open space ($k = -1$). The other symbols have the meanings previously mentioned: Λ is the cosmological constant, κ Einstein's gravitation constant, ρ the density, and p the pressure.

The Friedmann-Lemaître equations comprise static as well as evolutionary models. For example, the original static Einstein universe follows from $R' = R'' = 0$ and $R = \text{const} = R_E$. With $p = 0$ we obtain from equations (2.3) and (2.4) $\Lambda + \kappa \rho c^2 = (3k/R_E^2)$ and $\Lambda = (k/R_E^2)$. To satisfy these equations we must take $k = +1$ and are then led to $\Lambda = \Lambda_E = (1/R_E^2)$ and $\rho = \rho_0 = (2\Lambda_E/\kappa c^2) = (\Lambda_E c^2/4\pi G)$. It is only with this value of the cosmological constant that a static solution is possible for a matter-filled universe. If $\Lambda < \Lambda_E$ the universe will collapse (oscillating model) and if $\Lambda > \Lambda_E$ it will expand forever, asymptotically approaching a de Sitter state.

Ever-expanding models are also obtained for open and flat spaces if $\Lambda > 0$. This is seen from equation (2.3), which can then be written

$$3R'^2 = (-3k + \Lambda R^2 + \kappa \rho c^2 R^2)c^2. \tag{2.5}$$

Since the expression on the right side is positive if $k = 0$ or -1, the universe will always expand.

The de Sitter solution can also be obtained from the Friedmann-Lemaître equations, namely, as the solution for which $k = 0$ and $R'/R = \text{const}$. In that case, we must have from eqs. (2.3) and (2.4) that $R''/R = (R'/R)^2$ and, from the equality of the expressions on the right sides, $p = -\rho c^2$. To avoid a negative pressure (or a negative density), we are led to $\rho = p = 0$. Equation (2.3) now yields $R'/R = c\sqrt{\Lambda/3}$, the integral of which is the de Sitter solution $R(t) \sim \exp(ct\sqrt{\Lambda/3})$ in the form found by Lemaître in 1925.

A kind of de Sitter universe can be obtained also with $\Lambda = k = 0$ and $\rho \neq 0$, namely, if we retain $p = -\rho c^2$ as some kind of cosmic tension. From equations (2.3) and (2.4) we get $(R'/R)^2 - (R''/R) = 0$, which is the same as

$$\frac{d^2}{dt^2}(\ln R) = 0.$$

Integration gives $\ln R = at + b$, where a and b are constants, and then the exponential solution $R(t) \sim \exp(at)$. This type of model, with a nonvanishing density, corresponds to the steady-state universe, where $H = R'/R = a$ is a constant. The indicated derivation is, however, quite foreign to the steady-state theory.

It is of interest to note that several cosmological models, both standard relativistic and not, satisfy the relationship $G\rho/H^2 = \text{const}$, where the constant is not very different from unity. The relationship is satisfied by the Einstein–de Sitter model (constant $= 3/8\pi$), the Milne model, and the Dirac model, all of which are mentioned in later sections. Evidently it is also satisfied by the steady-state model, where each of the three factors (G, ρ, and H) is constant with respect to the period. However, in general the hypothesis does not apply to Friedmann-Lemaître models.

If we put $\Lambda = 0$ in equation (2.5) and introduce the Hubble parameter $H = R'/R$ we get the useful relation

$$kc^2 = H^2 R^2 \left(\frac{\kappa \rho c^4}{3H^2} - 1 \right) = H^2 R^2 \left(\frac{8\pi \rho G}{3H^2} - 1 \right).$$

The curvature depends on the sign of the quantity in the bracket, that is, the density. It is therefore convenient to define a critical density $\rho_c = 3H^2/8\pi G$, which happens to be the same as the density of the Einstein–de Sitter universe. With this definition,

$$kc^2 = H^2 R^2 \left(\frac{\rho}{\rho_c} - 1 \right).$$

In other words, world models with zero cosmological constant will be spatially closed if $\rho > \rho_c$; if $\rho = \rho_c$ they will be flat, and $\rho < \rho_c$ gives an open space. In modern literature the density parameter ρ/ρ_c is usually designated Ω.

From the Friedmann-Lemaître equations it follows that

$$\frac{d}{dt}(\rho c^3) + \frac{p}{c^2} \frac{dR^3}{dt} = 0, \tag{2.6}$$

which expresses energy conservation: in a closed world, the volume is $V = \pi^2 R^3$ and the energy $E = \rho c^2 V$. The equation can therefore be written as the familiar $dE = -pdV$. If, following Bronstein, it is assumed that Λ is not a constant, equation (2.6) will have to replaced by

$$\frac{d}{dt}(\rho c^3) + \frac{p}{c^2} \frac{dR^3}{dt} = -\kappa^{-1} R^3 \Lambda',$$

where $\Lambda' = d\Lambda/dt$. The corresponding energy equation does not conserve energy. Instead of $dE + pdV = 0$ the result is $dE + pdV = -\kappa^{-1}\pi^2 c^2 R^3 d\Lambda$ as suggested by Bronstein.

Friedmann-Lemaître models have a Robertson-Walker metric, which means that the line element can be written in the general form $ds^2 = c^2 dt^2 - R^2(t)d\sigma^2$, where, using polar coordinates, the space part is

$$d\sigma^2 = \frac{dr^2}{1 - kr^2} + d\theta^2 + \sin^2\theta d\phi^2.$$

Conversely, equations (2.3) and (2.4) can be obtained from the Robertson-Walker metric by substituting it into Einstein's field equations. The Robertson-Walker metric describes cosmological models which are homogeneous and isotropic, as explained by Robertson in his first derivation of 1929. A more general derivation, building on the cosmological principle, was independently given in 1935 by Robertson and Arthur Walker. (The cosmological principle will be explained in section 2.3.) Relativistic world models satisfying the Robertson-Walker (RW) metric are sometimes called FRW or FLRW cosmologies, the extra initials "F" and "L" referring to Friedmann and Lemaître, respectively.

2.2 THE PRIMEVAL ATOM

Cosmogonic Speculations

The idea of the universe originating in a singular event some finite time ago, that is, the big-bang idea, was first suggested as a scientific hypothesis in the early 1930s and then, in a revised and developed version, shortly after the end of the Second World War. But the essence of the idea goes far back in time insofar as it may be seen as an integral part of many religious world views. The entire Judeo-Christian tradition, in particular, builds upon the postulate of the world being created by God, although this creation need not be interpreted literally as a physical event. At any rate, we shall not be concerned here with purely religious and philosophical discussions of the origin or creation of the world, which belong to a tradition very different from that of the scientific cosmology that emerged in the first half of this century. Yet it may be interesting to mention some prescient speculations of the nineteenth century which in a rather surprising manner comprise some of the themes which were later taken up in cosmology.

One should not be surprised to learn that speculations about a big-bang origin of the universe predate the scientifically argued idea, whether these speculations had their roots in religious, philosophical, or scientific considerations. One remarkable example is provided by Edgar Allan Poe, the famous American poet and novelist, who in 1848 presented a cosmological vision which in some respects anticipated the essence of the later big bang theory as suggested by Lemaître. In his essay "Heureka" Poe imagined that the universe arose from the explosion of a singular state of matter "in its absolute extreme of Simplicity . . . absolutely unique, individual, undivided." He described the creation process as "one instantaneous flash." From the undifferentiated primordial atom followed the entire constitution and history of the universe: "This Oneness is a principle abundantly sufficient to account for the constitution, the existing phaenomena and the plainly inevitable annihilation of at least the material Universe." Poe argued that the atoms into which the original matter had fragmented would be diffused by means of radiation in such a way as to fill space homogeneously. The atoms were endowed both with a repulsive power and an "appetite for Unity" of which the latter would not only produce the celestial bodies by gravitational attraction, but also, in the end, force the atoms to coalesce back into the original state of oneness. Poe was thus led to a cyclical world model, a big bang followed by an expansion and a contraction. Although he repeatedly referred to astronomical authorities and in part dressed his vision in scientific terms, he made it clear that his essay was not to be considered an ordinary scientific treatise. Thus, after having identified the attractive force with Newtonian gravitation and the repulsive force with electricity, he added that "The former is the body; the latter the soul: the one is the material; the other the spiritual, principle of the Universe."[49]

Thirteen years later, in the fifth edition of his *Populäre Astronomie*, the German astronomer Johann Mädler called attention to Olbers's paradox, that is, how to explain the darkness of the night sky in a universe supposedly filled with an infinity of luminous stars. Mädler rejected the idea of interstellar absorption and instead suggested that, since the velocity of light is finite, "a *finite* amount of time has passed from the beginning of Creation until our day, and we, therefore, can only perceive the heavenly bodies out to the distance that light has travelled during that finite amount of time."[50] Mädler's suggestion of the universe being created a finite time ago agreed with Christian views, but ran against astronomical and physical orthodoxy. It left no impact on the science of his time and Mädler did not pursue the idea.

Poe was far from the only man of letters who took an interest in the great cosmogonical questions and sought to design a quasi-scientific scenario of the evolution of the entire universe. Later in the century the idea of a recurrent universe—one in which there is no unidirectional evolution on a very long time scale, but which develops cyclically and eternally—fascinated many people, both scientists and amateurs. Like the creation idea, this general idea goes far back in time and has its origin in primitive religion. Friedrich Nietzsche, the great and controversial German philosopher, found the recurrent universe philosophically attractive and believed that it followed from the laws of physics. The author of *Also Sprach Zarathustra* was led to a quasi-scientific vision of the universe no less grand, and no less speculative, than the vision presented by the author of *The Murders in the Rue Morgue*. Contrary to Poe, Nietzsche believed that the world had always existed and would always exist. He dismissed the idea of a created universe as superstition derived from religion and had no more patience with the idea of a final state in which the world had achieved thermodynamic equilibrium. His solution was to conceive the universe "as a definite quantity of energy, as a definite number of centres of energy." From this materialistic starting point he claimed that "it follows therefrom that the universe must go through a calculable number of combinations in the great game of chance which constitutes its existence." Nietzsche went on:

> In infinity, at some moment or other, every possible combination must once have been realized; not only this, but it must have been realized an infinite number of times. And inasmuch as between every one of these combinations and its next recurrence every other possible combination would necessarily have been undergone, and since every one of these combinations would determine the whole series in the same order, a circular movement of absolutely identical series is thus demonstrated: the universe is thus shown to be a circular movement which has already repeated itself an infinite number of times, and which plays its game for all eternity.[51]

The cyclical universe, as imagined by Nietzsche, was a popular theme in the late nineteenth century, when it was discussed also by some mathematical physicists, including Boltzmann and Poincaré. Although the steady-state universe is not cyclical, it is continually recurrent and may in a qualitative, philo-sophical sense be grouped together with universes of the cyclical type.

With the discovery of radioactivity, scientists were faced with a new source of energy which did not require any input from outside sources. Radioactive atoms spontaneously transmute into other atoms and in this way produce en-ergy as well as new matter. Could the phenomenon be of cosmological and not only of terrestrial significance? This seems to have been what the chemist Frederick Soddy suggested in 1904, shortly after he, in collaboration with Rutherford, had formulated the important law of radioactive decay. In the first textbook ever on radioactivity, he wrote: "[T]here is associated with the mate-rial structure of the atom an enormous store of energy which, in the majority of cases, remains latent and unknowable. . . . The internal movements of the atom must be highly irregular and cannot follow a definite sequence if the law of radioactive change is to hold good. . . . The atomic mass must be regarded as a function of the internal energy, and the dissipation of the latter into radio-activity occurs at the expense to some extent at least, of the mass of the sys-tem."[52] The suggestion of a mass-energy equivalence more than a year before Einstein's famous $E = mc^2$ formula may seem astounding, but in fact such equivalence was well known before Einstein, although in the more restricted sense of the electromagnetic world view. Soddy then proceeded: "Corre-spondingly, a sudden beginning of the universe—the time when present laws began to operate—is also fixed . . . It is necessary to suppose that the universe as a thing in being had its origin in some initial creative act, in which a certain amount of energy was conferred upon it, sufficient to keep it in being for some period of years." Again it is obvious that this was nothing but a spirited speculation, scarcely more scientific than Poe's and Nietzsche's except that it came from a recognized scientist. Yet the quotation merits attention as prob-ably being the first time that a physical mechanism responsible for the origin of the universe was suggested. That the suggestion should not be taken too seriously may be illustrated by comparing it with another of Soddy's discus-sions of "the revelations of radioactivity." In a series of public lectures deliv-ered at the University of Glasgow in 1908, he said: "The idea which arises in one's mind as the most attractive and consistent explanation of the universe in light of present knowledge, is perhaps that matter is breaking down and its energy being evolved and degraded in one part of a cycle of evolution, and in another part still unknown to us, the matter is being built up with the utilisa-tion of the waste energy. The consequences would be that, in spite of the incessant changes, an equilibrium condition would result, and continue in-definitely."[53] What Soddy proposed here was the very opposite of a big-bang theory, a kind of steady-state universe. Rather than hailing Soddy as a precur-sor of both big-bang and steady-state cosmology, we should take his visions for what they were, and were meant to be: speculations.

Although the idea of a universe with a sudden beginning was not discussed within the framework of relativistic cosmology in the 1920s (with the notable exception of Friedmann) it was occasionally mentioned, or hinted at, in other contexts. The case of James Jeans is noteworthy. In a popular book of 1929 he discussed under the heading "The Beginnings of the Universe" how far backward in time the evolution could be traced. Jeans argued that this time had to be finite and that the present matter thus had not existed forever. If we go back in time, say two hundred million million years, then "our next step back in time leads us to contemplate a definite event, or series of events, or continuous process, of creation of matter at some time not infinitely remote. In some way matter which had not previously existed, came, or was brought, into being."[54] Jeans discussed the possibility that matter was originally created by high-energy photons, which "might conceivably crystallise into electrons and protons, and finally form atoms. If we want a concrete picture of such a creation, we may think of the finger of God agitating the ether."

This may be seen as an anticipation of the big-bang universe, but also, for that matter, of continual creation of matter; Jeans's creation of matter could be both "a definite event" and a "continuous process." The fact that Jeans discussed the creation of matter and the beginning of the universe is remarkable; but it is also typical that this took place in a popular, noncommitted context. At any rate, Jeans admitted that his reference to God's finger was "crude imagery" and ended up with leaving the ultimate question of the creation of matter to philosophers and theologians: "Travelling as far back in time as we can, brings us not to the creation of the picture, but to its edge; the creation of the picture lies as much outside the picture as the artist is outside his canvas. . . . This brings us very near to those philosophical systems which regard the universe as a thought in the mind of the Creator, thereby reducing all discussion of material creation to futility."[55] So, according to Jeans, the understanding of the creation of the world just was not the business of scientists. A closer study of the literature of the 1920s will probably reveal other speculations of a similar kind, but there is no reason to assume that the mathematical cosmologists took them very seriously.

Early Physical Cosmology

As mentioned, the predominant tradition in relativistic cosmology was heavily oriented towards mathematics and had little contact with either physics or observational astronomy. In particular, atomic and subatomic theories of the constitution of matter were almost completely absent from the cosmological scene in the 1920s and 1930s. Lemaître's fireworks universe of 1931 may be seen as a first attempt to establish a cosmology that also took microphysical processes into account and made predictions of a physical nature, an attempt that was not, however, very successful until it was revived in a modified form by Gamow after the war. Even before Lemaître embarked upon this line of research there were a few attempts to consider the universe in terms of physics rather than to see it as merely a mathematical construct. Most of these attempts did not deal specifically with the universe at large, but rather with aspects of thermodynamics, stellar physics, and element formation; yet, at least in retrospect, they can be located in a cosmological context, being early examples of a trend which would reach a mature state only after the war.

Richard Tolman was professor of physical chemistry, first at the University of Illinois and from 1922 at the California Institute of Technology. This chemist with a strong interest in relativity and astronomy made a pioneering contribution to nuclear astrophysics in a paper of 1922. Tolman applied chemical equilibrium theory to the hypothetical reaction $4H \rightleftharpoons He$, i.e., the equilibrium between four hydrogen atoms and a helium atom (the latter assumed to consist of two extranuclear electrons and a nucleus with four protons and two electrons).[56] His purpose was to investigate if the relative abundances of the two elements, which soon would be recognized as the main constituents of the universe, could be explained in this manner. The result was disappointing in the sense that he found that even at very high temperatures and very low pressures hydrogen would combine almost completely to helium. This was, of course, contradicted by terrestrial, solar and stellar evidence, which shows hydrogen to be more abundant than helium.

The idea of considering the entire universe as a thermodynamical system subject to the ordinary laws of thermodynamics was entertained already in the nineteenth century, in connection with discussions of the first and second laws of thermodynamics. A new and more fruitful perspective was suggested by the German physicist Otto Stern, who in 1925–26 studied the conditions for matter and radiation to be in a state of equilibrium.[57] Stern was an expert in quantum theory and statistical mechanics and a former collaborator of Einstein's. Although Stern did not refer to a particular cosmological model, he assumed his calculations to hold for the entire universe, an assumption he found "very tempting." However, he was perplexed to find that the temperature needed for an equilibrium state was unrealistically high, and that for realistic temperatures all matter would disappear into radiation. Stern's work triggered several papers on the same subject, including the first applications of thermodynamics to definite cosmological models. Wilhelm Lenz, another German physicist, found the temperature of space to be related to the radius of the Einstein universe,[58] and in 1928 Tolman gave a much improved treatment in which he derived expressions for the energy and entropy of the same kind of world model.[59] This and other work in the same tradition was of course based on the assumption of a static, closed universe.

The problem that had occupied Tolman in 1922 was taken up by a Japanese physicist, Seitaro Suzuki, six years later.[60] Treating the helium-hydrogen equilibrium thermodynamically, he found that a considerable dissociation into hydrogen would require low pressures (about 1 atm) and very high temperatures (about 2×10^9 K), conditions that did not exist in the stars; with current estimates of the temperature and pressure in the interior of stars the degree of dissociation would be practically zero, again leading to the unacceptable conclusion reached by Tolman. However, contrary to his American colleague, Suzuki considered that the observed cosmic helium-hydrogen ratio might be explained on the equilibrium hypothesis "if the cosmos had, at the creation, the temperature higher than 10^9 degrees." This comment is remarkable in referring to the creation of the universe and in offering a possible cosmological explanation of the helium-hydrogen ratio, a subject that would become of central importance in cosmology only much later. It was, however, only a casual remark and Suzuki's work, published in a Japanese journal of physics, made in any case no impact at all on Western astronomy. Yet it deserves to be mentioned as probably the first example of a research program that later would revolutionize cosmology. The work of the Japanese physicist indicated the general program of "nuclear archaeology," that is, the attempt to reconstruct the history of the universe by means of hypothetical cosmic nuclear processes, and to test these by the resulting pattern of element abundances.

The works mentioned show that not all cosmology in the 1920s was relativistic theory, but that a few scientists attempted to adopt a more physical perspective in the study of the universe. These attempts assumed the world to be static and in thermodynamic equilibrium and they all led to disagreements with experience, to either a radiation-filled or a helium-filled universe. The reason was not only the assumption of a static universe but also, it turned out, the assumed equilibrium hypothesis. In 1931 the American chemists Harold Urey and Charles Bradley argued that the relative abundances of terrestrial elements could not be reconciled with the hypothesis of a thermodynamic equilibrium mixture, whatever its temperature.[61] This subject also would later be important in cosmology, in connection with nucleosynthesis of heavier elements (section 3.2). One might believe that the contradictory conclusions obtained by Tolman, Stern, Lenz, and others would make them question also the assumption of a static universe, but this was not the case.

When the expanding universe became a reality, Tolman immediately subjected it to thermodynamic analysis. In 1931 he considered in detail an expanding, radiation-filled universe and derived the dependence of the radiation density and entropy on the expansion.[62] Tolman's equations would later be used by Gamow and his collaborators in their big-bang model, but Tolman entertained no such ideas. He dealt with the subject in a general way, and had no reason to believe in an early universe actually dominated by radiation.

The Beginning of the World

In spite of the general euphoria over the expanding universe it was evident that the idea posed as many problems as it solved. One of these was the beginning of the expansion, a problem that was particularly emphasized by Eddington. Analyzing the various mathematical possibilities, he ended up by favoring as "the most attractive" the case in which the mass of the universe is equal to the mass of the Einstein universe. "There is at least a philosophical satisfaction in regarding the world as beginning to evolve infinitely slowly from a primitive uniform distribution in unstable equilibrium," he commented. Eddington also considered the possibility of a universe of mass larger than the

Einstein universe, but he rejected this possibility on the ground that "it seems to require a sudden and peculiar beginning of things."[63] In 1930 the world model of Eddington was in harmony with the one proposed by Lemaître in 1927. This so-called Lemaître-Eddington universe was an expanding universe, evolving gradually from an already existing preuniverse and asymptotically approaching a de Sitter world. It was thus a world without a proper beginning. The agreement between Eddington and his former student on this point did not last long.

Following the translation of Lemaître's 1927 work there appeared in the same issue of *Monthly Notices* an article by Lemaître in which he elaborated on various aspects of his theory of the expanding universe.[64] He derived the equations of his theory in a new way and examined the question of how the expansion originally started from a universe in equilibrium. This question had earlier been considered by Tolman and Eddington. Tolman argued that annihilation processes, i.e., conversion of electrons and protons into radiation energy, might explain the recession of the nebulae. The idea was criticized by Eddington, who suggested instead that formation of condensations in the Einstein world might be the mechanism responsible for the expansion.[65] Lemaître did not consider the annihilation hypothesis, but developed a theory which partly agreed with Eddington's idea. Lemaître's theory was based on what he called "stagnation," a sort of condensation process in which the total pressure diminishes. He concluded: "If, in a universe in equilibrium, the pressure begins to vary, the radius of the universe varies in the opposite sense. Therefore stagnation processes include expansion." He further considered the case in which the pressure of the equilibrium (p_0) suddenly, at $t = 0$, drops to zero as a result of an instantaneous stagnation. In this case he found that the time since the stagnation would have been infinitely long: "If p_0 tends to zero, t tends to infinity, the limiting case being the solution emphasized in our 1927 paper. As was pointed out by Eddington, such logarithmic infinities have no real physical significance."[66]

In March 1931 Lemaître, then, still worked with a model of the universe in which there was a beginning of the expansion but no creation of the world. At about that time he made another great conceptual innovation, introducing into cosmology for the first time the audacious notion of the beginning of the world in a realist sense. The step seems to have been inspired by an address which Eddington delivered in early January as president of the British Mathematical Association. The text of Eddington's address was published under the dramatic title "The End of the World: from the Standpoint of Mathematical Physics."[67] Eddington dealt in a popular fashion with one of his favorite themes, the role of universal entropy as an arrow of time. In accordance with earlier authors, such as Boltzmann and Kelvin, he discussed the "heat death" of the universe, that is, the state of maximum entropy and dissolution of the unidirectionality of time assumed to take place in the far future. Eddington's belief in an ultimate heat death was countered by H. Piaggio, who adopted Millikan's atom-building cosmology in order to argue that the universe is not like a watch that is always running down. "The process of creation may not yet be finished," he suggested.[68]

Eddington's discussion of heat death was not the only one in the first half of 1931. Tolman also discussed "the problem of the entropy of the universe," which he formulated as follows. If entropy is everywhere increasing toward a maximum, why has not the universe—supposed to be of infinite age—already reached this maximum state? Tolman showed that within the framework of relativistic thermodynamics classically entropy-increasing processes do not necessarily lead to an increase in entropy. He briefly mentioned another and much simpler answer, based on the assumption that the universe was created at a finite time in the past with sufficient available energy so that the entropy has not yet reached its maximum value. This might have been a reference to the new primeval atom hypothesis, but Tolman did not mention Lemaître's idea. As a modification he mentioned the Lemaître-Eddington assumption of "an infinite past during which the universe was in a quiescent metastable state of large available energy, and a disturbance at a finite time in the past initiated the process of degradation." However, Tolman had no sympathy at all for such assumptions, which he refused to take seriously. "These suggestions," he wrote, "depend too greatly on special *ad hoc* assumptions to be scientifically satisfying."[69]

Eddington did not find a finite-age universe satisfying either, but he nonetheless considered briefly what the state of the world would have been like if time were traced backward to a state in which the entropy tended to zero. It was not the first time that Eddington contemplated the question. He had done so three years earlier, when he came close to inventing the term "big bang." On this occasion he objected to "the implied discontinuity in the divine nature" and wrote that "As a scientist I simply do not believe that the Universe began with a bang."[70]

In his address of 1931, Eddington took up the subject anew. Would the universe of zero entropy correspond to "the beginning of the world"? In the words of Eddington: "Following time backwards, we find more and more organisation of the world. If we are not stopped earlier, we must come to the time when the matter and energy of the world had the maximum possible organisation. To go back further is impossible. We have come to an abrupt end of space-time—only we generally call it the 'beginning.'" Eddington answered that this question probably lay outside the range of scientific reasoning, but repeated that "philosophically, the notion of a beginning of the present order of Nature is repugnant to me."

This remark, taken rather out of its context, spurred Lemaître to state that the concept of the beginning of the world did not have to rest on personal or philosophical views. A few weeks after Eddington's address there appeared in *Nature* a letter entitled "The Beginning of the World from the Point of View of Quantum Theory."[71] In this letter Lemaître stated that "the present state of quantum theory suggests a beginning of the world very different from the present order of Nature." It is noteworthy that he argued for the primeval atom (a term not yet used) by means of quantum theory: "Thermodynamical principles from the point of view of quantum theory may be stated as follows: (1) Energy of constant total amount is distributed in discrete quanta. (2) The number of distinct quanta is ever increasing. If we go back in the course of time we must find fewer and fewer quanta, until we find all the energy in the universe packed in a few or even in a unique quantum." At that time Eddington attempted to link cosmology and quantum mechanics by means of his idiosyncratic interpretation of Dirac's wave equation of the electron. Lemaître, too, was fascinated by Dirac's equation but he did not believe it could be ascribed a cosmological significance and thus rejected Eddington's program.[72] His inspiration came from the current discussions of the philosophical implications of quantum mechanics and the problems of constructing a relativistic quantum theory of matter and fields.

In the early 1930s quantum theory was considered to be in a state of deep crisis, reflected in infinities turning up in quantum electrodynamics, the inadequacy of the theory in explaining the atomic nucleus and the cosmic radiation, and what appeared to be a violation of energy conservation in beta decay. In desperation, several physicists, including notables such as Bohr, Pauli, Schrödinger, and Heisenberg, suggested that the ordinary space-time continuum had to be revised in the quantum domain. In general, they believed that the time was ripe for drastic changes in the foundation of physics, possibly even giving up such time-honored laws as the strict conservation of energy and momentum. The way in which Lemaître phrased his letter to *Nature* reveals a close similarity to contemporary discussions among quantum theoreticians and suggests that the source of his primeval universe should be found as much in a quantum theoretical as in a cosmological tradition. His note was probably indebted to the views of Niels Bohr, who had recently argued that the concepts of space and time have only statistical validity.[73] Lemaître was deeply interested in these problems, which involved how to extend and interpret Heisenberg's indeterminacy principle.

Echoing Bohr, Lemaître claimed in 1931 that "in atomic processes, the notions of space and time are no more than statistical notions." He continued:

> If the world had begun with a simple quantum, the notions of space and time would altogether fail to have any meaning at the beginning; they would only begin to have a sensible meaning when the original quantum had been divided into a sufficient number of quanta. If this suggestion is correct, the beginning of the world happened a little before the beginning of space and time. I think that such a beginning of the world is far enough from the present order of nature to be not at all repugnant . . . we could conceive the beginning of the universe in the form of a unique atom, the atomic weight of which is the total mass of the universe. This highly unstable atom would divide in smaller and smaller atoms by a kind of super-radioactive process.[74]

With regard to the nature of the primeval atom, Lemaître naturally had to rely on speculations and metaphors derived from the meager contemporary knowledge of nuclear physics, which largely meant radioactivity. The hypothesis of a "unique atom" of gigantic atomic weight may seem rather fantastic, but it was a way of expressing a hypothetical state of the world which should not be taken too literally. Furthermore, speculations of the cosmic significance of transuranic elements were not uncommon at the time, when Jeans and others argued that stellar energy had its source in such elements of extreme atomic weights. Lemaître referred to this idea and it is possible that it inspired him to extrapolate it to the very extreme, an "atom" of the maximally conceivable atomic weight.

It seems that the source of Lemaître's big-bang hypothesis should be sought in a combination of an empirical fact, the existence of radioactive substances, and a theoretical insight, quantum-mechanical indeterminacy. As he later explained,[75] the idea arose when he thought about the conditions for radioactive elements still being with us. Didn't the very long half-lives of these elements, uranium and thorium, indicate that all elements had once been radioactive and that our present world was the nearly burned-out result of a previous radioactive universe? Was it a coincidence that the half-lives of uranium and thorium are comparable with the Hubble time? As far as the idea of a beginning of the world is concerned, one may assume that Lemaître was also influenced by his knowledge of Friedmann's paper, in which the possibility had first been discussed. But to formulate the idea of a radioactive beginning of the world in a satisfactory way, Lemaître had to avoid the Kantian antimony of beginning. This dilemma is based on determinism, according to which future states of a physical system can be inferred from some initial conditions. A deterministic explanation of a beginning will then have to refer to a more remote state as initial conditions, which is only to push the problem back in time. The problem ends in an infinite regress, that is, without a solution. This was where quantum mechanical indeterminacy came in. In a nondeterministic system the antimony will not arise and so Lemaître saw a way in which the world could have begun.

Lemaître was well aware of the speculative nature of his scenario of the beginning of the world and the problems which followed from it. Could our present world in all its colorful diversity really be the causal result of causes embedded in a single, undifferentiated quantum? Again he resorted to quantum mechanics: "Clearly the initial quantum could not conceal in itself the whole cause of evolution; but, according to the principle of indeterminacy, that is not necessary. Our world is now understood to be a world where something really happens; the whole story of the world need not have been written down in the first quantum like a song on the disc of a phonograph. The whole matter of the world must have been present at the beginning, but the story it has to tell may be written step by step."[76]

As a Catholic priest, Lemaître was, of course, aware that discussions about the beginning of the world could not, in the minds of most people, be separated from the question of God's creation of the world. He was at first inclined to include this aspect in his discussion, but then decided not to. In the typescript of the note of March 1931, there is a paragraph reading: "I think that everyone who believes in a supreme being supporting every being and every acting, believes also that God is essentially hidden and may be glad to see how present physics provides a veil hiding the creation."[77] The paragraph was crossed out by Lemaître, not because it did not represent his conviction, but because he found it unwise to introduce God in his purportedly scientific sketch.

As it was stated in the *Nature* note, Lemaître's suggestion had little in common with a proper scientific theory; it was an imaginative hypothesis more in line with Poe's speculations than with contemporary cosmology. But Lemaître soon found an opportunity to present a more elaborated and better argued version of his idea. At the meeting of the British Association for the Advancement of Science in October 1931 one of the sessions covered the subject "The Question of the Relation of the Physical Universe to Life and Mind"—a broad subject if there ever was one.[78] The session was organized by Herbert Dingle and included contributions from Jeans, Eddington, Milne, Millikan, Barnes, and de Sitter, among others. Apart from its scientific content, the discussion was also a public success of the first rank. It attracted a crowd of no less than two thousand. At first Lemaître was not invited to the meeting, but as a result of Eddington's intervention, it was arranged for Lemaître to deliver a talk following Jeans's opening address. At that time Jeans hesitated in accepting the expanding universe. "The concept of an expanding universe," he said, "may prove after all to be a false scent, and the truth may lie in some other direction." At the end of his talk Jeans therefore remarked that if some infallible oracle offered to give a yes or no answer to just one question, he would like the question to be, "Is the universe expanding at about the rate indicated by the spectra of the nebulae?" To Lemaître there was no doubt. He considered the cosmic expansion a scientific fact for which an oracle was not needed: "The expansion of the universe is a matter of astronomical facts interpreted by the theory of relativity, with the help of assumptions as to the homogeneity of space, without which any theory seems to be impossible." Lemaître, having recently moved from the expanding universe to the big-bang idea, would rather ask the oracle, "Has the universe ever been at rest, or did the expansion start from the beginning?" (But he added that he would really prefer the oracle to be silent "in order that a subsequent generation would not be deprived of the pleasure of searching for and of finding the solution.")

Jeans and other speakers at the meeting had emphasized the conflict between stellar theory and the new cosmology as far as the time needed for evolution was concerned (see section 2.4). Lemaître recognized the difficulty but believed that it could be resolved if a "fireworks theory" of evolution was introduced in cosmology. "The last two thousand million years are slow evolution," he declared; "they are ashes and smoke of bright but very rapid fireworks." He argued that the cosmic radiation was to be conceived as the remnants of the disintegration of the primeval superatom from which the stars were once formed. If such a view was adopted the time scale difficulty did not need to arise, Lemaître claimed. He pictured the evolution of the universe as follows: "At the origin, all the mass of the universe would exist in the form of a unique atom; the radius of the universe, although not strictly zero, being relatively small. The whole universe would be produced by the disintegration of this primeval atom." He further imagined that the primeval atom would first disintegrate into "atomic stars," atoms of weights comparable to those of the stars, and that the disintegration of these superatoms would produce the cosmic radiation and the ordinary matter now observed.

In January 1932, less than a year after he had suggested the primeval-atom hypothesis, Lemaître summarized his recent work:

> I found [in 1930], by applying a generalisation of a theorem of Birkhoff, that the formation of local condensations in an Einstein universe has no direct effect on the stability, but has an indirect effect, which I called "stagnation of the universe," which must include a rupture of the equilibrium in the sense of the expansion. Furthermore, I found that, in the hypothesis that the actual universe comes out from the rupture of equilibrium of an Einstein universe, the epoch of this rupture of equilibrium cannot be removed farther back than about 10^{11} years ago. A general conclusion of the theory of the expanding universe is that the time-scale of evolution is much shorter than was thought previously. This rules out of consideration most of the cosmical hypotheses and even the general "Leitmotiv" of these hypotheses which is the idea of Laplace of a primaeval nebula. . . . I proposed to replace the primaeval-nebula hypothesis, by a primaeval-atom hypothesis. An evolution from this starting-point would meet the requirement of the time-scale and furthermore would prove a natural explanation of the penetrating cosmic radiation.[79]

The Development of the Primeval-Atom Hypothesis

In another paper from the fall of 1931 Lemaître combined the purely qualitative idea of a primeval atom with a more precise, mathematical formulation taken from the nonstatic (Friedmann-Lemaître) field equations.[80] With a positive cosmological constant and the space-time curvature being zero or negative he found a model that expanded monotonically from a singular state, which he identified with the primeval atom. The model, subsequently known as the Lemaître model (not to be confused with his model of 1927), was characterized by having a beginning at $t = 0$ corresponding to the explosion of the primeval superatom. The universe starts expanding as in the Einstein–de Sitter case, $R(t) \sim t^{2/3}$, but as a result of keeping the cosmological constant different from zero, the expansion slows down. The universe then runs through a stagnation phase in which the expansion is temporarily halted, until it continues and then, for large values of t, the world approaches a de Sitter universe.

It should be emphasized that the Lemaître model has no natural connection to the idea of a primeval atom. This idea was grafted upon one particular solution of the relativistic equations for which $R = 0$ for $t = 0$, but it could equally well have been applied to other solutions with the same property, such

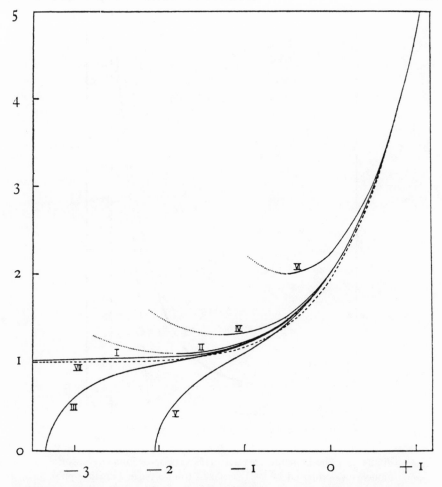

Figure 2.1. The graphical result of de Sitter's calculation of 1930 of various world models corresponding to different solutions to the Friedmann-Lemaître equations. The solutions III and V are examples of Lemaître models. The vertical axis is the scale factor R in terms of the constant Einstein radius. The horizontal axis measures the time as given by $\tau - \tau_0$, where $\tau = ct \sqrt{\Lambda/3}$ and τ_0 is a constant. *Source:* De Sitter (1930b), p. 213.

as the Einstein–de Sitter model. The Lemaître universe could therefore have been proposed as early as 1927 (or even in 1922 by Friedmann). In fact, the first reference to this kind of solution was due not to Lemaître, but to de Sitter, who in June 1930 included it among the various classes of solutions to the Friedmann-Lemaître equations (figure 2.1).[81] However, to de Sitter it was just a mathematical solution of no particular physical importance.

The model, represented in figure 2.2, remained Lemaître's favored one and was the first big-bang model of a physical nature to enter cosmology. In 1931 he described it as follows:

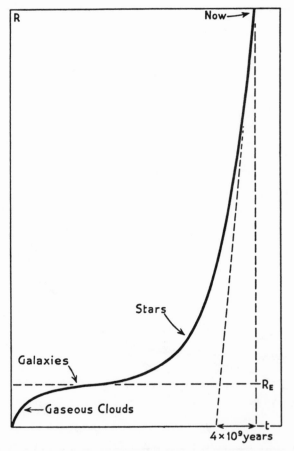

Figure 2.2. The Lemaître universe. This version of Lemaître's world model, first proposed in 1931, is taken from a paper Lemaître published in 1958. The present radius was assumed to be ten times the Einstein equilibrium radius. *Source:* Godart and Heller (1985), p. 98. Reprinted with the permission of the Pachart Foundation dba Pachart Publishing House.

The radius of space began at zero; the first stages of the expansion consisted of a rapid expansion determined by the mass of the initial atom, almost equal to the present mass of the universe. If this mass is sufficient, and the estimates which we can make indicate that it is indeed so, the initial expansion was able to permit the radius to exceed the value of the equilibrium radius. The expansion thus took place in three phases: a first period of rapid expansion in which the atom-universe was broken into atomic stars, a period of slowing-down, followed by a third period of accelerated expansion. It is doubtless in this third period that we find ourselves today, and the acceleration of space which followed the period of slow expansion could well be responsible for the separation of the stars into extra-galactic nebulae.[82]

In the same paper, Lemaître wrote poetically about "one of the most curious of the hieroglyphes of our astronomical library . . . the ultra-penetrating radiation, the cosmic rays." He suggested, as he had done at the meeting of the British Association, that the radiation was of cosmological origin and might date ten billion years back in time. The large value of the age of the world— some five times the Hubble time—was made possible by the choice of a non-vanishing cosmological constant: With $\Lambda = \Lambda_E$ (the Einstein value) the static Einstein universe follows, whereas a slightly larger value $\Lambda_E(1 + \varepsilon)$ corresponds to Lemaître models with different lengths of the stagnation phase. By adjusting the value of ε, any desired value of the stagnation era, and hence of the age of the world, can be obtained. The ten billion years mentioned by Lemaître was thus just one possible value.

Lemaître's reason for advocating the model bearing his name was motivated in particular by its ability to describe a world of an age longer than the Hubble time, thus avoiding conflict with geological and astronomical evidence. He often emphasized this pleasant feature and estimated that with realistic values of the cosmological constant "we may conclude that evolution during 15 to 20 [×] 10^9 years is within the realm of possibility."[83] Furthermore, the stagnation phase peculiar to this model was important to Lemaître because he considered it the era during which galaxies were formed. In a paper read before the (American) National Academy of Science in November 1933 he developed a theory of galaxy formation based on his new cosmological model.[84] In later publications also he returned to the problem on several occasions. Lemaître found that "in spite of the short duration of the cosmic evolution, this local instability [in the stagnation phase] seems to provide the explanation of the automatical formation of the nebulae and the nebular clusters."[85] He deduced that a cluster of nebulae, consisting of N galaxies, would typically have the radius $N^{1/3} \times 80\ 000$ light years.

As mentioned, Lemaître's cosmology rested on the assumption of a non-vanishing cosmological constant. Like his former teacher, Eddington, but for different reasons, he strongly believed in the necessity of this assumption, which he stuck to not solely in order to avoid the time scale difficulty. Far from being a "blunder" or a superfluous constant, he believed that it was a natural, indeed an indispensable, part of relativistic cosmology. In his 1933 address he suggested an interesting interpretation of the constant, namely, that it may be understood as a negative vacuum density. "Everything happens as though the energy *in vacuo* would be different from zero," he wrote, referring to general relativity applied to regions of space of extremely low density. He then argued that in order to avoid a nonrelativistic vacuum or ether—a medium in which absolute motion is detectable—a negative pressure $p = -\rho c^2$ must be introduced, the vacuum density being related to the cosmological constant by $\rho = \Lambda c^2/4\pi G$. This negative pressure is responsible for the exponential (de Sitter) expansion of Lemaître's universe during its last phase,[86] and it contributes a repulsive cosmic force of $\Lambda c^2 r/3$.

Lemaître remained committed to the cosmological constant throughout his life. He tried several times to convince Einstein of its necessity, but all that Einstein would admit was that a nonvanishing cosmological constant was helpful in solving the time scale dilemma. In a letter to Lemaître of 1947 he made it clear that his objections were aesthetically based: "Since I have introduced this term I had always a bad conscience. But at that time I could see no other possibility to deal with the fact of the existence of a finite mean density of matter. I found it very ugly indeed that the field law of gravitation should be composed of two logically independent terms which are connected by addition. About the justification of such feelings concerning logical simplicity it is difficult to argue. I cannot help to feel it strongly and I am unable to believe that such an ugly thing should be realized in nature."[87]

In his first presentation of the idea of the primeval atom Lemaître had already connected his hypothesis with the cosmic radiation, at that time a mysterious and largely unexplored area of physics. The content, energy, and origin of the radiation were disputed. Whereas Robert Millikan, one of the pioneers of cosmic-ray research, argued that the rays consisted of photons created by element formation in interstellar space, new data obtained by Arthur Compton and collaborators indicated that a large part of the radiation consisted of charged particles of extragalactic origin.[88] Lemaître was in favor of the latter hypothesis, which, he thought, might be interpreted within the primeval-atom framework. In the initial explosion superradioactive atoms were produced, and some of the particles (such as electrons, protons, and alpha particles) emitted from these atoms would presumably still be with us in the form of cosmic rays. He realized that the expansion of the universe would reduce the energy of the radiation, which therefore must have been enormous at the time shortly after the initial explosion. He therefore started a program of calculating the energies and trajectories of charged particles in the earth's magnetic field, intended as a test of his cosmological hypothesis. These complicated calculations, in part made jointly with the Mexican physicist Manuel Vallarta, made use of the differential analyzer, a recently completed mechanical (and, from 1935, electromechanical) analogue computer designed by the American engineer Vannevar Bush. Lemaître and Vallarta concluded in 1933 that their calculations, and Compton's discovery of an East-West effect, "give[s] some experimental support to the theory of super-radioactive origin of the cosmic radiation."[89]

Although Lemaître's hypothesis of 1931 was the first example of a big-bang universe, it was a universe evolving from a condensed material pre-universe, the primeval atom, and not from a singularity in space-time. In this respect the primeval-atom hypothesis may be considered only quantitatively different from the Lemaître-Eddington model. A cosmic singularity is a nonphysical notion insofar as neither space nor time exist, and the density is not only enormous, but infinite. Lemaître always emphasized that cosmology could and should be understood in physical terms, and he therefore denied that the beginning of the world could be represented by a true singularity or "annihilation of space" as he called it. However, he also realized that the unwanted initial singularity is not easy to get rid of if the Friedmann-Lemaître equations are kept. In 1931 he met with Einstein in Pasadena, and at Einstein's suggestion he studied a nonisotropic cosmological model in order to see if the singularity would disappear in this case. The result of the calculations was disappointing in the sense that Lemaître found that the singularity was unavoidable.[90] But neither he nor Einstein considered it a proof that an initial singularity was therefore a physical reality. "Matter has to find a way to avoid the annihilation of its volume," Lemaître declared. He believed that the field equations of general relativity would probably break down at extremely high densities and then make the singularity an artifact with no physical meaning. This was also Einstein's view, for in 1945 he wrote: "One may not . . . assume the validity of the equations for very high density of field and matter, and one may not conclude that 'the beginning of the expansion' must mean a singularity in the mathematical sense. . . . This consideration does, however, not alter the fact that 'the beginning of the world' really constitutes a beginning, from the point of view of the development of the new existing stars and systems of stars, at which those stars and systems of stars did not yet exist as individual entities."[91] This was a statement to which Lemaître fully subscribed.

An Old-Fashioned Cosmologist

With his innovations of 1930 and 1931 Lemaître suddenly became a scientific celebrity. In 1932–33 he was again in North America, where his presence aroused much interest. The fact that this mathematician and universe builder was also a Catholic priest contributed to the curiosity of the press. *Science News Letter* wrote about "[the] Belgian priest, Abbé Le Maitre, who teaches astronomy and mediates fruitfully in his monastery cell . . . [and] who pointed out that the universe, as postulated mathematically by Einstein, would collapse if anyone in it did so much as wave his arm."[92] His meetings with Einstein at the Mount Wilson Observatory in January 1933, and later the same year with Shapley at the Harvard Observatory, received particularly broad press coverage. Einstein, who had great personal sympathy for Lemaître, may have been an early convert to the big-bang theory. "This is the most beautiful and satisfactory explanation of creation to which I have ever listened," he reportedly said after Lemaître had presented his view at a meeting at the California Institute of Technology.[93]

The following year Lemaître received the Franqui Prize from the Belgian king, Leopold III. Einstein was one of the three sponsors for the prize, which carried with it about $33 000 and at the time was the world's highest scientific award other than the Nobel Prize. While at Harvard, Lemaître presented his new theory of galaxy formation, which, according to one account, involved a soapsuds universe rather than a shrapnel universe: "The suds, explained the Belgian scientist, are in disequilibrium, some whipped by a cosmic repulsive force (expressed by the constant lambda), some clutched at by the attractive force which earthlings know as gravity. While some bubbles swell and others contract, still others, are in a state of stagnation. Within some regions, where expansion is the rule, there are collapsing systems flying headlong away from one another. Also, in slowly collapsing regions are to be found a number of rapidly collapsing systems. Such a system is the Milky Way, the Galaxy to which Earth belongs."[94]

Press coverage apart, how did Lemaître's fellow cosmologists respond to this first version of the big-bang universe? Insofar as it was considered in its mathematical aspects only, i.e., as a solution to the Friedmann-Lemaître equations with $R \approx 0$ for $t = 0$, it was frequently mentioned together with other solutions of a similar kind, such as the Einstein–de Sitter solution. But it was not seen as particularly interesting, and was at any rate considered a mathematical and not a physical model. In an address to a philosophical forum in 1932, Robertson mentioned the possibility of an initial cosmic space-time singularity. However, he considered it neither a realistic nor an appealing possibility and preferred "emotionally more satisfactory" solutions such as the ever-expanding Lemaître-Eddington model.[95] Bishop Barnes was also sceptical. He considered Lemaître's hypothesis "a brilliantly clever *jeu d'esprit* rather than a sober reconstruction of the beginning of the world," a view which was undoubtedly shared by many cosmologists.[96]

As far as the physical aspects of the Lemaître model were concerned, including the exploding primeval atom, they were ignored in the scientific literature of the 1930s (but were sometimes mentioned in the popular literature). If there was a preference for a particular cosmological model, the Lemaître-Eddington model seems to have been the choice at least in the first half of the decade. Heckmann found this model attractive "because it allows the possibility of a world without catastrophic behaviour either in the past or the future,"[97] and Tolman, de Sitter, and Robertson also emphasized the advantages of the model. However, they all realized that there was a gap between the many nice mathematical solutions and the scanty observations of the one and only real universe. As de Sitter put it in 1930: "There is nothing in our observational data to determine the choice of any particular curve for the representation of the history of the universe. The selection must remain a matter of taste or of philosophical preference."[98] This was, of course, a rather pessimistic view, but there was little in the development of observational cosmology during the 1930s that gave reason for optimism.

Tolman, who collaborated with Hubble and strove to integrate theory and observations, admitted in 1934 that "the observational data . . . are . . . insufficient to make any decision as to the kind of temporal behaviour that should be ascribed to the model over long periods of time."[99] Yet he found it necessary to warn against philosophical preferences and "the evils of autistic and wish-fulfilling thinking" in cosmology. Among such prejudices he included the belief of the universe being created in the past: "We must be specially careful to keep our judgements uninfected by the demands of theology and unswerved by human hopes and fears. The discovery of models, which start expansion from a singular state of zero volume, must not be confused with a proof that the actual universe was created at a finite time in the past."[100] The beginning of the world was a radical concept, which was difficult to accept, and to which scientists had to accustom themselves.

Hubble, who only reluctantly entered the "shadowy realm" of cosmological theory, was in 1936 led to consider a big-bang model of Lemaître's type. In the Rhodes Memorial Lectures delivered at Oxford that year he concluded that *if* the universe was expanding, then a Lemaître universe with Λ = ca. 4.5×10^{-18} (light years)$^{-2}$ was the only possibility. "The model expands without reversal," Hubble explained. "The radius increases from zero to infinity. Past time is finite, future time is infinite. Comparatively recently, perhaps a thousand million years ago, the model started to expand from a small compact mass. The expansion was very rapid at first, but it has steadily slowed down to the rate we measure to-day."[101] However, Hubble in no way found this model attractive, among other reasons because it seemed to lead to a much too high density and a suspiciously small present size. He concluded that although it could not be ruled out, the Lemaître big-bang model was "rather dubious."

As for Eddington, he never accepted the primeval-atom hypothesis of his former student. In 1931 he had challenged Lemaître by calling big-bang ideas "repugnant," and he remained hostile to them. During the last ten years of his life, Eddington's main occupation was with neither astrophysics nor cosmology, but with an ambitious attempt to reformulate physics into a new "theory of everything."[102] In his last paper presented to the Royal Astronomical Society, in 1944, Eddington argued for a kind of Lemaître-Eddington universe with a Hubble constant of H = 572 km·s^{-1}·Mpc^{-1} and a cosmological constant $\Lambda = (\pi c^2 / 2\kappa M)^2$. Characteristically, he did not derive these values from comparison with observation, but from his "fundamental theory" claimed to unify cosmology and quantum theory. For the Einstein-like equilibrium state, Eddington obtained R_0 = ca. 302 Mpc and ρ_0 = ca. 1.2×10^{-27} g·cm^{-3}, and by comparison with astronomical data he concluded that "the present radius of the universe is between 1000 and 2500 megaparsecs, and most probably about 1500 megaparsecs."[103] With regard to the time elapsed since "the universe burst" Eddington argued that the equilibrium state was only uniform in a statistical sense. In that case the time scale would not be $t = \infty$ but rather $t = 9 \times 10^{10}$ years. This was not really the age of the world, but according to Eddington it was the time since condensations began to form.

Lemaître did not develop his cosmology much beyond the level it had attained by 1933. Between 1934 and the late 1940s he largely stopped working in cosmology, and what he published was mostly repetitions of earlier work. In 1946 he collected several of his cosmological articles in a book, *L'Hypothèse de l'Atome Primitif*, which made the primeval-atom hypothesis better known but contained nothing new.[104] In 1958, twenty-seven years after his introduction of the primeval atom, he gave a lecture at the Solvay conference which exclusively referred to his own work in the early 1930s and in which there was no sign of post-1940 cosmological advance.

In August 1961 Lemaître attended the Conference on the Instability of Systems of Galaxies at Santa Barbara. The sixty-seven-year-old father of the big-bang universe talked about his stagnation model with a positive cosmological constant. "It is realized," he wrote, "that this model, proposed some 30 years ago, may be considered an 'old fashioned cosmology.' This is not the place to discuss the strong prejudices that have arisen against it due to the reverence of an authority whose influence can only be compared to that of Aristotle in older times."[105] This was presumably an allusion to Fred Hoyle. The big-bang version which Lemaître sketched in 1961 scarcely differed at all from his 1931 version, and he referred to it in almost the same terms he had used thirty years earlier: "My cosmology can only be properly discussed when it is coupled with the primaeval atom hypothesis involving a beginning of the world, not only as a beginning of space, but also as a beginning of all multiplicity. It thus suppresses metaphysical questions about this beginning—the bottom of space-time and of all physics for which no origin can be conceived."[106]

It is ironic that Lemaître, the first scientist to emphasize the importance of introducing quantum ideas in the earliest universe, was so reticent, if not plainly uninterested, when Gamow took the step of transforming the big-bang idea into a version where quantum and nuclear physics could actually be applied and not only talked about. In his address to the British Association in 1931, Lemaître had emphasized that in order to decide whether his suggestion was a "wild imagination or [a] physical hypothesis," progress was needed in two areas, nuclear structure and the cosmic radiation. "We must wait," he said, "but we may trust the physicists that we do not have to wait too long." And, indeed, within two decades physicists had made tremendous progress in the two areas, demystified the cosmic radiation and produced a reliable, quantum-mechanical model of thermonuclear reactions. But in spite of the fact that Lemaître's appeal to the physicists was thus answered, he seems barely to have noticed the development. In his contribution to the Einstein volume of 1949 he merely repeated that "further progress in nuclear physics must be awaited" before the primeval-atom hypothesis could be turned into a proper theory. In what may have been an indirect response to Gamow's new theory (which he knew about), he wrote that "It is not very profitable to insist on [the precise nature of] the extreme physical conditions which arose at the very beginning."[107]

The strange fact is that the Belgian abbé showed a complete lack of interest in the new big-bang approach, although, in a sense, it vindicated his hypothesis. His collaborator Odon Godart recalled that he more than once suggested that Lemaître contact Gamow, but that Lemaître always refused.[108] His reticence seems to have been rooted in his general attitude to science, the ideals which he had adopted as a young man through his reading of Einstein. Lemaître's ideal of science continued to be the general theory of relativity, with its great conceptual beauty and unifying force. He could find nothing corresponding to this ideal in the development of nuclear and particle physics, which he found complicated and immature sciences; possibly practical, but certainly not beautiful. In conversations with Godart, he used to refer to the physics of elementary particles as "entomology." With such a conception it was not so strange after all that he ignored the new developments in big-bang cosmology.[109] Georges Lemaître died on 20 June 1966. Shortly before his death he was informed of the (1965) discovery of the cosmic microwave background radiation and its interpretation as the fossils of the big bang. Lemaître had expected such fossils, but in the form of cosmic rays and not microwave photons. Yet it must have pleased him to know that physicists had discovered something akin to the "ashes and smoke of bright but very rapid fireworks."

Lemaître on Religion and Science

Because Lemaître was a priest it is tempting to see his primeval-atom universe as a projection of his religious view of creation into a scientific context, that is, as ideologically motivated. Indeed, big-bang cosmology has often been suspected of concordism, and Lemaître of having suggested his hypothesis because of its agreement with Genesis. In 1939, the Swedish physicist Hannes Alfvén listened to Lemaître accounting for his theory at an astrophysics conference in Stockholm. Alfvén recalled that he felt "that the motivation for his theory was Lemaître's need to reconcile his physics with the Church's doctrine of creation *ex nihilo*."[110] During the controversy with the steady-state theory, this theme played a certain role and for this reason it is relevant to examine what Lemaître thought about the science-religion relationship.[111]

Lemaître's Christianity was far from fundamentalism. He had little but scorn for those who interpreted the Bible literally, as a scientific text. In connection with his stay in the United States in 1933 he elaborated on the subject: "Hundreds of professional and amateur scientists actually believe the Bible pretends to teach science. This is a good deal like assuming that there must be authentic religious dogma in the binomial theorem. Nevertheless, a lot of otherwise intelligent and well-educated men do go on believing or at least acting on such a belief. When they find the Bible's scientific references wrong, as they often are, they repudiate it utterly. Should a priest reject relativity because it contains no authoritative exposition of the doctrine of the Trinity?"[112] No, he should not. Salvation, not nature, is what religion is about and this is the reason why the doctrine of Trinity (which "is much more abstruse than anything in relativity or quantum mechanics") is part of the Bible. Furthermore: "The idea that because they [the writers of the Bible] were right in their doctrine of immortality and salvation they must also be right on all other subjects is simply the fallacy of people who have an incomplete understanding of why the Bible was given to us at all."

Lemaître's intellectual position can be characterized as epistemic optimism. In 1929 as well as on many later occasions he stated that God has given man intellectual faculties so as to be able to discover every aspect of the universe. Lemaître's God would hide nothing from the human mind, and consequently there could be no real contradiction between Christian belief and scientific cosmology. "The universe is not beyond human possibilities," he said in a talk delivered to the Catholic Institute in Paris in about 1950. "It is like Eden, the garden which had been placed at the disposal of man so that he could cultivate it and explore it."[113] This does not mean that Lemaître's cosmology was designed to fit theological views or that he used it in defending such views. On the contrary, he was always of the opinion that science and theology were separate fields which, though ultimately leading to the same goals, should not be mixed. At the Solvay Congress he made this crystal clear:

> As far as I can see, such a theory [of the primeval atom] remains entirely outside any metaphysical or religious question. It leaves the materialist free to deny any transcendental Being. He may keep, for the bottom of space-time, the same attitude of mind he has been able to adopt for events occurring in non-singular places in space-time. For the believer, it removes any attempt to familiarity with God, as were Laplace's chiquenaude or Jeans' finger. It is consonant with the wording of Isaias speaking of the "Hidden God" hidden even in the beginning of the universe. . . . Science has not to surrender in face of the Universe and when Pascal tries to infer the existence of God from the supposed infinitude of Nature, we may think that he is looking in the wrong direction.[114]

In spite of such unambiguous declarations, it was and still is often assumed that Lemaître's inspiration for, as well as justification of, his hypothesis has to be found in its similarity with the Christian view of the creation of the world. It may be tempting to suggest such an association, but it has no foundation in the sources.[115]

Contrary to many later cosmologists, Lemaître was careful to distinguish between the "beginning" and "creation" of the world, and never spoke himself of the explosion of the primeval atom in terms of the latter concept. He realized that the beginning of the world, the undisturbed primeval atom or initial singularity, might forever defy scientific analysis, but he denied that such lack of knowledge justified the entrance of religion. Many scientists as well as nonscientists have given in to the temptation of substituting science with God in the moment of creation. Lemaître always considered this a pseudosolution, scientifically as well as theologically illegitimate. Not only is "God" a nonsensical concept from the point of view of science (since it is beyond mathematical reasoning and experimental testing), but from the point of view of religion it would be all wrong to reduce God to a scientific hypothesis. Lemaître's emphasis on two separate levels of understanding, one scientific and another religious, did not imply that cosmology is irrelevant for religion. He believed that religious and philosophical values were important, indeed essential, to the scientist on a broader ethical level, but they should not interfere in his methods or conclusions. "To search thoroughly for the truth involves a searching of souls as well as of spectra," he said.[116]

2.3 COSMYTHOLOGIES

The 1930s witnessed a proliferation of cosmological ideas and models, many of which were opposed, in one way or another, to standard relativistic theory. Most of these heterodox theories were short-lived and failed to have any lasting effect on the cosmological scene, but in a few cases they attracted considerable interest as possible alternatives to the relativistic theories building on the Friedmann-Lemaître equations. Two, Milne's and Dirac's (and the extension of the latter by Jordan) had a lasting impact on the cosmological debate in Great Britain well into the 1950s, and formed part of the background of the later controversy between big-bang and steady-state cosmologies. The two theories accepted the expanding universe and pictured the universe as starting in a singularity-like state. They were thus, in spite of not building on the theory of general relativity, a kind of big-bang model. They also represent a peculiar trend in British cosmological thinking, a particular mixture of rationalism and empiricism, which was strong in the 1930s and 1940s, and the influence of which colored a good deal of the discussion in the 1950s also. In order to understand the later debate more fully it is essential to be aware of the earlier theories and the kind of cosmological reasoning they exemplified.[117]

Milne's Heresy

Edward Arthur Milne, born in 1896, was one of the leading astrophysicists in the period 1920–40 and an important figure in prewar cosmology. As a professor of mathematics at the University of Manchester and later, from 1929 to his death in 1950, the first holder of the Rouse Ball Professorship at Oxford University, he did fundamental work in the study of stellar atmospheres and structures in particular, which made him a fellow of the Royal Society at the age of only thirty years. Under the impression of the general excitement over the expanding universe, Milne began in 1932 to focus his interest on cosmology, a field in which he had previously shown almost no interest. He developed, first in 1933 and later in numerous other works, an original cosmological scheme very different from that of general relativity.[118]

Milne found the general theory of relativity to be mathematically as well as philosophically monstrous. It was his preference for common sense explanations and observationalist criteria that first made him embark on his ambitious program of reforming cosmology. Space was for him not an object of observation, but a system of reference, and thus could have no structure, curved or not. His own model of the universe supposed a flat, infinite Euclidean space and simple kinematic considerations. In his systematic and impressive 1935 presentation of his theory, *Relativity, Gravitation, and World-Structure*, he argued that all the fundamental laws of cosmic physics can be deduced from a very few principles, most of which may be obtained from analyzing the concepts used to order temporal experiences and to communicate them by means of optical signals. The physics arising from such considerations would naturally be restricted to distance and temporal relations; they could involve no initial appeal to dynamical or gravitational assumptions.

Milne managed to show that the expansion of the universe could easily be explained without Einstein's theory. He required only two postulates, Einstein's axiom about the constancy of the velocity of light, the basis of the special theory of relativity, and the so-called cosmological principle. According to this principle, which he first formulated in 1933, the world must appear the same to all equivalent observers, irrespective of their positions. Containing the postulates of large-scale isotropy and homogeneity, the cosmological principle occurred implicitly in Einstein's work of 1917 (and almost explicitly in 1931), but only rose to a basic principle with Milne. He originally called it the "extended principle of relativity" and formulated it as follows: "Not only the laws of nature, but also the events occurring in nature, the world itself, must appear the same to all observers, wherever they be, provided their space-frames and time scales are similarly oriented with respect to the events which are the subject of observation."[119] Applying the cosmological principle in interpreting Hubble's law, Milne found as a unique solution a uniformly expanding world model, i.e., H = const. The epoch of such a world is simply the Hubble time, $t_0 = T$. (Among the relativistic models a similar solution exists, but only in an empty universe; Milne's model was filled with matter.) Milne's theory had the advantage that it only allowed expansion, contrary to the relativistic theory which does not in itself distinguish between expansion and contraction.

Among the many results obtained by Milne in the mid-1930s was that the value of the gravitational constant depends upon the epoch. He found that Newton's "constant" varies with time according to $G = (c^3/M_0)t$, where c is the velocity of light and M_0 denotes "the apparent mass of the fictitious homogeneous universe," i.e., the mass of the universe according to those cosmological theories that adopted a curved finite space. "This means," Milne wrote, "that the Newtonian 'constant' of gravitation . . . should be proportional to the epoch t reckoned from the time-zero. It should be apparently increasing at the rate of one part in 2×10^9 per year."[120]

The term "apparently" is significant. Milne placed all of his results, including the mentioned equation, in a conventionalist perspective, and emphasized that the result does not imply that local gravitation, as in the solar system, varies in strength. Consequently, he did not think of his equation as verifiable. As another and more important illustration of the conventionalist character that Milne assigned to his equations, consider the time parameter t. This does not represent the usual dynamic time (τ), but what Milne termed the kinematic time (t). His analysis of time, lying at the heart of his project of reconstructing physics, was based on his wish to establish a rational system of timekeeping, without which, he felt, the universe would not be intelligible. Milne was able to show that in an ideal expanding universe, satisfying the cosmological principle, any two observers located at any two of the uniformly receding galaxies can calibrate their clocks in congruence. The common time of these observers is the kinematic time t. On the other hand, a set of observers, any two members of which are at relative rest, can also recalibrate their clocks and thus communicate rationally with each other. The time scale used in the latter case is the dynamic time τ, corresponding to the usual Newtonian time. This time scale is connected with t time by the rule $\tau = t_0 \log(t/t_0) + t_0$ or $d\tau/t_0 = dt/t$, where t_0 signifies the present epoch.

Milne's two time scales, further elaborated in cooperation with Gerald Whitrow and William McCrea, played an important role in later cosmological discussions. In Milne's practice, however, it was not very clear which processes followed t time and which τ time.[121] Nor was it very important, for according to him the choice of time parameters in various physical processes was largely a matter of convention: t time and τ time were particularly useful in cosmology, but in principle an infinity of different time scales could be established, each yielding different but equally valid equations for physical processes. The proportionality between G and t arises as a consequence of the arbitrary use of t time; in τ time, G reduces to a constant, the ordinary gravitational constant. Similarly, on the t scale there is a natural origin of time, $t = 0$, and the geometry is Euclidean; on the τ scale the universe remains stationary and hyperbolic, its history extending backwards to $\tau = -\infty$. But, as Milne liked to emphasize, the two descriptions of the universe "are merely two different descriptions of the same physical reality."[122] In the stationary version, Milne could not explain the redshifts as Doppler shifts, but he found that the wavelength of an atom would depend on τ time, and was in this way able to reproduce the universal redshift.

Was Milne's world model a big-bang universe? His answer seemed to be both yes (t time) and no (τ time), which was not a contradiction, but a result of his conventionalism. In 1933 he briefly discussed the "creation" at $t = 0$, to which he added the remark: "This question of creation has however only a philosophical importance. . . . What 'really happened' at $t = 0$ is not a scientific question."[123]

Milne's system of "kinematic relativity" diverged distinctly from traditional physics, although Milne maintained that he had merely followed to the end "Einstein's principle" of introducing only elements observable in principle. According to Milne, this principle implied that entities such as "laws of nature" and "curvature of space" had no objective existence, and hence that the general theory of relativity was unacceptable. Philosophically Milne subscribed to positivism, but of a peculiar brand that was heavily influenced by an almost unlimited confidence in the power of pure reasoning. For example, in a comment on whether the universe contains a finite or an infinite number of objects, he wrote: "I only point out here that whilst observation could conceivably verify the existence of a finite number of objects in the universe it could never conceivably verify the existence of an infinite number. The philosopher may take comfort from the fact that, in spite of the much vaunted sway and dominance of pure observation and experiment in ordinary physics, world-physics propounds questions of an objective, non-metaphysical character which cannot be answered by observation but must be answered, if at all, by pure reason; natural philosophy is something bigger than the totality of conceivable observations."[124]

Whatever his intentions, Milne's way of doing physics resulted in an extreme deductivism, the physical laws being compared with geometric theorems. As in geometry, empirical knowledge played only an insignificant role in kinematic relativity. In 1937 Milne described his ambitious programme in the following Cartesian way: "I have endeavoured to develop the consequences [of the cosmological principle] without any empirical appeal save to the existence of a temporal experience, an awareness of a before-and-after relation, for each individual observer."[125] Although he originally stressed that his kinematical approach was limited to world physics, in the late 1930s and 1940s he proceeded to apply it also to electrodynamics and even quantum theory. His attempt to reconstruct all of physics met with no success and was rejected even by those who found his cosmological program valuable.

Twenty years after Milne had started his work in kinematic relativity, it was largely forgotten, at least outside England, and today Milne's name appears in texts on cosmology only in relation to the uniformly expanding Milne model. However, his influence was very considerable in the period from about 1933 to 1948 and many British theoreticians regarded the kinematic approach as promising. From 1932 to 1940 there appeared about seventy papers related in one way or another to Milne's theory, which means that the theory had a predominant position in the period.[126] Among those working within the problem area defined by Milne's ideas were not only his collaborators Whitrow and McCrea, but also McVittie, Whittaker, Arthur G. Walker, T. Lewis, William A. Kermack, V. V. Narlikar, Martin Johnson, and, in the United States, Robertson. None of these theoretical cosmologists were willing to follow Milne all the way, but they took his work and approach seriously. It is worth noticing that A. G. Walker's derivation of the famous Robertson-Walker metric for homogeneous isotropic world models was based on Milne's work and not on general relativity. After having completed his Ph.D. under Whittaker in 1933, Walker devoted his talents to Milne's cosmology, and it was during this work that he derived the metric associated with his name.

In England, the impact of Milne's work extended far beyond the small communities of cosmologists and mathematical physicists. The biologist John B. S. Haldane found the theory interesting because of its thorough "historical" nature and because it pictured the universe as evolving from a singularity.[127] As a marxist, he referred approvingly to the theory as an expression of the fundamental dialectics of nature suggested by Friedrich Engels.[128] In early 1945 Haldane attempted to deduce from Milne's theory a comprehensive evolutionary picture of the world, admitting it to be the work of an amateur and probably "wildly speculative," which is was. Haldane did not hide that his preference for Milne's system was ideologically rooted, and that he saw it as a way of avoiding the choice between the traditional cosmologies, either with a beginning in the past or describing an eternal universe. "On the first hypothesis, why was it not created better; on the second, why has it not got better in the course of eternity?" he asked—a reasonable question for a European in January 1945. Haldane went on: "On neither theory have we very strong grounds for hoping that the world will be a better place a million, let alone a thousand, years hence, than it is today. But on Milne's theory the laws of nature change with time. The universe has a real history, not a series of cycles of evolution. Although, from one point of view, the past is infinite, life could not have started much before it did, or have got much further than it has at the present date. If this is so, human effort is worth while and human life has a meaning."[129]

This was not the first time that this theme entered cosmology, nor would it be the last. Haldane suggested that, since in Milne's model the radius of the world (in t time) varied as $R = ct$, there would at any time exist a maximum size of photons given by $\lambda_{max} \approx ct$ and thus a minimum frequency of the order t^{-1}. Going far back in time, say to $t = 10^{-72}$ s, photons would therefore have had enormous frequencies and energies, 10^{-72} Hz and 6.5×10^{45} erg; at the time $t = 10^{-92}$ s, the smallest photons would have had an energy corresponding to the mass of a galaxy! (But in τ time, where the universe is static, nothing would change.) Without mentioning Lemaître, Haldane pictured the universe as originating from one or a few such superphotons of almost infinite energy and sketched from this assumption the entire evolutionary history of matter and life.

Milne supported Haldane's grand cosmic speculation wholeheartedly.[130] One would think that the whole picture collapsed because of the arbitrary use of t time, but in 1945 Milne seems to have reached the conclusion that kinematic and dynamic time were not, after all, equally valid. He now found the τ scale "a concession to our Newtonian predilections," whereas "phenomena themselves" were best studied through the more important t scale. Milne's version of the big bang was this: "Just as the epoch $t = 0$ is a singularity in the mechanical t-history of the universe—an epoch at which the density was infinite—so the epoch $t = 0$ is a singularity in the optical history of the universe, namely, an epoch at which the frequency of radiation was infinite, because the wave-length had to be zero." He suggested that some of the original superphotons might have preserved a very high frequency and make up part of the present cosmic radiation. The Haldane-Milne hypothesis was short-lived and not well received in most scientific circles. Raymond Coutrez, an astronomer at the Royal Belgian Observatory, found the idea fascinating and praised Haldane's and Milne's "constructive and ingenious spirits."[131] A much more reserved reply came from another Belgian—Georges Lemaître—who questioned if Haldane's superphotons could be ascribed any physical meaning even in principle. Lemaître also used the opportunity to criticize Milne's idea of two time scales. He argued that all physical phenomena are so intimately interrelated that it is simply impossible to separate them into two different time categories.[132] As far as Lemaître was concerned, Milne's theory bordered on nonsense.

As Milne developed his program during the 1940s, it became increasingly clear that religion was part of it, namely, in the sense that his theory led to a creation of the world which he ascribed to God's supreme will. At a symposium in Brussels in 1947, Milne referred several times to his theory of kinematic relativity as if it reflected the thoughts and actions of God. For example, he used a strange theological argument to show that the creation of the universe had to have taken place as a singularity and that the created universe had to be infinite. Creation of an extended universe, such as the static Einstein universe, was said to be "a logical impossibility; something beyond the power of God himself." Thus, Milne explained: "However unnatural may seem the idea of creation at a point, it is the only form of creation which is free from logical contradiction."[133] And as far as the infinity of the universe was concerned, Milne offered the following argument: "If we attribute the act of creation to a Divine First Cause, then we can see in the infinite multiplicity of evolutionary experiments thus started the power and majesty of this First Cause. If evolution consists in the occurrence of mutations, there would be something little and, as it were, pettifogging in the application of the evolutionary process to a finite universe, in which only a finite number of evolutionary experiments could be practised. It would be to put Deity in a straitjacket. On the other hand, in creating an infinite universe, we can say that God has provided himself with the means of exhibiting and practising his own omnipotence."[134]

In the discussion following Milne's exposition, both Lemaître and Herman Weyl criticized Milne's cosmology. Lemaître made it clear that he disliked the system, which he found arbitrary and artificial. He must have felt Milne's use of religion rather embarrassing, but chose not to mention this aspect. Weyl was no less critical and found it "highly objectionable to infer creation by a divine power from the fact that . . . there is an absolute beginning of time $t = 0$."[135] After all, in Milne's system there was no beginning on the τ scale of time.

Finally, it should be mentioned that much of Milne's work in the 1930s was related to, and inspired other researchers to investigate, Newtonian models of the universe. Neo- or quasi-Newtonian cosmology was cultivated by many mathematicians and astronomers after 1934, when Milne and McCrea suggested a new way to apply Newtonian arguments to the expanding universe.[136] Among other results, they derived the Friedmann-Lemaître equations on such a basis. Neo-Newtonian cosmology was, in general, remote from physical theory and few scientists believed that the real universe obeyed the classical laws of mechanics. The occupation with neo-Newtonian cosmological models continued up through the 1950s, in many cases using ideas from general relativity, but was of more interest to the mathematician than the physicist. Since this kind of cosmology was of little relevance in the discussion of the real world picture, we shall not be further concerned with it.

Dirac and the Large-Number Hypothesis

Dirac's contributions to cosmology in the 1930s, limited to two articles, were less important than Milne's and attracted much less interest among cosmologists at the time they appeared. But they contained ideas which proved to be of considerable importance in the later cosmological discussion and which are still being discussed.[137] When thirty-four-year-old Paul Dirac first turned to cosmology in 1937, he was one of the world's leading theoretical physicists, a cofounder of quantum mechanics and a Nobel laureate of 1933 for his fundamental contributions to this field. His fame was particularly related to the relativistic wave equation of the electron (the "Dirac equation") of 1928 and his prediction, three years later, of the positively charged electrons, which were detected in 1932 and then became known as positrons. Dirac's interest in cosmology derived in part from Lemaître, who in 1933 had lectured on his primeval-atom hypothesis in Cambridge, and in part from Eddington's unorthodox attempt to bridge quantum physics and cosmology. Fascinated by the Dirac equation, Eddington interpreted it as describing, not an individual electron, but the structural relation of the electron to the entire universe (which he, contrary to Milne, described within orthodox general relativity). Developing this program during the 1930s, Eddington emphasized various numerical relationships between micro- and macrophysical constants of nature, including the "cosmical number," which is the number of protons in the universe, about 10^{79}. Eddington claimed to be able to deduce this number from pure theory, not only as an order of magnitude, but exactly. As he once wrote in a memorable introduction to a chapter in one of his books: "I believe there are 15,747,724,136,275,002,577,605,653,961,181,555,468,044,717,914,527,116, 709,366,231,425,076,185,631,031,296 protons in the universe, and the same number of electrons."[138]

In a brief note of 1937, clearly inspired by Eddington and Milne, Dirac suggested a reconsideration of cosmology based on the large dimensionless numbers that can be constructed from the fundamental constants of nature. He accepted a big-bang theory à la Lemaître and based his considerations on "current cosmological theories [according to which] the universe had a beginning about 2×10^9 years ago, when all the spiral nebulae were shot out from a small region of space, or perhaps from a point."[139] Dirac's fundamental postulate was the large-number hypothesis, namely, that all very large dimensionless numbers occurring in nature are interconnected. He paid particular significance to numbers of the orders of magnitude 10^{39} and 10^{78} [$= (10^{39})^2$] and this for the following reason. With a unit of time given by e^2/mc^3 the age of the universe, he observed, will be about 10^{39}, which is almost the same as the ratio between the electrostatic and gravitational forces between an electron and a proton, e^2/GmM. This suggested to him "some deep connexion in Nature between cosmology and atomic theory." If the numerical agreement is significant and charges and masses do not change with time (as Dirac assumed), the consequence is that the gravitational "constant" decreases with atomic time as $G \sim t^{-1}$.

Another consequence of the large-number hypothesis followed from the constant $\rho(cT^3)/M$, where ρ is the mean density of matter and $cT = c/H$ is the radius of the observable universe, the Hubble radius. Since the value of the constant—giving the number of particles in the universe—is about 10^{78}, and this is the square of the period in atomic time, Dirac concluded that the number of particles will increase with time according to the law $N \sim t^2$. Naturally, both of these suggestions—a gravitational constant decreasing in time and the spontaneous creation of matter—were highly unorthodox and conflicted with the general theory of relativity. The following year, 1938, Dirac decided that cosmic matter was conserved after all.[140] He now extended his considerations into a new cosmological model, which was infinite, spatially flat, and with a zero cosmological constant. However, in order to obtain these results, Dirac had to give up the hypothesis of matter creation, which thus survived for less than a year. He claimed that this step could be reconciled with the large-number hypothesis and maintained the idea of a decreasing constant of gravitation.

Dirac arrived at his cosmological model by observing that the Hubble time $T = R/R'$ and the reciprocal of the mean density of matter in the universe, if expressed in atomic units, are very large numbers of roughly the same order of magnitude, about 10^{40}. On account of the large-number hypothesis he therefore inferred that $\rho = k\, H(t) = k\,(R'/R)$ where k is a constant of the order of magnitude unity. Combining this equation with $\rho \sim R(t)^{-3}$ he got $R^{-3} \sim R'/R$ or $R(t) \sim (t + a)^{1/3}$. Dirac took the constant of integration a to be zero and thus arrived at a recession law of galaxies given by

$$R(t) \sim t^{1/3}. \tag{2.7}$$

It follows that the age of the Dirac universe is only $T/3$ and that the velocity of recession varies as $t^{-2/3}$. Dirac's cosmological model was clearly of the big bang type. As he wrote: "With this law of recession we still have a natural origin of time, namely the zero of the t in [2.7], when all the nebulae were extremely close together."[141] He did not explain what "extremely close" meant.

Dirac's adventure into cosmophysics had little impact on contemporary cosmology. After the war it was widely believed to have been disproved by arguments proposed by the Hungarian-American physicist Edward Teller in 1948.[142] Teller reasoned that if the gravitational constant had varied in accordance with Dirac's hypothesis, then the temperature of the earth in the past would be given by $\theta = \theta_0(t_0/t)^{2.25}$, where θ_0 is the present temperature, t the present age, and t_0 the age in the past. For the Cambrian era, some three hundred million years ago, this yields a temperature of about 110°C, which clearly conflicts with paleontological evidence of life on earth at that time. Consequently, Teller concluded that $G \sim t^{-1}$ must be wrong. His work was generally regarded as a mortal blow against Dirac's cosmology, although in fact it relied on assumptions which could be (and later were) questioned.

For a long period Dirac left the field of cosmology. He only returned to his theory in the early 1970s, at a time when relativistic big-bang theory had obtained a paradigmatic status. Although Dirac dropped the idea of matter creation in 1938, it was taken up by a few other physicists and astronomers who felt fascinated by the large-number hypothesis. Subrahmanyan Chandrasekhar adopted Dirac's reasoning and proposed that the numbers of particles in stars might increase in time as $t^{3/2}$ and those in galaxies as $t^{7/4}$. Other applications of Dirac's idea were made by the astrophysicist D. S. Kothari and by the physicist Frederick Arnot.[143] However, most astronomers and physicists received Dirac's unorthodox theory with silence, if not embarrassment. Yet, in the long run, Dirac's speculations proved more influential than Milne's much discussed theory. The lasting impact of the theory, it turned out much later, was the large-number hypothesis, not the idea of a varying gravitational constant nor the particular universe model suggested. The general idea of a fundamental interconnection between the large combinations of natural constants proved to be a source of constant fascination in postwar cosmology. It appealed to big-bang and steady-state cosmologists alike.

A Debate on Cosmophysics

The kind of rationalistic cosmophysics expounded in different versions by Eddington, Milne, and Dirac enjoyed considerable popularity in England, but was strongly opposed by more empirically minded scientists and philosophers. Herbert Dingle, in 1937 a forty-seven-year-old astrophysicist at the Imperial College, University of London, was among the first to launch a counterattack against the unbalanced a priori methods of what he called the "modern Aristotelians." Dingle specialized as a young man in astronomical applications of spectroscopy. In 1932 he spent a year at Caltech, where he met Tolman and also Einstein and became interested in theoretical cosmology. Dingle made one of the first systematic studies of anisotropic universes, arguing that the assumption of spatial homogeneity was nothing but "a working hypothesis, valid so long as it does not conflict with observation or with theoretical probability."[144] This view contrasted with Milne's new cosmological principle, which Dingle categorically dismissed. Greatly interested also in the history and philosophy of science, Dingle had recently published a book defining science as organized common sense,[145] and he now felt this empiricist view threatened by the theories of Milne, Eddington, Dirac, and their epigones.

Dingle accused the rationalistic camp of perverting the proper or "Galilean" method of science, which Milne saw as an unsophisticated empirical inductivism. His scathing criticism was aimed primarily against Milne's theory, and especially the cosmological principle, which, he felt, was nothing but pseudo-science. To admit general principles not derived from experience would be the death of experimental philosophy, Dingle declared: "Milne and Dirac . . . plunge headlong into an ocean of 'principles' of their own making. . . . Instead of the induction of principles from phenomena we are given a pseudo-science of inveterate cosmythology, and invited to commit suicide to avoid the need of dying. If anyone is uncertain about the place of imagination in science, let him compare Lord Rayleigh's discovery of argon with Dirac's discovery of the contemporary creation of protons which, according to *The Times*, 'alters fundamentally our ideas of the structure of the universe and the nature of time.' "[146] Dingle's strongly worded objections caused a heated debate in *Nature* that engaged many of Britain's most prominent scholars, among them the astronomers McCrea, Whitrow, and R. A. Sampson, the physicist Charles G. Darwin, the mathematician and geophysicist Harold Jeffreys, the biologist J. B. S. Haldane, and the philosopher G. Dawe Hicks.

In reply, Milne denied Dingle's charges of mysticism and extravagance, which, in fact, were unfair. Milne's ambition was rather to reconstruct physics so as to open it to the common person, and he often declared his distaste for the sort of obscurity that he claimed to find in Jeans. But with regard to rationalism, Milne saw no reason to bow: "The universe is rational. By this I mean that given the mere statement of *what* is, the laws obeyed can be deduced by a process of inference. There would then be not two creations [one of matter, the other of law] but one, and we should be left only with the supreme irrationality of creation, in Whitehead's phrase. . . . Laws of Nature would then be no more arbitrary than geometrical theorems. God's creation would be subject to laws not at God's further disposal."[147]

Dingle made it clear that there was more at stake than a few mathematical physicists' idiosyncrasies. What really alarmed him were matters of science policy and cultural standards, "the general intellectual miasma that threatens to envelop the world of science." Alluding to the dark clouds over intellectual life in Nazi Germany and the Soviet Union, he warned against the authoritarian tendencies that he saw represented in cosmophysics: "The criterion for distinguishing sense from nonsense has to a large extent been lost: our minds are ready to tolerate any statement, no matter how ridiculous it obviously is, if only it comes from a man of repute and is accompanied by an array of symbols of Clarendon type. . . . There is evidence enough on the Continent of the effects of doctrines derived 'rationally without recourse to experience.' To purify the air seems to me an urgent necessity." Dingle had earlier criticized modern theoretical physics for being esoteric and remote from sound scientific reasoning, features he traced back to an excessive mathematization of physics. In 1934 he objected to the theoretical physicists' portrayal of mathematics "as the magic wand of the few instead of the concentrated reason of all . . . its very name has become a mental opiate, and elementary fallacies which a generation ago would have been detected by the most ordinary of thinkers, now deceive the acutest minds, which lie bemused under its spell."[148]

The responses to Dingle's criticism can only have confirmed him in his belief in the necessity of an intellectual crusade against cosmophysics. Most participants distanced themselves from his extreme empiricism, although few explicitly sided with Milne's equally extreme rationalism. At any rate, the rationalistic and deductivistic trend continued to enjoy popularity in Britain during the years of the war. It was part of a zeitgeist particular to British intellectual life. As mentioned, Haldane exploited Milne's theory in his evolutionary speculations in a way that must have made Dingle's blood boil. And Frederick Arnot, a Scottish physicist, developed his own elaborate version of cosmophysics, very much based on the ideas of Milne and Dirac.[149] In the 1950s Dingle would again be on the warpath against those he considered heirs of the cosmophysicists of the 1930s.

Jordan's Empirical Cosmology

The German theoretical physicist Pascual Jordan was another of the founders of quantum mechanics who turned to cosmology in the late 1930s. Born in 1902, the same year as Dirac, Jordan had been a pioneer in the early phase of quantum electrodynamics, but in the 1930s he turned toward other aspects of science, including biology and the philosophical foundations of physics. When Jordan came across Dirac's cosmological theory, he immediately adopted it and developed it along his own lines. As he wrote in 1952, referring to the question of continuous creation of matter: "I am the only one who has been ready to take Dirac's world model seriously, which even its originator has partly abandoned, and to reconsider its more precise formulation."[150]

Jordan took Dirac's theory as his point of departure, but combined it with elements of Eddington's numerological reasoning, according to which numerical agreements between dimensionless numbers were significant also if the numbers were not large in Dirac's sense. Jordan found the mass of the universe to be continually increasing, as had Dirac in 1937, but he argued that this increase did not violate energy conservation. Drawing on an argument previously suggested by the Austrian-American physicist Arthur Haas,[151] Jordan argued as follows. The value of the Einstein gravitational constant $\kappa = 8\pi G/c^2$ is about 10^{-27} in cgs units, which, together with the values of the radius of the visible universe R and its average density ρ, suggests that $R = (\kappa\rho)^{-1/2}$. The estimated value of the mass of the universe then leads to the suggestion that $R \sim \kappa M$, which can also be written $\kappa M^2/R \sim M$ or $GM^2/R \sim Mc^2$. The right side is the total mass-energy of the universe, while the expression on the left is the negative potential energy of the universe. According to Jordan, the mass increase was thus compensated by an increase in negative potential energy, leaving the total energy of the universe unchanged (namely, equal to zero).

Jordan suggested that the mass increase did not take place incrementally, particle per particle, but that entire stars were formed in this way.[152]

His theory was a strange version of big-bang cosmology. In the sense that his universe had a beginning in time, when it started expanding from a singularity, it was a big-bang model; but the mass of the universe was not there originally, it was formed along with the expansion, and so there was no explosion. It was a big bang without a bang. As Jordan explained in 1944, the history of the universe could be traced back to a time equal to one atomic time unit, about 10^{-23} s.[153] At that time the universe would have a radius of one atomic length, 2×10^{-23} m, and consist of only one pair of newly created neutrons. As space expanded and the neutrons separated, the change in gravitational energy would be balanced by the creation of new matter. Ten seconds after this tiny "big" bang the universe would have grown to the size of the sun with a mass less than that of the moon; at that time it would consist of about 10^{12} protostars with an average mass of 10^9 kg.

Jordan's general view of cosmology differed from that of his British counterparts. As a convinced positivist, he emphasized that there was nothing speculative or a priori about numerological reasoning based on the constants of nature. He wanted to establish a cosmology based on what could be isolated as purely empirical facts, freed from the opaque mathematical hypotheses of standard theory. Somewhat naively he thought that it was possible "to distinguish quite clearly between what are *observational facts*—and as such *independent of any theory*—and what are the results of theoretical condiderations." The various relations between the dimensionless constants he regarded as "mere reformulations of facts of experience, freed from hypotheses." Far from being cosmythology, as Dingle would have it, Jordan called his version of Dirac's theory "empirical cosmology."[154]

Because of the war Jordan's cosmological speculations did not receive much immediate attention, but the extended versions he presented after 1945 played some role in the cosmological debate. Collaborating with German mathematical physicists, Jordan formulated his ideas in an extended form of general relativity, where the gravitational "constant" appeared as a local scalar field quantity depending on time and space.[155] This theory and its claimed applications to cosmology and geophysics were cultivated by a small school of German researchers, but outside Germany it seems to have attracted only modest interest. It did, however, exert an influence on Robert Dicke's thinking in the 1960s. Jordan's post-1948 cosmology differed from both the steady-state theory and the ordinary relativistic evolution theories with G = const. Because of the extra degree of freedom in the κ field, Jordan's theory allowed an even larger variety of cosmological models than the Friedmann-Lemaître theory. One of these was identical with the Dirac model, and another with the linearly expanding Milne model, but there were numerous other possibilities. With regard to flexibility and lack of uniqueness, Jordan's theory of cosmology was the very opposite of the steady-state theory.[156]

2.4 THE TIME SCALE DIFFICULTY

The idea of a universe of a finite age, whether of the big-bang type or not, was originally seen as a revolt from a conceptual point of view. As we have noted, Lemaître's theory was met with considerable resistance because of its foundation in the notion of the beginning of the world. Gradually, however, people grew accustomed to the strange notion, which at the end of the 1930s was no longer seen as particularly shocking, especially if separated from speculations about the physical origin of the universe. But in addition to the conceptual problem, there was from the very start an uneasiness about the related problem of the age of the universe as compared with the age of its constituents. The problem is a logical one, derived from the very concept of the universe. For obvious reasons, any acceptable finite-age cosmological model must satisfy the criterion that the age of the universe is greater than or equal to the age of any component (earth, stars, etc.), where "age of the universe" refers to the present phase of expansion, a comment which is relevant for oscillating models. A model for which the inequality is definitely violated must be ruled out, not because it clashes with experience, but because it clashes with logic.

With the acceptance of Friedmann-Lemaître cosmology in the 1930s the notion of the age of the universe obtained a firmer foundation in the sense that it could now be related to a measurable quantity, the Hubble constant, and actually be calculated for specific models. The result was discomforting, leading to the paradox that is often called the time scale difficulty, that is, what seemed to be a violation of the inviolable inequality stated above. It became "the nightmare of cosmologists," as historian of astronomy John North has aptly phrased it.[157] To realize the nature of this nightmare we recall that according to the Friedmann-Lemaître theory the Hubble time (the inverse of the present Hubble constant) is in most cases of the same order of magnitude as the age of the universe, in the following denoted t_0. Indeed, the Hubble time is often referred to as the age of the universe, although this is an unfortunate way of speaking since it is only in special (and unrealistic) cases that equality is secured. As an illustration of the time scale difficulty, consider the Einstein–de Sitter solution for which the age of the universe is $2/3H = 2T/3$. The age is smaller than the Hubble time, which in the mid-1930s was taken to be about 1.8 billion years.

The problem of a universe of age about 1.2 billion years arises most clearly if compared with the age of the earth, a quantity more reliably known than the ages of celestial objects. The use of radioactivity as a dating method goes back to the early part of the century, but it took many years until the method was perfected to a degree that it could yield reasonably reliable estimates. Compared with 1.2 billion years, the figures obtained were large. In 1921, H. N. Russell used analyses of thorium and uranium minerals to estimate the maximum age of the earth's crust to be approximately four billion years.[158] Eight years later Rutherford found from comparison of the abundances and half-lives of the two uranium isotopes (U-235 and U-238) that the upper limit of the age of the earth was 3.4 billion years.[159] The accepted value of the age of the earth in the 1930s, based on radiometric methods, was between two and three billion years.

Now the Einstein–de Sitter model is just one solution among many, but in most cases the time scale difficulty remains. In general, the age of the universe can be computed from the time variation of the scale factor (R', R''), but will depend on the value of the cosmological constant and the mean density of the universe; and, of course, on the value of the Hubble constant. For $\Lambda = 0$, as was usually assumed, any nonzero density value will result in an age smaller than the Hubble time, $t_0 < H^{-1} = T$. The larger the density, the smaller the age. Only in the unrealistic case of $\rho = 0$ will we have $t_0 = T$, and even that will be of no help.

The problem appeared in an even more drastic form if the Hubble time was compared with the ages of the stars and galaxies, as was usual in the astronomical discussions of the 1930s. In a series of works starting in 1922, James Jeans had studied the dynamics of star systems, assuming the stars to be in a kind of statistical equilibrium in much the same manner as the molecules of a gas. He reached the conclusion that the age of the galaxies had to be enormous, between one thousand and ten thousand billion years. "The time for the condensation of . . . nebulae to form stars . . . is of the order of 10^{13} years," he wrote in 1928. "A dynamical investigation has shown that the time necessary to break up the formation of the less massive stars would again be of the order of 10^{13} years."[160] This so-called long time scale enjoyed general respect, in part, no doubt, derived from Jeans's authority in matters of stellar dynamics. Although some astronomers realized that 10^{13} years might be too high a value and that Jeans's calculations were not to be taken for granted, the consensus was that the ages of stars and galaxies were several orders of magnitude higher than the age of the earth. Naturally, this made the value of the Hubble time embarrassingly small.

The time scale difficulty was often mentioned in the early papers on the expanding universe and became a standard ingredient in the first phase of big-bang cosmology. At the British Association meeting in 1931 the problem was emphasized by de Sitter. He could see no way out of the dilemma except suggesting that perhaps the expansion of the universe and the evolution of stellar systems were unconnected processes which only happened to take place side by side. The following year he elaborated on this suggestion, forced upon him because "there is no way out of the dilemma." De Sitter was deeply worried. "I am afraid," he wrote,

> all we can do is to accept the paradox and try to accomodate ourselves to it, as we have done to so many paradoxes lately in modern physical theories. We shall have to get accustomed to the idea that the change of the quantity R, commonly called "the radius of the universe," and the evolutionary changes of stars and stellar

systems are two different processes, going on side by side without any apparent connection between them. After all the "universe" is an hypothesis, like the atom, and must be allowed the freedom to have properties and to do things which would be contradictory and impossible for a finite material structure.[161]

Other cosmologists shared de Sitter's worries but believed that the paradox might be solved without admitting a universe with contradictory properties. It was a main reason for the popularity of the Lemaître-Eddington model, where the paradox does not need to arise. Oscillatory, nonsingular models were also thought of as a possible explanation. The British astronomer Harold Spencer Jones judged the difficulty to be "one of the outstanding problems of astronomy at present," but suggested that if the universe is oscillating then "the paradox of the stars being apparently older than the Universe is avoided."[162]

De Sitter seems to have preferred two kinds of models at the time. Remarkably, neither of them was the Einstein–de Sitter model suggested in 1932, but only, it seems, as an academic possibility. The Dutch pioneer cosmologist preferred either a universe which "may have been practically stationary at or very near its minimum size for an infinite time before starting to expand;" or a universe which "may have contracted during an infinite time and, after passing through a minimum a few thousand million years ago, started to expand again."[163] In 1933 he was more inclined toward a version of the second model, where the universe contracted to a point, with all the galaxies passing simultaneously through the singular point with the speed of light. He suggested that this "short, but very vigorous" critical event took place 3–5 billion years ago. According to de Sitter, the stars would survive the critical event and their ages thus be much longer than the age of the universe based on the recession of the galaxies.[164]

Richard Tolman pointed out that all cosmological models with a singular state of beginning were idealized to such a degree that the very concept of age might lose its meaning. Perhaps the discrepancy was a result of the use of homogeneous models and illegitimate extrapolations of the assumed singular state? In conformity with de Sitter he warned against identifying the beginning of the expansion with the beginning of the universe, a concept which Tolman found difficult to accept in 1934: "The difference between the time scales for stellar evolution and nebular expansion suggests that no definiteness could now be attached to any idea as to *the* beginning of the physical universe. Indeed, it is difficult to escape the feeling that the time span for the phenomena of the universe might be most appropriately taken as extending from minus infinity in the past to plus infinity in the future."[165]

Hubble recognized in the time scale problem an argument against the relativistic theory of the expanding universe. In 1936 he concluded that the universe, if expanding at all, was decelerating. This implied that "The time-scale is not lengthened; on the contrary, it is materially shortened [compared to T]. . . . The maximum permissible span appears to be of the order of 1500 million years, but the true value might lie anywhere between the maximum and half the maximum."[166] Six years later Hubble repeated his view of the "contradictions or at least . . . grave difficulties" that the time scale problem represented to relativistic cosmology. He concluded that "either the measures are unreliable or red shifts do not represent expansion of the universe."[167]

Both de Sitter and Tolman accepted the long time scale, but at the time of Tolman's comment this estimate had begun to lose its authority and was giving way to a much shorter time scale. The first one to question Jeans's conclusion seems to have been Ernst Julius Öpik, who in a semipopular article of 1933 reviewed existing data for terrestrial, meteoritic, stellar, and galactic ages. Öpik was born in Estonia in 1893 and did important work there and in the Soviet Union. After the war, in 1948, he went to the Armagh Observatory in Ireland, where he remained for thirty-three years. In his article of 1933, Öpik reached the conclusion that "The combined evidence . . . , once we assume that the recession is real, all this evidence points to an age of the stellar universe of the same order of magnitude as the currently accepted age of the solar system: not much more than 3,000 million years."[168] Jeans had not taken into consideration the rotation of galaxies (only established in the late 1920s) and in 1934 the Harvard astronomer Bart Bok argued that the neglected tidal forces would invalidate parts of Jeans's argument. Other astronomers soon discovered other weaknesses in the arguments for the long time scale, and in a Royal Astronomical Society discussion of 1935 the whole problem was reexamined.[169] Although Jeans maintained the long time scale of 10^{12}–10^{13} years, most of the other speakers, including Eddington and Milne, followed Bok in advocating a short time scale of 3–5 billion years. The growing suspicion of flaws in Jeans's arguments was strengthened when the Soviet astrophysicist Victor Ambarzumian pointed out in 1936 that the observational data of double stars did not, contrary to Jeans's contention, support the long time scale hypothesis.[170] The new, much shorter time scale won general support during the following years and by 1940 Jeans's estimate was discarded by most astronomers.[171]

The replacement of the long with the short time scale did not really solve the time scale problem, though. After all, the age of the universe as based on the Hubble constant was still too low and still smaller than the age of the earth. But in the wake of the new consensus among astronomers there followed a shift in attitude, in which the term "age of the universe" was considered suspicious because of its assumed metaphysical status. Astronomers preferred instead to speak of the "cosmic time scale" as something that allowed them to establish reference posts in the past rather than to date the present epoch according to an absolute beginning of time.[172]

Of course, neither the short time scale nor astronomers' change of terminology made the time scale difficulty disappear. It was still there and continued to cause concern. One possible way out of the problem would be to question the value of the Hubble constant, but the confidence in Hubble's data prevented scientists from following this route. That the value might be questionable had already been indicated in 1931 by the young Dutch astronomer Jan

Oort, who from a critical reexamination of galactic data tentatively concluded that the Hubble constant was only about 290 km·s^{-1}·Mpc^{-1}, corresponding to a Hubble time of about 3.4 billion years.[173] However, Oort's analysis was unable to shake the confidence in the standard Hubble value and it was twenty years before the value changed. It rested on the Cepheid method and then on the zero point in the period-luminosity function as determined by Shapley about 1916. With regard to this calibration, Hubble concluded in 1936 that "further revision is expected to be of minor importance."[174] As late as 1949, Tolman found it "highly improbable" that observational changes in the value of the Hubble constant would lead to a resolution of the time scale problem, and in the same year George Gamow concluded that "this discrepancy is certainly beyond the limits of errors in the astronomical measurements of the red shift."[175] Following a systematic discussion of the relationship between the Hubble time and the age of the universe in Friedmann-Lemaître models, the German physicist Helmut Hönl similarly dismissed a drastic revision of the Hubble constant as "very unlikely."[176] Hönl found it more reasonable to follow Dirac and Jordan in assuming that the law of radioactive decay had changed in time, in which case radiometric datings could not be trusted. The theoreticians were in for a surprise.

In 1948 Warren Weaver, director of the Rockefeller Foundation, asked his friend Tolman about the current status of the problem of the age of the universe, and in response Tolman produced a careful review which was published posthumously.[177] Made at the same time as the steady-state cosmology was formulated in England—but this was unknown to Tolman—the memorandum provides interesting insight into how a mainstream relativist looked upon the problem and the associated notion of the big-bang universe. In accordance with his view of 1934, and with the consensus among astronomers, Tolman adopted a cautious attitude. He preferred to put "the age of the universe" in brackets, and to speak of a cosmic time in the meaning of either the time since which gravitational interaction had taken place or the time which had elapsed since the beginning of the recession of the nebulae. "I see at present no evidence against the assumption that the material universe has always existed," he wrote. "This [notion of cosmic time] carries for me no implication that the universe was created without previous past history at the time of any such event." Although Tolman accepted Hubble's data, and was led to a value for the cosmic time of only 1.24 billion years, he did not find the discrepancy damaging to relativistic cosmology. As to the possibility that the paradox might indicate a failure in the general theory of relativity, Tolman summarily dismissed it, because, as he argued, without general relativity cosmology would be open to all kinds of "unbridled fancy." He had in mind such ideas as tired-light explanations of the redshifts and the theories of Milne and Dirac.

Tolman could not resolve the time scale difficulty, but he believed that it was not a proper paradox and that modifications of relativistic models might well lead to an answer. As an example he mentioned an investigation which recently had been made by Guy Omer on world models including inhomo-geneities. Omer, who did his work as a Ph.D. student under Tolman, exam-ined select classes of such models. With appropriate values of the mass den-sity, the curvature, and the cosmological constant he was able to arrive at an age for his model universe of 3.64 billion years. "[There is] a very real flexi-bility of the general theory of relativity. It is not necessary to invent any new physical principles to rationalize the currently accepted observational data," he concluded.[178] Tolman agreed and stated optimistically that it was possible to introduce "sufficient flexibility so that we do not need to expect trouble as to the time scale." Although neither Tolman's nor Omer's remarks were di-rected against the steady-state theory, which they did not know about at the time, their remarks could well have been. The "very real flexibility" seen as a virtue by the American cosmologists was seen quite differently in Cambridge, England. Indeed, this difference in view came to be a major ingredient in the controversy which raged throughout the 1950s.

In some contrast to the situation in the early 1930s, the general attitude toward the time scale difficulty among relativistic cosmologists in the late 1940s was, as exemplified by Tolman's, relaxed. Astronomers and physicists recognized that there was a difficulty, but they did not see it as a real paradox, and it did not create any real sense of crisis with regard to the future of relativ-istic cosmology. They could always fall back on the flexibility of the Fried-mann-Lemaître equations, which provided sufficient room to avoid any glar-ing inconsistency between data and theory. As mentioned, one way was to rely on some suitable inhomogeneous model. Another way, known for some time, was to reintroduce the cosmological constant in the homogeneous mod-els. For models with $\Lambda > 0$ and positive acceleration, such as the Lemaître-Eddington model, the Hubble time is shorter than the age of the universe, and then the time scale problem can be avoided. This was pointed out by Gamow, for example, who in 1949 showed that an open universe with a positive cos-mological constant could be brought into agreement with observations; or, rather, he showed that with appropriately chosen values of the curvature and the cosmological constant, any age could be obtained.[179] The attitude prevail-ing among many cosmologists at the time of the emergence of the steady-state theory was expressed by Whitrow, the British astronomer and relativist: "The discrepancy may be due to an overestimation of the age of the earth or to the fact that in the remote past the nebulae were moving more slowly so that the universe has taken a somewhat longer time to expand to its present state. . . . All the available evidence points to the universe having expanded from an initial state of maximum concentration a few thousand million years ago."[180]

Yet not all relativists were willing to exploit as shamelessly as Gamow, Tolman, and Whitrow the flexibility of the relativistic equations to solve the time scale problem. Einstein was one of the few skeptics. Having eliminated the cosmological constant in 1931, he firmly refused to reintroduce it because it would spoil the "logical economy" of the theory. With this position the time scale problem became serious, as Einstein frankly emphasized in 1945. Had it only been a contradiction between the long time scale and the age of the universe, he would gladly have sacrificed the theories of stellar evolution, but Einstein recognized that comparison with the age of the earth resulted in a genuine dilemma. As he wrote: "The age of the universe . . . must certainly exceed that of the firm crust of the earth as found from the radioactive minerals. Since determination of age by these minerals is reliable in every respect, the cosmologic theory here presented would be disproved if it were found to contradict any such results. In this case I see no reasonable solution."[181] It is remarkable that Einstein, the father of relativistic cosmology and a believer in the supremacy of theory over (claimed) observational facts, was apparently willing to give up the theory because of the time scale discrepancy, a step most other relativist cosmologists did not even consider. But then the "if" in Einstein's conclusion may be important: he was not, after all, convinced that the time scale difficulty was an incontrovertible fact.

The unsatisfactory state of the time scale difficulty would play a considerable role in the emergence and early discussion of the steady-state theory. It was also of importance to the alternative cosmological schemes of the 1930s, such as Milne's and Dirac's, which, as models of the expanding universe, had to face the same difficulty. (For those alternatives that did not admit the expansion of the universe, there was, of course, no difficulty.) In Milne's system no time scale discrepancy appeared in spite of the age of his uniformly expanding universe being equal to the Hubble time; on the dynamic time scale (the τ scale) there is no origin of the universe, and the age of the universe is infinite. Milne could therefore claim that his theory solved "the paradox into which contemporary physics is led in discussing the age of the universe."[182] In Dirac's cosmology the age of the universe was even smaller than in the Einstein–de Sitter case, only about 0.7 billion years, or one-third of the age of the earth's crust. Dirac was aware of the difficulty, but did not regard it as menacing: "This does not cause an inconsistency," he claimed in 1938, "since a thorough application of our present ideas would require us to have the rate of radioactive decay [on which the age of the earth was based] varying with the epoch and greater in the distant past than it is now."[183] He probably had in mind that an application of two time parameters à la Milne would give the stated result.

Gamow's Big Bang

THE BIG-BANG UNIVERSE is usually associated with the Russian-American physicist George Gamow and not with Lemaître, the Belgian abbé. Although the idea of a big bang was proposed years before Gamow became interested in the subject, and although he knew about the works of Friedmann and Lemaître, his contributions were original and very different from those of earlier cosmologists. First and foremost, Gamow was a nuclear physicist, neither a mathematician nor an astronomer. Influenced by his own and others' work in nuclear physics, he conceived the early universe as a nuclear oven in which the elements constituting our present universe were once cooked. This was a persistent theme in Gamow's cosmology. Although the theme was not entirely new, it was only with Gamow that it was developed systematically. He thereby provided cosmology with a new perspective, linking the science of the universe intimately to nuclear physics. But it was a link not appreciated, indeed scarcely noticed, by mathematical cosmologists in the period. By conceiving cosmology as a branch of high-energy nuclear physics, Gamow—first alone and later with his collaborators Ralph Alpher and Robert Herman—not only created a cosmological theory which appealed more to physicists than to astronomers and relativists; he also created an American niche in cosmology, which in its spirit differed distinctly from the one cultivated in Europe. The niche received widespread attention only after 1965, at a time when a new generation had taken over in cosmology, but it was founded by Gamow.

Gamow's big-bang model described a universe born from the nuclear reactions in a primordial superdense state of nucleonic matter. According to the refined version, the early universe was dominated by radiation, and Gamow, Alpher, and Herman predicted that the remnants of the primordial radiation would still exist in the form of a low-temperature microwave background, which was finally detected more than fifteen years later. In spite of several successes and a generally promising development, the early big-bang theory failed to gain general recognition. In fact, work on the theory stopped for more than a decade and it was only revived about 1964, at a time when its originators were no longer active in cosmological research. In order to understand this peculiarity, the evaluation of the big-bang theory's strengths and weaknesses must be seen in the larger framework of the state of cosmology in the 1950s, which includes the existence of rival cosmologies such as the steady-state theory.

3.1 NUCLEAR PHYSICS AND STELLAR ENERGY

Gamow's route to the big bang had its roots in earlier efforts to explain two all-important problems in astrophysics: What is the source of the energy irradiated from the stars? How are the chemical elements that make up our world formed? These questions occupied a central position in the new astrophysics, which from the 1920s was able to draw upon atomic and nuclear physics, but they remained for a long time restricted to the stars. In our context, nuclear astrophysics is not of importance in itself, but only insofar as it paved the way for extending the perspective from a stellar to a cosmological level.

Early Ideas

Speculations about both the formation of chemical elements and the source of stellar energy—not necessarily seen as related problems—can be traced far back in the nineteenth century, but it was only with the discovery of radioactivity that they acquired some measure of plausibility. With radioactivity came the disintegration theory and the composite atom, and then the possibility of element synthesis as part of the cosmic evolution. Johannes Stark, a German physicist and early advocate of these views, tied stellar energy production together with element-forming processes: "As the transformation of atoms in some elements is still going on, it may be supposed that there was a time when our chemical atoms did not exist in the present amount, while other types of matter were more common. In the later change of the arrangement of the positive and negative electrons, or in the genesis of the present chemical atoms, a very large amount of the potential energy of their electrons was transformed to kinetic energy. . . . It is reasonable to suppose that the temperature of the sun and stars is partly due to the genesis of chemical atoms."[1] Writing in 1903, Stark had no idea of either the atomic nucleus or the quantum structure of the atom. Although his speculations were shared by several other physicists, they were considered to be nothing but speculations, and at the time it could hardly have been otherwise. It was only with the new atomic theory of Ernest Rutherford and Niels Bohr of 1911–13 that a rational foundation for studying the problems was established.

Among the first scientists to suggest ideas of the formation of atoms in stars (or elsewhere) within the new framework of atomic structure was the American chemist William Harkins in a work of 1917. Harkins had a weakness for numerological speculations and was not taken very seriously among European physicists. But scattered around in his many works there were important insights. At a time when the nuclear atom was quite new, Harkins argued that the observed relative abundance of the various elements reflected their degree of stability and that this was a nuclear property, i.e., depending on the nuclear charge (Z) and not on the atomic mass (A): "The variation in the abundance of elements as found would seem to be the result of an atomic evolution . . . [and] related to the atomic number."[2] Considering the nuclei to be composed of electrons, alpha particles, and protons—a name not yet invented—he further hypothesized that "one of the first steps in the formation of a complex atom is the change of hydrogen into helium." This process, $4 \ ^1H \rightarrow \ ^4He$, was also considered two years later by the French chemical physicist Jean Perrin, a Nobel laureate of 1926.[3] Notice that there is no problem with the number of electrons in the scheme: although the helium atom contains only two extra-nuclear electrons, its nucleus was believed to consist of a system of four protons and two electrons.

Whereas neither Harkins nor Perrin was interested in explaining the energy produced in the stars, this was precisely the problem, or rather one of several problems, that occupied Eddington's fertile mind at the time. This problem, and the one relating to the sun in particular, had been the subject of investigation by several of the most eminent physicists of the nineteenth century. According to the theory of Hermann von Helmholtz and Kelvin (William Thomson), the source of solar heat was gravitational, arising from a gradual contraction of the sun. Even an imperceptible annual contraction, they calculated, would be able to account for the energy irradiated. However, the theory allowed the sun a lifetime of at most one hundred million years, a figure in sharp contrast with the much longer time scale of the earth required by the evolution theories held by biologists and geologists. The result was a famous dispute over the age of the earth, in which Kelvin attacked the virtually indefinite time scale of the geological uniformitarianism espoused by Charles Lyell and his followers.[4] The dispute concerned not only the geological time scale but also the one required by Charles Darwin's new theory of biological evolution. By and large, Kelvin's physically based time scale was considered more reliable than the one argued by the evolutionists, a result due in part to the great authority of Kelvin and the physicists. However, about 1890 the mass of evidence accumulated by the geologists made it increasingly difficult to believe in the physicists' much shorter time scale.

When radioactivity was discovered at the end of the century it was realized that the Helmholtz-Kelvin theory was probably not necessary, and for a short period it was believed that the sun and the stars might be made up of radioactive material. As early as 1903, Rutherford and Soddy pointed out that radioactivity might have important consequences for "cosmical physics" in general and the solar energy problem in particular.[5] Four years later, in an address delivered before the Royal Astronomical Society of Canada on 3 April 1907, Rutherford returned to the matter. He now speculated that at the enormous temperature of the sun ordinary matter might become radioactive, perhaps with the same intensity as radium (but evidently with a different lifetime). If so, he wrote, "this new source of heat would allow the sun to shine for a much longer period than the older theory allows." Admitting that it was only a speculation, Rutherford nonetheless insisted that it deserved to "be taken seriously into consideration in coming to a decision of the probable duration of the sun's heat, and consequently of the time for habitation of our globe."[6] However, this idea also failed to account for a sufficiently long time scale, and so the problem remained unsolved. The answer, it seemed to Eddington and a few other astrophysicists about 1920, lay in subatomic energy released by the loss of mass in nuclear processes according to Einstein's famous $E = mc^2$ formula.

As early as 1917, the same year that Harkins ventured his suggestion, Eddington speculated that in order to avoid the short time scale of the contraction theory, one might postulate that "the star has some unknown supply of energy, a slow process of annihilation of matter (through positive and negative electrons occasionally annulling one another)."[7] The process conjectured by Eddington was $p^+ + e^- \rightarrow \gamma$, where γ denotes energy of some form, for example, a gamma quantum; p^+ denotes a "positive electron," a name sometimes used for the hydrogen nucleus until the 1920s, when the name "proton" became established (it was introduced by Rutherford in his Bakerian Lecture of 1920). Although annihilation processes between positive and negative electrons later became part of physics, Eddington's conjecture has no foundation in physical theory; protons and electrons cannot annihilate. However, this became known only much later, and in the 1920s the suggested process was seen as reasonable, if hypothetical.

Eddington repeated the proton-electron annihilation hypothesis in 1919, but now found the collision frequency, even in stars of high density, to be too small to account for the necessary energy output. As an alternative he speculated that the energy might be liberated by the instantaneous annihilation of a proton, if only because "in the present confusion of ideas it is not unwelcome to have a possible explanation of the difficulties."[8] He did not comment on the obvious disagreement between the hypothesis and the firmly established law of charge conservation.

The following year, in a brilliant address given to the British Association for the Advancement of Science at its annual meeting in Cardiff, Eddington reviewed the problem, referring to the Helmholtz-Kelvin theory as "an unburied corpse" that deserved to be treated with no more respect than Archbishop Ussher's notorious date for the creation of the world. (James Ussher, archbishop of Armagh, proposed about 1640 that Creation had taken place in 4004 B.C.—on 23 October, to be precise.) Eddington suggested that the real source of stellar energy might lie in nuclear processes such as the formation of helium from hydrogen, and held that "what is possible in the Cavendish Laboratory [to make atomic nuclei react] may not be too difficult in the sun."[9] Neither Eddington nor the few other astronomers and physicists who entertained such ideas could account for the assumed formation process, but it was attractive because it supplied the stars with a much longer time scale than previous theories. The fusion hypothesis also had a sounder physical basis, resting as it did on Francis Aston's recent experiments with his mass spectrograph. According to Aston's measurements, the mass of a helium nucleus was nearly 1% less than that of four hydrogen nuclei and so the reaction $4H \rightarrow He$ would be followed by a considerable release of energy.

Six years later, in his classical *Internal Constitution of the Stars*, Eddington surveyed the problem in greater detail. He discussed the annihilation of protons and electrons into pure radiation energy, but at the time he considered this possibility speculative and favored the hydrogen-to-helium mechanism, although he recognized the problems with this possibility too: "How the necessary materials of 4 mutually repelling protons and 2 electrons can be gathered together in one spot, baffles imagination," he wrote. Clearly, the particles must have enormous velocities and so the interior of the stars must be extremely hot, perhaps hotter than allowed by astronomical knowledge. But this was an argument that did not shake Eddington. In a famous statement, the prophetic nature of which would be realized a quarter of a century later, he wrote: "The helium which we handle must have been put together at some time and some place. We do not argue with the critic who urges that the stars are not hot enough for this process; we tell him to go and find *a hotter place*."[10]

Digressing for a moment, it was not all scientists who believed that extreme temperatures are needed to produce helium from protons. Friedrich Paneth, an eminent Austrian chemist and pioneer in radiochemistry (and later a Nobel laureate), was one who did not. As a professor of chemistry in Berlin he tried in 1926 together with his assistant Kurt Peters to synthetize helium at room temperature by absorbing molecular hydrogen in the metal palladium; in this so-called occlusion process large amounts of the gas will be absorbed and subjected to a very large internal pressure. The two Berlin chemists reasoned that under these circumstances the collision frequency might be large enough to make some helium. Remarkably, they did detect small amounts of helium, but half a year later they retracted their claim, which they showed was due to experimental errors. Inspired by the experiment of Paneth and Peters, and undeterred by their failure, the Swedish chemist and industrialist John Tandberg took up the challenge. In order to force more hydrogen into the palladium, he used electrolysis. Believing that he had succeeded where Paneth and Peters had failed, Tandberg applied in 1927 for a patent for a "method to produce helium and useful energy." The Swedish patent authorities rejected the application, and the whole story would have been forgotten had it not been for the controversy over cold fusion more than sixty years later.[11] Suffice to say that from all we know, Eddington was right: fusion of hydrogen into helium requires very hot places indeed.

Nuclear Theory Enters Astrophysics

The breakthrough in an understanding of stellar energy came as a result of the new quantum mechanics, which was founded in 1925–26 by Werner Heisenberg, Paul Dirac, Max Born, Pascual Jordan, Erwin Schrödinger, and others. Originally quantum mechanics was thought of as a general theory of atomic structure, that is, applicable in particular to the electron system of atoms. The atomic nucleus was terra incognita, and it was uncertain whether quantum mechanics would apply also to this enigmatic part of the atom. That it did was first shown in 1928. In that year, George Gamow gave a quantitative explanation of alpha radioactivity by showing how an alpha particle, believed to preexist in the nucleus, can penetrate through the electrostatic potential barrier in spite of having less energy than that needed classically to escape the nucleus.[12] The explanation was founded directly on the new wave mechanics, Schrödinger's mathematically more manageable version of Heisenberg's quantum mechanics. Gamow's theory of the tunneling effect, and the equivalent one proposed independently by Edward Condon and Ronald Gurney in the United States, created much interest in the physics community. The German physicist Max von Laue pointed out that the equations in Gamow's emission process are reversible and thus might be supposed to be useful also in understanding the building up of elements by nuclear reactions.[13] Laue did not elaborate the suggestion, but later the same year, 1929, two young physicists did just that in a work which counts as one of the pioneering contributions to nuclear astrophysics.

The Austrian Friedrich ("Fritz") Houtermans was twenty-five years old when he met Gamow in Göttingen in the summer of 1928. He became a close friend of the one-year-younger Russian physicist and acquainted with the new theory of alpha decay. Houtermans, who was a communist, went in 1935 to the Soviet Union, where he worked at the Physico-Technical Institute in Kharkov. He fell prey to the purges in 1937, but after arrest and more than two years in prison with continuous interrogation and torture, he was extradited to the Gestapo in Germany.[14] In the peaceful year of 1928, Houtermans went from Göttingen to Berlin, where he met Robert d'Escourt Atkinson, a thirty-year-old British astronomer and physicist, who after studies in Oxford was doing graduate work in Germany. Contrary to Houtermans, Atkinson was familiar with the astrophysical literature and knew about Eddington's estimates of the temperatures in the interior of stars. In discussions between the two they realized that although the inverse tunneling effect might not be detected in laboratory experiments, it might furnish the key to understanding stellar nuclear processes. They wrote to Gamow about the idea and during the Christmas holidays Houtermans, Atkinson, and Gamow went to Austria to ski and discuss the work. After having made a couple of errors—which fortunately canceled out—they decided that the theory was in reasonable agreement with stellar data.[15]

Atkinson and Houtermans applied the inverted Gamow theory to the interior of a star with temperature 4×10^7 K and density 10 g·cm^{-3}. Under these conditions they found that the probability of alpha particles entering even a light nucleus was vanishingly small. Proton-nucleus reactions, on the other hand, were found to take place at an appreciable rate governed by the expression $\exp(-Z/v)$, where Z is the atomic number of the target nucleus and v the velocity of the incoming proton.[16] Making use of the Maxwell distribution, which for a given temperature gives the number of particles moving more rapidly than a certain velocity, Atkinson and Houtermans derived a general expression $W(Z, T)$ relating the cross section (reaction probability) to the temperature and atomic number: W increases with T and decreases with Z. For the mentioned values of temperature and density they found that the average lifetime for protons reacting with helium-4 would be eight seconds, whereas for neon-20 it would rise drastically to one billion years. They suggested that the source of stellar energy might be a transmutation of four protons into an alpha particle, the process taking place not by the improbable four-particle collision but by the consecutive capture of protons by a light nucleus and the subsequent expulsion of an alpha particle.

The Atkinson-Houtermans theory was a promising beginning of nuclear astrophysics, but at first it attracted little attention.[17] It built, of course, on the inadequate knowledge of nuclear physics at the time; in particular, in agreement with the universally held view, it presupposed the nucleus to consist of protons and electrons. About 1930 physicists became increasingly aware that somehow electrons did not belong to the nucleus, but with protons and electrons the only known elementary particles they were unable to suggest another model of the atomic nucleus. This, and the related problem of explaining the continuous spectrum of beta rays, led some physicists to speculate that perhaps energy conservation is not strictly valid in nuclear transformations. The idea was introduced by Niels Bohr, who indicated that in this way the problem of stellar energy might also be better understood. He never explained exactly how, but his advocacy of the idea is illuminating with regard to the uncertain state of stellar and nuclear physics at the time. In an unpublished manuscript from the summer of 1929, Bohr wrote: "Indeed, if as ordinarily assumed, the reversal of radioactive processes takes place in the interior of celestial bodies, it would depend on the temperature whether energy in the mean would be gained or lost through the capture of an electron by a nucleus and its subsequent expulsion as a β-ray. Perhaps in considerations of this kind we can find an explanation of the energy source which according to present astrophysical theory is wanting in the sun. . . ."[18]

Wolfgang Pauli, to whom Bohr sent his manuscript, was less than happy about the suggestion. "Let this note rest for a good long time and let the stars shine in peace!" he wrote.[19] Bohr followed the advice, but maintained his view for at least three years. It was well known through informal communication channels and from various addresses given by Bohr and his associates. Although primarily meant to solve problems in microphysics, the hypothesis of energy nonconservation was occasionally referred to also in the astrophysical literature. Tolman, for one, considered it important enough to include it in his 1934 textbook.[20]

Bohr's idea seems to have been influential especially among the young Leningrad physicists Gamow, Landau, and Bronstein, who did much to propagate and develop it. For example, in a paper completed in February 1931,

Lev Landau argued that, theoretically, stellar masses larger than 1.5 solar masses would collapse gravitationally into a singularity. Since such "ridiculous tendencies" are not in fact observed, he concluded that the stars might possess high-density "pathological regions" in which the laws of quantum mechanics fail. "Following a beautiful idea of Professor Niels Bohr's," he wrote, "we are able to believe that the stellar radiation is due simply to a violation of the law of energy, which law, as Bohr has first pointed out, is no longer valid in the relativistic quantum theory."[21] Landau's younger colleague, Matvei Bronstein, suggested applying Bohr's vague idea not only to the stars, but to the entire universe. In accordance with Landau, he hypothesized that "The only source of stellar radiation is the violation of the law of energy going on in the nucleus [of stars]." Moreover, he argued that the cosmological Friedmann-Lemaître equations had to be modified in accordance with Bohr's idea of energy nonconservation, namely, to include a time-dependent cosmological constant.[22] The fascination with energy-nonconserving processes and the assumed breakdown of quantum theory faded out about 1932, in part as a result of the discovery of the neutron (which made the problem of intranuclear electrons disappear), and in part because Bohr's idea proved to be incompatible with the theory of general relativity. From a sociological point of view, the episode illustrates the enormous authority that Bohr's views held at the time, not least among the young physicists.

Before the mid-1920s it was generally assumed that the stars have approximately the same chemical composition as the earth, with iron being the most abundant element. The abundances of elements in the stars were first estimated from spectral data by Cecilia Payne (later Payne-Gaposchkin), who in 1925 was led to the suggestion that hydrogen and helium were the dominant constituents of stellar atmospheres. This idea went against the established view, and Payne was persuaded by H. N. Russell to discount her result and conclude that her calculated abundances of hydrogen and helium were "improbably high, and [are] almost certainly not real."[23] However, some years later Payne's original suggestion was vindicated. Between 1928 and 1930 works by Albrecht Unsöld, William McCrea, and Russell demonstrated that hydrogen is by far the most abundant element in the sun's atmosphere. This important result at once indicated that hydrogen may also be the predominant element in the interior of the sun and most other stars, a suggestion that Eddington had made earlier. Eddington's conjecture was confirmed by the Danish astrophysicist Bengt Strömgren, who in 1932 calculated that the hydrogen content in the interior of stars must be about one third by weight, meaning that hydrogen is the most common element in the stars (namely, in terms of number of atoms).[24] This again implied that hydrogen probably was of special significance in stellar nuclear processes, as already suggested by Atkinson and Houtermans in 1929, if for other reasons.

With the new knowledge of hydrogen's predominant role in stars, Atkinson gave in 1931 a greatly expanded, but largely qualitative, version of his earlier work with Houtermans. Atkinson, who at this time had moved to the United States, addressed his exposition to astronomers and not physicists. He assumed that "in its initial state any star, or indeed the entire universe, was composed solely of hydrogen."[25] Without the still unknown neutrons and deuterons, helium could not be built up directly from protons, but Atkinson devised a cyclic model in which helium was formed by disintegration of unstable nuclei, an elaboration of the idea already suggested in the 1929 paper. Atkinson attempted to explain the abundances of the entire range of elements (except helium and hydrogen) by proton capture processes, but his results, although suggestive, were not very satisfactory.

In 1932 nuclear physics experienced its annus mirabilis with the first high-voltage accelerators and the discoveries of the neutron, deuterium (the hydrogen-2 isotope), and the positron. Shortly thereafter, in 1933, Enrico Fermi introduced his important theory of beta radioactivity, which incorporated yet another new (but hypothetical) particle, the neutrino suggested by Pauli a couple of years earlier. The accelerator experiments made by the Cavendish physicists John Cockcroft and Ernest Walton in 1932 confirmed the theory of Gamow, Houtermans, and Atkinson by showing that a proton of relatively low energy could penetrate into a lithium nucleus, splitting the compound nucleus exothermically into two helium nuclei: $^7Li + p \rightarrow {}^4He + {}^4He + 17$ MeV. At the Cavendish other thermonuclear reactions were studied in 1933–34 with the recently discovered deuterons as projectiles.[26] (The deuteron, consisting of a proton and a neutron, is the nucleus of deuterium.) By bombarding heavy ammonium chloride with accelerated deuterons, the exothermic reactions $d + d \rightarrow {}^3He + n$ and $d + d \rightarrow {}^3H + p$ were obtained. However, although the individual reactions were energy producing, energy loss made these early experiments massively energy consuming as a whole. It was evident that fusion energy was a matter for the stars, not the laboratory.

The new knowledge, theoretical and experimental, changed the scene in astrophysics also, and provided new possibilities of stellar nuclear reactions. A work of T. E. Sterne, a British astrophysicist working at Harvard University, illustrates how rapidly the new microphysical insight was transferred to astrophysics.[27] Sterne developed an equilibrium theory for the formation of elements in the stars. Completed in March 1933, his work included a discussion of the role not only of neutrons, but also of the recently discovered positrons. At that time physicists still discussed whether the neutron was a real elementary particle and whether the positron was identical with the positive electron predicted by Dirac in 1931 (as assumed by Sterne). Atkinson was also quick to make astrophysical use of the still infant elementary-particle physics. For example, in 1936 he pointed out the significance of the reaction $p + p \rightarrow d + e^+$, where e^+ denotes a positron.[28] Atkinson argued that with a mixture of deuterons, protons, and neutrons, a starting point was provided for subsequent reactions supposedly leading to the synthesis of heavier elements, including helium.

In the late 1930s, nuclear astrophysics was developed into an advanced and successful theory by, in particular, a small group of American nuclear physicists with no previous knowledge of astronomy. They set a pattern to be repeated and enhanced in later stages of cosmology and astrophysics: the successful entrance into these fields, traditionally seen as branches of astronomy, by young experts in quantum mechanics and theoretical nuclear physics. Foremost among these was George (or Georgii) Antonovich Gamow, the Russian-born nuclear physicist whose theory of alpha decay had served as the foundation for the first application of quantum mechanics to stellar element synthesis. Gamow was born in Odessa in southern Ukraine on 4 March 1904. As a young boy he came to question the truth of the Christian dogmas he was taught in school. Could the wine and bread served during Communion really turn into the blood and flesh of Jesus Christ? George decided to check the dogma scientifically, and examined the supposedly transubstianted wine and bread under his small microscope. He could find no sign of transubstantiation. "I think this was the experiment which made me a scientist," he recalled.[29] Gamow went to high school during the 1917 Bolshevik revolution and the subsequent civil war, but was too young, and too uninterested, to take part in the tumultuous military and political events. After a year at the local university he went to Leningrad, then still Petrograd (St. Petersburg), to study physics. Leningrad was the center of theoretical physics in the young Soviet Union, and Gamow thrived in the intellectual atmosphere, together with fellow students and friends such as Lev Landau, Victor Ambarzumian, Matvei Bronstein, and Dmitri Iwanenko, and the five-years-older Vladimir Fock, all of whom were quickly on their way to becoming important theoretical physicists. In 1923–24 Gamow attended lectures by Friedmann, who had recently extended the solutions of Einstein's cosmological field equations to cover nonstatic cases also. The subject of Friedmann's lectures, general relativity, interested Gamow greatly and left him with an early interest in relativistic physics. His first scientific paper, coauthored by Iwanenko, was an attempt to formulate Schrödinger's new wave mechanics in accordance with five-dimensional relativity theory.[30]

The intellectual atmosphere cultivated by the young theoretical physicists in Leningrad was part of the larger, international physics atmosphere. The new generation of theoretical physicists was bright and young, and they mastered quantum mechanics and relativity. They were also inexperienced and socially and philosophically unsophisticated, but in the excitement over the revelations of the new physics this was considered a virtue rather than a failure. The intellectual climate has been described by Carl Friedrich von Weizsäcker, in 1932 a twenty-year-old physics student and member of the international new-generation physics gang. "It was very difficult not to be senile after having lived thirty years," Weizsäcker recalled in an interview of 1963. "I feel that the general attitude was just an attitude of . . . an immense 'Hochmut,' an immense feeling of superiority, as compared to old professors of theoretical physics, to every experimental physicist, to every philosopher, to politicians, and to whatever sorts of people you might find in the world, because we had understood the thing and they didn't know what we were speaking about."[31] The somewhat childish attitude so vividly recollected by Weizsäcker was very much the attitude of Gamow and the Leningrad group, too.

In 1928–31, after having graduated from the University of Leningrad, Gamow went on a fellowship to the West to visit the centers of quantum physics in Göttingen, Cambridge, and Copenhagen. It was while attending a summer school in Göttingen in 1928 that the twenty-four-year-old Russian invented the mechanism for expulsion of alpha particles and thus made himself a name in international physics. The Belgian physicist Léon Rosenfeld, who met Gamow in Göttingen in the summer of 1928, recalled him as "a Slav giant, fair haired and speaking a very picturesque German; in fact he was picturesque in everything, even in his physics."[32] Traveling to Western Europe as often as possible, Gamow quickly became a leading authority in the new and exciting field of nuclear theory, a field in which he was a pioneer and wrote the first monograph ever. However, about 1932 the political climate hardened in the Soviet Union and Gamow was faced with increasing difficulties in getting permission to go abroad. In addition, he felt his work and security threatened by the ideological emphasis on correct Marxist-Leninist thinking and the associated resistance against modern physics in some quarters. In 1932 local party officials requested his employer, the State Radium Institute in Leningrad, to give information about the young physicist. In reply the Radium Institute answered that "[Gamow] has kept away from politics and social activity. In his behaviour, he is relatively undisciplined and is a typical representative of the literary-artistic bohemia."[33] This may sound innocent enough, but even apolitical bohemians could be potential targets of the rising political campaign. As a result of the worsened situation, Gamow decided to leave his fatherland, which he did in 1933, after being permitted to accept an invitation to the Solvay congress that year. In 1934 he arrived in the United States to take up a position as professor at George Washington University.

Gamow thrived in the United States and traveled widely to attend conferences and give lectures on nuclear physics. In June 1936 he visited Stanford University, where he was interviewed by the local newspaper. Or rather, since the reporter "was overcome by the warmth of the day," Gamow wrote his own interview. It went as follows:

Adam, the first man, didn't know anything about the nucleus but Dr. George Gamow, visiting professor from George Washington University, pretends he does. He says for example that the nucleus is 0.00000000000003 feet in diameter. Nobody believes it, but that doesn't make any difference to him.

He also says that the nuclear energy contained in a pound of lithium is enough to run the United States Navy for a period of three years. But to get this energy you would have to heat a mixture of lithium and hydrogen up to 50,000,000 degrees Fahrenheit. If one had a little stove of this temperature installed at Stanford, it would burn everything alive within a radius of 10,000 miles and broil all the fish in the Pacific Ocean.

If you could go as fast as nuclear particles generally do, it wouldn't take you more than one ten-thousandth of a second to go to Miller's where you could meet Gamow and get more details.[34]

Since 1935 Gamow's research focused increasingly on astrophysics, a subject to which he had already contributed while in the Soviet Union. Together with his friend Landau, he published in 1933 a brief paper on stellar temperatures.[35] In America, Gamow took an interest in almost all branches of this science, and contributed to many of them, but he was particularly fascinated by the prospect of explaining the relative abundances of elements by means of nuclear processes. This program was not Gamow's invention, but he gave it a new turn by emphasizing the role of neutrons in nuclear building-up processes. In a lecture delivered at Ohio State University in the summer of 1935 he first proposed that neutron capture might be the basic mechanism in the formation of heavier nuclei in stars.[36] Inspired by Fermi's contemporary experiments bombarding elements with neutrons, Gamow believed that what was possible in Fermi's laboratory in Rome might not be too difficult in the sun (to paraphrase Eddington). He suggested vaguely that "The neutrons which can be ejected from the nuclei of light elements by collisions with protons may stick to the nuclei of different heavy elements thus securing the possibility of the formation of still heavier nuclei."

A more precise version of this general idea was offered in a review completed two years later.[37] Gamow now based his idea on the process $p + p \to d + e^+$. With a large enough number of deuterons, the process would produce neutrons according to $d + d \to {}^3\text{He} + n$. The neutrons would then react with a nucleus of mass number A and atomic number Z, producing a beta-radioactive isotope with mass number $A + 1$ and the same Z; as a result of the decay, the isotope would transform into a nucleus with unchanged mass number but with atomic number $Z + 1$, i.e., an element with a higher position in the periodic table—just as in Fermi's experiments. With such a chain of consecutive captures of neutrons Gamow hoped to build up all the elements from an original stellar state of pure hydrogen. However, the deceptively simple program was difficult to harmonize with the models of nuclear reactions in the stars which originated at the time. There were not enough neutrons and so Gamow's scheme did not work.

Gamow was not the first to propose a theory of element formation based on neutron capture. Half a year earlier a British physicist, Harold Walke, suggested a much more detailed theory on the same basis. The theory is not well known, and seems to have had very little impact, but it deserves a place in the history of nuclear astrophysics. Like Gamow, Walke took his inspiration from current laboratory experiments: "The atomic physicist, with his sources of high potentials and his discharge-tubes, is synthesizing elements in the same way as is occurring in stellar interiors, and the processes observed, which result in the liberation of such large amounts of energy of the order of million of volts indicate how the intense radiation of stars in maintained and why their temperatures are so high."[38] Walke started with an initial state of the universe consisting of a dilute neutron gas. Neutrons, not hydrogen, were considered the source of all the elements. Condensations in the gas would cause frequent collisions and "as a result hydrogen is produced, this element being the first to evolve and appearing at a quite definite stage in cosmic development." Walke, who argued that the proton was really a neutron-positron composite, imagined that hydrogen was produced in energetic neutron-neutron collisions, for example, $n + n \to n + n + e^+ + e^- \to n + p + e^-$, which might also result in a deuteron and an electron. The deuterons would then form helium-4 and also helium-5 ("which has not yet been detected, but it is most probably stable"). The heavier elements were believed to be formed by neutron capture and subsequent beta decay. Walke suggested definite processes all the way up to neon-20. Although most of these were plain wrong (which was not evident at all in 1935), his general line of argument was prescient. Moreover, contrary to other contemporary works on element formation, Walke's was in part of a cosmological nature. However, he seems not to have been acquainted with, or found it relevant to refer to, cosmological models. His theory might have furnished an astrophysical link to Lemaître's neutronic primeval-atom hypothesis, but this is not what happened. Cosmology and nuclear astrophysics remained separate for yet another couple of years.

In his 1935 Ohio address, Gamow also considered a mechanism of element formation different from the neutron capture hypothesis. Gamow's alternative was based on a develoment of Landau's theory of stars, according to which most stars included a core of superdense "neutronic" matter of nuclear density, i.e., about 10^{12} g·cm^{-3}. The idea of "neutron stars"—stellar bodies made up of tightly packed neutrons—had been proposed by Baade and Zwicky the previous year as a possible explanation of supernovae.[39] Gamow believed at the time that all stars were composed in this way, and that the gravitational energy liberated in the contraction to densities of the order of the atomic nucleus might constitute the source of stellar energy. This was a return to the old and long-discarded contraction theory of Helmholtz and Kelvin, but based on densities unimaginable in the nineteenth century. A similar kind of modernized Helmholtz-Kelvin theory was proposed by Landau, who in 1938 found that the transformation to a neutronic state of only 2% of the sun's mass would be enough to account for the solar energy during a period of two billion years.[40] Gamow further suggested that the neutronic, stellar nucleus might play a role in the formation of elements, as follows: "The eruptive processes from its [the stellar nucleus's] surface will throw out the small pieces of nuclear substance which coming into the outside layer of the star will immediately disintegrate giving rise to the nuclei of different stable and radioactive elements."[41] Although Gamow soon abandoned this idea, it resurfaced seven years later in a different, cosmological context.

Bethe and Stellar Energy Processes

Together with his friend and colleague Edward Teller, Gamow organized in the spring of 1938 a conference on "Problems of Stellar Energy-Sources" in Washington, D.C.[42] This was the fourth of a series of annual conferences on theoretical physics arranged jointly by the George Washington University and the Department of Terrestrial Magnetism of the Carnegie Institution of Washington. The purpose of this important series of conferences was to establish an informal forum of a small and select number of physicists. The structure and idea of the conferences were modeled on the conferences held at Bohr's institute in Copenhagen, which Gamow knew so well. In a letter to Bohr of 1 June 1935, Gamow mentioned explicitly that the first Washington conference was to take place in accordance with the "Kopenhagener Geist," and that it was the first example of such a discussion conference in American physics.

The 1938 conference was attended by thirty-four scientists, both astrophysicists and nuclear and quantum physicists. The first group included Chandrasekhar and Strömgren, the second Merle Tuve, John Neumann, George Breit, and of course Teller and Gamow. Among the invited theoretical physicists was also Hans Albrecht Bethe, "who on his arrival knew nothing about the interior of stars but everything about the interior of the nucleus."[43] Thermonuclear reactions in stars was the central topic of the Washington conference and reflected Gamow's own research interests. Teller once referred to the subject as "Gamow's game," but it was Bethe, not Gamow, who turned out to be the game's key player.[44]

Bethe was born in 1906 in Strasbourg, then spelled Straßburg and part of the German empire. He studied under Sommerfeld in Munich, where he obtained his Ph.D. in 1928, and afterwards he held temporary positions at various German universities. Recognized as one of the brightest theoretical physicists of his generation, he did important work in quantum theory in the early 1930s. Being half of Jewish descent he fled Germany in 1933, first to England and two years later to the United States. Bethe settled at Cornell University in Ithaca, New York, where he quickly established himself as the foremost authority in nuclear theory. During the Washington conference, he became interested in the problem of explaining the energy of the sun, a problem he saw and approached as a problem in nuclear physics. At that time, Bethe was still a German citizen, but three years later, in 1941, he became naturalized as a citizen of the United States.

At about the same time, one of Gamow's and Teller's former students, Charles Critchfield, proposed a scheme based solely on reactions between protons, and together with Bethe he developed the idea into the first quantitative model of solar energy production.[45] The two physicists analyzed in precise, quantum-mechanical terms the reaction $p + p \rightarrow d + e^+$ earlier considered by Atkinson but only now taken to be the fundamental process in the stars. The deuterons formed in this way were assumed to react with protons in such a way that the end result of the "proton-proton cycle" would be helium:

process 1: $p + p \rightarrow d + e^+$,

process 2: $d + p \rightarrow {}^3\text{He} + \gamma$,

process 3: ${}^3\text{He} + {}^4\text{He} \rightarrow {}^7\text{Be} + \gamma$ followed by ${}^7\text{Be} \rightarrow {}^7\text{Li} + e^+$,

process 4: ${}^7\text{Li} + p \rightarrow {}^4\text{He} + {}^4\text{He}$,

net process: $4p \rightarrow {}^4\text{He} + 2e^+$.

Twenty-one years after Harkins and Eddington had first conjectured the hydrogen-to-helium process, Bethe and Critchfield could present it in a fully developed version which agreed remarkably well with observations. Thus, with current estimates for the hydrogen content and temperature of the interior of the sun (35% by mass; 20×10^6 K) they calculated the energy production to be 2.2 erg·g^{-1}·s^{-1}. Observations said 2.0 erg·g^{-1}·s^{-1}.

In later versions of the scheme a neutrino follows together with the positron, leaving the net process as $4p \rightarrow {}^4\text{He} + 2e^+ + 2\nu$. This was known in 1938, but at the time the hypothetical neutrino was viewed with some suspicion and not usually included in the reaction schemes. However, although most physicists believed that the neutrino would probably never be detected because of its extremely feeble interaction with matter, they nonetheless believed that the particle existed. In his 1939 theory, Bethe discussed briefly the role of neutrinos in the inverse beta process $p + \bar{\nu} \rightarrow n + e^+$ (where $\bar{\nu}$ is an antineutrino).

While working out the details of the proton-proton cycle Bethe was occupied also with another idea for producing helium from hydrogen, which he had first thought of during the Washington conference. According to Gamow's colorful narrative, Bethe solved the solar energy problem while returning by train to Cornell from the conference: "Taking out a piece of paper, he began to cover it with rows of formulas and numerals, no doubt to the great surprise of his fellow-passengers. . . . and as the Sun, all unaware of the trouble it was causing, began to sink slowly under the horizon, the problem was still unsolved. But Hans Bethe is not the man to miss a good meal simply because of some difficulties with the Sun and, redoubling his efforts, he had the correct answer at the very moment when the passing dinner-car steward announced the first call for dinner."[46] Gamow's account is perhaps not entirely reliable, as he himself admitted in the preface to *The Birth and Death of the Sun*: "It is perhaps best to warn the reader against giving too great credence to such minutiae in the following pages as . . . the relationship between Dr. Hans

Bethe's famous appetite and his rapid solution of the problem of solar reaction."[47] Whatever the minutiae, Bethe did solve the problem with amazing speed. Less than half a year after the Washington conference, he had acquired a thorough knowledge of astrophysics and performed detailed calculations of cyclic reactions of protons with carbon and nitrogen nuclei.[48] Bethe's explanation of the energy production of the sun—the ultimate demystification of the celestial body once thought sacred and divine—took place 144 years after the famous astronomer William Herschel had seriously suggested that the sun was inhabited "by beings whose organs are adapted to the peculiar circumstances of that vast globe."[49] Yes, progress does take place in science.

The essence of the CN cycle (sometimes called the Bethe-Weizsäcker cycle) as suggested by Bethe in 1938 was the catalytic action of carbon-12:

process 1: $p + {}^{12}C \rightarrow {}^{13}N + \gamma$ followed by ${}^{13}N \rightarrow {}^{13}C + e^+$,

process 2: $p + {}^{13}C \rightarrow {}^{14}N + \gamma$,

process 3: $p + {}^{14}N \rightarrow {}^{15}O + \gamma$ followed by ${}^{15}O \rightarrow {}^{15}N + e^+$,

process 4: $p + {}^{15}N \rightarrow {}^{12}C + {}^{4}He$,

net process: $4p \rightarrow {}^{4}He + 2e^+$.

The end result of the CN cycle is thus the same as that of the proton-proton cycle, but the two processes differ in their lifetimes and dependences on temperature. Bethe did not explain the origin of the carbon needed as catalyst for his reactions; it was just assumed to be there and could not, in fact, be built up from hydrogen. We shall not be concerned here with the details of Bethe's paper, except to repeat that it built on detailed calculations supplied with values of cross sections determined experimentally. The work relied intimately on laboratory physics, and, conversely, it demonstrated to experimental nuclear physicists that their work might have an astrophysical relevance. For example, at Caltech's Kellogg Laboratory nuclear physicists were studying the reactions between protons and carbon nuclei and other elements, not at first with an eye to astrophysics. But, as William Fowler later recalled: "Bethe's paper told us that we were studying in the laboratory processes which are occurring in the sun and other stars. It made a lasting impression on us."[50] Bethe's work started a process in which stellar energy was partly turned into a laboratory science, a process which after the war was extended also to cosmology. This observation goes a long way in accounting for the fact that early big-bang cosmology was completely dominated by American nuclear physicists.

The main result of the 1939 theory was the CN cycle, which Bethe argued was the principal energy source for main-sequence stars, including the sun. Having calculated the temperature dependence of both cycles, he concluded that for temperatures below 16 million degrees the proton-proton cycle would dominate, for higher temperatures the CN cycle. Turning the problem around, he calculated that the CN cycle, in order to give the energy produced by the sun, would require a central temperature of 18.5 million degrees—in excellent agreement with the value of 19 million based on astrophysical models of the sun. The work of Bethe and other researchers at the end of the 1930s was hailed as a great success and a breakthrough in nuclear astrophysics. Even before its appearance in the *Physical Review* in the spring of 1939, Bethe's theory was widely applauded by astronomers. According to Russell, it was "the most notable achievement of theoretical astrophysics of the last fifteen years."[51] Twenty-eight years later Bethe was awarded the Nobel Prize for his work.

The astronomical consequences of the CN cycle were quickly explored, including a nuclear physical interpretation of the Hertzsprung-Russell diagram suggested by Gamow. According to him, the evolutionary meaning of the diagram was that stars evolve upward along the main sequence as they slowly deplete their content of hydrogen. In a survey in the fall of 1939, where he first suggested this interpretation, Gamow concluded optimistically that "the problem of stellar energy sources and the main features of stellar evolution can be considered at present as practically solved."[52] As was usual for Gamow, he was a bit too optimistic. According to two young British astronomers, Fred Hoyle and Raymond Lyttleton, the optimism was unfounded. They objected to the "embarrassing" feature that there was no room for synthesis of heavier elements in the theory. This would seem to imply "that the stars can no longer be regarded as the building place of the heavy elements," a conclusion which Hoyle and Lyttleton dismissed.[53]

A few weeks after the Washington conference, a symposium on a related subject was held at the University of Notre Dame, South Bend, Indiana. Arranged by Arthur Haas, an Austrian physicist who had settled in the United States, the symposium on "The Physics of the Universe and the Nature of Primordial Particles" was attended by more than one hundred scientists. Among the speakers were A. H. Compton, G. Lemaître, H. Shapley, C. D. Anderson, A. E. Haas, W. D. Harkins, and G. Breit.[54] Lemaître was at the time a visiting professor at Notre Dame. "Founder of the theory of the expanding universe, Canon Lemaître's place in the scientific firmament is alongside that of Einstein and Richard C. Tolman, of the California Institute of Technology," the press release stated. "These three are the great leaders in science's most abstruse investigations, the geometrics of space, time and matter."[55]

Although the Notre Dame symposium did not result in specific results, as the Washington meeting did, the very content and structure of the symposium is interesting. It must have been one of the very first conferences where cosmology was a major theme. Furthermore, the subject—cosmology *and* particle physics—has a strikingly modern sound. It reflected an American interest in physical cosmology (or cosmic physics) different from the mathematical cosmology which dominated in Europe. Whereas contemporary British cosmology was essentially a mathematical (in part, an epistemological) study of an abstract universe, either a relativistic one or the one advocated by Milne, in the United States cosmology was seen as connected with physical problems related to stellar energy and, especially, cosmic rays. This was indeed the mold of Gamow's later revival of the big-bang theory, although in his program cosmic rays played a subordinate role compared to the problem of element formation. If only in retrospect, the two 1938 conferences marked a new phase in cosmology.

Weizsäcker's Cosmological Sketch

Bethe was not concerned with the synthesis of heavier elements. On the contrary, he declared that this problem was unsolvable within the framework of ordinary stellar ovens. "Under present conditions, no elements heavier than helium can be built up to any appreciable extent," he wrote. "Therefore we must assume that the heavier elements were built up *before* the stars reached their present state of temperature and density. No attempt will be made at speculations about this previous state of stellar matter."[56] The last remark could have been a reference to a work in *Physikalische Zeitschrift* appearing a week after Bethe submitted his paper.[57] The author of the German work was Carl Friedrich von Weizsäcker, one of the few Europeans active in nuclear astrophysics. Weizsäcker had obtained his Ph.D. in Leipzig in 1933, where he worked with the famous Heisenberg on problems in quantum theory. A specialist in quantum and nuclear physics, Weizsäcker had an interest in astronomy since his childhood. He had browsed through the German edition of Eddington's *Internal Constitution of the Stars*, and at Bohr's institute in Copenhagen he had discussed astrophysics with Bengt Strömgren. When Weizsäcker wrote his important paper in 1938, he was only twenty-six years old. Working at Berlin's Kaiser-Wilhelm Institut für Physik, he benefited from discussions with Ludwig Biermann, a German astronomer at the Berlin University Observatory.

In June 1938 Gamow attended a conference in Warsaw on "New Theories in Physics." On his way back to the United States he visited Berlin and met with Weizsäcker, with whom he discussed problems of nuclear astrophysics. Weizsäcker learned from Gamow that Bethe was investigating the same nuclear processes as he was, the proton-proton and CN cycles. However, the works of the two German physicists were not only independent, they differed considerably. First of all, Weizsäcker's work, in which he suggested the CN cycle half a year before Bethe, was largely based on qualitative arguments and lacked the detailed calculations that made Bethe's work a full-blown scientific theory. Moreover, whereas Bethe focused on energy production, Weizsäcker considered this "the unproblematical part of the theory"; and whereas Bethe did not deal with the origin of the elements and refrained from "speculations about this previous state of stellar matter," his colleague in Berlin had no such reservations. This makes Weizsäcker's paper, in spite of being much more primitive than Bethe's (from a scientific point of view), the more interesting one from a cosmological perspective.

In 1937 Weizsäcker had endorsed the "wider synthesis hypothesis," that is, the attempt to build up heavier elements from a process starting with proton-

proton reactions, but in his paper of 1938 he renounced this view. He now argued, as Bethe did independently, that it was impossible to synthesize heavier elements in stars and, therefore, that "It is quite possible that the formation of the elements took place before the origin of the stars, in a state of the universe significantly different from today's." However, contrary to Bethe, Weizsäcker found it tempting "to draw from the frequency of distribution of the elements conclusions about an earlier state of the universe in which this distribution might have originated." He realized this to be a problematic program, not only because of the technical difficulties but also because it related to a state of the universe of which no direct empirical knowledge was available and to which the present laws of nature could not be uncritically extrapolated. He took the sensible position "to presuppose, first of all hypothetically, the permanence in time of our natural laws and . . . be prepared for the appearance of error when this theory is compared with historical documents still available today." The program outlined by Weizsäcker aimed at inferring the state of the unknown, early universe from present conditions. He relied on crude calculations based on the assumption that the original distribution of elements was in thermodynamical equilibrium and tried to estimate from the present distribution the original temperature and density. Relying in part on previous methods of this kind, first used by two German physical chemists, Ladislaus Farkas and Paul Harteck,[58] Weizsäcker was led to the tentative conclusion of the early universe having "a temperature of an order of magnitude of that which would come about through the complete transformation of the nuclear binding energy into heat [about 2×10^{11} K]. The accompanying density is likewise already in the neighbourhood of the density of the nucleus."

Weizsäcker argued qualitatively that the gross distribution of the elements was obtained under these extreme conditions, and that subsequently the temperature and density would decrease, with new nuclear reactions taking place that would account for the fine structure. At the end of his paper, he addressed more directly, and more speculatively, cosmological questions. He imagined "a great primeval aggregation of matter perhaps consisting of pure hydrogen" collapsing under the influence of gravity to form the extreme conditions under which element formation would take place. Admitting that the size of this primeval aggregation was purely a matter of speculation, he invited his readers to imagine "the entire universe as known to us combined in it." He saw this speculation as justified by the observation that the energy released in the original formation of the elements would impart to the nuclei a velocity of the order of magnitude of one-tenth of the velocity of light: "At approximately this speed the fragments of the [primordial] star should fly apart. If we ask where today speeds of this order of magnitude may be observed, we find them only in the recessional motion of the spiral nebulae. Therefore, we ought at least to reckon with the possibility that this motion has its cause in a primeval catastrophe of the sort considered above."

One might suspect that Weizsäcker was inspired by Lemaître, with whose theory his picture has an obvious affinity, but this seems not to have been the case. Rather than referring to the Belgian abbé, Weizsäcker had resort to Edward Milne's demonstration of 1933 that the Hubble relation follows kinematically for a collection of particles—galaxies—moving in empty space. "My proposal differs from Milne's current theory in that it proposes a specific cause for the expansion . . . it appears to me rather a superiority of the new proposal that it makes superfluous the difficult-to-prove assumption that the presently known part of the universe has qualitatively the same structure as the whole [the cosmological principle?]." Apart from this reference, Weizsäcker seems not to have been influenced by Milne's theory of kinematic relativity.

Weizsäcker's paper of 1938 deserves to be recognized not only as a pioneering contribution to our understanding of why the stars shine, but also as an interesting attempt to provide the origin of the universe with a physical explanation. Weizsäcker's picture was that of a big-bang universe, although he imagined the original superdense state to have been preceded by a diluted state of hydrogen atoms. This picture has much in common with Lemaître's primeval-atom hypothesis; for example, both scientists pictured the original state as a compressed universe of nuclear density which then exploded. Insofar as Weizsäcker's theory is placed in a cosmological tradition, it was an extreme case of physical cosmology. What separates it from Lemaître's theory, and also from the later theory of Gamow, is, first of all, that it did not consider the geometric structure of the world at all. In particular, Weizsäcker did not refer to general-relativistic models and did not try to combine his nuclear-historical sketch with the geometrical history of the universe as given by the Friedmann-Lemaître equations. In this sense, it was only half a big-bang hypothesis.

Weizsäcker seems to have meant his theory, insofar as it referred to the universe at large, as no more than an occasional adventure into cosmology. He referred briefly to it in the summer of 1939, in a physico-philosophical discussion of the entropic theory of time. In this connection he argued that documents of a much earlier state of the universe are still with us in the form of, for example, the recession of the galaxies and the abundance distribution of the chemical elements. But Weizsäcker did not mention the superdense state he had contemplated the previous year. He now found it likely that "for about 10^{10} years ago the matter of the universe consisted of thinly distributed hydrogen at rest, with a constant density and zero absolute temperature."[59] At any rate, Weizsäcker soon became occupied with new developments in nuclear physics—the fission of uranium—and did not return to either cosmology or nuclear astrophysics. During the war he did important work on cosmogony, the science dealing with the formation of the solar system, but this was of no direct relevance to cosmology proper. Weizsäcker was never profoundly interested in cosmological models and did not take much interest in them. His interests were in the energy sources of the stars and the origin of the planetary system, and he felt that cosmological model making was not sufficiently secured in known physical laws to merit serious attention.[60]

Weizsäcker's big-bang speculation received some attention among German astronomers and physicists during the war. His work was also well known in Great Britain and the United States, but there it was seen in a more narrow perspective, as simply a theory of stellar energy production. In Germany, the cosmological aspect was discussed by the physicists Pascual Jordan, F. G. Houtermanns, and J. H. Jensen, and the astronomers A. Unsöld, H. Kienle, and O. Heckmann, who all interpreted Weizsäcker's theory as including a *Urexplosion*—a primeval explosion or big bang.[61] However, as pointed out by Heckmann, the mechanism suggested by Weizsäcker was not really sufficient to account for the largest known redshifts. According to Weizsäcker's hypothesis, the kinetic energy of the receding galaxies should approximately be given by $\varphi Mc^2 = 1/2\ Mv^2$, where M is the total mass and φ is the fraction of mass transformed into energy by the fusion of four hydrogen nuclei into a helium nucleus. With $\varphi = 0.01$, this gives a maximum recessional velocity of $0.14c$, considerably less than the $0.23c$ which Heckmann derived from the data of the weakest galaxies. In spite of the apparent discrepancy, Heckmann found a big-bang model à la Weizsäcker attractive because it promised an understanding of the abundances of chemical elements.[62]

Gamow's interest, like Bethe's, focused on nuclear reactions in existing objects, the stars. He was aware of the possibility of a cosmological origin of the elements at a time some two billion years ago as suggested by Weizsäcker, but at first he seems not to have taken it very seriously. In a review article from the summer of 1938 he wrote: "Since the physical conditions in this epoch, at a very long time ago, are highly hypothetical, a broad area for speculations about the origin of stars and their properties 'in statu nascendi' is opened up here."[63] However, having mentioned the possibility, he shelved it. Yet the idea of a prestellar, dense universe seems to have continued to occupy a corner of Gamow's mind. In a popular book written in 1939 he wrote that "a long, long time ago" the universe consisted of a primordial gas of extreme density and temperature which gradually decreased as a result of expansion.[64] Referring to Weizsäcker, who had argued that uranium and thorium must have been built up under such extreme conditions, Gamow now found that the scenario of a superdense and superhot early universe was supported by "good physical reality." We may conclude that at this time, in late 1939, Gamow entertained the idea of a big-bang universe, at least in a vague sense.

Although Gamow's primary interest at the time was in astrophysics, and not in cosmology, he was already acquainted with relativistic cosmology before the war (recall that he was a former student of Friedmann). This is shown by a paper he wrote with Teller, his colleague at George Washington University, on the gravitational condensation of diffuse matter into galaxies in 1939.[65] The two physicists considered the problem cosmologically, i.e., within the framework of an expanding universe, and concluded that galaxy formation was only possible in the past, when the scale of the universe was about six hundred times smaller than today. There was nothing particularly remarkable in this conclusion—a similar one had been obtained by Lemaître in 1934—but it is worthy of notice because it shows Gamow's familiarity with cosmology. In their use of relativistic cosmology, Gamow and Teller relied on Tolman's monograph of 1934, a work which remained Gamow's standard source when he turned seriously to cosmology some years later.

Around 1940 it was not uncommon for astrophysicists to refer to the possibility that the formation of elements (or other phenomena, such as galaxy formation) might have taken place in a much earlier state of the universe where matter was compressed to, say, nuclear densities. Neither was it uncommon to connect this hypothetical state with the expansion of the universe, that is, to conceive of it as something like Lemaître's primeval atom. But with the exception of Weizsäcker, such references were brief and casual and they rarely mentioned Lemaître's hypothesis. Not only did physicists and astronomers find the idea highly speculative (which it was), they also saw no advantage in a cosmological approach to element formation compared to the stellar approach. So long as they could justify their belief in the heavier elements being built up in stars they saw no reason to replace the stellar ovens with an even hotter and denser one. It was the realization that the stars were unable to produce a wide range of elements in the right ratios that convinced some astrophysicists of the necessity to focus on a prestellar formation and then opened up for them the possibility of a big-bang universe.

3.2 THE ULTIMATE NUCLEAR OVEN

The steps taken by Gamow in the 1940s toward the big-bang theory were a natural continuation of existing research, which was given a new, cosmological direction. The unplanned result was a new world picture, essentially the picture of the world we still believe in. If one looks for a definite moment of revelation in which the big-bang idea was first realized, one will look in vain. There was no such first and crucial event, no sudden insight. Instead there was a series of events through which the new nuclear-physical big-bang model gradually emerged. The formative phases of this process were almost entirely the work of Gamow, but the subsequent development of his ideas into a proper theory of the early universe was as much the result of work done by his collaborators Alpher and Herman.

The Revival of the Primeval Atom

America's entrance into the war had a serious effect on physics in the country, which to a large extent became directed toward pressing military needs. But although some nuclear research was classified and the volumes of the *Physical Review* and the *Astrophysical Journal* became markedly slimmer, the changed conditions did not mean a stop to pure research. Perhaps because of his Russian origin, Gamow was not cleared to work in the Manhattan Project or related projects, but during the war years he was a consultant for the navy in matters of high explosives. His duties left him with enough time to continue his work in astrophysics and extend it to the realm of cosmology.

The conclusion reached by Weizsäcker and Bethe, that heavier elements cannot be produced in appreciable amounts by either the proton-proton or the CN cycle, seemed to indicate that these elements had a nonstellar origin, perhaps in an earlier state of the universe. The program outlined by Weizsäcker and some earlier researchers was to infer the physical conditions of this hypothetical state by assuming that the nuclear species were originally in a state of dynamical equilibrium. On this basis it might be possible to estimate the temperature and density necessary to reproduce an element distribution corresponding to the one known empirically.

The method can schematically be presented as follows. Consider a triplet af isotopes of the same element X, with mass numbers A, $A + 1$, and $A + 2$, for example, the magnesium isotopes of masses 24, 25, and 26. In equilibrium with a neutron gas, the isotopes are related as ${}^A X + {}^1 n \rightleftharpoons {}^{A+1} X$ and ${}^{A+1} X + {}^1 n \rightleftharpoons {}^{A+2} X$. Leaving out some finer details, it follows from statistical equilibrium theory—the law of mass action known from chemistry—that

$$\frac{n(A)n^*}{n(A+1)} = G(T) \exp\left(-\frac{E_{A+1}}{kT}\right) \tag{3.1}$$

and

$$\frac{n(A+1)n^*}{n(A+2)} = G(T) \exp\left(-\frac{E_{A+2}}{kT}\right), \tag{3.2}$$

where $n(A)$ denotes the concentration of A nuclei (etc.), n^* the concentration of neutrons, k Boltzmann's constant, and T the absolute temperature. $G(T)$ is a known function of T, and E_A is the binding energy of a neutron in a nucleus of mass A, which is an experimentally known quantity. From the two equations it follows that

$$\frac{n(A)}{n(A+1)} \frac{n(A+2)}{n(A+1)} = \exp\left(-\frac{E_{A+2} - E_{A+1}}{kT}\right).$$

Since the binding energies are known, the equilibrium temperature can thus be found if the abundance ratios on the left side are also known, and then the neutron concentration n^* can be obtained from one of the equations (3.1) or (3.2). Again, one can argue the other way around and use values of T and n^* to find the abundance ratios and then compare these with the ones observed. One can always find values that reproduce a particular ratio, say that between Mg-25 and Mg-24, but to make sense, all of the ratios must be reproducible from the *same* set of (T, n^*) values, at least approximately. Since the equilibrium must be established in a finite time, the temperature must be high. The kinetic energy kT must be comparable with the binding energy per nucleon, that is, $kT \approx 1$ MeV or $T \approx 10^{10}$ K. Because the equilibrium approach rests on thermodynamical reasoning, it has the advantage that it can be followed in the absence of detailed knowledge of the nuclear processes. All that is needed is essentially the binding energies of the isotopes involved.

In order for the program to succeed it was necessary to have a reliable knowledge of the relative abundance of isotopes over most of the periodic system. Such knowledge first became available in 1937 when the Swiss-Norwegian geochemist Victor Goldschmidt published an extensive survey based on terrestrial, meteoritic, and astrochemical measurements painstakingly carried out for many years.[66] Goldschmidt realized that his data probably reflected the cosmic creation history of the elements and attempted to find correlations with nuclear structure (figure 3.1). At the end of 1941 Chandrasekhar and his student Louis Henrich at the Yerkes Observatory used a more advanced version of this method to derive T and n^* for the prestellar stage of the universe, when the cooking of heavier elements was assumed to have taken place.[67] With $T = 8 \times 10^9$ K and $n^* = 10^7$ g·cm^{-3}, they found a promising agreement with Goldschmidt's data for elements up to argon, but for the heavy elements the success turned into something like a disaster. In the words of Chandrasekhar and Henrich: "It is found that the physical conditions under which we would predict anything like the observed relative abundances of the elements beyond oxygen and up to potassium (say) will be wholly inadequate to account for any appreciable amounts of the heavy nuclei. . . . [T]he conclusion is inescapable that to predict anything like the observed relative abundances of the heaviest nuclei . . . we need distinctly different conditions from those indicated by the relative abundances of the isotopes of a single element."[68] As a way out of the dilemma they suggested that the different elements were formed during successive stages of the prestellar state, the heavy ones at a temperature of 10^{10}–10^{11} K and a density near the nuclear density, and the lighter ones under the less extreme conditions following the original expansion. During the expansion the ratios of the heavy elements were assumed to be frozen up. Once again, if only hesitatingly, the problem of element formation was brought into contact with evolutionary cosmology.

The partial failure of Chandrasekhar and Henrich was taken up at the eighth Washington Conference on Theoretical Physics, held on 23–25 April 1942. The subject of the conference, suggested by Gamow, was "The Problems of Stellar Evolution and Cosmology." It was attended by twenty-six physicists and astronomers, including Atkinson, Chandrasekhar, Critchfield, Gamow, Teller, Pauli, and Svein Rosseland; but, as noticed in the report of the conference, several of the scientists invited to take part "could not do so because of urgent unexpected demands of their national-defense problems."[69] Indeed, in the spring of 1942, four months after Pearl Harbor and the official declaration of war on Japan, Italy, and Germany, most physicists had other obligations than accounting for the formation of the elements. The Manhattan Project, recommended to President Roosevelt in March, was on its way, and there was

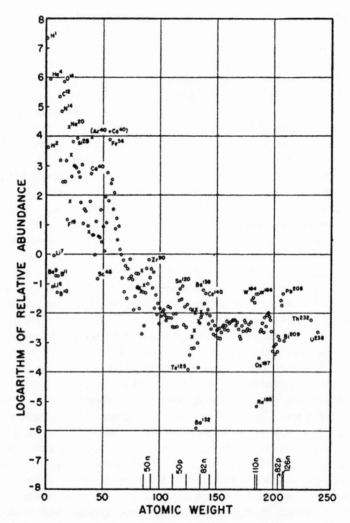

Figure 3.1. The relative cosmic abundances of the elements, according to Goldschmidt's compilation of data from 1938. The points indicate the abundance ratios relative to silicon, which on the logarithmic scale is taken to be 4. *Source:* Reproduced from the version given in Alpher (1948), p. 1578.

a general mobilization of scientists for defense purposes. Probably for this reason, the 1942 conference was the last in the series of Washington conferences on theoretical physics during the war years.

The Washington meeting is an important event in the history of cosmology because it was here agreed that a big-bang universe was necessary in order to account for the existence of the heavy elements. In discussions of the Chandrasekhar-Henrich results, it was concluded that the equilibrium mechanism suggested a few months earlier was not tenable. Goldschmidt's abundance curve shows that, apart from local variations, for atomic weights larger than about 100 all elements have nearly the same abundance. It was this feature that the equilibrium theory could not explain and which indicated that the failure might lie in the equilibrium hypothesis itself. For according to this hypothesis the relative abundance of an isotope $[n(A)/n(A + 1) = x$, say] must follow an expression of the form $\log x = - (E_{A + 1}/kT) - \log n^*$, which follows from equation (3.1). Since the binding energy of a neutron is approximately proportional to the atomic weight, for any given temperature a relation of the form $\log x \approx -A$ must therefore exist. Whereas this fits well with the lighter elements, it disagrees violently with the leveling off in the diagram for heavier elements.

According to the report of the conference, written by Gamow and J. A. Fleming of the Carnegie Institution's Department of Terrestrial Magnetism: "It seems, therefore, more plausible that the elements originated in a process of explosive character, which took place at the 'beginning of time' and resulted in the present expansion of the universe." In itself, this conclusion was only a minor advance compared with the suggestion made by Weizsäcker four years earlier, but it was better founded and a bit more precise in its linking of the origin of elements with a cosmic big bang. And whereas Weizsäcker's suggestion had been a kind of private speculation, the conclusion of 1942 was shared by many of the participating scientists, if not all of them.[70] It shows that the big-bang picture was gaining momentum at least among nuclear physicists. Yet it shared with Weizsäcker's scenario its qualitative nature and foundation in nuclear-physical considerations of an unspecified sort. What reactions took place in the hypothetical "process of explosive character" was still a matter of pure speculation.

Undeterred by speculative scenarios, Gamow suggested half a year after the Washington conference the possibility that the original superdense nuclear matter consisted of, or immediately after its explosion resulted in, a series of highly unstable superheavy atoms.[71] These hypothetical nuclei, which he suggested to be several times heavier than uranium, would then break up into smaller parts. Although ordinary fission processes result in only two fragments, Gamow saw no reason why the superheavy nuclei should not fragment into many smaller nuclei spread over the entire periodic system. He realized, of course, that there was not the slightest evidence for either his superheavy elements or multiple fission, but nonetheless concluded that "this possibility is not entirely excluded." He must soon have realized that it was, in fact, excluded, or at least implausible, for he never returned to comparing the big bang with an enormous atomic bomb of the fission type. In spite of its speculative nature and brief lifetime, the hypothesis deserves to be remembered because it was the first attempt to give a physical picture of the big bang.

As far as can be judged from the conference report and Gamow's subsequent paper, there still was no attempt to connect nuclear physics with relativistic models of cosmological evolution. All the same, at the 1942 conference Gamow clearly endorsed a big-bang picture and suggested that the gross material of the present world is the result of what happened some two billion years ago in a highly compressed primeval state. This was a major conceptual change, but at the time it seems not to have been considered as such by either Gamow or other physicists. They were working within an established tradition on a clearly defined problem area, the origin of the elements, and when they realized that the site of nuclear cooking could not be stellar they felt forced to choose the only other alternative they could see, the hypothetical big bang. Conceptually this was a step of paramount importance, but from the point of view of Gamow it was just a natural, if somewhat speculative, extension of a research program he had pursued for years. Moreover, it was not an entirely new idea. Not only had it been ventured explicitly by Lemaître and Weizsäcker, and implicitly by Chandrasekhar and Henrich, but in his work with Teller on galaxy formation Gamow had himself come close to the big-bang idea.

At any rate, it took more than three years until Gamow developed the idea of 1942, a lapse of time which should primarily be sought in external circumstances—there was, after all, a war going on—but which also supports the suggestion that he did not see the 1942 big-bang version as particularly important. The first evidence of a breakthrough is probably contained in a letter to Bohr of 24 October 1945, congratulating him on his sixtieth birthday. The letter suggests that Gamow resumed work on big-bang cosmology in part as a response to the recently ended war and the new world order with a strong and aggressive Soviet Union. It is a different side of Gamow we meet in the letter, from which I quote an extensive part.

It would be so nice if the end of the war would mean the return to the peacefull life as some fifteen years ago when we have been drinking hot chokolade at one of your previous birthdays paa [at] Blegdamsvejen. But somehow I do not feel this way at the present moment, and it looks to me more as the eve of a great Deluge comming from the East which is bound to engulf the free man on the Earth. Sorry! I didn't start this letter to develope pessimistic truths and this is just the mood . . . But it would be realy so much nicer if one could begin to work again on pure science without the heavy clouds hanging in the air! That is what I am trying to do at present studying the problem of the origin of elements at the early stages of the expanding universe. It means bringing together the relativistic formulae for expansion and the rates of thermonuclear and fission reactions. One interesting point is that the period of time during which the original fission took place (as estimated from the relativistic expansion formulae) must have been less than one millisecond, whereas only about one tenth of the second was avaliable to establish the subsequent thermodynamical equilibrium (if any) between different lighter nuclei. I am planing to have our next conference here in spring on that problem and the other problems on the borderline between nuclear physics and cosmology. It is such a pitty that you are not here to participate in it! I would love still better to come to visit you in

Copenhagen even though this means comming too close to the red line crossing the Europe. As soon as communications will come to a more normal state I think I will do that. We are living here quietly out in the woods on what is now being considered the safe distance from downtown Washington, my boy is ten years old and knows quite a bit about atoms, nuclei and stars. Though his spelling is just as bad as mine (heredity of course!).[72]

In the fall of 1945 Gamow thus had the essential idea of a revised big bang model building on a combination of nuclear physics and the Friedmann-Lemaître equations. Apparently he still believed that the big bang started with fission processes. The Ninth Washington Conference on Theoretical Physics was held from 31 October to 2 November 1946, with Bohr as one of the participants. However, the subject was not cosmology and nuclear physics, as originally planned by Gamow, but "The Physics of Living Matter."

At the time of the ninth Washington conference Gamow had modified the idea he mentioned to Bohr and developed it into a publishable form. On 13 September 1946 Gamow sent a brief paper to *Physical Review* which some-times has been hailed as the foundation of modern big-bang cosmology.[73] The essential progress in that paper was that Gamow combined two perspectives, neither of which was new, but which together served as the foundation for much of the further development. He repeated the conclusion obtained at the Washington conference and then made the decisive step to connect it with the relativistic theory of the expanding universe, very much in the same manner that he and Teller had discussed in their joint work of 1939. As in this work, Gamow took his starting point in the general Friedman-Lemaître equation for the time dependency of the scale factor. He wrote it as

$$\frac{dR}{dt} = \sqrt{\frac{8\pi G}{3}\rho R^2 - \left(\frac{\text{const}}{R_0}\right)^2},$$
(3.3)

where R_0 describes the present curvature of space. Introducing the proper distance of any linear dimension l, the equation can also be written

$$\frac{dl}{dt} = \sqrt{\frac{8\pi G}{3}\rho l^2 - \left(\frac{cl_0}{R_0}\right)^2},$$
(3.4)

where l_0 is the unit of proper distance. Now consider with Gamow a cube with sides of length l containing 1 g of matter. From the value of the mean density of matter (10^{-30} g·cm^{-3}) it follows that $l_{now} \approx 10^{10}$ cm. The Hubble constant is $H = (dl/ldt) = 1.8 \times 10^{-17}$ s^{-1}, from which follows a rate of expansion $(dl/dt)_{now} \approx 1.8 \times 10^{-7}$ cm·s^{-1} and, from equation (3.4), $R_0^2 \approx -2.9 \times 10^{34}$ cm^2. According to Gamow, the imaginary value of R_0 showed the universe to be open, that is, expanding without limits.

In the distant past, when the universe was compressed to a density of the order 10^6 g·cm^{-3}, as supposedly needed for element formation, equation (3.3) still holds. Using the new value of ρ and the same value of R_0^2 gives $l \approx 0.01$ cm and $dl/dt \approx 0.01$ cm·s^{-1}. That is, the volume is reduced by a factor of 10^{24} and the rate of expansion increased by a factor of 10^5. Gamow summarized: "This means that the epoch when the mean density of the universe was of the order 10^6 g cm^{-3}, the expansion must have been proceeding at such as high rate, that this high density was reduced by an order of magnitude in only one second . . . [therefore] the conditions necessary for rapid nuclear reactions were existing only for a very short time."[74] After having ruled out the possibility of early equilibrium processes, Gamow imagined the earliest universe to consist of a comparatively cold (but thick) gaseous soup of neutrons. This first version of the big bang was cold because the high rate of expansion would thin out the soup before an appreciable amount of neutrons could have decayed. At that time the neutron lifetime was not known very accurately—Gamow was satisfied with taking it to be "presumably of the order of magnitude of one hour." As a result of the expansion the cold neutron soup or gas was imagined to coagulate into larger and larger neutronic complexes, which by subsequent emission of beta particles would turn into the known chemical elements. That is, Gamow thought of a kind of very large atomic nuclei originally made up only of neutrons, a picture not unlike the one proposed by Lemaître in 1931. "The decrease of relative abundance along the natural sequence of elements," Gamow wrote, "must be understood as being caused by the longer time which was required for the formation of heavy neutronic complexes by the successive processes of radiative capture." Apart from the radiation emitted in nuclear reactions, Gamow did not include electromagnetic radiation in his picture.

As we have seen, many of the elements of Gamow's early version of the big bang theory were in existence in the late 1930s. It is natural to speculate that the idea of conceiving the primeval universe as a neutronic state was inspired by Gamow's interest in stars with neutron cores. Such superdense objects had been discussed by several researchers, including Landau, Gamow, Baade, Zwicky, Oppenheimer, and Robert Serber. Although purely hypothetical, neutronic stellar bodies of very high density had acquired a certain respectability. Gamow returned to the subject in 1939, when he suggested that during the gravitational contraction of heavy stars nuclei and free electrons would transform into neutrons.[75] The time-reversed picture is an expanding core of densely packed neutrons which transform into complex nuclei—a kind of local big bang.

In his arguments for the big-bang theory, in both his scientific and popular works, Gamow presented equilibrium theories as hopelessly inadequate and outdated. As early as 1942 he claimed that the attempt to understand the formation of elements by equilibrium processes in stars "has been entirely abandoned," and in 1946 he stated as a matter of fact that "the *only* way of explaining the observed abundance-curve lies in the assumption of some kind of unequilibrium process taking place during a limited interval of time."[76] This was, however, a distortion of the facts. Most specialists outside Gamow's small group did not think of equilibrium theory as a failure, and denied that it was necessarily so. This kind of theory continued to be developed in still more sophisticated versions by a great many scientists, especially in Europe. To mention only a few examples, the Italian-Brazilian physicist Gleb Wataghin argued that elements could be formed in equilibrium at a temperature of about one billion degrees, and in 1946 the Swedish physicist Oskar Klein and his collaborators Göran Beskow and L. Treffenberg used the powerful method of Gibbsian statistical mechanics to provide partial support for the conclusion.[77] However, although the Stockholm physicists obtained a satisfactory agreement up to about copper, they were unable to account for the relatively large amounts of heavier elements. Yet they did not consider this a failure of the equilibrium theory, as Gamow did. Klein, Beskow, and Treffenberg concluded optimistically that the discrepancies were "hardly to be regarded as a difficulty for the theory but rather as a most interesting problem for its further development."[78]

Moreover, if heavy element synthesis seemed impossible in the interior of ordinary stars, one could assume the locale of nuclear buildup to be in special types of hot stars or taking place during special phases of the evolution of stars, where temperature and density were very high. Following such an approach, G. van Albada in the Netherlands and Fred Hoyle in England were able to obtain a reasonable agreement with observations for the heavy elements also.[79] Or one could argue, as did a British physicist, that whereas the heavier elements are built up by neutron capture, the lighter ones are formed in thermodynamic equilibrium.[80] The point is that Gamow's conclusion was far from universally accepted and remained contested. In a careful review of 1950, the Dutch physicist Dirk ter Haar concluded that "at present the equilibrium theory offers a better solution for the problem how to account for the observed abundances of the chemical elements than the [big bang] theory."[81] Ter Haar saw no need for an explosive universe. He repeated his conclusion of the superiority of the equilibrium method three years later.[82]

The explosive universe was not Gamow's only preoccupation with cosmology in September 1946. He also wanted to explain the rotation of galaxies, and for this purpose he suggested that the entire visible universe is provided with an angular momentum, i.e., that it rotates around some center far beyond the reach of telecopes. In a letter to Einstein he suggested that such a rotating universe might be better suited to describe the big bang than the ordinary one. However, in order to describe it relativistically, anisotropic solutions of the field equations had to be found, and this posed a problem. Modestly describing himself as "a very bad mathematician," Gamow wrote:

> Instead of the "spherical space" we will have an "ellipsoidal space" with the masses of the universe rotating arround the two "poles." I do not think that anybody ever tried to obtain the line-element for such anisotropic expanding universe, and

I wonder if you have ever thought about it? It would have the advantage that, unlike in ordinary simmetrical solutions, the radius of the universe would not go through zero in the beginning of time (even though $\Lambda = 0$). For each value of the "total angular momentum of the universe" we would have the maximum density and temperature at the extreem cotracted state. It is important to remember that in order to explain the present relative abbundance of chemical elements one must agree that in "the Days of Creation" the mean density and temp. of the Universe was 10^7 gm/cm^3 and $10^{10\circ}$K. If the solution for the anisotropic (rotating) expanding universe will be found the above conditions will permit to calculate universal angular momentum, and to predict the amplitude in the variations of radial velocities of galaxies to be observed by Hubble and Humasson. Don't you think it is very exiting?[83]

Einstein did not find the idea "exiting" at all. In a critical reply to Gamow's hypothesis, he questioned whether it had any meaning to ascribe to the universe as a whole an angular momentum, a "quantity [which] has no meaning independent of the choice of the coordinate system."[84] Gamow seems to have abandoned the idea, but anisotropic or rotating models of the universe were later taken up by several mathematical cosmologists, first by the mathematician (Einstein's colleague at the Institute for Advanced Study in Princeton) Kurt Gödel in 1949.[85] Gödel found the kind of solution Gamow had asked for, but not one that could have any physical interest or might have appealed to Gamow: apart from having no redshift, the Gödel universe has many pathological features that make it of mathematical and philosophical interest only. For example, it is theoretically possible to make a round trip in time—travel to the past or future and back again—which of course implies all kinds of problems of causality. Gödel's model universe is fascinating in lending scientific support to what seems to be science fiction, but it is definitely not the kind of universe you and I live in.

Alpha, Beta, Gamma, Delta

In 1946, the same year that Gamow first put some flesh on the explosion model of the universe, he started supervising a Ph.D. student working on primordial nucleosynthesis.[86] The son of émigré Russian Jews, Ralph Alpher was born in Washington, D.C., in 1921. Inspired by the popular writings of Jeans and Eddington, he became in his teen years interested in science generally and astronomy particularly. In 1940, as an undergraduate, he started working with the navy, and five years later he received his master's degree in physics from George Washington University, where he had followed evening courses. During his period as a navy employee he worked on the protection of ships against magnetic mines and on submarine detection by airborne magnetometers. In 1944, Alpher obtained a position at the Applied Physics Laboratory of the Johns Hopkins University, where he first worked with guided missile systems and on rocket studies of the cosmic radiation. After having spent most of a year on an ill-fated doctoral dissertation—the results of which, he discovered to his chagrin, had been published by another physicist—Alpher decided to examine the formation of elements in the primeval universe. This was a topic where he was unlikely to meet competition. At that time he had only a limited knowledge of cosmology and nuclear theory, most of which he had learned in courses given by Gamow. In relativity and cosmology the main source was Tolman's 1934 monograph, and he also studied Pauli's famous encyclopedia article and Weyl's book. Nuclear theory was learned through Gamow's textbook, supplemented by with extended reading of articles, including Bethe's review articles. Alpher also studied a wide range of works in astrophysics and fundamental physics, including Eddington's posthumously published *Fundamental Theory*.[87]

The thesis topic was suggested by Gamow, who at the time had formulated his cold big bang scenario and was now eager to have Alpher examine how elements could be formed in agreement with this model. Alpher was soon discussing his ongoing work with Robert Herman, his neighbor at the Johns Hopkins laboratory. The two would soon join forces in pioneering work in cosmology. Herman's family background was similar to Alpher's; his parents were Russian Jews, who had arrived in New York in 1910, where Robert was born four years later. After undergraduate studies at City College, New York, he entered Princeton University in 1936 and there received his Ph.D. in physics in 1940. Two years later he joined the new Applied Physics Laboratory, where he did militarily related work and remained until 1955. Contrary to the younger Alpher, Herman was already a seasoned researcher when he came to cosmology. In 1948 he had published twenty-eight scientific works, most of them in areas of chemical physics such as molecular spectroscopy, which was also the topic of his dissertation. Also contrary to Alpher, Herman was well acquainted with relativity and cosmology from his time at Princeton, where he had studied these subjects under H. P. Robertson.

In Gamow's 1946 big-bang version, elements were supposed to be built up by neutron capture. In fact, in 1946 Gamow did not refer directly to neutron capture processes, but to an unspecified "coagulation-process" forming large neutral complexes, i.e., a kind of polyneutrons. But he seems to have decided soon thereafter that the essential process was the capture of neutrons by protons and other nuclei. To transform the idea into a quantitative theory, empirical knowledge of such processes was needed. It was therefore fortunate that precise data on the cross sections of fast neutrons captured by a wide range of elements had just appeared. The data, obtained by Donald J. Hughes, a physicist from Argonne National Laboratory, were determined with an eye to materials suitable for the construction of nuclear reactors.[88] Hughes presented his data in a ten-minute paper at a meeting of the American Physical Society with Alpher in the audience; what Alpher came to think of was their use, not in nuclear reactor design, but in the design of the universe. He therefore requested Hughes to send him the full set of data, which Hughes did.

When Alpher studied the data he saw in them a key to understanding the cosmic distribution of elements. The exciting feature was that Hughes's cross-

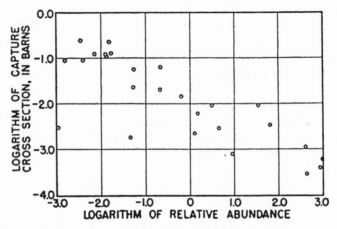

Figure 3.2. Alpher's correlation of Hughes's neutron capture cross sections with Goldschmidt's data for the relative abundances of nuclei. *Source:* Alpher (1948), p. 1581.

section data were correlated with the relative abundance of the elements (figure 3.2), an observation which indicated that Gamow's idea was basically correct and might be developed into a quantitative theory. Alpher completed his Ph.D. work in the summer of 1948, having continuously discussed its progress with Gamow, who naturally was greatly interested in it. At that time the twenty-seven-year-old Alpher was a novice in the world of science. He had only written one research paper, dealing with missile dynamics. The dissertation generated large public interest because of its unusual subject and was widely reported in the newspapers. The weekly *Science News Letters* reported the amazing discovery that the creation of the world had taken place two billion years ago "minus the 300 seconds needed to get the process started."[89] The political association—the constructive big bang as compared with destructive nuclear weapons—was noticed, and so was the religious association. Not everyone found it right to discuss scientifically the creation of the world, the domain traditionally reserved for God. Alpher received letters from people assuring him that they would pray for his soul and for his redemption.[90]

Like Gamow, Alpher was careful not to connect his work on big-bang theory with theological questions. He was a scientist and considered cosmology a subject of science, not of religion or philosophy. As a teenager he had lost faith in Jewish belief and come to doubt if God existed. Later in life he tended toward Unitarianism and described himself as "an avocational cosmologist who is also an agnostic with atheistic tendencies."[91] Opposed to dogmatism and fundamentalist tendencies, Alpher believed that scientific reasoning alone could reveal the secrets of nature; his work with and approach to cosmology were for him a kind of, or substitute for, religious experience. Maybe it would one day be necessary to introduce God in big-bang cosmology, he thought, but in that case the necessity would not follow from the Bible, but from scientific evidence. "Meanwhile," he wrote in 1983, "it seems to me to be undignified and unworthy to hide the *sum total of ignorance* in a god-concept." Contrary to what was argued by some opponents of the big-bang theory in the 1950s, this theory did not imply any divine action; at least, this was the view of Gamow and Alpher, who were both opposed to the notion.

Before the completion of the dissertation, Gamow and Alpher could already publish the first version of an improved big-bang model.[92] Although deeply occupied with explaining the early development of the universe, Gamow was at the time also busy with more mundane work. As a consultant for the Applied Physics Laboratory he was involved in military research, especially in studying the means for long-range navigation related to the needs of the airforce.[93] His memorandum on this subject appeared on 13 February 1948, a few days before he sent his and Alpher's joint paper to the *Physical Review*.

The paper by Gamow and Alpher is generally known as the "$\alpha\beta\gamma$ paper." Quite apart from its content, the paper has a famous history, which has made it part of the physics folklore. As it appeared in *Physical Review*, it was co-authored by Bethe, although he did not, in fact, contribute at all. Gamow recalled the story behind the $\alpha\beta\gamma$ terminology as follows: "In writing up the preliminary communication of this work, I was unhappy that the letter β was missing between α and γ. Thus, sending the manuscript for publication in *Phys. Rev.*, I put in the name of Hans Bethe (in absentia) between our names. This was planned as a surprise to Hans when he would unexpectedly find his name as co-author and I was sure that, being my old friend, and having a good sense of humor he would not mind. What I did not know was that at that time he was one of the reviewers for *Phys. Rev.* and that the manuscript was sent to him for evaluation. But he did not make any changes in it except to strike out the words "in absentia" after his name, thus endorsing the idea and the results."[94] Gamow even tried to make Herman change his name to Delter so the $\alpha\beta\gamma$ theory would become an $\alpha\beta\gamma\delta$ theory. Although Herman resisted the temptation, in a later paper Gamow referred to "the neutron-capture theory of the origin of atomic species recently developed by Alpher, Bethe, Gamow and Delter," a reference which must have caused some confusion.[95] Bethe, who was not foreign to this kind of fun, took the joke with good humor as Gamow expected. As he later told Alpher and Herman, "I felt at the time that it was rather a nice joke, and that the paper had a chance of being correct, so that I did not mind my name being added to it."[96] The last word on the matter was added by Gamow in 1960: "There was . . . a rumor that later, when the $\alpha\beta\gamma$ theory went temporarily on the rocks, Dr. Bethe seriously considered changing his name to Zacharias."[97]

Almost equally as important as the joke behind the name of the $\alpha\beta\gamma$ paper was its content. The theory, which in part built on Alpher's still uncompleted dissertation, was preliminary and programmatic, but it did give a new picture of the earliest universe and indicated the route to be followed in further research. The very early universe was now described as a hot, highly compressed neutron gas which at some time ($t = 0$) started decaying into protons and electrons. What caused the neutrons to decay about two billion years ago was left as unanswered in this model as it was in Lemaître's. And so was the question of the origin of the primordial gas, which was just assumed as a starting point. On some occasions Gamow speculated that the explosion—the combined initial expansion and decay—might have been the result of an earlier "hypothetical universe collapse" which had squeezed the free electrons into the protons to form neutrons.[98] Gamow apparently liked the idea of a grand cosmic one-cycle process, from infinite rarefaction over a superdense state toward a new state of infinite rarefaction. However, this hypothesis, rather similar to the one entertained by Weizsäcker in 1938 (and vaguely by de Sitter in 1933), was of no importance to the theory and was never mentioned by the more cautious Alpher and Herman. In 1954 Gamow concluded that although our expanding universe was presumably the result of an earlier contraction, "from the physical point of view we must forget entirely about the pre-collapse period and try to explain all things on the basis of facts which are no older than five billion years—plus or minus five per cent."[99]

With the spontaneous decay of neutrons, protons would be formed, and some of these would combine with remaining neutrons to form deuterons; from these nuclei heavier elements were assumed to be synthesized by successive neutron capture and beta decay. In such a process the rate of increase of concentration of nuclei of mass number A must be equal to the difference between the rate of the buildup $(A - 1) \rightarrow A$ and the rate of further transformation $A \rightarrow (A + 1)$. Writing $n(A)$ for the relative numbers of nuclei and $\sigma(A)$ for the capture cross sections, the differential equation governing the formation processes can therefore be written as

$$\frac{dn(A)}{dt} = \Phi(t) \left[\sigma(A - 1)\, n\, (A - 1) - \sigma(A)n(A) \right] \tag{3.5}$$

with A running from 1 to about 240. $\Phi(t)$ is a factor representing the frequency of nuclear collisions at a given density and temperature; since both of these quantities decrease as a result of the expansion, so will $\Phi(t)$. By integrating the equation, $n(A)$ can be found for different mass numbers and a theoretical abundance curve thus obtained. A preliminary result was communicated in the $\alpha\beta\gamma$ paper, and a more detailed discussion included in a paper Alpher submitted in early July, which was based on his dissertation. For the initial neutronic state of the world Alpher introduced the term "ylem," which according to Webster's Dictionary, where he had come across it, was an obsolete noun for "the primordial substance from which the elements were formed."[100] Gamow liked the word and used it frequently, inevitably leaving his readers with the impression that he had invented it (and perhaps himself believing so).

In his first computations, which disregarded the expansion, Alpher did not have access to electronic computers. In order to avoid carrying out the integration over more than two hundred consecutive differential equations, he di-

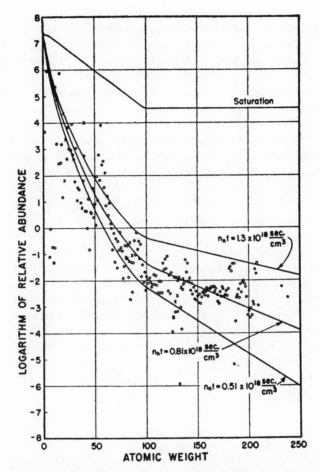

Figure 3.3. Alpher's comparison of relative abundances computed by the neutron capture theory with Goldschmidt's abundance data. The best fit is for $n^*t = 0.81 \times 10^{18}$ sec·cm⁻³. The "saturation" curve represents the relative abundances had the neutron capture process continued for an indefinite period, i.e., ignoring cutoff due to neutron decay and expansion of the universe. *Source:* Alpher (1948), p. 1585.

vided the interval between $A = 1$ and $A = 240$ into twelve sections and performed the calculations for each of these. Although a crude approximation, the result did not differ much from later, more exact calculations. The result is shown in figure 3.3. As is seen, a generally good agreement with observations was obtained, or rather made possible as a result of fitting the theoretical curve to the data. With a suitable value for the quantity n^*t, where n^* is the concentration of neutrons and t is the period of element formation, the agreement was made possible.

With the value of n^*t indicated by the abundance curve, it was possible to estimate the physical conditions of the ylem. To do so the relativistic theory of the expanding universe was introduced by means of the same equation (3.3) that Gamow had used two years earlier. By integration Alpher found the density to vary with the time according to $\rho = 1.3 \times 10^6 \, t^{-2}$ with t in seconds and ρ in g·cm^{-3}. The ylem of Gamow and Alpher was a physical proto-universe, for which reason the singularity occurring at $t = 0$, where the density becomes infinite, had to be avoided or, rather, disregarded. They assumed that the initial explosion of the ylem, and the corresponding initial formation of elements, only started at a time t_0, when the density had the finite value ρ_0. With a temperature of $T \approx 10^9$ K, corresponding to the dissociation energy of deuterons, Alpher found the initial conditions $t_0 \approx 2.5 \times 10^4$ s and $\rho_0 \approx 1.3 \times 10^{-3}$ g·cm^{-3}.[101]

However, these values were based on the assumption that there is only matter in the primordial universe, as assumed by equation (3.3). In the spring of 1948 Alpher and Gamow realized that the early universe was in fact dominated by radiation, not matter: with a temperature about 10^9 K, the radiation density as given by the Stefan-Boltzmann law $\rho_r = (a/c^2)T^4$ is much larger than the matter density. In their papers from the summer of 1948, Alpher and Gamow separately made the important innovation of a radiation-filled universe, which, Alpher showed, reduced the starting time of element formation to about 250 s. In his slightly earlier paper, Gamow considered the cosmological model for an expanding radiation-filled universe of the form

$$\frac{dR}{dt} = \sqrt{\frac{8\pi G}{3}\rho_r R^2}. \tag{3.6}$$

The result follows from equation (3.3) with $\rho = \rho_r + \rho_m \approx \rho_r$ and ρ_r much larger than the curvature term involving R_0. Gamow discussed briefly the time dependence of ρ_r and ρ_m (matter density) in such a model, where $\rho_r \sim t^{-2}$ and $\rho_m \sim t^{-3/2}$. As Gamow noted, it follows that the domination of radiation over matter will decrease over time, and at a certain period the two densities will be equal. He estimated that this would have taken place at about $t = 10^9$ years, when the universe had cooled to about 10^3 K, and suggested that under these conditions galaxies would begin to form. Although Gamow's numerical estimates were wrong, it was the first time that the significance of the crossover or decoupling time was noticed.

At that time, Gamow knew Einstein well. During the war Einstein served as a consultant for the High Explosive Division, the same U.S. Navy branch where Gamow did part-time work. Because of his status and advanced age Einstein was allowed to stay in Princeton and Gamow acted as courier, periodically going from Washington to Princeton where he presented the various military-technical problems to the famous physicist. "Einstein would meet me in his study at home, wearing one of his famous soft sweaters, and we would go through all the proposals, one by one. He approved practically all of them, saying, 'Oh yes, very interesting, very, very ingenious,' and the next day the admiral in charge of the bureau was very happy when I reported to him Einstein's comments."[102] Later, when working on the big-bang theory, Gamow sent a copy of the manuscript of one of his articles to Einstein. The father of relativistic cosmology answered: "I am convinced that the abundance of elements as function of the atomic weight is a highly important starting point for cosmogonic speculations. The idea that the whole expansion process started with a neutron gas seems to be quite natural too. The explanation of the abundance curve by formation of the heavier elements in making use of the known facts of probability coefficients seem to me pretty convincing."[103]

Einstein was not the only notability whom Gamow thought should know about the new big-bang picture of the universe. In 1951 he sent a copy of a popular article to the pope, Pius XII, which was received through the apostolic delegate, the archbishop of Laodicea. According to the archbishop, the paper "was presented to the Holy Father who read it with satisfaction and who looks forward to the publication of your book on 'The Creation of the Universe.'" A few months later the secretary of state of the Vatican City informed Gamow that the pope had received the book from its author and was eager to read it.[104] When Gamow sent these works to the pope it was not because Gamow was a Catholic or had any special relations to the Catholic church. But he knew that the pope had recently endorsed the big-bang universe in a controversial statement, apparently unaware of the recent developments in the United States (see further in section 5.3).

Further Developments

The preliminary developments of the $\alpha\beta\gamma$ theory from the summer of 1948 were to a large extent the collective result of discussions between Gamow, Alpher, and Herman, and it is difficult to decide more exactly which result was due to whom. Gamow spent part of the summer at Los Alamos working on the hydrogen bomb project, for which he had been cleared; another part he spent at Ohio State University, where he prepared a larger paper on the big-bang theory for *Nature*. In this paper he restated his and Alpher's earlier results and presented graphically the earliest history of the universe in the first version of what he would later refer to as "the divine creation curve" (figure 3.4). Gamow was particularly fascinated by the possibility of using the knowledge to calculate the condensations out of which galaxies were assumed to be formed. For the masses and diameters of the original galaxies he obtained the impressive formulas

$$M = \frac{2^{31/8} 5^{7/4} \pi^{5/4} e \hbar^{5/4} \epsilon^{5/4}}{3^{17/8} m^{15/4} c^{5/4} G^{7/4}} = \text{ca. } 2.7 \times 10^7 \text{ sun masses,}$$

$$D = \frac{2^{45/8} 5^{1/4} \pi^{7/4} e^3 \hbar^{3/4} \epsilon^{15/4}}{3^{27/8} m^{29/4} c^{35/4} G^{5/4}} = 13.000 \text{ light years.}$$

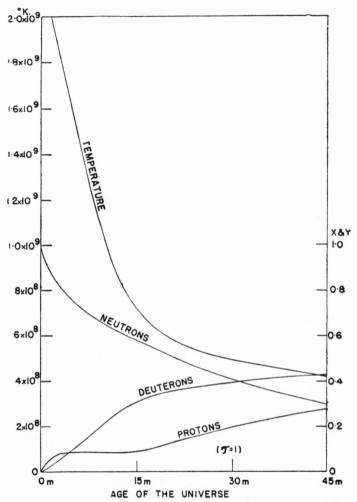

Figure 3.4. Gamow's "divine creation curve" as of October 1948. The variation of the concentrations of neutrons, protons, and deuterons with time is shown on the right scale, and the temperature decrease is shown on the left scale. *Source:* Gamow (1948b), p. 681. Reprinted with the permission of Macmillan Magazines Limited.

These were expressions that would surely have delighted Eddington, had he lived to see them. The fact that the estimate for galactic masses differed from the one accepted by astronomers by a factor of about one hundred did not spoil Gamow's pleasure in observing that "masses and sizes comparable to those of stellar galaxies can be expressed in terms of fundamental constants, and the basic quantities of nuclear physics."[105] To a reporter from Science Service he said that "I am quite excited about these new results on the evolution of the universe. It seems to me to be the best piece of work I have done since the theory of radioactive decay 20 years ago."[106]

Alpher and Herman seem to have been less impressed. When they received Gamow's manuscript they found several errors in it, but Gamow felt that it was too late to change his manuscript and instead urged his colleagues to write a note to *Nature* with their corrections.[107] The note, appearing in the November 13 issue, was more than just a correction of Gamow's too hasty calculations. As a result of their recalculation of the matter density, Alpher and Herman were led to conclude that the formation of condensations required taking into account both matter and radiation in the expanding universe. That is, instead of using either (3.3) or (3.6), they now integrated the full equation

$$\frac{dR}{dt} = \sqrt{\frac{8\pi G}{3}(\rho_r + \rho_m)R^2 - \left(\frac{\text{const}}{R_0}\right)^2}. \tag{3.7}$$

As a result they found the crossover time to be about 10^7 years, about ten times smaller than Gamow's result. What is more important, they also corrected Gamow's value of $T = 340$ K and added the remark, "The temperature of the gas at the time of condensations was 600° K, and the temperature in the universe at the present time is found to be about 5° K."[108] This remark, apparently of no particular significance, was later to be seen as highly interesting, almost (but only almost) worth a Nobel Prize.

The calculations of Alpher and Herman, and the way in which they had obtained the 5 K value, were made clear in a subsequent article in which they also elaborated on Gamow's galaxy formation theory.[109] They concluded, however, that this problem was scarcely worth investigating in detail. The formation of a galaxy required some initial instability and this was just what the Soviet physicist Evgeny M. Lifshitz had recently proved could not take place in a relativistic expanding universe.[110] A physical mechanism for the formation of galaxies was therefore missing and for this reason Alpher and Herman could not share Gamow's enthusiasm. With regard to the physical conditions of the early universe, they pointed out that the product of ρ_r and $\rho_m^{-4/3}$ would remain constant during the expansion. That is,

$$\rho_{r,0}\,\rho_{m,0}^{-4/3} = \rho_r\rho_m^{-4/3}, \tag{3.8}$$

where the quantities to the left denote initial densities and those to the right are, for example, present values. They now reasoned as follows. In the relation (3.8), ρ_m is known from astronomy to be about 10^{-30} g·cm^{-3}, and $\rho_{r,0}$ and $\rho_{m,0}$, the densities at the formation of elements, can be estimated from the density and temperature conditions necessary for nucleosynthesis. They took $\rho_{m,0}$ to be ca. 10^{-6} g·cm^{-3} and the corresponding radiation density to be ca. 1 g·cm^{-3}. The present radiation density can then be found. Alpher and Herman calculated in this way ρ_r = ca. 10^{-32} g·cm^{-3}, and from the Stefan-Boltzmann law they arrived at the result that this "corresponds to a temperature now of the order of 5 K."

In the early universe, where radiation dominates, Alpher and Herman showed by integrating the cosmological equation (3.7) that the radiation density would vary as

$$\rho_r = \frac{3}{32\pi G}\frac{1}{t^2} = 4.48 \times 10^5 t^{-2}\text{g}\cdot\text{cm}^{-3}$$

and the temperature as

$$T = \left(\frac{3c^2}{32\pi Ga}\right)^{1/4}\frac{1}{\sqrt{t}} = 1.52 \times 10^{10}t^{-1/2}\text{ K,}$$

where a is the constant in the Stefan-Boltzmann law. Concerning the temperature of 5 K they explained, "This mean temperature for the universe is to be interpreted as the background temperature which would result from the universal expansion alone. However, the thermal energy resulting from the nuclear energy production in stars would increase this value."[111]

For the variation in time of the temperatures and densities, Alpher and Herman produced an improved version of the "divine creation curve" (figure 3.5). They found (as had Gamow previously) that in the early, radiation-dominated era expressions of the form $\rho_m \sim t^{-3/2}$ and $\rho_r \sim t^{-2}$ would hold, whereas in the later, matter-dominated era they got $\rho_m \sim t^{-3}$ and $\rho_r \sim t^{-4}$. In the same two eras the scale factor and Hubble constant would relate to the time as $R \sim t^{1/2}$ and $H = (2t)^{-1}$ (radiation domination), and $R \sim t$ and $H = t^{-1}$ (matter domination). As to the geometry of the universe, they maintained the earlier conclusion of an open, hyperbolic universe. However, for Alpher and Herman, as for Gamow, the geometry and expansion type of the universe were of minor importance. Their concern was with the physical conditions of the early universe, concentrating on the first few minutes and extending at most to the crossover time. They chose, wisely, to ignore the initial singularity and the original explosion itself. Alpher and Herman agreed with Einstein that the field equations would probably break down very near $t = 0$, but, as they wrote, "since we do not concern ourselves with the 'beginning' this difficulty is obviated."[112] Neither they nor Gamow liked the term "big bang" because it emphasized the original explosion rather than the evolutionary feature of the model. In 1968, shortly before his death, Gamow said in an interview: "I don't like the word 'big bang'; I never call it 'big bang', because it is a kind of cliché."[113]

In addition to their publications, technical and popular, the three Washington physicists also publicized their ongoing research in various lectures. For example, they gave a joint paper at the Chicago meeting of the American Physical Society taking place 26–27 November 1948. From the abstract we

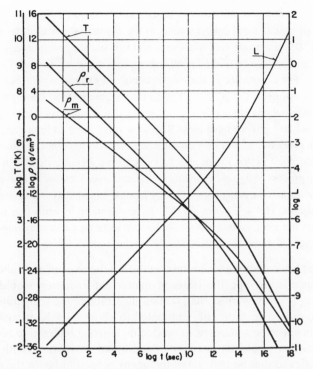

Figure 3.5. One of Alpher and Herman's "divine creation curves" from 1949. The figure shows the expansion of the universe (variation of proper distance L with time) and the time dependence of the matter and radiation densities (ρ_m, ρ_r) as well as the temperature T. *Source:* Alpher and Herman (1949), p. 1093.

learn that Gamow, Alpher, and Herman had by then made a study of the formation of deuterium under the conditions of an expanding universe with a matter density varying as $\rho_m = 7.2 \times 10^{-3} \, t^{-3/2}$. As to the medium-heavy elements, "impressed with the large relative abundances of the elements in the vicinity of iron, we are led to suggest that they may have been caused by the thermal evaporation of particles from the various nuclear species, the relative rates depending on the binding energy per particle."[114]

Contrary to Alpher and Herman, Gamow was strongly interested also in the problem of galaxy formation. He considered it an integral part of the new big-bang cosmology and had already investigated the problem before the war, in his work with Teller. The two physicists decided that the time was ripe for a new attack and in early 1949 they joined forces with Nicholas Metropolis and the mathematician Stanislav Ulam, whom they knew from the hydrogen bomb project. At first they arrived at the same result as Lifshitz, namely, that perturbations cannot give rise to condensations. But if the effects of radiation were taken into account (which Lifshitz had not done), things appeared more promising. On 29 January 1949 Gamow reported to Alpher:

Last week we (Gamow, Metropolis, Teller and Ulam) lead a vigorous attack on the problem of gravitational condensations in the expanding space in the presence of radiation. The results are very interesting and I should say revolutionary. First of all no condensations can be formed at all unless ρ_0 (in $\rho = \rho_0/t^{3/2}$ formula) is 0.1 or still higher. . . . For all smaller ρ_0 the kinetic energy of expansion will simply tear the condensations apart. For $\rho_0 \geq 0.1$ there exists a time-interval (near $t = 10^5$ years) where condensations are possible. . . . It seems that it may give the galactic masses as high [as] 10^{42} g if we have luck. Radiation plays quite an important role in the formation process. . . . You may be surprised that I do not object against such high ρ_0. But the point is that Fermi and Turkevich have recently studied in details the formation process of lightest elements with actual cross sections and have found that it will not work at the densities we usually assume unless there is a strong resonance for the reaction $He^4 + T$ below 400 kV. Experimentally it seems that such a resonance does not exist, so that one would have to increase the density quite considerably to go through the He-bottleneck.[115]

Gamow and his collaborators found that during the expansion of the universe the difference in the specific heats of matter and radiation would result in matter being slightly warmer than radiation. Although the temperature difference at the crossover time only amounted to about 0.01 K, it was found to have significant effects, namely, in producing a strong interaction between the gas particles that would act as a kind of additional gravitation. With this force Gamow believed that he had found the agent needed to begin the condensation process. However, further calculations showed that the interactions were not sufficient to provide a mechanism for condensation, and Gamow was forced to give up the idea.[116]

At the end of 1948, big-bang cosmology had experienced a drastic development and had now, if still only in outline, turned into a proper theory with quantitative estimates of how the universe had evolved in time. The justification of the new picture of the hot, radiation-filled early universe was primarily its agreement with the empirical abundance curve. However, so far the model was rather primitive, and no detailed calculations of nucleosynthesis had been made. As we have seen, the big-bang model grew out of Gamow's earlier attempts to understand element formation in a prestellar context. Although this context was different from that of Lemaître's primeval-atom hypothesis, qualitatively the Gamow big bang shared some features with Lemaître's idea. It is therefore natural to ask if Gamow and his coworkers were inspired by the Belgian cosmologist. The answer seems to be no. During the period when Alpher prepared his dissertation he studied systematically all relevant papers on cosmology, including those of Friedmann and Lemaître, and read Lemaître's recently published book (in French).[117] Undoubtedly Gamow was aware of the literature also. But the tradition in which the American physicists worked was very different from that of Lemaître, who was never very interested in element formation. Gamow did not need Lemaître to get to the big bang, and when the idea first occurred its further development followed from its rationale, the formation of elements. It is thus not surprising that Gamow, Alpher, and Herman saw no reason to refer to Lemaître's work in their scientific papers.

3.3 COSMOLOGY AS A BRANCH OF PHYSICS

With the pioneering work on the big-bang theory in 1948, the door was opened for further developments. These took place during the following five years, when several American nuclear physicists took up the challenging problems of the early universe and Alpher and Herman developed the big-bang scenario into a full-grown scientific theory. However, the development was not a string of successes, and after a period of four years the big-bang program came to a halt. It included the prediction of a cosmic microwave background, first made in 1948, but ignored by both physicists and astronomers. How was the big-bang theory developed in the period 1948–53? Why was it abandoned at a time when great progress had been made?

Polyneutrons and Improved Calculations

Not all nuclear-physical models of element formation were based on the hot primordial gas assumed by Gamow, Alpher, and Herman. In 1948–49 Maria Goeppert Mayer and Edward Teller, then at the University of Chicago, suggested that whereas the light elements may have been formed by thermonuclear equilibrium reactions, the mechanism of heavy-element formation was very different.[118] Polish-born Maria Goeppert had come to the United States in 1930 after marriage with an American chemist (hence her name Mayer), and was in 1946, at the age of forty, attached to Chicago's new Institute for Nuclear Studies. The starting point of Mayer and Teller was a condensed fluid of cold nuclear matter consisting of or having a large excess of neutrons. This primordial hypothetical object, a "polyneutron," differed from Gamow's somewhat similar speculation of 1942 in not comprising the entire mass of the universe, but having a mass less than that of a star. The early universe would thus have had to comprise a multitude of polyneutrons, the origin of which Mayer and Teller did not account for. The polyneutron was assumed to break up in a kind of fission process, first to very heavy fragments with an excess of neutrons. The breakup of the polyneutron was not an ordinary fission, however, but the formation of small droplets which would break off from the surface of the polyneutron. After a series of nuclear reactions involving beta decay and neutron evaporation, heavy stable elements would be formed from the droplets. Mayer and Teller found that this mechanism was able to lead to a distribution of heavy isotopes ($Z > 34$) in rough agreement with observations.

The polyneutron theory was first discussed in public during the Eighth Solvay Congress taking place from 27 September to 2 October 1948, as always in Brussels.[119] The Solvay report was prepared jointly by Teller and Mayer, but only Teller was invited to participate. In his report on the formation of elements in the universe, Teller summarized the current state of the subject and carefully discussed various schemes to explain the origin of elements. Apart from the new suggestion of polyneutrons, he also discussed the equilibrium theories and the latest version of the big-bang neutron capture theory of Alpher, Herman, and Gamow, including several of their works still in press. Among other objections to the simple Gamow theory, Mayer and Teller pointed out that it was unable to explain the existence of stable isobars, i.e., nuclei with the same mass number but different atomic numbers. (For example, molybdenum-92 and zirconium-92 are isobars, and Gamow's theory could only account for the latter.) The formation of uranium was also problematic, they argued. For these and other reasons they found it useful to suggest the polyneutron alternative, which they "excused as a radical attempt to clear up the non-equilibrium processes which must have played an essential part in the formation of elements."[120]

The starting objects of Mayer and Teller were made of a stuff similar to Gamow and Alpher's ylem, and they referred to it by that name—"a word introduced by Gamow." Alpher's paper had not yet appeared, and so the ylem was introduced in public for the first time at the Solvay Congress, the name being credited to Gamow and not to Alpher. During the subsequent discussion, the polyneutron theory was discussed by Rudolf Peierls and Klein. Peierls found the theory unsatisfactory because it did not provide a mechanism for the compression of enough matter into a "bottle." And even if such a mechanism could be found, he said, "a legitimate question is that of the origin of the bottle." Teller replied: "For the time being I prefer what I hear whispered here in my neighbourhood, that the bottle was made by God."[121]

As far as the starting point was concerned, the polyneutron hypothesis was neither more nor less plausible than the hot big-bang neutron gas. But compared with the Gamow-Alpher-Herman theory it had the methodological disadvantage of explaining only the formation of the heavy elements and not addressing the light ones. An attempt to unite the two theories was made by W. Band, an American physicist.[122] He suggested that the lighter elements were formed in a hot, nuclear gas à la Gamow, which then separated into two phases: one which became the light elements, and one which became the polyneutrons out of which the heavier elements were formed.

The polyneutron theory seems not to have been considered a serious alternative to the big bang and was, at any rate, short-lived. In 1952 the theory was reexamined by the distinguished British (German-born) physicist Rudolf Peierls and his collaborators K. S. Singwi and D. Wroe, who extended it into a cosmological theory related to the big-bang model.[123] Contrary to Mayer and Teller, the British physicists considered polyneutrons in the context of the expanding universe. They argued that polyneutrons could have been formed cosmologically from the original cold nuclear matter. Under these conditions the expansion will be governed by the material pressure only, and the density will decrease in time according to

$$\rho = \frac{1}{6\pi G} \frac{1}{t^2} = 7.94 \times 10^5 t^{-2} \mathrm{g} \cdot \mathrm{cm}^{-3}.$$

Peierls, Singwi, and Wroe showed that the expansion would result in a tension which would fragment the nuclear matter and after about 10^3 s (the half-life of the neutron) produce polyneutrons. At that time the density of the universe would have been drastically reduced so the formed polyneutrons would, in effect, decay in a vacuum. Although the theory of Peierls and his collaborators was more satisfactory than the Mayer-Teller theory, its conclusion was that the polyneutron idea could not work because it would result in a world consisting mostly of heavy elements. Since the real world has a large abundance of light elements over the heavy ones, Peierls and his collaborators concluded that the polyneutron theory was incorrect. No more was heard of polyneutrons, yet the theory of Mayer and Teller was not without consequences. It stimulated Mayer to make a closer study of nuclear structure and propose the shell theory of nuclei, which in 1963 made her the second female Nobel prizewinner in physics (the prize was shared with the German physicist Johannes H. D. Jensen, who independently suggested the shell model).

Meanwhile, Alpher and Herman, as well as a few other nuclear physicists, continued to develop and refine the big-bang model of element formation. Among the young American physicists who joined the program was J. Samuel Smart of the University of Minnesota, who in 1948 had completed his Ph.D. thesis on cosmological nucleosynthesis under Critchfield. Smart investigated various aspects of the neutron capture theory not dealt with by Alpher and Herman, such as the effects of nuclear stability and varying density conditions.[124] The original neutron capture calculations of Alpher and Herman assumed for simplicity a static universe, i.e., $\Phi(t) = $ const in equation (3.5). The reason was practical, grounded in the overwhelming calculational work which would otherwise be required to integrate the rate equations. This was the period in which the first electronic computers, based on radio valves, became available for nonmilitary research, and in 1951 Alpher and Herman got the opportunity to use the new machines. They first made calculations with MADDIDA, a computer developed by North American Aviation, and with the SEAC digital computing machine of the National Bureau of Standards they were able to solve their equations with the expansion of the universe explicitly included. The results did not differ drastically from those obtained earlier, but showed that the initial matter density had to be five times as large as that found previously to represent the abundance distribution of elements.[125] Armed with the most recent value of the half-life of the neutron,

12.8 min, Alpher and Herman concluded that the best agreement with the data was obtained with an initial temperature of 1.3×10^9 K and a density of ca. 9×10^{-4} g·cm^{-3} at a time about 140 s after the beginning of the expansion (figure 3.5). Until that period only decay of neutrons was assumed to take place. The composition of the initially purely neutronic ylem was found to have changed to 88% neutrons and 12% protons at $t \approx 140$ s.

The assumption that the ylem consisted only of neutrons simplified the picture of the earliest universe, but it was realized that it was merely an assumption. In early 1950, Chushiro Hayashi of Nanikawa University in Japan reconsidered the question and suggested that at the very high temperatures in the initial phase of the expansion, other processes than radioactive neutron decay should also be taken into account. Specifically, he considered an initial temperature of about 10^{12} K, corresponding to a time after the big bang of the order of 10^{-4} s.[126] He argued that even if only neutrons and photons were initially present, electron-positron pairs would be created from the electromagnetic field. At the high temperatures, induced beta processes such as $n + e^+ \rightarrow p + \bar{\nu}$ and $n + \nu \rightarrow p + e^-$ would have a major effect on the ratio between protons and neutrons (ν and $\bar{\nu}$ denote a neutrino and an antineutrino, respectively). In the thermal equilibrium existing at the highest temperatures, the proton-to-neutron ratio would be given by $n_p/n_n = \exp[(m_n - m_p)c^2/kT]$ and thus would be larger than 1. This ratio would become frozen when the universe cooled and equilibrium was no longer possible, and Hayashi calculated that this required a ratio between the numbers of neutrons and protons at the beginning of element formation of around $n/p = 1/4$. In other words, the ylem would be far richer in protons than assumed by Gamow, Alpher, and Herman, namely, 50% by weight corresponding to equilibrium.

Hayashi, who did not consider the formation of the heavier elements, found this a promising result because it seemed to explain the main content of the universe, hydrogen and helium. Whatever the detailed reactions that led to helium in the early universe, the net process must be $2n + 2p \rightarrow {}^4$He, he argued; when all the neutrons have been used to produce helium there will still be 60% of the initial protons and one would thus expect a present hydrogen-to-helium ratio of 6:1. Given the crudeness of the calculations Hayashi found the predicted value to agree satisfactorily with the rather uncertain observational data. Hayashi was the first of the rather few physicists who took an interest in the work of Alpher and Herman, and his contribution was important in indicating a new insight into the isotopic composition of the early universe. Hayashi later became professor of physics at the University of Kyoto and did important work on stellar formation and other topics of astrophysics.

Alpher and Herman found Hayashi's approach interesting, but also pointed out that it was an oversimplification and that the neutron-proton ratio of 1:4 failed to yield a sufficient number of heavier elements. Among other things, Hayashi had used in his calculations a neutron half-life of 30 min, very different from the 13 min currently established experimentally; this alone meant that his conclusions had to be modified. All the same, Hayashi's work inspired

Alpher and Herman to refine their calculations and incorporate new reactions, which they did in a paper of 1953 which marks the climax of the Gamow-Alpher-Herman big-bang theory.[127] The work was done in collaboration with James Follin, a colleague at the Applied Physics Laboratory who previously had assisted Alpher and Herman in their calculations. It provided a detailed and comprehensive analysis of the early universe and brought big-bang theory to a state that would not change significantly over the next fifteen years.

The Alpher-Herman-Follin work made sophisticated use of the most recent advances in nuclear and particle theory. For example, relativistic quantum statistics was used instead of classical statistics; the role of gravitation quanta (gravitons) was discussed; different theoretical models of the neutrino were taken into account; and processes involving both muons and pions (then called μ and π mesons) were considered. In a digression on the assumed homogeneity of the early universe, Alpher, Herman, and Follin remarked that in this state the universe would consist of causally unconnected parts because the horizon distance ct would be much smaller than the radius of the universe. This seems to be the first time that the basis for the "horizon problem," so much discussed later, was mentioned explicitly. The problem, which attracted interest only in the 1970s and was eventually solved with the advent of inflation cosmology, concerns how large-scale uniformity could be formed in an early universe consisting of parts with no causal connection.

With the theory of Alpher, Herman, and Follin, the big-bang scenario was pushed back to a time of only 10^{-4} s after the initial explosion and continued until the decoupling time some 100 000 years later (the detailed calculations were carried on until about 600 s). During this period the universe was found to expand by a factor of 10^8 and the temperature to drop from almost 10^{13} K to 10^3 K. The earliest universe was no longer the ylem made up of neutrons, or neutrons plus radiation and a few protons, as imagined a few years earlier. It now consisted of photons, neutrinos, and electrons (positive as well as negative), and also muons, the heavy electrons discovered in 1937; only a very small part was made up of nucleons. After a few minutes' expansion most electrons would have vanished by annihilation, producing an environment of mainly photons and neutrinos. The most important events in the early history of the universe, as found by Alpher, Herman, and Follin, are shown in table 3.1. The three physicists found that while neutrons and protons would at first exist in almost equal numbers, the neutron-to-proton ratio would decrease with the expansion. They calculated the ratio to lie between 1:4.5 and 1:6 at the time when nucleosynthesis started, the precise value depending on the half-life of the neutron and whether or not the neutrino was taken to be identical to the antineutrino. Using the same argument as Hayashi, they found a minimum value for the hydrogen-to-helium ratio between 7:1 and 10:1, broadly consistent with astronomical data. The figures correspond to a weight percentage of helium between 29% and 36%. Alpher, Herman, and Follin thereby supplied the big-bang theory with a useful basis for testing and further calculations, but it would be eleven years before the test became effective.

TABLE 3.1
History of the Alpher-Herman-Follin Universe

Temperature (K)	Time	Events and Particle Content
$> 10^{12}$	$< 6 \times 10^{-5}$s	Doubtful validity of field equations
10^{12}	6×10^{-5}s	Thermodynamic equilibrium; radiation density 10^{-26}g·cm^{-3}; photons, neutrinos, electrons
10^{11}	8×10^{-3}s	Neutrinos frozen in; photons, neutrinos, electrons
2×10^{10}	0.2 s	Electron-positron annihilation starts; neutrinos, electrons
5×10^8	600 s	Electron-positron annihilation; nucleogenesis; photons, neutrinos
3×10^8	30 min	Nucleogenesis essentially complete
170	10^8y	Mass density 10^{-26} g·cm^{-3}; galaxy formation

The Mass Gap Problem

The $\alpha\beta\gamma$ approach was to build up the elements sequentially by neutron capture, but Gamow and Alpher recognized at an early stage that this was not enough, and that detailed thermonuclear reactions between the light elements would also have to be included. Apart from the technical difficulties in such calculations they also required knowledge of thermonuclear reaction rates, most of which either did not exist or were still classified in late 1948. In the fall of that year, Enrico Fermi, together with his junior colleague at Chicago University, the chemist Anthony Turkevich, attended a colloquium by Alpher on his new work. Fermi at once became interested in the problem and decided to attack it with the use of the most recent reaction rates available to him. At that time Fermi was already interested in some areas of astrophysics, especially in the primary cosmic radiation. Neither was cosmology a new field for him. For example, in informal evening lectures to graduate students at the end of 1946 he covered elementary aspects of stellar evolution, structure of white dwarfs, energy production in supernovae, and also general relativity and cosmology.[128] Fermi's interest in element formation also seems to date from before the Gamow-Alpher theory. Together with Turkevich he proposed an explanation for the formation of higher-charged isobars, but the work was not published.[129] Further insight into Fermi's interest in topics of cosmology may be obtained from a series of lectures he gave in Rome in October 1949. One of these was a careful review of theories of the origin of the chemical elements, where Fermi paid particular attention to the hot big-bang theory of Gamow and Alpher (mentioning the $\alpha\beta\gamma$ paper and Bethe's role in it).[130]

In early 1949, Gamow reported to Alpher: "Fermi and Turkevich have recently spent quite a lot of time on it [the problem of the formation of light elements], and have apparently obtained very satisfactory results (with the exception that they get too much helium)."[131] Fermi and Turkevich never pub-

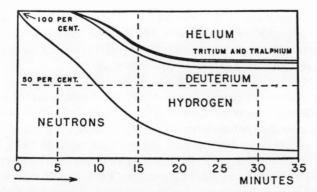

Figure 3.6. The result of the calculations of Fermi and Turkevich concerning the chemical composition of the first half-hour of the universe. Tralphium is the helium-3 isotope. Starting with a purely neutronic ylem, more than half of it is transformed into hydrogen and about 24% by weight into helium, in fair agreement with contemporary empirical knowledge. *Source:* Gamow (1952b), p. 68.

lished their work, presumably because they failed in their attempt to build up elements heavier than helium, but they communicated it to Gamow and his colleagues in Washington, through whom it became generally known.[132] Fermi and Turkevich first considered the building up of isotopes of hydrogen and helium from an ylem consisting of neutrons at $t = 0$. They adopted the radiation-filled universe of Gamow and Alpher, and assumed the nucleon density at $t = 1$ s to be 1.7×10^{-3} g·cm^{-3}. From this follows the nucleon density at later times, which decreases as $t^{-3/2}$. Since the only important reaction during the first 300 s would be decay of neutrons, they started their calculations at $t = 300$ s, at which time the primordial gas would consist of neutrons and protons in the ratio 7 to 3 (the half-life of the neutron was taken to be 11.5 min). Fermi and Turkevich considered twenty-eight possible nuclear reactions between neutrons, protons, deuterons, tritons (hydrogen-3 nuclei), and helium-3 nuclei. The latter were called tralpha particles by Gamow, who invented the name tralphium for helium-3; the names seem not to have been adopted by other physicists.

Among the reactions studied by Fermi and Turkevich was the natural beta decay of neutrons and tritons as well as thermonuclear processes such as $n + d \rightarrow t + \gamma$, $p + t \rightarrow {}^{4}\text{He} + \gamma$ and $d + {}^{3}\text{He} \rightarrow {}^{4}\text{He} + p$. By leaving out some of the twenty-eight processes that have small reaction rates and integrating the resulting rate equations, Fermi and Turkevich found the composition of the early universe to vary in time as shown in figure 3.6. They stopped their calculations at $t = 2000$ s, at which time the formation of elements heavier than helium would begin to be important. The results, based on detailed calculations and real reaction rates, were encouraging. The predicted amount of deuterium exceeded the observed one by a factor of 150, but this was only to be expected because most of the deuterons would be consumed in subsequent processes leading to heavier elements. For the number ratio between hydrogen and helium Fermi and Turkevich found H/He = ca. 6.7, in approximate agreement with the value estimated from observations.

In spite of this success, the ambitious work of Fermi and Turkevich failed in its broader aim, to account quantitatively for the existence of the lighter elements. The reason was the apparently insuperable difficulty in synthesizing elements heavier than helium from the particles existing five minutes after the big bang. As already pointed out by Bethe in his work of 1939, nuclei of atomic weights 5 and 8 do not exist. The absence of such nuclei was presumed in 1937 and in 1939 it was almost a certainty, but not yet a definitely established fact; ten years later it was, and the problem was then how to bridge the gaps at these two masses. In order to build up elements heavier than helium, nuclei with $A = 5$ and 8 cannot be used and some other mechanism than neutron capture therefore had to be found. If this was not possible the entire program of building up elements from the ylem would seem to fail. The mass gap problem was clearly recognized by Alpher in 1948; he vaguely suggested that it might be solved by considering unspecified reactions between deuterons and tritons.

When Fermi and Turkevich were faced with the problem at atomic weight 5, they realized that it was not so easily solved. They just were unable to find a simple process bridging the gap and, at the same time, giving a sufficient amount of matter. As the best nonsimple candidate they suggested the process $t + {}^{4}\text{He} \rightarrow {}^{7}\text{Li} + \gamma$, which, they found, would proceed at a sufficiently high rate if there existed in lithium-7 a resonance state at a certain energy (see the letter from Gamow to Alpher quoted in section 3.2). Unfortunately, no such resonance was known experimentally and they therefore had to look for other solutions. One possibility, suggested by Turkevich, was to change the initial conditions of the process by assuming a more dense universe with a different ratio between protons and neutrons. This increased the production of lithium, but not enough. Yet the difficulty was far from being seen as an impossibility. In June 1949 Gamow wrote to his "Dear children" (Alpher and Herman): "Tonny Turkevich found the way to build the 'chain bridge' across the 'mass-five cravasse' [*sic*], and is calculating now if sufficient trafic can go through. It is ${}_{6}\text{C}^{10} + {}_{1}\text{T}^{3} \rightarrow {}_{3}\text{Li}^{6} + {}_{4}\text{Be}^{7} + 0.002$ m.u."[133]

Turkevich's suggestion was based on an idea first proposed by Eugene Wigner, the distinguished nuclear theoretician, namely, to bridge the mass gap by a kind of cyclic or catalytic reaction.[134] According to quantum mechanics a very small amount of material can leak through the mass-5 gap and Wigner suggested that these "seed nuclei" could serve to bridge the gap. For example, with carbon-10 as a seed, the lithium-6 and beryllium-7 nuclei might capture neutrons and become a pair of carbon-10 nuclei. The process could then begin again with one of the nuclei serving to bring a new triton across the mass gap. The particular mechanism suggested by Turkevich turned out not to work, however. In his Rome address Fermi reviewed the difficulties, which he considered to be serious indeed: "One is left with the sad conclusion that this theory is incapable of explaining the way in which the elements have been formed, which after all was to be expected."[135] Although Fermi concluded that the theory was wrong, he also praised it for its precise hypotheses and considered it much more satisfactory than other theories—he thought of the equilibrium theories—and hoped that the mistakes might be corrected within the theory's framework.

Gamow, Alpher, and Herman for a time considered other versions of the seed mechanism as possible solutions. For example, Alpher and Herman speculated that perhaps the explosion of the ylem would result not only in individual nucleons but also in some aggregates of neutrons which might serve as seeds. Or perhaps the desired process could be accomplished by means of a photoelectric effect caused by the gamma photons? Yet another suggestion was the possibility of self-heating, i.e., the nuclear gas might temporarily become much hotter than the radiation and thus greatly increase the thermonuclear reaction rates. This possibility was suggested by Gamow, who handed over the problem of calculating the nuclear reactions under these conditions to his student, Arthur Carson. The complicated numerical calculations, made on electronic computers, seem to have led to nothing.[136]

By 1953, after many ingenious suggestions and the involvement of some of the world's most eminent nuclear physicists, the mass-gap problem was still not solved. How did the failure affect the big-bang program? According to most retrospective accounts, it was a stumbling block that hindered further progress, but this evaluation does not seem to agree with the contemporary views of the physicists involved. Everybody recognized the problem to be a serious one, but Gamow and his small group did not consider it unsolvable, just unsolved. After all, because the more or less simple ideas had failed it did not mean that the mass-gap problem was beyond the power of physics. Gamow concluded that "*So far* no . . . [satisfactory] reaction has been found, which may be due simply to a deficiency in our information about the various isotopes which may be involved."[137] Alpher, Herman, and Follin followed Gamow in suggesting that the solution might lie in more detailed and realistic calculations. "To resolve this and other difficulties in the theory," they wrote in 1953, "it will apparently be necessary to remove many of the simplifying restrictions." The belief that the mass-gap problem could be solved within the big-bang framework was not wishful thinking in the situation of nuclear physics at the time. Although discouraging, the problem did not suggest that the theory was wrong in any fundamental sense. Gamow and his group had reasons to be moderately optimistic. As Alpher and Herman rather modestly concluded in 1953: "It is the belief of the authors, and perhaps not without prejudice, that the theory of element formation principally by neutron-capture reactions in the expanding universe may suffer from fewer difficulties than other theories. In addition, this theory has as a minimum provided a fit to the general trend of the relative abundance data with relatively few adjustable parameters within the framework of what at present appears to be a reasonable cosmological theory."[138] We shall return to the mass gap problem and other aspects of nucleosynthesis in section 6.2.

Cosmic Background Predictions

The universe is filled with radiation, and the cosmic microwave background is only one among other radiation backgrounds. The first calculation of the temperature of space, based on the energy of starlight, was performed by Eddington in 1926.[139] From an estimated value of the total stellar energy flux received on earth and the Stefan-Boltzmann law he found the result 3.2 K. Some years later Ernst Regener studied the ionization caused by the cosmic radiation and concluded that intergalactic space, where starlight is negligible, was filled with a background of temperature 2.8 K.[140] The temperatures found by Eddington, Regener, and other researchers are close to the temperature of the cosmic microwave background. This is remarkable, but it is important to be aware that both starlight and cosmic rays are entirely different radiation backgrounds from the microwaves predicted in 1948 and discovered in 1965.

As mentioned, Alpher and Herman first calculated the present temperature of the decoupled primordial radiation in 1948, when they reported a value of 5 K. Although it was not mentioned either then or in later publications that the radiation is in the microwave region, this follows immediately from the temperature: for blackbody radiation the dominating wavelengths are given by Wien's displacement law, $\lambda_{max} T = 0.29$ cm·K, which for $T = 5$ K corresponds to a peak wavelength of 0.06 cm. In their detailed calculation of 1949, Alpher and Herman made it clear that what they had called "the temperature in the universe" the previous year referred to a blackbody distributed background radiation quite different from the starlight. But they also mentioned that "the thermal energy resulting from the nuclear energy production in the stars would increase this value [of 5 K]," thereby perhaps creating the impression that the universal background radiation might not be observationally separable from starlight.

Alpher and Herman repeated the $T = 5$ K value in their paper of 1949, but the following year they obtained $T = 28$ K.[141] The reason for the discrepancy had nothing to do with the theory, which remained unchanged, but was simply the result of Alpher and Herman's wish to use in their calculations the most recent observational data. Unfortunately they picked up a recent suggestion by the German astronomer Alfred Behr, who believed he had evidence for a matter density of 10^{-29} g·cm^{-3}, and with this value, ten times the one used earlier, they ended up with 28 K.

Whereas the approach of Alpher and Herman was based on the big-bang model, from which they deduced the present background radiation as separate from stellar radiation, it seems that Gamow's ideas of the matter were different and rather less clear. In an undated letter to Alpher from either 1948 or

1949 he wrote: "The space temperature of about 5° K is explained by the present radiation of stars (C-cycle etc.). The only thing we can tell is that the 'residual temperature' from the original heat of the universe is *not higher* than 5° K, but it could be as close to zero as one likes. Thus one cannot use 5° K and ρ_{rad}(now) $= 10^{-32}$ as the observed boundary conditions."[142] This indicates that Gamow had not understood, or did not agree with, the reasoning of Alpher and Herman; according to them, the cosmic background radiation temperature could not be as small as one likes. That Gamow considered the radiation in another perspective than did Alpher and Herman is also indicated by his published references to it. He first mentioned the radiation in 1950, when he suggested a temperature of 3 K, but without any argument at all. Apparently he considered the value an observational one, and the fact that it happens to coincide with what was found fifteen years later must be considered accidental.

The next time Gamow mentioned the subject, in *The Creation of the Universe*, he got 50 K! The wrong result was obtained by simply inserting the age of the universe (he used 10^{17} seconds or ca. three billion years) in the equation (3.8). This is obviously not valid since the relationship only holds in the early, radiation-dominated period. Gamow's comment that the value "is in reasonable agreement with the actual temperature of interstellar space" adds to the suspicion that he understood the background radiation in a different way from Alpher and Herman. (There is no reason to assume the cosmic background radiation to be of the same order as the measured temperature of space.) The confusion only increased with Gamow's two latest estimates, appearing in 1953 and 1956, when he obtained the values 7 K and 6 K. In neither of the cases did he use or refer to the Alpher-Herman approach, but relied on rough calculations of a somewhat obscure kind which did not relate to the primordial nucleosynthesis.[143] Whatever the soundness of Gamow's calculations, he was quite clear about the residual radiation left over from the superdense state of the world. As he pointed out in 1956, referring to the 6 K radiation, "the residual heat found at present in the universe is comparable with the heat provided by nuclear transformations in stars."[144]

The prediction of the cosmic background radiation was first made in 1948 and it was repeated at least seven times until 1956. It was thus well known, but made no impact at all. The reason for the neglect of the prediction is not entirely clear, but some contributing factors may have been the following. First, the different and in part conflicting derivations by Gamow and by Alpher and Herman may have confused many readers. Was the radiation of cosmic origin only or did it include also stellar radiation? Calculations of the temperature of space were old, and it would not have been clear to many readers that this was something different. Second, the lack of explicit reference to its range of wavelength may have masked the fact that it is a microwave background radiation, and thus physicists occupied with microwave techniques (such as Robert Dicke and Charles Townes) may have ignored it as irrelevant. Third, the wide range in temperatures, between 3 K and 50 K, cannot have increased the credibility of the prediction. Fourth, the predictions appeared in a nuclear-physical rather than a standard astronomical context, which may have made them less known among astronomers. It is interesting to speculate what would have happened if the prediction had also been published in the *Astrophysical Journal*. Perhaps it would then have attracted the interest of, say, McKellar (see below)? Fifth, even if the prediction was understood, it may have seemed of theoretical interest only; neither Gamow, Alpher, nor Herman suggested that it might actually be detected, and if so, how. They did not state, for example: "From our theory follows the existence of a cosmic blackbody-distributed radiation in the microwave range, corresponding to a temperature of about 5 K. Although such a radiation may be difficult to detect by present-day technology, it would be of interest to look for it."[145] Had they phrased their prediction something like this, there is reason to believe that the subsequent history would have been different. Wordings are important.

In fact, when Alpher and Herman made their prediction in 1948, traces of the radiation had already been observed, albeit unwittingly. To account for the observation of a new interstellar absorption line of $\lambda = 3874.6$ Å, the Australian astrophysicist Andrew McKellar suggested in 1940 that it was due to absorption by rotating cyanogen molecules (or, more correctly, radicals, CN).[146] From this hypothesis he predicted that there might exist yet another line due to the excitation of a cyanogen molecule in its ground state to a second rotational state of $\lambda = 3580.0$ Å. The measured intensity of an absorption line depends on the number of molecules in the excited state relative to that in the ground state, namely, on $n_1/n_0 = \exp(-\Delta E/kT)$ where ΔE is the energy of the excited state. Since $\Delta E = hc/\lambda$ and the wavelength was known from laboratory experiment to be 2.6 mm, McKellar could determine the excitation temperature to be about 2.3 K. The precise value would depend on the intensity of the predicted line relative to the observed, and in his paper McKellar suggested values between $T = 0.8$ K and $T = 2.7$ K. The prediction was verified the following year by his compatriot Walter Adams,[147] but at the time nobody thought of interpreting the apparent existence of a thermal bath cosmologically. Yet McKellar and Adams had actually detected an instance of the cosmic microwave background radiation. It is not surprising that their result did not receive much attention. After all, they had only reported the existence of a rotational temperature of 2.3 K for cyanogen molecules in a restricted part of the universe, which could easily be explained without invoking cosmology. Furthermore, there was at that time no theoretical reason for expecting a cosmic background radiation at all. Even if the existence of such radiation had been suspected, it was impossible to establish at that time whether the value represented the noise temperature of the CN receiver or had its origin in a source outside the molecular cloud. In 1948, seven years and a world war later, the discovery of McKellar and Adams was either forgotten or relegated to the position of just another result of cosmic molecular spectroscopy. It was mentioned by the physical chemist Gerhard Herzberg in a classic text on spectroscopy, and must thus have been known to thousands of scientists. But, not unreasonably, Herzberg referred to the temperature of 2.3 K as having "of course a very restricted meaning."[148]

What if the idea of connecting the Alpher-Herman prediction with the McKellar-Adams observation had occurred to someone? Might the cosmic background radiation then have been searched for and found in, say, the late 1940s? It appears that the technology of microwave astronomy at that time had almost, but not quite, reached the level at which such a detection might have taken place. In fact, in 1946 Robert Dicke and coworkers at MIT tested equipment that could detect a cosmic microwave background of intensity corresponding to about 20 K in the microwave range.[149] However, they did not refer to such a background, but only to a "radiation from cosmic matter." Also, this work was unrelated to cosmology, and is only mentioned because it suggests that by 1950 detection of the background radiation might have been technologically possible, and also because of Dicke's later role in the discovery.

An Almost Aborted Research Program

For the sake of brevity I refer to the early American big-bang program as the Gamow program. As we have seen, this program was based on the conceptual innovations introduced by Gamow in the early and mid-1940s; it reached maturity in the late 1940s, and culminated in 1953 with the work of Alpher, Herman, and Follin. Contrary to earlier workers in cosmology, Gamow and his collaborators considered cosmology to be intimately linked with nuclear physics and gave priority to the latter field. Being physicists by training and spirit, they saw the early universe as something which could be dealt with by ordinary methods of physics. It was surely a difficult problem, but nonetheless a problem which did not differ qualitatively from other problems of nuclear physics. If Bethe could explain the mysteries of the sun by using nuclear theory, they saw no reason why the mysteries of the early universe could not be likewise explained. The Gamow approach was thoroughly pragmatic, reflecting the general attitude among the majority of American physicists. This pragmatic credo had been a characteristic element in American physics since the 1930s. It included the doctrine that "A theoretical physicist . . . does not ordinarily argue about philosophical implications of his theory," as the leading quantum physicist, John Slater, expressed it in 1938.[150] It was no less significant in the postwar period, when it had rather been reinforced by the wartime experiences common to most physicists. The sole purpose of theoretical physics, according to this view, was to devise theoretical models that could be used to predict the outcome of future experiments. It would seem that cosmology would be the worst imaginable candidate for such a pragmatic view of science, but this was not how Gamow and his collaborators looked at it. Theoretical models of the early universe could not be tested in the laboratory, but they could be used to predict certain present conditions, which then could be tested by the ordinary means of physics. In other words, they could be used for making inferences rather than for predictions in the ordinary sense.

Gamow, Alpher, and Herman were confident that advanced calculations with input of nuclear-physical laboratory data would give the right answer, and they saw no need to introduce new principles or discuss the conceptual state of cosmology at any length. For them, the electronic computer was more important than philosophically oriented conceptual innovations. In this respect their attitude was similar to that of many other American physicists of the same generation and with the same background as Alpher and Herman. For example, the physicists who developed the new quantum electrodynamics—Julian Schwinger and Richard Feynman in particular—adopted a conservative approach in the sense that they accepted the received formulation of quantum mechanics and relativity. They had no sympathy for revolutionary solutions or philosophically appealing arguments of a radical nature. Their emphasis was, like Alpher and Herman's, on numbers obtained from calculations. But whereas the numbers that the quantum theoreticians came up with could be, and were, tested experimentally, there was no correspondingly direct way to test the calculations of the nuclear cosmologists.

The approach of Gamow and his collaborators focused rather narrowly on the question of element formation, which was seen as both the rationale and the crucial test of cosmology. Because the abundance distribution of the elements was known empirically, and the possible nuclear reactions were limited by empirical knowledge gained from laboratory experiments, theirs was an empirical cosmology. And for this reason they had little interest in or respect for the more deductive approach that had traditionally ruled in cosmology.

In accordance with their pragmatic approach, Gamow and his collaborators rarely commented on methodological issues. But there is little doubt that they considered cosmological theories such as those of Milne, Jordan, and the steady-state theoreticians with distrust and lack of appreciation. As the prevailing attitude among American quantum physicists has been described as an engineering attitude, so did Gamow conceive of himself as a cosmo-engineer. At a conference in Denver, Gamow distinguished between two schools of cosmology, which he labeled postulatory cosmology and factual cosmology. According to the first school—he no doubt had steady-state cosmology in mind—"one asks oneself what the properties of matter and radiation should be in order to obtain philosophically desired cosmological models." In factual cosmology, on the other hand, "we accept the physically established laws governing matter and radiation and look for cosmological models which are derived on the basis of these laws and are consistent with astronomical observations." Naturally, this was the approach favored by Gamow and also by Alpher and Herman. It is revealing that Gamow drew an analogy between the approach of the engineer and that of the factual cosmologist: "When an engineer wants to design an automobile, a jet plane, or a spaceship, he starts with the well-known physical and chemical properties of the materials he uses and looks for the arrangement of these materials which would satisfy his purposes.

Similarly, a physicist looks for the universe which will satisfy known laws, be these obtained by direct experiment or theoretical deductions."[151] This rather profane cosmology-as-engineering view was as foreign to the British school of cosmology as the steady-state theory was to the Americans.

It is evident that the Gamow program was physical rather than astronomical in the traditional sense. Although it had important astronomical implications, and although Gamow had a broad interest in and knowledge of astrophysics and astronomy, the program was primarily oriented towards nuclear physicists. This is shown not only by the scientists who participated in the program—none of whom were astronomers—but also by the publications in which the results were reported. With one exception (the semipopular *Vistas in Astronomy*) no articles on big-bang theory appeared in astronomical journals. All the research articles appeared in either physics journals, mostly in the *Physical Review*, or in *Nature* (where they could be published quickly). It might have seemed natural if Gamow, Alpher, and Herman had also used the leading *Astrophysical Journal,* but they did not. Their approach fitted well into the pages of the *Physical Review*, and Gamow was also concerned that Chandrasekhar, who was one of the editors of the *Astrophysical Journal*, would find the big-bang model too speculative.[152]

In the early 1950s the Gamow big-bang theory had developed into a minor and apparently thriving industry. Although well known also in Europe, it was cultivated almost exclusively by American physicists. The theory was disseminated not only through the scientific literature, but also through many meetings and conferences, by way of Gamow's popular accounts, and by articles in general journals and newspapers. For example, at an American Physical Society meeting in 1951, Alpher and Herman presented their neutron capture theory and included a discussion of the predicted cosmic background radiation; and Gamow and Alpher gave separate addresses on aspects of the theory at a conference on the abundances of the elements held at the Yerkes Observatory in the fall of 1952.[153] The following summer an important symposium on astrophysics was arranged at the University of Michigan; among the participants were Walter Baade, Allan Sandage, Edwin Salpeter, Vera Rubin, and Margaret and Geoffrey Burbidge. Gamow gave a series of lectures on the big-bang theory, including the synthesis of elements, principles of general relativity, and the formation of galaxies.[154] The theory received wide public notice through Gamow's *The Creation of the Universe*, and was frequently mentioned in the press, if not always with the greatest insight. Thus, in an account of the Gamow-Alpher-Hermann theory of 1951, *Time* reported that "Followers of Gamow agree that the universe is still expanding, as a result of the original explosion of the ylem. What they find harder to explain is why the earth should happen to be at the exact center of the great expansion."[155]

The big-bang core group was small, consisting mainly of Gamow, Alpher, and Herman. But in addition there was a larger group of physicists who, in some way or other, contributed to the field. The population until 1953 includes the following scientists, all of whom have previously been mentioned:

Gamow, Alpher, Herman, Smart, Follin, Hayashi, Berlin, Wigner, Fermi, Turkevich, Carson, Bethe, Band, Teller, Mayer, Peierls, Singwi, and Wroe. Of these, the last five worked with the polyneutron theory, which related to, but was not a proper big-bang theory. Although Wigner, Turkevich, and Carson did not publish on the theory, they performed calculations within its framework. Bethe is included, in part because of his official affiliation with the $\alpha\beta\gamma$ theory, in part because of his general interest in it (for example, he sat on Alpher's dissertation defence committee in 1948).

The Gamow program was successful in the sense that it was able to attract, if only for a while, considerable attention among American nuclear physicists. With the exception of Hayashi, no one outside America contributed to the big-bang theory. Furthermore, it is noteworthy that the physicists involved in the program were East Coast or Chicago physicists with, in most cases, personal connections to Gamow. There seems to have been no interest among physicists at, e.g., Caltech. Whereas the Gamow program may be said to have been in a healthy state in the early 1950s, interest soon faded, and after 1953 it was only Alpher, Herman, and Follin who sought to carry it on. The endeavor met with little success. Although Gamow wrote a few papers on the theory in the 1950s, he seems not to have been very interested in developing it and was, at any rate, soon attracted by other areas of science. Inspired by the discovery of the structure of DNA in 1953, he became increasingly interested in biology and engaged himself in work on the nucleotide molecules making up DNA. With or without Gamow, given the fact that the big-bang theory was resurrected about 1965 in much the same form as it had twelve years earlier, the low interest after 1953 is puzzling. With the Alpher-Herman-Follin paper of that year the theory came to an almost complete stop. The three physicists sought to revive interest in it, and gave several presentations of new calculations of light-element formation at scientific meetings, but to no avail. As Alpher and Herman later tersely remarked: "There seemed to be little interest in the results at the time."[156]

It is difficult to say why the big-bang program was disregarded for so many years, and that during a period when the general opinion among astronomers was in favor of an evolving universe with a finite age. Undoubtedly the failure in bridging the mass gap cooled the interest among physicists, although, as we have seen, there was no compelling reason to give up looking for a solution. Moreover, even if it was accepted that the gap could not be bridged, the big-bang theory might well be correct. As Gamow conceded in 1954, perhaps the heavier elements did not have a cosmological origin, but were the later results of nuclear reactions in the interior of special kinds of stars.[157] The general impression is that many of the physicists who had been engaged in the problem (e.g., Wigner, Fermi, and Turkevich) were not seriously interested in the big-bang theory and did not feel committed to it. Their interest was in nuclear physics, and they may have considered the mass-gap calculations little more than an interesting exercise in that science. Unable to find a solution, they skipped the matter and went on to other, more important matters. It may be tempting to put the blame for the ebb in interest in big-bang cosmology on the steady-state theory, but this would be wrong; the steady-state theory did not enjoy general support, and in America it was not even taken seriously by most astronomers. On the other hand, the success of Hoyle and his collaborators in 1957 in explaining the formation of the elements by stellar processes undoubtedly had an adverse effect on the big-bang program. If elements could be formed in the interior of stars, the rationale of the big-bang theory might seem problematical. From the point of view of the nuclear physicists, the processes taking place in the stars were no less interesting than those assumed to have taken place in the early universe.

The big-bang theory was closely associated with Gamow, which possibly had an adverse effect on the attractiveness of the theory in the 1950s. Gamow lost active interest in the theory and lived a scientifically more isolated life after he went to the University of Colorado in 1956. He was always fascinated by new ideas, but when the ideas had been formulated and understood and the consequences had to be worked out in detail, his interest waned. R. D. Richtmyer, a Colorado physicist who knew Gamow well, recalled that his colleague "shied away from any subject when original ideas were no longer important and technical elaborations were accumulating instead." There were good nuclear physicists and astrophysicists at the University of Colorado, "but apparently they did not seek out Gamow, nor he them." Gamow's style of doing physics estranged him from the new generation of physicists. Richtmyer suggested that "nuclear physics and astrophysics, in both of which he had played dominant roles, no longer appealed to him when million-dollar computers, elaborate theories, and volumes of data were involved."[158] What Richtmyer did not mention was Gamow's heavy consumption of alcohol, which sometimes led to embarrassing scenes at meetings and conferences and which undoubtedly contributed to the decline of his status in the physics community. Vera Rubin was a Ph.D. student of Gamow's when she attended the 1953 Michigan summer school on astrophysics. She recalled that Gamow finished "1/2 bottle of something" during discussions with Baade. "Gamow embarrassed me by his behavior—sleeping during lectures, asking 'stupid' questions (fundamental, I would now call them), but he understood the astronomy before anyone else."[159]

As Alpher and Herman have themselves suggested, it may also have had an effect that the two were neither astrophysicists nor pure nuclear physicists. In the mid-1950s they changed to scientific careers in private industry, Alpher at General Electric's Corporate Research and Development and Herman at General Motors Research Laboratories. In these positions they continued for a period to work with cosmology, but their main work was, of course, in other fields. Alpher worked in a variety of areas which included shock waves, plasma physics, fluid dynamics, and technological forecasting. Herman did research in solid state and chemical physics and also pioneering work in transportation and traffic science. In this branch of applied (but also fundamental) science, he developed an important statistical theory together with Ilya Prirogine, the famous Belgian physicist and Nobel Prize winner in chemistry. However, even in American science, with its close links to industry, there was (and is) a clear divide between industry and academia, and it is quite possible that the industrial affiliations of Alpher and Herman made many academic physicists look upon their work with either distrust or indifference. How seriously can one take a theory proposed by a reputed alcoholic and two scientists who have defected from academia to industry?

Internal competition in American physics may have been yet another factor that prevented the Gamow-Alpher-Herman theory from obtaining more than a temporary foothold. The year 1948 saw the emergence of the big-bang theory of the universe, but it was an event which caused no sensation at a time when American physics boomed and other developments seemed much more interesting.[160] In the same year that the $\alpha\beta\gamma$ theory appeared, the Pocono Conference on quantum electrodynamics took place, featuring Richard Feynman's and Julian Schwinger's brilliant presentations of their new ideas. A few months earlier the giant Berkeley 184-inch synchrocyclotron produced the first artificially made mesons (pions), the new elementary particle supposed to be the carrier of nuclear forces. This case of big physics was matched by big astronomy in the shape of the 200-inch Hale telescope on Mount Palomar, which was dedicated in June 1948. Solid state physics also experienced a spectacular boom with the recently invented transistor. In this situation of expanding frontiers and fascinating new possibilities, Gamow's big-bang program had difficulty in attracting much interest and was probably seen as of little reward by most physicists. Relying as it did on classical relativistic cosmology, and then on general relativity, it appealed to a branch of theoretical physics which was not in vogue at the time and of which most nuclear physicists had no knowledge. As argued by Sylvan Schweber, the great advances in nuclear, particle, and solid state physics during the 1950s created a pack effect with most physicists working on successful theories and ideas. The attraction to dominant research programs with a strong empirical appeal left other areas impoverished. General relativity was one of them, big-bang cosmology another. It is only in retrospect that we can see that the big-bang theory was one of the really big events in physics in 1948, a big bang in a dual meaning.

Alpher and Herman were not typical young American physicists in the postwar period. This period saw a drastic trend toward specialization in general, and toward nuclear and particle physics in particular. General relativity was rarely studied by the new generation of physicists, and to master both nuclear physics and general relativity was highly unusual; it seemed inefficient, unrelated to career, almost un-American. The lack of interest in cosmology and general relativity was reflected in the placement of courses in these fields. In most of the large American universities, general relativity was taught within the mathematics departments, not the physics departments. The barrier separating the trend-setting nuclear and particle physicists from the dusty cosmologists naturally implied that cross-disciplinary research programs such as that of Gamow, Alpher, and Herman had difficulty recruiting new people.

Many cosmologists have recalled the low status of their field in the 1950s. Robert Dicke, who took up gravitation theory in the mid-1950s after a career in quantum theory, recalled that he was isolated. "It's a puzzle to me how cosmology got so separated off from the rest of physics," he remarked. "Here is all that matter in the universe, and it doesn't seem to bother anybody. It's here, but where did it come from? Questions of this kind just weren't asked [during the 1950s]."[161] Well, they were asked, but few listened.

If the early big-bang cosmologists did not speak the language of the nuclear and particle physicists, neither did they speak that of the astronomers. It may have been a problem for the big-bang theory that it was primarily seen as a somewhat speculative nuclear-physical theory, which was foreign to most astrophysicists and to almost all astronomers. It was a theory without a clear connection to existing astronomical disciplines. Although the formation of elements through nuclear reactions had long been an established subject of astrophysics, most scientists were suspicious when it came to a theory which dealt with the *creation* of matter. Their suspicion was not diminished by the fact that the big-bang theory placed the creation two billion years ago. From 10 to 12 September 1953 a conference on nuclear astrophysics took place in Liège in Belgium with about sixty participants, including Fowler, Salpeter, Bondi, Gold, and ter Haar. Alpher, Herman, and Follins contributed with a brief summary of their new work, and the big-bang theory was also discussed by ter Haar. Most participants ignored it. The Belgian astrophysicist Evry Schatzman remarked: "One should not introduce the idea of creation in the elaboration of the theories of formation of the elements. The problem is to study under which conditions the actually observed abundance of the elements have been produced, and not to invent a state of the universe completely different from the one of its actual state."[162] The last sentence was a repetition of an objection often raised by the steady-state physicists, but Schatzman had no sympathy for either the big-bang or the steady-state theory. He and most other astrophysicists at the time simply did not find it reasonable to mix cosmology and astrophysics.

The decline in interest in the big-bang program was only temporary. After a ten-year hibernation it was revived and turned into a shape soon known as standard big-bang cosmology. Although it was a new generation of physicists and astronomers which was responsible for this development, the essential part of the new cosmology was the one worked out by Gamow, Alpher, and Herman around 1950. That they had mixed feelings about the way in which the revival took place is another matter.

The Steady-State Alternative

IN THE LATE 1940s cosmology was still a very small business, characterized by a lack of institutional support and very little disciplinary and conceptual unity. The rather few people who worked with cosmological problems were astronomers, mathematicians, and physicists who in no way considered themselves as cosmologists and for whom the science of the universe at large was only a part-time occupation. To the extent that there was a consensus view, most researchers probably favored some version of relativistic cosmology with a universe of finite age, but such models were rarely discussed as realistic candidates for the universe. With the exception of Gamow and his group, the physical big bang was not an accepted part of cosmology, relativistic or not. What most astronomers could agree upon was not that the universe had once exploded from a primeval state, but that it was now in a state of evolution—expanding—and had been so for a very long period. The evolutionary feature did not merely imply expansion, but also that the large-scale features of the universe changed systematically over time, with the distances between galaxies becoming ever larger and the universe therefore becoming more dilute. It was realized, of course, that if the universe is ever expanding it may be ascribed a beginning in time in agreement with the Friedmann-Lemaître equations, and thus be said to have been created. But most astronomers preferred to neglect what may seem to be a natural consequence of the evolutionary, relativistic worldview.

At the same time as Gamow and Alpher reformed big-bang theory with the $\alpha\beta\gamma$ paper, there appeared in Cambridge, England, a diametrically opposed cosmological theory, first referred to as the "new cosmology" but soon generally known as steady-state cosmology. The basic features of this theory were contained in two interrelated postulates: First, that the universe has always and will always look the same to any observer, whatever his location in space and time; this is the perfect cosmological principle; secondly, that matter is continually created throughout the universe, emerging spontaneously out of nowhere. Apart from being an important cosmological theory in its own right, the steady-state theory provoked a major controversy in cosmology by questioning standard assumptions of the evolutionary, relativistic theory. This had two consequences, which together contributed significantly in changing cosmology from a mathematical game to a scientific theory. By offering a radically new world picture, steady-state theory forced astronomers and physicists to think more deeply and critically about the foundations of cosmology; and because the alternative had direct observational consequences, different from those of the evolution theory, it was an important factor in the emergence of new observational methods and practices. The observational cosmology that transformed the field in the late 1950s and early 1960s was related to the attempts to test which of the two rival cosmologies corresponded best to the real world.

4.1 STATIONARY UNIVERSES AND CREATION OF MATTER

Although the originators of the steady-state theory worked out their model of the universe independently of earlier suggestions, it is of interest to recall that essential components of the theory can be found in the literature on cosmology even before the Second World War. It is not well known that there existed an earlier kind of steady-state cosmology which differed not only from the later Cambridge version but also, and even more so, from the works of Hubble, Tolman, Lemaître, and other mainstream cosmologists. Readers who prefer to think that science develops linearly, or that outsider theories have no place in the annals of science, are advised to skip the present section (and perhaps the rest of the book).

MacMillan's Universe

In 1958, at a time when the steady state theory was much alive and the center of a heated controversy, the American physicist Richard Schlegel called attention to a much earlier theory suggested by a former colleague of his at the University of Chicago, William Duncan MacMillan.[1] As Schlegel pointed out, this theory anticipated to some extent the steady-state theory. MacMillan first expounded his hypothesis in 1918, when he speculated that the radiant energy from the stars was partly absorbed in the ether and there converted into potential energy associated with new matter. The atoms thus formed in space might disappear again by being freed of the energy by which they were organized. MacMillan suggested that stars, although they were formed and eventually dissolved, were permanent forms of physical existence: "There is no necessary limit to its [a star's] age, and though the star itself may rise and fall, the universe as a whole is not necessarily altered. The singular points may change their positions and their brilliancy, but it is not necessary to suppose that the universe as a whole has ever been or ever will be essentially different from what it is today."[2]

Seven years later MacMillan developed his ideas at a symposium of the American Mathematical Society. Without claiming that they necessarily reflected the actual state of the world, he set up a number of cosmological postulates, one of which was the homogeneity of the universe and another that "the universe does not change always in any one direction. . . . It is like the surface of the ocean, never twice alike and yet always the same."[3] The way in which the stationary state was maintained was, he suggested, that radiant energy from the stars was absorbed in "the fine structure of space" from where it sooner or later would reappear in the form of new atoms. MacMillan repeated his view of 1918, that it was unnecessary "to suppose that the universe as a whole has ever been or ever will be essentially different from what it is today," a statement which can reasonably be seen as a version of the later perfect cosmological principle. However, written as it was years before the expanding universe became a reality, this was not a particularly controversial statement at a time when the static universe was universally accepted. Where MacMillan's view differed from standard cosmology was in its denial of the gradual and irreversible degradation of energy and matter, of the increase of entropy which eventually would leave the universe dead and structureless.

Although MacMillan's views may seem surprisingly modern in some respects, his cosmological scenario represented a reaction against the more recent developments in physics and astronomy. Politically and scientifically, MacMillan was a conservative who rejected Einstein's theory of relativity and believed in the scientific validity of common sense. His universe was Newtonian, and he explicitly endorsed the classical notion of an absolute time which was independent of space. The Chicago astronomer had no confidence in mathematical models of the universe, relativistic or not, and declared in 1927 that "The exclusive use of mathematics is a dangerous thing in cosmology."[4] His suggested cosmological model was designed to lend support to a cosmic optimism which he felt was threatened by the world view of modern physics. "The haunting fear of a general stellar death is gone," he wrote in 1920, referring to his own theory. "The forbidding picture of the galaxy as a dismal, dreary graveyard of dead stars fades away from our sight; and in its stead we see an indefinite continuation of our present active, living universe with its never-ending ebb and flow of energy."[5]

MacMillan was an early advocate of the idea that stellar energy consists in the transformation of matter into radiation by some kind of annihilation process, the same idea which Eddington had discussed a little earlier. Due to the enormous gravitational pressure in the interior of a star, the atoms would dissociate into electrons and nuclei, and these would occasionally annihilate. In this way the mass of the star would diminish in time if it was not—so MacMillan hypothesized—being balanced by the accretion of dust particles gathered by the star through its journey through space. "The stars feed upon it [the nebulosity] in exactly the same sense that the cattle feed upon the grass of the fields," he explained in 1923.[6] But if this process of maintaining the stellar masses and luminosities continued, eventually the interstellar space would be swept clean of its rarefied dust atmosphere. This contradicted MacMillan's fundamental postulate of an eternal, unchanging universe, and so he had to come up with a suggestion to regenerate small amounts of matter in space. The answer was a kind of pair creation, to use a later term. The mechanism proposed by MacMillan for this hypothetical process was obscure and relied more on nineteenth-century physics than on the atomic physics of the 1920s. Rather than just producing quanta of radiation energy, the annihilation between electrons and nuclei was said to result in a "unit" with no electromagnetic or gravitational characteristics. Exactly what this unit was supposed to be is not clear, but MacMillan spoke of it as being different from radiation energy: it was a carrier of such energy, but in itself not a quantum of energy. He speculated that "perhaps they constitute the units of that unsubstantial something which we call the ether."[7] In free space the units, of ether or whatever, might on rare occasions undergo a process of substantialization, be separated into the electrons and nuclei of which they were once formed. In this way MacMillan believed he had constructed an indefinite steady state for both stars and the space between them.

In other writings MacMillan did not mention the idea of dust accretion, but he kept to the idea of creation of matter in interstellar space, presumably in the form of hydrogen atoms. His scenario had the advantage that it solved the much discussed problem of the heat death of the universe. In MacMillan's stationary universe there was no reason to assume that the entropy would always increase and leave the universe as a dead soup of radiation. This was an important motive for the cosmology suggested by the American astronomer. When the expansion of the universe was discovered, MacMillan belonged to the minority who questioned the standard interpretation of the redshift and sought to avoid the conclusion of an expanding universe. He suggested a "tired-light" hypothesis according to which the photons lost energy on their travels through space, and speculated in accordance with his earlier stated view that perhaps the emitted energy "disappears into the fine structure of space and reappears eventually in the structure of the atom."[8]

Some of MacMillan's ideas replayed themes which can be found much farther back in time. They had their origin in mid-nineteenth-century physics, when the cosmic implications of the laws of thermodynamics were first discussed. If these laws predict an irreversibly degenerating universe there ought to be, some physicists felt, a mechanism balancing the dissolution in order to keep the universe static and eternal. As early as 1852, at the meeting of the British Association in Belfast, the British physicist and engineer William Rankine speculated that "the world, as now created, may possibly be provided within itself with the means of reconcentrating its physical energies, and renewing its activity and life." Rankine's idea was to conceive the universe as consisting of compartments in any of which the radiant energy is either dissipated or "reconcentrated into foci . . . [with] a store of chemical power being thus reproduced at the expense of a corresponding amount of radiant heat." He believed that there was some evidence for this, for, as he wrote, "some of the luminous objects which we see in distant regions of space may be, not stars, but foci in the interstellar ether."[9]

Speculations similar to Rankine's abounded in Victorian and Edwardian England. Shortly before the First World War, the eminent British geophysicist Arthur Holmes repeated that there probably existed a mechanism which could replenish the energy lost by the sun and others stars. He succinctly put the problem as follows: "If the development of the universe be everywhere towards equalisation of temperature implied by the laws of thermodynamics, the question arises—why in the abundance of time past, has this melancholy state not already overtaken us? Either we must believe in a definite beginning . . . or else we must assume that the phenomena which we have studied simply reflect our limited experience."[10] Holmes, like many others, tended toward the latter alternative. There is no known evidence that MacMillan was inspired by Holmes and earlier physicists, but neither the problem MacMillan wrestled with nor the solution he came up with was entirely original. To mention but one more example, the French metaphysician Henri Bergson proclaimed in 1907 that "the universe is not made, but is being made continually. It is growing, perhaps indefinitely, by the addition of new worlds."[11]

Many of the cosmogonical discussions of the nineteenth and twentieth centuries were modifications of earlier geological discussions concerning the age and history of the earth. In the late eighteenth century there was a dispute between catastrophic and uniformitarian views; according to the former, cataclysmic processes (such as mighty floods and earthquakes) had once shaped the crust of the earth, whereas the uniformitarianists held that all geological changes were the cumulative effects of physical agencies operating also today. To some of the uniformitarianists this implied that the world was eternal. It is easy to see the parallel with the later division between big-bang and steady-state cosmologies. For example, a British geological author held in 1789 a kind of steady-state view, including a version of the perfect cosmological principle. The author, George Toulmin, asserted that "A succession of events, something similar to what is continually observed, has *ever* taken place; nature having, through an eternal period of duration, acted by laws fixed and immutable." Furthermore: "Thus does nature support herself! And she will *ever* continue to support herself in the same manner as we now behold her. . . . And as the vegetables flourish and decay, and men receive the spark of life, with which they triumph for a time, then fading die, or change the mode of their existence and felicity, so they have *ever* done, and *ever* will do."[12]

Millikan, Atom-Building, and Cosmic Rays

Although MacMillan's speculations made no impact on his astronomer colleagues, they did find a receptive ear in Robert Millikan. The famous experimentalist was at the University of Chicago until 1921 and knew MacMillan well. It was there that Millikan had established his reputation in physics by measuring the charge of the electron and by confirming Einstein's photoelectric law experimentally (but without accepting Einstein's quantum-theoretical explanation of the law). In 1915 MacMillan discussed his ideas of atom building by condensation of radiation with Millikan, who later paid tribute to his former colleague as the "investigator [who], on the theoretic side, is the foremost representative of the idea of the development of cosmic energy by the process of atom building."[13] In his studies of the cosmic radiation, which started after Millikan went to Pasadena to build up the California Institute of Technology, he developed MacMillan's view into what can reasonably be called the MacMillan-Millikan theory of steady-state cosmology.

The penetrating radiation, in Europe known as the *Höhenstrahlung* (high-altitude radiation) was discovered about 1912 by the Austrian physicist Victor Hess, who recognized the existence of a permanent ionizing radiation of extraterrestrial origin. His results were confirmed and extended by measurements made in 1913–14 by Werner Kohlhörster, a German physicist. After the war, when balloon experiments were resumed, Kohlhörster explicitly linked the *Höhenstrahlung* to cosmology, concluding that "the radiation from above is a phenomenon the origin of which is to be sought in the cosmos."[14] In accordance with speculations suggested by Walther Nernst, Kohlhörster believed that the radiation might be a kind of radioactivity emitted by a hypothetical transuranic *Urmaterie*, or primordial matter, distributed in space.

At that time the nature of the cosmic radiation was still a mystery, and the field so little explored that Millikan, who had started measurements in 1922, could claim in 1925 that he had discovered cosmic rays. His first experiments of 1922–23 indicated, in fact, that no such rays existed, and Millikan at first concluded that the penetrating radiation was of terrestrial origin. But he soon corrected his mistake, and in 1925 he claimed to have incontrovertible evidence for a radiation coming from the depths of space. In America, the supposedly new radiation was often referred to as "Millikan rays."[15] Experiments with balloon-borne electrometers conducted with his student G. Harvey Cameron made Millikan conclude that the cosmic rays were energetically inhomogeneous and electrically neutral, consisting of distinct bands of high-energy photons.[16] The energies of these bands were found to correspond to the binding energies of atomic nuclei, from which finding Millikan concluded that cosmic-ray photons arose from nuclear building-up processes in the universe. In an article of 1928, Millikan and Cameron phrased their conclusion as follows: "The observed cosmic rays are the signals sent out through the ether announcing the continuous creation of the heavier elements out of the lighter."[17] Millikan rejected as incompatible with his data the idea of Jeans and Eddington that the process responsible for the cosmic rays was proton-electron annihilation.

To indicate the reasoning of Millikan and Cameron, consider the assumed fusion of four protons into a helium nucleus, i.e., the process $4p^+ + 2e^- \rightarrow {}^4\text{He}$ (recall that this was years before the neutron was discovered and realized to be a nuclear constituent). During such a building up of a nucleus there will be a mass defect, the mass of the helium nucleus being 0.029 atomic mass units smaller than the sum of its constituents (Millikan and Cameron's value). The loss in mass corresponds by Einstein's equation $E = mc^2$ to an excess energy, which Millikan assumed revealed itself in the form of a cosmic-ray photon. Using the relationships $E = h\nu$ and $\nu\lambda = c$, where ν is the frequency and λ the wavelength, he found the theoretical wavelength of the photon to be 0.000 46 Å; this he transformed to an absorption coefficient in water (μ) by means of Dirac's new quantum-mechanical formula for Compton scattering. The result was an absorption coefficient of $\mu = 0.30$ per meter of water. Since this coincided almost exactly with one of the values he and Cameron had found experimentally, Millikan interpreted the cosmic-ray band as the result of a cosmic formation of helium. Using similar reasoning, he interpreted other bands as indications of the building up of oxygen, silicon and iron out of hydrogen and helium.

In the paper of 1928 Millikan extended his cosmic-ray studies into a grand cosmological scheme, the rudiments of which had been in his mind since his youth. The essence of Millikan's scheme was as follows. Space is filled with a tenous gas of electrons and protons which are the building blocks of all matter. These will occasionally combine into heavier nuclei, thereby producing the cosmic-ray photons which Millikan dramatically described as "the birth cries of the elements."[18] With regard to these cosmic atom-building processes Millikan concluded that they did not take place step by step, but in single acts, the heavier elements being formed directly from protons and electrons (and, in some cases, helium nuclei). He argued, moreover, that the processes did not take place in stars, but on the contrary "only in the interstellar or intergalactic spaces where densities and temperatures are essentially zero."[19] Once formed, the atoms were supposed to aggregate into larger gravitational systems and eventually into stars.

But this was not all, for if it was it would in the end seem to lead to a dead universe, namely, when the building blocks were used up, and it was as essential for Millikan as it was for MacMillan to avoid this consequence, to have a creative and evolutionary universe. The solution was to postulate that electrons and protons were regenerated by the photons emitted by the stars: "These building stones are continually being replenished throughout the heavens by the condensation with the aid of some as yet wholly unknown mechanism of radiant heat into positive and negative electrons."[20] In this way Millikan had constructed a grand cosmic cycle and avoided the universal heat death which he intensely disliked, for philosophical, social, and religious as much as for scientific reasons. The cost was a somewhat forced explanation and a violation of the second law of thermodynamics. But Millikan argued that this was justified on a cosmological scale and concluded that "we are able to regard the universe as in a steady state now, though a state not satisfying the condition of microscopic reversibility."

The kind of universe favored by Millikan contrasted with, and can in part be seen as a reaction to, the cosmological ideas presented by James Jeans in the 1920s.[21] According to Jeans, young stars consisted mainly of transuranic elements which would spontaneously transform into radiation, in part by ordinary radioactive decay, but mostly by annihilation. In addition to electron-proton annihilation processes, annihilation of entire atomic nuclei was also supposed to take place, a conclusion which Jeans asserted was inevitable. The cosmic processes were unidirectional, from the complex to the simple, and governed by the second law of thermodynamics. Contrary to most other ideas of evolution, Jeans held that the world started with complex, transuranic atoms which gradually decomposed into radiation and lighter atoms. He believed that the transuranic elements making up most of the stars had a "greater capacity for the spontaneous generation of radiation by annihilating themselves [while] the lightest elements survive for longest the disintegrating action of time." What happened, as Jeans wrote in 1928, was that "the universe is melting away into radiation," occasionally leaving pockets of dead ashes such as terrestrial matter.[22] According to this pessimistic view, life was an accident and the universe determined to become dead in the far future. It was a view to which scientists such as MacMillan, Millikan, and Nernst (and many others) strongly objected.

Millikan's cosmic scenario was fully consistent with MacMillan's, and Millikan acknowledged his debt to his former colleague, whom he presented as the pioneer whose theoretical views he had now proved experimentally.[23] On his side, MacMillan saw Millikan's cosmic-rays theory as "striking confirmation" of his own hypothesis.[24] The theory made headlines in American newspapers, which praised its spirit of robust optimism, so much needed at a time marked by the economic recession. Thus, according to the *New York Times* on the last day of 1930: "Neither drought nor flood nor financial depression nor any other terrestrial ill can stay the cosmic optimism of the science that not only has such practical applications, but that has faith in a continuing creation and that cooperates with 'a Creator continually on the job.'"[25]

However, the appraisal of the news media did not extend to support from astronomers or physicists. Atom-building processes in interstellar space, with protons coalescing into helium nuclei, were considered by the Soviet physicist Georgii Pokrowski, who referred favorably to Millikan's views.[26] So did an American chemist, S. B. Stone, who suggested one of the earliest theories of the origin of elements by nuclear processes.[27] But these were exceptions. In England, the physicist Edmond Stoner severely criticized Millikan's cosmic theory and pointed out that the probability of atom-building processes in interstellar space was vanishingly small. Contrary to Millikan, Stoner concluded that the universe is changing irreversibly and that the source of cosmic rays was probably annihilation of matter.[28] About 1931 it was realized that the experimental foundation was less than satisfactory. Cosmic-ray energies turned out not to be distributed in bands; the maximum energies were much larger than any atom-building process could account for; and the cosmic radiation did not only, and perhaps not even mainly, consist of photons. Millikan and his coworkers admitted that annihilation processes had to be taken into account in addition to the atom-building processes; but even then the modified theory could not stand up to the rapid progress in cosmic-ray physics.

By the mid-1930s, Millikan's cosmophysical theory had silently disappeared from the scene of physics. Millikan became more cautious with regard to the central question of retransformation of radiant energy back into atoms—"a question upon which there never existed any experimental evidence of any sort," he now declared (thereby contradicting his own earlier statements). Although Millikan continued to raise objections to "the dogma of the heat death," he abandoned his earlier position with regard to linking cosmic-ray physics with cosmology. As he wrote in 1935: "The cosmic rays up to the present have no significant bearing upon the question as to whether the universe is ultimately running down or whether there are processes which keep it in equilibrium. These are questions which at present must be left to the philosopher and metaphysician."[29] Yet the metaphysical foundation on which he had originally based his theory, an evolving universe revealing the Creator's continual activity, remained essential to Millikan's spiritual outlook and view of science.[30]

Both MacMillan and Millikan associated a universe without some kind of continual creation with a mechanical, atheistic, and materialistic world view which modern science had made obsolete. It was not mere rhetoric when Millikan and Cameron ended their paper of 1928 by pointing out that "modern philosophers and theologians have also objected [to a running-down universe] on the ground that it overthrows the doctrine of Immanence and requires a return to the middle-age assumption of a Deus ex Machina."[31] This was a valid objection for Millikan too. At the end of 1930, in his address as retiring president of the American Association for the Advancement of Science, Millikan emphasized that his hypothesis of hydrogen being replenished out of radiation was "a little bit of experimental finger-pointing in that direction," namely, "to allow the Creator to be continually on his job."[32] In spite of the disagreements between Millikan and Jeans, their metaphysical commitments were not all that different.

Millikan had no doubt that the world was created by God, and his and MacMillan's steady-state theory was believed to be consonant with Christian belief. In consideration of the similarity between the MacMillan-Millikan view of the universe and the later steady-state theory, it is interesting to note that the latter theory was often associated with atheism, and big-bang theory with theism. Obviously, Millikan would not have accepted such associations. Neither would all theologians, as is shown by the case of William Ralph Inge, dean of St. Paul's and a well-known author of theological and cultural texts. In a book of 1933 Inge dealt at length with the second law of thermodynamics, this "naive deistic doctrine that some billions of years ago God wound up the material universe, and has left it to run down of itself ever since."[33] Inge favored a perpetual universe with an eternal Creator and believed that there must be in the universe some hitherto unknown agency that counterbalances the growth in entropy: "Are we sure that there is no creation (say) of hydrogen atoms out of radiation? A discovery of such a balance between creation and destruction would be extremely welcome to most of us. It would end the necessity for believing in the creation of the universe in Time. It would satisfy our very natural feeling that a perpetual continuance of the universe would be in more accordance with what we may imagine to be the will of God than its temporary existence and final annihilation."[34] No wonder that Inge referred to Millikan's theory of cosmic rebuilding processes as "very attractive."

According to Millikan and MacMillan, the question was not whether the world was created or not, but rather whether it was created once and for all or if creation took place continually. This question, a classic in natural philosophy since the famous dispute between Newton and Leibniz, also occupied Reginald Kapp, a professor of electrical engineering at University College, London. In a book of 1940, significantly entitled *Science versus Materialism*, Kapp argued against the materialistic universe, much as Millikan had done earlier.[35] On qualitative and philosophical grounds, Kapp distinguished between what he called "once-upon-a-time-theories" and "at-any-time-theories," the first being a reference to cosmology of the big-bang type. Kapp argued that the at-any-time-theory, in which creation of matter continually occurred together with disappearance of matter, was philosophically preferable, but his discussion was restricted to a very general level. After the advent of the steady-state theory, Kapp claimed that his book anticipated this theory's idea of matter creation.

From Test Tube to Stars, and Beyond

The steady-state view of the universe favored by MacMillan and Millikan was independently developed in Europe by the German chemist Walther Nernst, one of the founders of modern physical chemistry and electrochemistry, and a Nobel Prize winner of 1920 for his so-called heat theorem (or third law of thermodynamics). Considering that the MacMillan-Millikan theory and Nernst's cosmophysical theory had much in common, it is remarkable that the Americans did not refer to the German chemist, who, on his side, never mentioned the works of MacMillan and Millikan.

Nernst became interested in astrophysics during the First World War and cultivated the subject in his own, unorthodox way until the end of his life in 1941. Cosmology may seem a strange subject for a chemist to take up, but Nernst was not the only prominent physical chemist who turned from the test tube to the stars. So did other distinguished physical chemists, including Richard Tolman, Jean Perrin, and Svante Arrhenius. As we have seen in previous chapters, chemists—and especially physical chemists—were active in several areas of physical cosmology and played a considerable role in the field of element formation in particular. In the 1910s the behavior of chemical reactions at very high and very low temperatures was a hot topic in physical chemistry, which also included the study of radioactive processes. It was not so unnatural to extend such interests to deal also with the unknown cosmic processes. Furthermore, since the mid-nineteenth century there had existed in chemistry a tradition in "cosmochemistry," where chemists speculated about the formation and disappearance of elements on a cosmic scale. As shown by the works of William Crookes and Norman Lockyer, for example, not all chemists were narrow empiricists. Cosmology was not necessarily a foreign subject to the chemist.[36]

Most likely, Nernst's interest in cosmology received inspiration from Arrhenius, the great Swedish physical chemist (Nobel laureate in 1903) who was an old friend of Nernst. Unwilling to accept the universal heat death, Arrhenius searched in the early part of the century for mechanisms that would obliterate it. He believed he had found in radiation pressure a mechanism which allowed the possibility that "the cosmic development can take place in a continuing cycle, where there is no trace of any beginning or end, and by means of which an ever undiminished prospect can be maintained even for life."[37] Combining the effects of radiation pressure with a speculative hypothesis of galactic collisions, Arrhenius devised a scheme in which these dissipative forces compensated the universal gravitation. New celestial bodies would be formed at the expense of those degrading, and as a result the heat death could be avoided and a stationary universe secured eternally. To Arrhenius and many of his contemporaries it was an intellectual necessity to establish a cosmology, however qualitative and speculative, that secured eternal cosmic evolution in an infinite, self-perpetuating universe. Only such a universe would be comprehensible, they claimed, and accord with their philosophical views. Arrhenius's cosmophysical ideas were widely discussed, if more among laypersons and astronomically interested scientists than among professional astronomers. In Germany, in particular, they stimulated much speculation about the evolution of the universe.

Nernst first thought of a mechanism for maintaining energy equilibrium in the universe in 1912, like Arrhenius in connection with an attempt to avoid the continuous degradation of energy and matter.[38] Ever since 1886, when he first became acquainted with Boltzmann's gloomy prediction of a cosmic heat death, he had denied believing in this alleged consequence of the second law of thermodynamics.[39] Nernst was strongly opposed to any *Götterdämmerung des Weltalls*—a universal Armageddon—and, as he later wrote, his work in astrophysics was wholly motivated by his desire to find an alternative.[40] In his important work in physical chemistry, Nernst consistently avoided using the concept of entropy, which he considered unnecessary and viewed with some suspicion. Presumably this attitude was colored by his strong views on the heat death with which entropy was often associated.

In his lecture of 1912 Nernst indicated a way of saving the world from the heat death, namely, by assuming that the ether, supposed to be the end product of radioactive decay, could attain all possible kinds of configurations. "In this way," he explained, "an atom of a chemical element (most probably even a high-atomic element) will be recreated from time to time." Nernst realized the speculative basis of his hypothesis, which he presented as "less as an attempt to establish a new cosmical world-picture than an illustration of our theme, that is, the thermodynamic approach." And yet nine years later he had developed the idea into just such a cosmical world picture, now justifying it by the new quantum theory. In a booklet of 1921 he suggested that the ether, far from being an inert medium, was a huge energy reservoir with its energy being stored in the form of the zero-point energy known from quantum theory [where the energy of a harmonic oscillator is $E = h\nu(n + 1/2)$, $n = 0, 1, 2, \ldots$, i.e., different from zero even in the lowest quantum state]. The existence of an ether filled with enormous amounts of zero-point energy was first discussed by Nernst in 1916, in connection with an attempt to explain the reaction rates of chemical processes.[41]

In 1921 Nernst speculated that fluctuations of this hidden energy would occasionally, over long spans of time, form configurations out of which radio-active atoms would be created. Eternal recycling of radioactivity would prevent the heat death and secure the stationary universe to which Nernst was metaphysically committed: "Our eyes need not, in the far future, have to look at the world as a horrible graveyard, but at a continual abundance of brightly shining stars which come into existence and disappear."[42] In later publications between 1928 and 1938, Nernst developed these ideas into a cosmological view of a stationary universe with creation of matter, which in several respects showed striking similarity with that advocated by MacMillan and Millikan.

A view somewhat similar to that held originally by Nernst was independently proposed in 1921 by Emil Wiechert, a recognized German theoretical physicist and pioneer of geophysics.[43] Wiechert was opposed to Einsteinian relativity and stuck to the ether worldview, which he claimed constituted a cure against the "materialism" of the theory of relativity (a common theme in German physics in the early 1920s). Like Nernst, he speculated that ether-matter transmutations might continually take place in the depths of space and in this way provide a cosmic cycle making the heat death avoidable. The views of Nernst and Wiechert reflected certain trends in the zeitgeist of the early Weimar Republic, including a conception of the laws of nature as being merely statistically valid. According to Nernst, laws of nature were idealizations of empirical facts and not of an absolute nature; just as Boltzmann had shown the second law of thermodynamics to be only statistically valid, so, he claimed, were other laws of nature. Energy conservation was believed to be no exception.[44]

Nernst based his cosmological view on what he called the "principle of the stationary condition of the cosmos," according to which the universe does not change in its large-scale features.[45] The transformation of stellar matter by radioactive processes, such as those taking place in the stars, could therefore not be a one-way process but had to be balanced by the formation of new matter. Nernst regarded his principle an "intellectual necessity," an a priori hypothesis which could not be subjected to direct experimental tests and the acceptance of which was, in the end, "a matter of taste." Much the same attitude was held by MacMillan, who emphasized the postulational nature of his system. "It is a matter of taste," he wrote, "and *de gustibus non est disputandum* [taste cannot be discussed]."[46]

With regard to the energy production in stars, Nernst accepted neither the idea of building-up processes à la Millikan nor the annihilation and fusion processes which at the time were advocated by Eddington in particular.[47] Instead of this "highly fantastic hypothesis" he postulated that the source of energy was radioactivity, namely, due to the disintegration of one or more hypothetical transuranic elements with a relatively short lifetime and large decay energy. As mentioned, his early speculations of an *Urmaterie*—a form of primordial matter consisting of hyperradioactive, transuranic elements formed by the ether—was adopted by his compatriot Werner Kohlhörster in his cosmic-ray research in the 1920s. Nernst even urged chemists to search for "this most important element" which, in his view, might well exist also on earth. When Fermi and others later reported the production of transuranic elements artificially, Nernst was encouraged and considered it a support of his conjecture.[48] It may seem rather fantastic to postulate stars consisting of transuranic elements, but the hypothesis had a certain credibility at the time, when such elements were purely conjectural. As mentioned, Jeans suggested the same thing.

The energy irradiated by the hyperradioactive nuclei would in part be absorbed by the ether, and out of the energy-enriched ether new hyperradioactive nuclei would be formed, either coalescing into new stars or floating around in space. Nernst realized that such a scenario, and his principle of a stationary universe, seemed incompatible with fundamental physical laws such as the second law of thermodynamics. Furthermore, if matter could be created at all, according to established physical knowledge—or instinct—it would presumably be in the form of elementary particles and not as elements of extremely high atomic weight. Therefore: "The investigator of Nature stands before the choice, either to relinquish the principle of the stationary condition, which above was characterized as an intellectual necessity, or to assume a possible continuous failure of the two above mentioned natural laws."[49] Nernst chose the latter alternative.

Nernst's reference to the ether as an active medium for the interaction between energy and matter was indebted to a view of the ether he had held for many years. It was, like the ether itself, an unorthodox and unwelcome concept at a time when mainstream physics had relegated the ether to the status of a historical mistake, or merely a metaphor for the transmission of electromagnetic signals. But for Nernst the ether continued to be a vital and essential part of the physical world, and he was not alone in referring to the ether as a real medium. In the early part of this century electrons or electromagnetic fields were often seen as "intermediate between ether and ponderable matter," as the French physicist Jean Becquerel expressed it in 1908.[50] The previous year, another Frenchman, the amateur physicist Gustave LeBon, had given an elaborate account of the "dematerialization of matter" in which the ether was of central importance. LeBon wrote about a hypothetical primordial condensation "effected at the beginning of the ages by a mechanism totally unknown to us, [from which] are derived the atoms, considered by several physicists . . . as condensation nuclei in the ether." Furthermore, "if we know very little about the ether, we must, however, consider it certain that the greater part of the phenomena in the universe are the consequences of its manifestations. It is, no doubt, the first source and the ultimate end of things, the substratum of the worlds and of all beings moving on their surface."[51] This was a common view at the time and one with which Nernst felt a great deal of sympathy.

In the same year that LeBon philosophized about the dematerialized ether, Nernst speculated that the same imponderable medium might consist of what he called "neutrons"—weightless combinations of positive and negative electrons. In his classical textbook on theoretical chemistry he wrote: "We will assume that neutrons, like the light ether, are everywhere, . . . a space filled with these molecules [neutrons] will be weightless, electrically nonconducting, but polarizable, i.e., possess such properties which physics usually claims for the aether."[52] Such an ether, filled with energy and capable of creating matter, was also an essential part of Nernst's cosmological hypothesis.

The theory of the expanding universe in general, and the big-bang theory in particular, were clearly incompatible with Nernst's cosmological view. In the 1930s, when this theory had won wide acceptance, he therefore sought to develop his stationary world as an alternative to the expanding universe. In 1937, probably referring to Lemaître, he concluded that "the hypothesis of the 'exploding universe', which has no place according to my views, is now excluded."[53] In the new version the sources of stellar energy were either nuclear disintegration by high-speed collision with nuclear particles or the radioactivity from the transuranic elements he had conjectured earlier. He pictured the ether as consisting of zero-mass neutrons, these being transformed into real, ponderable neutrons by the absorption of radiation energy into the ether's pool of zero-point energy. In this way neutrons would be created throughout the universe so that interstellar space would include a rarefied gas of neutrons mixed with electrons and protons arising from the disintegration of the radioactive neutrons. From this cosmic particle gas heavier elements would be formed, including as a dominant component highly radioactive elements. Nernst found his hypothesis justified by measurements of the energy and particle content of the cosmic radiation. According to Nernst, cosmic rays consisted of neutrons and their decay products, protons and electrons, together with gamma photons from the cosmic radioactive elements.

In the mid-1930s any plausible cosmological theory would have to account for the observed linear relationship between redshifts and distances of the galaxies. Nernst denied, of course, the Doppler interpretation of the redshift and noted with satisfaction that this was an interpretation that even Hubble now seemed to question. As an alternative to the Doppler interpretation, and one consistent with his principle of a stationary universe, Nernst assumed that the photons irradiated by the galaxies would continually lose energy on their journey through space. The energy would be absorbed by the ether (and hence disappear) and secure a constant low temperature of intergalactic space, which Nernst calculated as 0.75 K.[54] It was an easy matter in this way to obtain the redshift-distance formula without galactic recession. Assume that a photon loses its energy according to $dE/dt = -HE$, where H is a constant and $E = h\nu$

is the photon energy. This yields ln $(v_0/v) = Ht$ or $(v_0 - v)/v = Ht$ if the decrease in frequency is small compared to the frequency $(v_0 - v \ll v)$. Since $t = r/c$, we have $\Delta v/v = Hr/c$, which is the empirically confirmed Hubble law. Thus, as Nernst first suggested in 1935, the cause of the linearity is not that the universe expands, but simply that it takes more time for the photons to reach the earth the farther away the galaxy is located. The Hubble constant H was not to be considered a cosmical constant, but a "quantum decay constant" giving the decay rate of photons.[55]

The particular cosmological hypothesis proposed by Nernst did not arouse much attention. Nobel laureate and grand old man of physical chemistry as he was, astronomers chose to ignore him. His views were well known in Germany, though. For example, in their previously mentioned works of 1925 and 1926, both Stern and Lenz referred to Nernst's theory. And Max von Laue, in the first suggestion ever of using quantum mechanics in the study of element formation, believed that "The hypothesis which Nernst has introduced into cosmical physics, i.e., that radioactive elements under favourable conditions are re-created, thus receives considerable support."[56] Later, in 1937, the then seventy-three-year old Nernst gave an account of his views at the 32nd Annual Meeting of the German Astronomical Society, held in Breslau.[57]

Outside Germany, Nernst's cosmophysical theory was received with silence. Eddington, one of the few who did respond to Nernst's hypothesis, rejected it. Among other things, he objected to Nernst's "antievolutionary" claim that matter was created in the form of complicated and unstable elements of large atomic weight. "Personally," wrote Eddington, "when I contemplate the uranium nucleus consisting of an agglomeration of 238 protons and 146 electrons, I want to know how all these have been gathered together."[58] This was a criticism not only of Nernst, but also of Jeans. Contrary to Nernst, Millikan, and MacMillan, Eddington dismissed the idea that a hypothetical renewal of matter might prevent the ultimate heat death of the universe: "Sub-atomic energy extends the life of the universe from millions to billions of years; other possibilities of rejuvenation may extend it from billions to trillions. But unless we can circumvent the second law of thermodynamics—which is much the same as saying unless we can make time run backwards—the universe is steadily getting nearer to an ultimate state of uniform changelessness."[59]

There is little doubt that Eddington disagreed with the kind of universe advocated by MacMillan, Millikan, and Nernst, and that he did so not only for scientific, but also for aesthetic and religious reasons. He had "no great desire" that the second law of thermodynamics should succeed in averting the final running down of the universe, as he wrote in 1928. His arguments against cyclic conceptions of the universe—"I am no Phoenix worshipper," he admitted—were also arguments against the stationary Nernst-MacMillan universe: "I would feel more content that the Universe should accomplish some great scheme of evolution and, having achieved whatever may be achieved, lapse back into chaotic changelessness, than that its purpose should be banalised by continual repetition. I am an Evolutionist, not a Multiplicationist. It seems rather stupid to keep doing the same thing over and over again."[60]

On this matter, Jeans agreed. In his best-selling *The Mysterious Universe*, published in 1930, he discussed the conception of a cyclic universe. Jeans did not refer to oscillating relativistic models with this term, but to the view of an imperishable universe "in process of being built, not out of the ashes of the old, but out of the radiation set free by the combustion of the old." Jeans admitted that Millikan's scheme—he didn't mention either MacMillan or Nernst—was "entirely logical and self-consistent," but dismissed it on the ground that it violated the second law of thermodynamics.[61] The previous year he had dismissed also the Phoenix universe (or "fancy"), and that with the same kind of argument as Eddington. "It is hard to see," wrote Jeans, "what advantage could accrue from an eternal reiteration of the same theme, or even from endless variations of it."[62] The cosmological view of Nernst, and also that of MacMillan and Millikan, reflected a deep, philosophically based opposition to the notion of heat death as well as to any notion of the universe having a birth and a death. A universe of finite age was totally unacceptable to Nernst, who in 1938, in a discussion with the forty-eight-years younger Weizsäcker, insisted that this view betrayed the very foundation of science and could not possibly be true.[63]

The cosmological hypotheses of Nernst, MacMillan, and Millikan had in common a picture of the universe as being in a steady state with the disappearance of matter balanced by a creation of new matter. The creation hypothesis was introduced as a means of avoiding a running-down universe, and not, as in the later steady-state theory, to counteract the thinning out of the universe, which would otherwise follow from the universal expansion. The latter phenomenon was ignored by Millikan and denied by MacMillan and Nernst, a fact which presumably contributed to the astronomers' lack of interest in the theories. Moreover, the early steady-state hypotheses were qualitative scenarios which completely ignored what was usually considered cosmology at the time, viz., the geometric structure of the universe and its evolution in time. One will look in vain for discussions of the metrical properties of space in the works of Nernst, MacMillan, and Millikan. Given the nature of this kind of theory, it is understandable that what little scientific interest there was in it before the war diminished to zero after 1945. It is significant that what appears to be the only attempt to revive the Nernst-MacMillan universe took place in the pages of a philosophical journal.[64]

Speculations on Ether and Matter

One of the remarkable features of the cosmological schemes of the Nernst-MacMillan type was that they postulated some kind of creation of matter. It is difficult to judge what this was taken to mean, more precisely, but in most cases the authors referred to some form of radiation energy converted into matter. They did not, like the later steady-state theoreticians, discuss true creation-out-of-nothing processes in which the law of energy conservation is violated. Nernst's version came close to this audacious notion, but then he would not have accepted the ether as nothing.[65] Whether creation out of nothing or by way of radiation energy, cosmic creation of matter was an unorthodox idea.

Long before the law of energy conservation had been contemplated, the permanence of matter was accepted as a scientific-philosophical doctrine. John Dalton, the founder of chemical atomism, wrote in 1808 that "We might as well attempt to introduce a new planet into the solar system, or to annihilate one already in existence, as to create or destroy a particle of hydrogen."[66] Although progress in physics in the twentieth century indicated that hydrogen atoms can in fact be destroyed—transformed to other particles or to energy— it remained an almost uncontested doctrine that atoms cannot be created in any proper sense. Yet the idea was not so unorthodox that it was not suggested by a few chemists and physicists.

Scientific ideas of the cosmic creation of matter can be traced back to the 1880s, when speculations about the cosmic genesis of elements were widely discussed, especially among British chemists. An interesting example is William Crookes's address to the British Association for the Advancement of Science in 1886, in which he suggested a far-ranging evolutionary scheme for the formation of elements out of a hypothetical primordial matter, the protyle. In one of the many visionary passages from this address, given long before relativistic energy-mass equivalence became known, we read: "Let us start at the moment when the first element came into existence. Before this time matter, as we know it, was not. It is equally impossible to conceive of matter without energy, as of energy without matter; from one point of view the two are convertible terms. Before the birth of atoms all those forms of energy which become evident when matter acts upon matter could not have existed— they were locked up in the *protyle* as latent potentialities only. Coincident with the creation of atoms all those attributes and properties which form the means of discriminating one chemical element from another start into existence fully endowed with energy."[67] According to Crookes, who rejected the cosmic heat death, the universe was in continual creation, with genesis of matter taking place perpetually in space as a result of radiant energy transforming into protyle. In Crookes's vision, we can thus (with good will) see anticipations of parts of both big-bang and steady-state cosmologies.

Many of the suggestions in the early part of the twentieth century were like those of Nernst, associated with what remained of the etherial worldview. For example, Oliver Lodge speculated vaguely in the early 1920s about some relationship between radiation pressure and the birth of matter.[68] As early as 1903 he mentioned the role possibly played by radiation pressure in "the cosmic scheme." This scheme appears permanent, but in reality, Lodge claimed, it is in constant flux: "The resolution of an atom into its electron constituents, and the aggregation of these constituents into fresh atoms, are both perfectly thinkable."[69]

Lodge was a firm believer in the reality of the world ether, which for him was inextricably connected with his Christian faith in the immortality of the soul. He speculated that matter might be created out of the ether and believed that such a process would be necessary in order to restore a stationary universe; without matter creation, the radiation pressure of the stars would result in an increasing dilution of matter. According to Lodge's scenario, the world once consisted of an undifferentiated, all-extensive substance—the ether. Out of the ether's energy, or rather its dynamic fluctuations, electrons and protons were formed as "knots" in the ether and eventually, through a long evolutionary sequence, there appeared atoms, molecules, gaseous clouds, and galaxies. The stars would shine their energy into the confines of space with, in the end, the universal heat death as a result. But Lodge was no more willing to accept the heat death than were McMillan and Nernst. "Are the operations in time, of which we have been speaking, really a sequence, or are they a co-existence?" he asked. "Or may we suppose that there is a recuperative process at work, the formation of matter as well as its destruction?"[70] According to Lodge, this was indeed what one had to suppose: "It may even be that fresh matter, in the form of electrons and protons, can be generated by radiation, apart from its photoelectric effect on already existing matter. . . . I would urge that creation is a continuous process, not going on once for all and then stopping, but continuing now, and always continuing: that what we are confronted with is not really a succession, a series, a beginning and an ending, a past and a future, —but in some sense an eternal Now."[71] Lodge was thus led to an idea of a steady-state universe very similar to the views held by McMillan, Millikan, and Nernst at about the same time. Contrary to these scientists, Lodge did not present his speculations as a scientific hypothesis, but admitted that they were part of a larger worldview, where religion and psychic evolution were the components of principal interest.

Although Jeans saw no escape from the heat death, and in this respect differed from many other cosmoscientists, ideas of creation of matter appeared also in his view of the universe. Thus, in 1928 he conjectured that "the centres of the nebulae are of the nature of 'singular points,' at which matter is poured into our universe from some other, and entirely extraneous, spatial dimension, so that, to a denizen of our universe, they appear as points at which matter is continuously being created."[72] Jeans's conclusion appeared in connection with an attempt to explain why so many galaxies have spiral arms. His words may sound odd to a modern reader, but they should probably be understood in the context of the older ether hypotheses of Karl Pearson and others, according to which the world ether (and then, latently, matter) might flow out of singular points or "ether squirts." It was a view familiar to Jeans, who many years earlier had offered an explanation of radioactivity along the same line.[73] Jeans also discussed whether the universe was of finite age or if it was perhaps in a kind of steady state with neither beginning nor end. He considered it "not impossible" that matter could be created continually: "We are free to think of stars and other astronomical bodies as passing in an endless steady stream from creation to extinction, just as human beings pass from birth to the grave, with a new generation always ready to step into the place vacated by the old."[74]

By 1948 the idea of creation of matter was far from new. In addition to the already mentioned speculations, recall that Dirac suggested the idea in 1937 (as creation out of nothing) and that Jordan took it over the following year. The idea also briefly appeared in Milne's cosmology. In 1946, and again in 1948, Milne examined the structure of spiral galaxies according to his unorthodox theory of kinematic relativity.[75] His conclusion was in general agreement with the speculations of Jeans, cited above. According to Milne, "The centre of each nebula, before the nebulae separated from one another by the expansion of the universe, has indeed been a singular point, where matter was created at some supra-sensual event, which is the origin of time for our *t*-scale."[76]

More on Creation

A few works in mathematical cosmology also discussed the possibility of matter creation, mostly in connection with examinations of the de Sitter universe (which is characterized by being devoid of matter!). Attempting to account for the observed distribution of nebulae, Tolman considered in 1929 what he called the "hypothesis of continuous formation," i.e., that nebulae are continuously formed at such a rate that their concentration remains constant. Tolman discussed the possibilities of continual conversion of radiation into matter as well as the annihilation of matter into radiation, but it is unclear if he thought of matter creation as an elementary process. In any case, he decided that although the possibility of matter creation should not be ruled out, it "appears to have little inherent probability." As he wrote: "Science has not yet provided us with any clearly worked-out mechanism by which such a formation of nebulae could take place, and an appeal to special acts of creation is not within the province of our discussion."[77] In the late 1930s, a group of Japanese mathematical physicists also, the so-called Hiroshima school, was led to the idea of continual creation of matter in their attempt to formulate a "world geometry" which would apply to both quantum mechanics and cosmology.[78] This kind of cosmology attracted no interest in the West. Not only were many of the contributions unknown to Western researchers, they were also highly abstract and apparently of no relevance to the real universe.

In none of the mentioned works were there any serious attempts to explain how matter was created, i.e., to base matter creation on a physical mechanism. The authors either ignored the question or stated that the processes might be due to "an exchange between ether and matter" (Nernst) or "some as yet wholly unknown mechanism" (Millikan and Cameron), or be "a subject of further investigation" (Dirac). The only one who actually suggested a mechanism was not looking for matter creation and, when he came across it, wanted to get rid of it. Erwin Schrödinger, the father of wave mechanics, had in about 1936 become interested in Eddington's program of unifying quantum mechanics and cosmology. He spent the academic year 1938–39 at the University of Ghent and also gave lectures in Louvain and elsewhere in Belgium. In Louvain he met Lemaître, whom he came to know well and with whom he had many discussions about cosmology. Schrödinger investigated the proper vibrations of quantum waves in closed world models in the hope of deriving in this way the mass spectrum of elementary particles as consequences of the structure of space-time.[79] In 1939 he extended this program to expanding universes satisfying a Robertson-Walker metric with a time-dependent radius of curvature $R(t)$.

The result of the investigation, which Schrödinger found to be "alarming" as well as "of outstanding importance," was this: "With particles it would mean production or annihilation of matter, merely by the expansion, . . . the alarming phenomena (i.e. pair production and reflexion of light in space) are not connected with the *velocity* of expansion, but would probably be caused by *accelerated* expansion."[80] Schrödinger examined the solutions of the quantum-mechanical Klein-Gordon wave equation which describes massive scalar particles. With $R = $ const, the solutions would be separate plane waves with positive or negative frequencies (or energies), but during the expansion Schrödinger found that "There will be then a mutual adulteration of positive and negative frequency terms in the course of time, giving rise to [pair production]." Interpreted by means of Dirac's hole theory (which Schrödinger did not use), the phenomenon can be explained as follows.[81] In the vacuum state, the sea of negative-energy particles is completely filled with particles of energy $E = -h\nu$. A rapidly expanding space provides sufficient gravitational energy to lift one of the particles out of the sea to a level of positive energy, which is exactly the mechanism of pair production: a hole, or antiparticle, is left in the negative-energy sea and a positive-energy particle of $E = h\nu$ is created. Solely as a result of the expansion of the universe, a pair of particles (such as electron-positron) has been formed out of the vacuum, and that without violating energy conservation.

The following year Schrödinger extended his investigation to cover also the Dirac equation in an expanding universe,[82] but he now judged the implied "alarming" possibility of matter creation to be of little relevance to the real universe; in most cases, he noticed with satisfaction, there would be no mixing of positive and negative frequencies and thus no accumulated pair production. At any rate, Schrödinger seems to have decided that the phenomenon was not, after all, of "outstanding importance" and he simply ignored it in his later work, in which there are no references to it. Even to those of his colleagues at the Dublin institute who worked with related problems, he did not mention his past preoccupation with matter creation.[83] This, and the fact that the suggestion appeared in a Dutch periodical shortly after the outbreak of the war, may account for the total neglect of Schrödinger's discovery that a varying gravitational field can create particles. More than any of the earlier suggestions of matter creation, Schrödinger's came close to the idea that was later examined by steady-state theoreticians. But they, too, did not know about his work.

4.2 A CAMBRIDGE TRIO

The modern steady-state theory appeared in 1948, in two separate papers. The authors were three young physicists from Cambridge University, who had been acquainted for several years, but who had only recently become interested in problems of a cosmological nature. To a considerable extent the theory was the result of collaborative work and was indebted to the discussions between the three. The steady-state theory challenged traditional ideas about the evolution of the universe and quickly became a subject of controversy among physicists and astronomers. As a result of Hoyle's popular exposition of 1950, the theory also became known to a wider audience, which made it even more controversial.

Young Fred Hoyle

Of the three founders of the steady-state theory, Fred Hoyle was the senior and the one best known in the astronomical community. He was born in 1915 in the small village of Gilstead, near Bingley, Yorkshire. Young Fred did not like school and later felt that he did not get much out of his years at Bingley Grammar School. "The tragedy of my life is that I never had a teacher who pushed me, from the age of 6 to the age of 18," he said in 1989; "Until I went to Cambridge, I was just allowed to drift, more or less."[84] At the time he entered Bingley Grammar, he had developed an interest in science. He read the family's only book on science, a chemistry textbook, and did chemical experiments in the kitchen, much to the dismay of his mother. One can understand Mrs. Hoyle's protests, for Fred's master experiment was the preparation of phosphine—an explosive, poisonous, and extremely evil-smelling gas. At the age of twelve he borrowed from the local library Eddington's *Stars and Atoms* (published that year), which made a great impression on him.[85]

After having finished at Bingley Grammar, Hoyle planned to study chemistry at Leeds University, but, as it happened, he was admitted in the fall of 1933 to Cambridge University on a Yorkshire scholarship. At first he wavered between physics and mathematics, which in the Cambridge terminology at the time corresponded to experimental and theoretical physics. He decided on theoretical physics (applied mathematics) and prepared for the mathematics tripos examination, in which he did very well, and won, in 1936, the Mayhew

Prize for the best performance in theoretical physics. Two years later he won another prize, one of two Smith's research prizes. As a graduate student Hoyle was first assigned Rudolf Peierls as supervisor. Peierls was a leading German-born quantum theoretician who had been Pauli's assistant and in 1933 settled in England, as a Jew unwilling to return to what had then become the Third Reich. When Peierls left Cambridge in 1937 to take over a new chair at the University of Birmingham, Hoyle was without a supervisor. He continued consultations with Peierls, however, and regularly hitch-hiked between Cambridge and Birmingham. Hoyle didn't mind working on his own, but it was arranged that the young Maurice Pryce, only two years older than Hoyle, should become his supervisor for a period. Later, in 1939, Hoyle was transferred to the famous Dirac, who usually took very few students, but made an exception in Hoyle's case. Hoyle had followed Dirac's lectures on quantum mechanics and was impressed by the way in which Dirac insisted on the priority of mathematical equations before physical interpretation.[86]

Hoyle's first research work was a theoretical investigation of the nature of beta-radioactive processes, which he conceived of as taking place through intermediate states in the daughter nucleus. In this way he was able to account better for the experimental data and to criticize the standard theory of beta decay due to Emil Konopinski and George Uhlenbeck (which was a modification of the theory originally proposed by Enrico Fermi in 1933). The work led to a paper in the *Proceedings of the Royal Society* of 1938, a collaboration with Peierls and Bethe, and a brief association with the experimental physicists at the Cavendish Laboratory.[87] Peierls next suggested that Hoyle take up an examination of wave equations for hypothetical particles with spin quantum number larger than one-half. Dirac's relativistic wave equation of 1928 described particles with half-integral spin (such as electrons), but with the discovery of new elementary particles it became natural to look for a generalization of the wave equation to cover particles with arbitrary spin. Dirac had himself proposed such a generalization in 1936, and when meson theory became part of physics slightly later, the topic became of central interest in theoretical physics. Pryce, Hoyle's friend and supervisor, was among those who studied the subject. In the spring of 1938 Hoyle completed a manuscript on wave equations for particles with spin larger than one-half. It was apparently intended as part of a dissertation, but the work was withdrawn shortly before publication and never appeared. Pauli, who stayed with Dirac in Cambridge in March 1938, had seen the manuscript and later received a copy from Hoyle. Pauli was at the time engaged in the very same kind of work together with his research student Markus Fierz, and although he found Hoyle's manuscript interesting, he also found it flawed on a number of points and inferior to Fierz's. "Unfortunately I cannot see how something of Hoyle's work can be saved," he wrote Peierls in May 1938; "Both of the two final sections of his summary are the opposite of the truth, and in the second section I would substitute 'simplified' with 'complicated'." And to the Dutch-American physicist George Uhlenbeck: "Hoyle's considerations ... were quite wrong and he has to withdraw his work from publication (this is more Peierls' fault than Hoyle's)."[88]

The failure with the work on wave equations was naturally a disappointment to the ambitious Hoyle, but it far from ruined his career possibilities as a theoretical physicist. In the spring of 1939 he was elected a fellow of St. John's College and obtained an 1851 Exhibition Fellowship which made him materially well off. In the same year, he published a large work on quantum electrodynamics, in which he proposed a new method to avoid the infinities which for years had plagued this much-discussed theory.[89] Hoyle suggested providing the electron with a kind of structure and spatial extent through its interaction with the radiation field. It was one of the first examples of what later would be known as nonlocal field theories, which attracted much attention after the war. Although Hoyle announced his intention to publish a continuation of the work, it never appeared. At that time, the summer of 1939, he had decided to leave quantum and nuclear theory and instead become an astronomer.

This was a serious and risky change to make in the career of a young scientist with no formal education in astronomy, but for various reasons Hoyle was convinced that theoretical physics was not his future. In 1985 he wrote: "My generation was the unluckiest of the past half century. We were too late to receive anything but crumbs from the rich table of the years around 1926 [when quantum mechanics was discovered], we were too early for quantum electrodynamics, and very much too early for quarks. Additionally from 1939 to 1945 we lost six wartime years."[90] His unsuccessful competition with Fierz can only have strengthened his feelings that he might never become a first-class quantum theoretician. And anything less than first class was incompatible with Hoyle's ambitions.

In one of his recollections, Hoyle added another reason for his dissatisfaction with quantum physics and his decision to change to astronomy. Having mastered the mathematical methods of quantum mechanics, he also wanted to understand the theory's conceptual basis. It led him into a state of intellectual crisis when he realized that the standard Copenhagen interpretation of quantum mechanics was not only inadequate, but could not possibly be the correct way of understanding quantum phenomena. According to Bohr and the majority view, quantum mechanics gave an exhaustive description of physical reality, and Heisenberg's uncertainty principle was to be understood as one of the conditions for obtaining knowledge of the physical world. In the spring of 1938, young Hoyle constructed a thought experiment which, he believed, showed that quantum uncertainties could occur even in systems large compared to the observer; thus they could not be explained, as the Copenhagen orthodoxy insisted, as interference caused by the observer.[91] Unknown to Hoyle at the time, his challenge was somewhat similar to Schrödinger's famous cat paradox, first published in 1935 but not much discussed before the war.[92] Hoyle decided that the philosophical foundation of quantum mechanics was less than satisfactory, and discovered that apparently no one else bothered about these kinds of conceptual difficulties. According to Hoyle, the unsatisfactory situation motivated him to leave quantum mechanics at least for a time. "This [crisis] caused me a year later to quit theoretical physics for astronomy. My brain patterns were seriously disturbed, and I could not see how to reshuffle them in a satisfactory way. . . . This was the question I failed to resolve, the question which led me in 1939 to leave theoretical physics for astronomy, thinking with youthful idealism to be entering a more rational subject."[93] It was Hoyle's first move away from orthodoxy in science, but not his last.

Hoyle's conversion to astronomy was assisted by his friend, the astronomer Raymond Lyttleton, who after two years at Princeton University had returned to Cambridge, where he was a fellow of St. John's College. Lyttleton, whose special interest was in theories of planetary formation, directed Hoyle's attention to the problem of how moving stars pick up interstellar matter. This problem had been treated in Eddington's *Internal Constitution of the Stars*, but Lyttleton believed Eddington's result gave a much too small accretion rate. The result of Lyttleton's and Hoyle's calculations was an important paper which demonstrated the importance of accretion of interstellar gas in the evolution of stars.[94] Other works with Lyttleton continued this line of research, and yet other papers, with Lyttleton or alone, dealt with the formation of the solar system, the constitution of stars, and stellar nuclear processes. One of the early papers was a rather sharply formulated criticism of Gamow's theory of stellar evolution, which Hoyle and Lyttleton found totally inadequate.[95] They criticized Gamow for overestimating the importance of nuclear physics and ignoring the dynamical considerations that, in their opinion, were much more important. What matters in the present context is that Hoyle was very successful in his move from theoretical physics to astronomy, and that he very quickly established himself as one of Great Britain's leading theoretical astronomers. Given the conditions—that Hoyle had no previous background in astronomy and that he was occupied with war-related research between 1940 and 1945—his productivity was amazing. Between 1939 and 1947 he wrote twenty-five astronomical research papers, of which ten were coauthored by Lyttleton, one written with Bondi, and one with both Lyttleton and Bondi. One of the few subjects the prolific and versatile Hoyle did not work on was cosmology.

Fred Hoyle had married at the end of 1939, and shortly thereafter he and his wife Barbara (née Clark) had to move from Cambridge. There was a war going on, and England's, and Hoyle's, prime concern was now the war against the apparently victorious Hitler Germany, and not the interior of stars. In the efforts to fight the enemy, radar research was given top priority, and many of the best British scientists were drawn into this crucial area of military research. Hoyle was one of them. He began work on aspects of this topic, primarily intended for the Royal Navy, in 1940 within the Admiralty Radar Group. At first installed in a small hut on a field near Portsmouth, Hoyle and his collaborators developed a new radar system better suited for ships. After he became leader of the theoretical section of the Admiralty Signal Establishment, in 1942, the section was transferred to Witley, between Portsmouth and London. It was there that he first met with two new members of the section, Hermann Bondi and Tommy Gold.

From Vienna to Cambridge

Hermann Bondi belonged to a Jewish family which had moved from Germany to Vienna in 1884, where Hermann was born on 1 November 1919. His father was nonreligious, and Hermann's meeting with members of the Viennese orthodox Jewish community left him with a feeling toward religion that he later described as "direct hostile."[96] He developed from an early age an antireligious view, seeing religion as associated with intolerance. In school Hermann quickly demonstrated a remarkable talent for mathematics and physics and contemplated following a career in science. In 1936 Eddington happened to be in Vienna, and the sixteen-year-old schoolboy managed to meet the famous astronomer. Hermann was greatly impressed. During his last year in school, the self-confident young man decided that Austria was too much of a scientific backwater to live up to his ambitions, and he managed to be admitted as a foreign student at Trinity College, Cambridge University. This was quite unusual, but the acceptance went through when Hermann's mathematical abilities were judged very favorably by Abraham (or Adolf) Frankel, an eminent mathematician at the Hebrew University in Jerusalem, who was a relative of Hermann. When Hermann Bondi arrived in England in the fall of 1937, it was thus as a foreign student of his own will, not as a refugee. (The racial laws in Austria were only introduced after the *Anschluss* in March 1938.) Bondi recalled: "It was not push that brought me, but pull. It was because I wanted to be here, and I never had any doubt that this is where I wanted to live for the rest of my days."[97]

   Bondi liked England and did well as an undergraduate student of mathematics. However, with the outbreak of the war, and with Austria part of the new pan-German Reich, his career was temporarily halted. As a citizen of a hostile nation, Bondi was interned in May 1940 and only released fifteen months later, in August 1941. The letters of recommendation that he obtained from Cambridge University speak of a student of unusual abilities. According to Eddington, he was "a mathematical student of great brilliance and promise . . . very highly thought of, and likely to do work of importance." Bondi's teacher in mathematics, Abram Besicovitch, wrote that when Bondi arrived from Austria his mathematical training was very inadequate. "In fact during his first term I could not give him any individual teaching because his knowledge of mathematics was inferior to that of the weakest of my pupils. But by the beginning of the next term it was the opposite picture. . . . [T]he ability displayed by him leaves me in no doubt that if he is given an opportunity he will develop into a distinguished original mathematician."[98] Bondi spent most of his time in internment in a camp in Quebec, under conditions which in many ways were more comfortable than those many people faced in England. Shortly after Bondi arrived in Canada in July 1940, his parents, who had also come to England from Vienna, settled as immigrants in New York. They obtained a United States immigration visa for their son, but because Hermann had arrived by a military ship (and thus not paid for his ticket) he was not allowed to cross the border. Helene Bondi, Hermann's mother, tried to get Einstein to intervene, which he did.[99] However, nothing came of the efforts and so Hermann Bondi became a British citizen and not an American. He was keen to return to the England he had come to like, and it was only for a brief period, when return looked problematical, that he considered following his parents' wishes.

On the first night of the internment, when the interned people had to bed on a concrete floor at a place somewhere near Cambridge, Bondi found himself next to another young Austrian who had also studied at Cambridge and whose parents (it later turned out) had even known Bondi's parents back in Vienna. Thomas Gold, born 22 May 1920, was the son of a wealthy Austrian-Jewish businessman and a German, non-Jewish mother. Like Bondi, he was brought up without any kind of religious education or influence. In 1930 the family moved to Berlin, where Thomas went to school without distinguishing himself except in mathematics. After having changed to a Swiss boarding school, Thomas became interested in science and confident about his own abilities. At that time he had no particular interest in astronomy, but he read several of Jeans's and Eddington's books which influenced and stimulated him.[100] After Thomas Gold had finished school in 1937, he went to England, where his parents had settled. He wished to study physics, but on his father's advice he signed up for mechanical sciences (engineering) at Cambridge University after having learned the Latin which was demanded for entering the university. Compared with Bondi, and also Hoyle, he was slow and thorough, and (Bondi recalled) a bit of a playboy, without the burning academic ambitions which characterized the other two. Gold had not come very far with his studies when he was interned.

Contrary to Bondi, Gold recalled the stay in Canada, and especially the voyage, as a rough time. When the ship arrived, Gold and most of the other people aboard had dysentery and felt depressed, but circumstances eventually improved. During the boring period of internment, Bondi and Gold came to know each other well. Bondi used the time in teaching Gold and a few others various subjects of physics, and in this way Gold became exposed also to Bondi's general view of science. "He also talked to me quite a lot about mathematical physics," Gold recalled, "and especially about the outlook he always had—an outlook that went quickly to the heart of a problem and placed it in the proper context."[101] Two of the three later steady-state scientists had come together, still unaware of the presence of Fred Hoyle.

After their release from internment, Bondi and Gold resumed their separate studies. Gold, who was released three months before Bondi, obtained his B.A.

in the summer of 1942 and was then hired by Hoyle's radar group fresh from the university. "Gold came to me with a reputation of being difficult," Hoyle recalled,[102] himself being subject to the same reputation. Bondi, too, joined the group, but slightly earlier and only after having cultivated his academic career for a while. The recruitment of Gold to the group was in part due to the recommendations of Bondi and Pryce. Bondi's specialty was classical dynamics, and with the distinguished mathematician and geophysicist Harold Jeffreys as his supervisor, he managed to make his debut in the world of pure research before he was drawn into militarily oriented work.[103]

As it happened, Bondi and Gold rented a little house in the village of Dunsfold, near Witley. Hoyle lived with his wife and child almost fifty miles away. Having no car, he had to bicycle six miles to the nearest railway station every morning and evening, an unpleasant sport during windy mornings in the wintertime. "I grew to hate the bicycle with a deep, bitter hatred, and have in fact never ridden one since that time," Hoyle wrote in 1965.[104] To avoid the inconvenience of commuting with bicycle and train, Hoyle frequently stayed with Bondi and Gold in their three-bedroom house. The result of the encounter between the three was a lasting friendship and long discussions about physics and astronomy. "I acquired," Bondi recalled in 1990, "my first taste there for domestic engineering and for cooking, under Tommy's guidance, but he and I and Fred, when he was with us, spent all our time discussing scientific questions. Fred's enormously stimulating mind, his deep physical intuition, his knowledge of the most interesting problems in astronomy, all combined to give me an outstanding scientific education in the few hours left after a hard day's work."[105] The atmosphere in the house at Dunsfold has been vividly characterized by Gold:

> In the evening Fred would typically walk around and with great emphasis say: "Well what could that Hubble observation mean? Find out what it could mean!" He would continue along this line, sometimes being rather repetitious, even aggravating, drumming away at particular points without any obvious purpose. At other times Fred would have Bondi sit cross-legged on the floor, then sit behind him in an armchair and kick him every five minutes to make him scribble faster, just as you might whip a horse. He would sit there and say: "Now come on, do this, do that," and Bondi would calculate with furious speeds, though *what* he was calculating was not always clear to him—as on the occasion he asked Fred, "Now, at this point do I multiply or divide by 10^{46}?"[106]

Of the trio, Gold was the perceptive listener whose physical imagination could bring new perspectives into the discussions; Bondi was the disciplined and sharp mathematical mind; and Hoyle was the dynamo, the versatile, undisciplined, and intuitive renaissance scientist. "Very often, with his papers, as with his conversation," Bondi recalled about Hoyle, "I found it very irritating that he'd obviously used entirely illegitimate arguments. And then when I tried to get the thing right, I came to the same conclusion as he had come to."[107]

During the wartime years Hoyle continued his astronomical studies and developed an interest in the physics of supernovae. In the fall of 1944 he went on an official tour to the United States, where he visited MIT and the naval headquarters in San Diego as part of his work with naval radar. Armed with a letter of introduction from H. N. Russell, he used the opportunity to go to the Mount Wilson Observatory, located some one hundred and twenty miles away from San Diego. At the Observatory he spent a weekend with the astronomer Walter Baade, who provided him with reprints and fresh information about novae, a subject which Baade had pioneered some years earlier. The meeting inspired Hoyle to take up the study of element formation in stars and novae, a subject in which he soon became a leading expert and with which his name will forever be associated. "So it came about that I left Pasadena," he recalled, "having for the first time an idea of just how exceedingly high stellar temperatures and densities might become during the late stages of stellar evolution."[108] The way home went via Montreal, where he met with his old friend Maurice Pryce, whose work there was related to the efforts to produce a nuclear bomb.

Careers and New Horizons

After the war had ended, Hoyle, Bondi, and Gold returned to academic activities. Gold finally obtained his master's degree in 1945, and in 1947 he became a fellow of Trinity College. His scientific work took place at the Cavendish Laboratory and at Cambridge University's Zoology Laboratory, where he worked for the Medical Research Council between 1947 and 1949. In 1948, when the steady-state theory appeared, Gold was completely unknown in the astronomical community, and his only publications dealt with a subject which all astronomers and most physicists would have found peripheral—the physics of the inner ear.[109] Gold analyzed the working of the ear from the perspective of signal processing and the electronic techniques that he had become thoroughly familiar with in his work with radar during the war. This was a perspective uncommon to the zoologists and medical doctors traditionally occupied with the subject. As Gold recalled forty years later: "I had to address myself to an audience of otologists—the doctors and the medical people who deal with hearing—the only ones who were doing any kind of research in this field. The mismatch was obvious: it was completely hopeless. There was no common language, and of course the medical profession just would not learn what it would take to understand the subject. On the other hand, they sure made their judgments about the matter, without having any basis at all. So it just essentially forced me out of the field."[110] If medical doctors rejected Gold's physicalistic model of hearing, physicists just ignored it. Appearing in the biological section of the *Proceedings of the Royal Society*, Gold's work was probably better known to physiologists than to physicists. Indeed, whereas it was abstracted in the *Psychological Abstracts*, one looks for it in vain in the *Physics Abstracts*.

When Bondi returned to Cambridge in the summer of 1945, he still had his research fellowship and was appointed an assistant lecturer by the Faculty of Mathematics. At that time Bondi and Gold were still stateless subjects, but by the end of 1946 they both became naturalized as British citizens. During the years Bondi worked for the Admiralty, one of his duties had been to develop a satisfactory theory for the functioning of the magnetron, the special vacuum tube which was the central component of the radar sender. This work, and the discussions with Hoyle, drew him into astrophysics and inspired him to make a detailed study of the dynamics of the accretion processes which had previously been introduced by Hoyle and Lyttleton. The work earned him a fellowship at Trinity in 1943 and resulted in what he later judged to be his first substantial paper. The paper, written jointly with Hoyle, was a mathematically rigorous investigation of how stars accrete interstellar matter, including a calculation of the accretion rate in an idealized case.[111] Other papers between 1944 and 1948 dealt with hydrodynamics, the dynamical theory of the earth's core (with Lyttleton), and the accretion theory applied to the sun (with Lyttleton and Hoyle). In 1948 Bondi was appointed a university lecturer, which is a permanent position. He had at that time made his entry into theoretical astronomy and made himself known as a promising applied mathematician.

Hoyle continued his work in various branches of astrophysics and theoretical astronomy, either on his own or in collaboration with Lyttleton and Bondi. One of his important works was an investigation of the formation of heavy elements in stars by means of thermonuclear reactions taking place at extremely high temperatures, a line of investigation spurred by his meeting with Baade in 1944.[112] Basing his work on statistical methods rather than on actual nuclear processes, Hoyle found values of temperature and density that gave the required abundances of the elements after cooling and expansion. He suggested that the necessary, very high temperatures and densities indicated supernovas as the sources of element formation. This work became well known and was often referred to by Alpher, Herman, and the other American physicists who at the time were working out the foundation of big-bang cosmology.

Although Hoyle had been elected a fellow of the Royal Astronomical Society as early as 1940, on the suggestion of Eddington, he was widely seen as a physicist intruding in astronomy and not as a real astronomer. His style, impatience, drive, and prolific rate of publishing were not appreciated by all astronomers. When his and Lyttleton's first paper was submitted to the Society's *Monthly Notices*, it had to be withdrawn after discussions with the council, which failed to persuade the two authors to change their manuscript according to the suggestions of the referees. If Gold had a reputation of being difficult, so had Hoyle. According to the historian of the Royal Astronomical Society, the episode "led to an unfortunate lack of confidence between the authors and the Council that was to endure for several years."[113] After the war, when Hoyle's productivity peaked, he had difficulties getting his papers published as quickly and smoothly as he wanted. Many astronomers within the society felt that his papers were too many and too speculative, and that he was not careful enough in acknowledging his debts to other authors. It was only in about 1947, when Hoyle was elected to the council and William McCrea had become its secretary (1946–49), that the situation improved. But even then relations between Hoyle and the astronomical community were somewhat tense, and Hoyle's reputation as a maverick did not substantially change.

There may have been good reasons for the skeptical and sometimes hostile attitude that some astronomers held regarding Hoyle. With Hoyle's ego and uncompromising ambitions, it is understandable that he was not universally popular. What matters here is that the tensions existed. Hoyle himself felt that his career was deliberately impeded by conservative astronomers afraid of losing their power and status and unable to appreciate and understand theoretical astrophysics. The feeling only strengthened with the advent of steady-state cosmology, which Hoyle felt was treated unfairly by many astronomers. The distrust between Hoyle and the astronomical community would be an important theme in the history of the steady-state theory.

The close relationship that had existed during the war between Hoyle, Bondi, and Gold continued after 1945. Hoyle could not afford to buy or rent a house in Cambridge and therefore lived with his family in the country, south of the town. Since Bondi had rooms in Trinity College, Hoyle spent a good deal of his time there. When Bondi married in the fall of 1947, he and his wife moved to a flat near Trinity and Hoyle continued his frequent visits, often joined by Gold. Even Bondi's marriage remained within the closed Cambridge circle; his wife, Christine Stockman, was an astrophysics research student, supervised by Hoyle. It was in these circumstances that the steady-state theory germinated. Until then, cosmology had never been an important part of the freewheeling discussions between the three physicists, none of whom had any particular knowledge of or interest in the field. None of their publications until 1947 dealt with cosmological matters. After 1945 Bondi had become interested in the general theory of relativity, and about 1947 he was asked by the Royal Astronomical Society to prepare a review article on the state of cosmology. It is perhaps surprising that Bondi, who had little previous experience with the subject, was requested to write the review (and not, for example, McVittie, McCrea, Whitrow, Temple, or Walker—Great Britain had no shortage of theoretical cosmologists). At any rate, Bondi accepted and started reading about cosmology. At that time, 1946 or early 1947, neither Bondi, Gold, nor Hoyle seems to have been acquainted with even the most important literature in the field. For example, it was only when preparing for the review that Bondi came across H. P. Robertson's seminal review article of relativistic cosmology from 1933, and it was also only then that Hoyle studied the article.[114] Another work that Bondi studied and benefited from was Otto Heckmann's book of 1942. This was the most thorough treatment of cosmology at the time, but, being published in German during the war, it was rare and not well known among British astronomers.[115] At that time cosmology began seriously to occupy the minds of the three physicists and become an increasingly important part of their discussions. Bondi demonstrated his newly acquired mastery of the technical details of relativistic cosmology in an important paper on inhomogeneous solutions to Einstein's field equations, completed in the summer of 1947.[116]

At the time Bondi's review appeared in *Monthly Notices*, he and his two colleagues had decided that the best model for the universe was probably a stationary world with continuous creation of matter. Although the idea of a steady-state universe was not mentioned in the article, Bondi's work reflected the state of thought which characterized the slightly later steady-state theory. The review paper can in some respects be seen as a methodological prologue to the steady-state theory.[117] Considering the fact that Bondi was an expert mathematician, the qualitative and philosophical nature of the paper is remarkable. It contained only very few equations (five, to be exact) and was, even in a British context, unusually critical with respect to the methodological foundation of cosmology. On the other hand, it also followed a tradition in British cosmology in which qualitative and philosophical considerations were given high priority. Whether supporting Milne's kind of cosmology or not, British workers in cosmology generally agreed that the study of the universe had to be based on certain a priori principles and could never become an ordinary empirical science. As George Temple, a theoretical physicist at King's College, summarized in a review of 1939: "A purely empirical basis is wholly inadequate for cosmological studies. The interpretation of observations necessarily requires conceptual and speculative elements."[118] Bondi's review after the war confirmed this view, which was also taken over into the steady-state theory the same year. It was, of course, a view which did not appeal to pragmatists (such as Gamow) and positivists (such as Jordan), but in England it enjoyed general acceptance. It remained a subject of discussion among British cosmologists well into the 1960s.

Of direct relevance to the later steady-state theory was Bondi's introduction, in which he distinguished between two different approaches to the study of cosmology, which he called the "extrapolatory" and the "deductive" approaches. In the first approach, characteristic of relativistic cosmology, the physicist extrapolates ordinary terrestrial physics to form a comprehensive theory (the general theory of relativity), which is then assumed to be valid for the entire universe; the consequences of this assumption are then compared with measurements. The alternative, which Bondi saw realized in Milne's kinematic relativity, is to start from a small number of cosmological axioms or postulates and from these to deduce the corresponding physical theories.

Bondi did not comment on the relative merits of the two approaches, but emphasized that, whatever the approach, any cosmological theory had to account for one very fundamental assumption, viz., the repeatability of experiments. "Experimental science is based on the assumption that the repetition of an experiment will reproduce the original results, and indeed the realm of experimental science is defined by this criterion," Bondi wrote in his introduction. "Our assumption of repeatability is meaningless unless we make the hypothesis that the place and time of performing an experiment have no direct influence on its outcome. It is the fundamental assumption of all experimental science on which the structure of cosmology rests and which, as we shall see, in turn requires a suitable cosmology." After reflecting on this theme he formulated what he called the homogeneity assumption, which is a version of the cosmological principle. In Bondi's formulation: "An observer situated in a nebula and moving with the nebula will observe the same properties of the universe as any other similarly situated observer at any time."[119] But, as Bondi noticed, the only stable relativistic world model which is both homogeneous in space and stationary in time is the de Sitter model, and this model does not describe the real universe, since it assumes a world in which there is neither matter nor radiation. Within the framework of relativistic cosmology, the stationary condition must therefore be dropped, leading to the ordinary cosmological principle, which agrees with the expansion of the universe inferred from measurements. Viewed with hindsight, Bondi's formulation comes close indeed to the perfect cosmological principle, but in his paper it was not presented as such, but merely as the old cosmological principle associated with the static models of the universe of the 1920s.

4.3 EMERGENCE OF THE STEADY-STATE THEORY

Motives and Inspirations

During the discussions Hoyle, Bondi, and Gold had about cosmology, they agreed that relativistic cosmology was not as solidly founded as most astronomers seemed to believe, and eventually they were faced with the possibility of an unchanging, expanding universe with creation of matter, that is, the steady-state universe. Exactly how and when this idea turned up is unknown, but it seems certain that the basic idea came from Gold, and that at first it was meant as nothing but an academic possibility. Hoyle and Bondi found it an interesting, but not very serious, suggestion and believed that they could easily disprove it. As Bondi recalled: "Fred and I said: 'Ach, we will disprove this before dinner.' Dinner was a little late that night, and before very long we all saw that this was a perfectly possible solution to the question."[120] When they discovered that the steady-state scenario seemed able to accommodate all the observations they could think of, they became really interested in investigating the suggestion. They now concentrated on developing the consequences of what at this stage was merely a loose idea. According to Hoyle's recollections, the idea of a dynamic yet stationary universe may have originated when the three physicists in 1946 went to the cinema. The movie was

> a ghost-story film, which had four separate parts linked ingeniously together in such a way that the film became circular, its end the same as the beginning. I have not been able to trace the name of the film but, drawing on a remote corner of my memory, I think it was called *The Dead of Night*. Tommy Gold was much taken with it and later that evening he remarked, "How if the universe is constructed like that?"

One tends to think of unchanging situations as being necessarily static. What the ghost-story film did sharply for all three of us was to remove this wrong notion. One can have unchanging situations that are dynamic, as for instance a smoothly flowing river. The universe had to be dynamic, since Hubble's red-shift law proved it to be so, but if the universe could be unchangingly dynamic, . . . From this position it did not take us long to see that there would need to be a continuous creation of matter.[121]

The charming story is hardly to be taken very seriously, although, like all origin myths, it undoubtedly has some foundation in fact. Bondi and Gold remembered the movie, but did not believe it had any connection with the origin of the steady-state idea. Moreover, whereas Hoyle placed the origin in 1946, Bondi's memory was that it was well into 1947 before the idea was seriously discussed.[122] Even less credible as an origin event is the story, again recounted by Hoyle, that a search for a lost item influenced the cosmological thoughts of Bondi and himself. The two were drinking tea in front of the fire. Then, "One of us dropped a small article, I have forgotten exactly what. I think it was a screw or a nail, but it may have been a pencil sharpener or some such item. It was of no particular consequence in itself, but at first somewhat idly we looked around to retrieve it. When at first we failed we thought little of it and simply started a more intensive search. We worked at it for over an hour, finally more or less dismantling the whole room. But the article was simply not to be found. A simple time reversal of this situation leads to the creation of matter."[123] The incident, which Hoyle probably only meant as a good story, would not be worth mentioning had it not been taken seriously by the American sociologist of science Lewis Feuer in a paper of 1977, in which he views it in the light of psychoanalysis.[124]

There is, in fact, no reason to have recourse to such simplistic and colorful origin myths, for the emergence of steady-state theory can well be accounted for without the help of external inspirations. During the discussions of cosmology that took place in Bondi's flat in Cambridge, he, Gold, and Hoyle soon decided that the standard evolutionary theory was not satisfactory. It is difficult to decide with certainty what were their objections, and to distinguish between the motives that stimulated the discovery of the steady-state alternative from the reasons subsequently presented as justification. From the recollections of the authors and their original papers on the subject, it seems reasonable to assume that at least four objections to the evolutionary theory contributed to their dissatisfaction. Two of these were concerned with the status of the laws of physics, and two with the nature of relativistic cosmology. None of them were entirely original.

The widely accepted finite-age expansion cosmologies of the Friedmann-Lemaître type operated with one initial creation of matter, a big bang. The creation event itself was not discussed, and it was generally supposed that it was beyond physical explanation. If the laws of physics only came into existence with or shortly after the event, how could it possibly be explained? Hoyle, in particular, disliked the notion of a singular creation event which forever would be beyond scientific understanding. Furthermore, from Mach's principle one would expect the laws and constants of physics to be influenced by the cosmic environment, and in an evolving universe this would seem to imply that the laws changed in time; but then, as Bondi stressed in his review article of 1948, there would be no good reason for the repeatability of experiments, the very basis of experimental science. The time scale difficulty, generally admitted as a weakness of the relativistic evolution models, undoubtedly played a role too. The difficulty could be avoided, as it was in the Lemaître-Eddington and Lemaître models, for example; but these were solutions which did not appeal to Bondi, Gold, and Hoyle, who considered them to be "a bit of a fiddle" and who developed "a considerable dislike" for the models because they depended on fine-tunings of the cosmological constant.[125] They were struck by the lack of uniqueness in the relativistic expansion models and felt repelled by the fact that these models could accommodate almost any observation, and hence had little real predictive power. The dissatisfaction with the logical structure of relativistic cosmology seems to have been an important motive for Bondi and Gold in particular.

The steady-state theory was not, as is sometimes claimed, proposed just in order to avoid the time scale difficulty. If this had been the case, it would have been an ad hoc hypothesis. But although the difficulty was not of primary importance in the formation of the theory, naturally it played a role, and in the subsequent debate it entered as an important element. As we have seen, in 1948 it was generally realized that the age of the solar system compared with the Hubble time constituted a real problem for relativistic cosmology. However, most astronomers did not find the time scale difficulty serious enough to justify abandoning relativistic models. The three Cambridge physicists' skeptical attitude to relativistic cosmology made them view the time scale difficulty as more grave than those who were committed to the relativistic cause. "This discrepancy," wrote Bondi in his review paper of early 1948, "is very real and awkward, although there are various means of avoiding it."[126] He referred to the introduction of the cosmological constant, as used by Lemaître, Eddington, and Gamow, but this was a use he found thoroughly objectionable from a methodological point of view. When he and Gold published their steady-state theory later the same year, they did not fail to point out that the time scale difficulty was absent in their theory, but remained in (almost) all relativistic models of the expanding universe: "In our theory there is . . . no difficulty whatever in taking the age of our galaxy to be anything indicated by local observations (such as $5 - 8 \times 10^9$ years), although T [the Hubble time] is much shorter."[127] A passage to the same effect appeared in Hoyle's first publication on the steady-state theory.[128]

The Theory Takes Shape

The three having agreed upon the unsatisfactory state of affairs in orthodox relativistic cosmology, Gold came up with his idea of continual creation of matter as a means to keep the universe stationary in spite of its expansion. This may have been in the winter of 1946–47. The idea of matter creation was not new at the time, and one may ask to what extent Gold and his two friends were inspired by earlier authors on this topic. There is no reason to assume that they were inspired by, or had even heard of, the steady-state versions of Nernst, MacMillan, and Millikan. These speculations belonged to the past and were largely forgotten in 1947. Even in the 1930s they were considered unorthodox and ignored by most cosmologists. Neither is there any reason to assume that Bondi, Gold, or Hoyle was acquainted with the 1939 work of Schrödinger or with the speculations of Lodge and Kapp. On the other hand, they knew about Jeans's remark quoted in section 4.1 and cited by Hoyle in his 1948 paper on steady-state cosmology.[129] They must also have come across Milne's remark about matter creation in his paper of 1946, but, brief and linked to Milne's theory of kinematic relativity as it was, it hardly left much of an impression. Dirac's (and Jordan's) views are more likely to have been of some influence, but with regard to matter creation it is again unlikely that the idea came from Dirac. Although he had introduced the idea in his note of 1937, in Dirac's only major work on cosmology, the one of 1938, matter was assumed to be conserved.

On 20 December 1946, slightly before or at about the time that Gold came up with his suggestion of continual creation of matter, Hoyle gave a paper at a meeting in Birmingham before the Physical Society of London. The subject of the talk was "On the Formation of Heavy Elements in Stars," a topic Hoyle was working on at the time, and on which he would later publish a paper with the same title. During the discussion he was asked by Peierls, his former supervisor, how certain it was that hydrogen was the primary element out of which the other elements were formed. And if hydrogen were the primary element, Peierls continued, then where did the hydrogen come from?[130] Hoyle could not answer the question, of course, but it triggered a critical phase in his thinking, and presumably made him more receptive to Gold's idea, whether it was suggested before or after the Birmingham meeting. In a general way, Hoyle's earlier work, partly done in collaboration with Lyttleton, was conducive to the steady-state idea. Their work on the formation of planets, stars, and galaxies was indirectly based on a self-perpetuating universe, and they had never felt any need to consider a primeval universe radically different from the present one. Hoyle's work on the formation of elements in stars and novae can only have strengthened his belief that no such hypothesis was needed. Lyttleton was very much part of the Hoyle-Bondi-Gold circle, but he played no role in the discussions leading to the steady-state idea. All the same, Lyttleton was at the time considered to be involved in the new cosmology, which sometimes was referred to as the Hoyle-Lyttleton-Bondi-Gold universe.[131]

Although it seems futile to search for direct influences, in a general and indirect sense Bondi, Gold, and Hoyle were undoubtedly influenced by earlier and contemporary workers. Whereas mainstream relativist cosmologists such as Robertson, Tolman, and Lemaître mainly acted as (mostly anonymous) targets of criticism, the views of Milne and Dirac influenced to some extent the way in which the steady-state theory was formed. According to Hoyle,

Dirac's large-number hypothesis was "an important thread" for the three in their cosmological discussions.[132] In most respects, the steady-state model contrasted sharply with Dirac's cosmology, in which not only the universe but also the gravitational constant changes in time. Yet the large-number hypothesis was considered important by many steady-state advocates. For example, it occupied a prominent position in McCrea's exposition of 1950, in which he found it favorable to the steady-state theory that it could be brought into conformance with the large-number hypothesis.[133]

At the time when the steady-state theory appeared, Milne was almost alone in defending his system of kinematic relativity. The steady-state theoreticians did not accept Milne's cosmology, which they found to be artificial and too far removed from experimental testing. Yet Bondi and Gold recognized Milne's system to represent some of the methodological qualities they thought were important in cosmology. This relatively positive evaluation was Bondi's rather than Gold's, however. Gold did not know Milne's theory well and what he knew did not appeal to him. In particular, he disliked the idea of the two time scales.[134] In his influential textbook of 1952, Bondi praised kinematic relativity from a methodological point of view, but the praise went to the theory as an abstract discipline, not as a valid description of the real universe. In Bondi's view, cosmology should not be regarded either "as a minor branch of general relativity or as a branch of philosophy and logic."[135] The last reference was presumably to Milne's system. Yet Bondi also admired this system, which, he stated in 1948, "certainly deserves permanent credit for showing what a powerful tool kinematics can be and how far-reaching are the implications of a cosmological principle."[136] What appealed to Bondi (but not to Hoyle) was the deductive nature of Milne's theory, and his insistence that cosmology was not a branch of general relativity. Milne's construction of a cosmological theory based on definite principles rather than on an extrapolation of terrestrial physics worked to some extent as a methodological model for Bondi and Gold.

The view that Hoyle, Bondi, and Gold had arrived at in the spring of 1947 was merely a loose idea that the universe might possibly be in a steady state, with creation of matter providing a constant density. In order to formulate it into a publishable theory it needed to be developed quantitatively and supplied with arguments of empirical support. This caused a dead stop for most of a year. Hoyle's problem was the matter creation, which seemed to imply violation of energy conservation; since Hoyle wanted to formulate a field-theoretical version in close conformance with the general theory of relativity, this blocked progress for some time. Moreover, during 1947 Hoyle was occupied with many other matters besides cosmology. One of them was solar physics, on which subject he was commissioned to write a book in the authoritative Cambridge Monographs series. The work, the first of many books from Hoyle's pen, kept him busy for some months.[137]

Hoyle's solution to the cosmological creation problem, which came in late 1947, was that with a proper definition of energy conservation, the disagreement between general relativity and steady-state creation of matter could be avoided. The ideas of Bondi and Gold followed a different route from those of Hoyle. They were not interested in formulating a steady-state theory in the field-theoretical version desired by Hoyle, and had no objections against accepting energy nonconservation on a cosmic scale. Their problem was to deduce from their general assumptions quantitative results of an empirical nature. They were not happy at all about Hoyle's approach and felt that they "were left a little high and dry." But about December 1947 Bondi discovered that it was possible to account very well for the number counts of galaxies on his and Gold's theory. And when he also realized that the geometry of the steady-state space had to be a de Sitter metric, the road was open for writing a paper. At that time Hoyle worked independently of Bondi and Gold and there seems to have been a certain rivalry between them. In February 1948 Hoyle completed in a rush a first version of his paper, which, he recalls, was finished at 2:30 A.M. He wanted to add Gold as a coauthor, but Gold declined and explained that he would rather go ahead with Bondi. When Hoyle's paper appeared it was under his name only, but with full acknowledgment of Gold as the originator of the steady-state idea.

Realizing that Hoyle was about to finish his paper, Gold and Bondi speeded up their efforts. However, they could not complete their work before Hoyle, who had already presented his paper at a seminar in the Cavendish on 1 March 1948, in front of notables such as Dirac and Heisenberg. A couple of days later he submitted it to the *Proceedings of the Physical Society*, but publication proved more difficult than Hoyle had thought. The manuscript was rejected by the Physical Society, whose secretary replied in courteous words: "[We] have now regretfully decided that the Proceedings is not the most suitable medium of publication, especially in view of the acute shortage of paper which is forcing us to reject papers we would otherwise be glad to publish. The Committee suggests that you submit the paper to the Royal Astronomical Society."[138] A somewhat offended Hoyle (who did not accept the excuse of shortage of paper) then sent the manuscript to the *Physical Review* in the United States. However, the editors of this journal wanted it so drastically reduced in size that Hoyle declined and withdrew the work. Only on the third try did he succeed, when he submitted it to the *Monthly Notices* about 1 August. Hoyle had apparently wanted to avoid this journal, perhaps fearing that astronomers would be less appreciative of his work than physicists, and also recalling the troubles he had previously had with this journal in obtaining rapid publication. If this was the case, he was wrong, for the Royal Astronomical Society gave the paper a quick and positive response. This was due in large measure to the society's secretary, McCrea, who from the very beginning was favorably inclined to the steady-state theory. Because of the delay of Hoyle's paper, the one of Bondi and Gold appeared shortly earlier. It was received by the *Monthly Notices* on 14 July, three weeks before Hoyle's. As a secretary McCrea was responsible for the society's publications, and he decided to referee the Bondi-Gold paper himself. McCrea was enthusiastic and had the paper published with no comments. Bondi and Gold knew Hoyle's paper, which he had shown them in March, and they referred to it in the belief that it would be published before theirs.

Even before the appearance in print of the steady-state theory, it was known to many people. As mentioned, Hoyle had lectured on his version in March 1948, and on 15 June Gold gave a talk to the Kapitza Club on "A Stationary Universe."[139] In the summer of 1948, Hoyle and Bondi and his wife participated in the Seventh General Assembly of the International Astronomical Union in Zurich. The meeting, which took place 11–18 August, was the first general assembly of the International Astronomical Union after the war and was attended by about four hundred astronomers from thirty-one countries. Among these were leading astrophysicists and cosmologists, including Albrecht Unsöld from Germany, Oskar Klein from Sweden, and Lemaître from Belgium. Encouraged by McCrea (who already knew the theory), Bondi talked to a number of senior astronomers about his and Gold's theory, including the distinguished Swedish astronomer Bertil Lindblad, the new president of the International Astronomical Union. Lindblad found the idea interesting and worthy of exploration, a response that encouraged Bondi. Most likely, Hoyle also talked informally to some of the participants about his version of the steady-state theory.

4.4 Two Steady-State Papers

The Hoyle Version

Hoyle presented the aim of his paper as formulating a cosmology which avoided the various difficulties of standard relativistic theories; and to do so without leaving the framework of general relativity and without making use of the cosmological constant. Among the unsatisfactory features of the ordinary models, he mentioned the time scale problem and "aesthetic objections" to the notion of a universe created in the past. "For it is against the spirit of scientific enquiry to regard observable effects as arising from 'causes unknown to science', and this is in principle what creation-in-the-past implies."[140] Although Hoyle introduced from the very beginning what he called the "wide cosmological principle," which was his name for Bondi and Gold's perfect cosmological principle, he was careful not to introduce either this or the idea of matter creation axiomatically. They were concepts following as consequences of his theory, not presuppositions for it.

The ordinary field equations of general relativity entail energy conservation and so cannot be used to describe a steady-state universe. In order to make the equations accommodate creation of matter, Hoyle replaced the cosmological term Λg_{mn} with a symmetric tensor of nonvanishing divergence, the creation tensor C_{mn}. He thus arrived at the alternative equations

$$R_{mn} - \frac{1}{2} g_{mn} R + C_{mn} = -\kappa T_{mn}. \tag{4.1}$$

Like Bondi and Gold, Hoyle used the word creation rather than more neutral terms such as formation or origin. They did so with no ulterior motive, just to express the existence of matter where none had been before, but it soon turned out that creation was an explosive term which invited much controversy. Hoyle found creation an appropriate term because pair creation, where pairs of particles and antiparticles are formed by radiation, was a term already in widespread use among quantum physicists.[141]

The effect of the creation tensor was spontaneously to create matter, which might possibly be generated in the form of neutrons, Hoyle guessed. Associated with the creation tensor was a vector field parallel to a geodesic at each point of the homogeneous and isotropically expanding universe. The field was written

$$C_m = \frac{3c}{a} (1, 0, 0, 0), \tag{4.2}$$

where a is a constant. Hoyle showed that the solution of the field equations (4.1) would be given by a metric of the form $ds^2 = c^2 dt^2 - R^2(t)(dx^2 + dy^2 + dz^2)$, i.e., a space of zero curvature. The distance function $R(t)$ satisfies the differential equations $2RR'' + R'^2 + 3c/a\ RR' = 0$ and $3R'^2 = \kappa\rho c^2 R^2$, where a is the same constant as in (4.2) and ρ is the average density of matter. If, in these equations, the zero of time is chosen so that $R = 1$ for $t = 0$, it follows that $R(t) = \exp(ct/a)$ and

$$3/a^2 = \kappa\rho. \tag{4.3}$$

In polar coordinates the metric can then be written

$$ds^2 = c^2 dt^2 - \exp(2ct/a)[dr^2 + r^2 (d\theta^2 + \sin^2\theta\ d\varphi^2)].$$

This is a metric of the de Sitter type, formally with the constant a replacing the quantity $\sqrt{3/\Lambda}$. However, contrary to de Sitter's original metric, the steady-state metric is valid for a universe filled with mass. As to the physical meaning of the constant a, it is related to the mass density by equation (4.3). Hoyle showed that a light signal emitted from a distance larger than a would never reach the observer, meaning that the quantity could be interpreted as the radius of the observable universe.

It follows from equation (4.3) that the density is independent of the time, and that this is also the case with the mass of the observable universe $M_{obs} = 4/3\ \pi a^3 \rho = 4\pi a/\kappa$. The matter which is created by the presence of the tensor field C_{mn} compensates for the loss of matter passing out of the observable universe. Hoyle showed that in this model there is a redshift corresponding to the recessional velocity

$$v = r\frac{c}{a} \exp\left(\frac{ct}{a}\right) = r R(t)\frac{c}{a},$$

where $rR(t)$ is the absolute distance from the galaxy to the earth. By comparison with the Hubble relation, he was thus led to the identification

$$c/a = H \text{ or } a = c/H = cT, \tag{4.4}$$

where H is now a true constant, the Hubble constant inferred from the redshift observations. With this identification, Hoyle had deduced a definite value for the density of matter in the universe, $\rho = 3H^2/\kappa c^2 = 3H^2/8\pi G$, which follows from equations (4.3) and (4.4). Estimating the numerical values of the quantities appearing in his model, he obtained $a \approx 1.8 \times 10^{27}$ cm, $\rho \approx 5 \times 10^{-28}$ g·cm^{-3}, and $M_{obs} \approx 1.2 \times 10^{55}$ g. The predicted mass density was thus much larger than the mean density of stars and galaxies estimated from observations (about 5×10^{-31} g·cm^{-3}). Hoyle did not consider the discrepancy a problem for the theory. It just showed, he argued, that only one-thousandth of the cosmic matter was in a condensed state. In any case, the small density value was not a problem for the steady-state theory specifically. The mean density of the universe predicted by the Einstein–de Sitter model happens to be exactly the same as in the steady-state model.

As in the earlier cosmologies of Nernst, MacMillan, and Millikan, there was no heat death in Hoyle's model. He argued that although the entropy increases locally, the creation of matter prevents a global increase of entropy towards a maximum value. The influence of numerological considerations à la Eddington and Dirac was evident in the last part of the paper, in which Hoyle observed that with k denoting the range of nuclear forces the dimensionless combination $a/3k$ would be of the order of magnitude 10^{39}. Since this is a large number in Dirac's sense, it suggested the following invariant relationship between gravitational, electromagnetic, and nuclear constants:

$$\frac{a}{3k} \approx \frac{e^2}{GmM}.$$

Hoyle belived that this might be of deep significance and that it was a further argument in favor of his theory. A similar positive evaluation of the large-number coincidences was contained also in the work of Bondi and Gold, who found it satisfactory that according to the steady-state theory, such coincidences were permanent and not (as in Dirac's theory) the accidental result of a certain phase of the cosmic history. The enduring influence of Dirac's large-number philosophy on the views of Hoyle, in particular, is evident from the many publications over the next decades in which Hoyle returned to the significance of very large dimensionless numbers.

The Bondi-Gold Version

In its style and approach, the paper written by Bondi and Gold was very different from Hoyle's. It started with a lengthy discussion of repeatability in science, very much along the same lines as in Bondi's earlier review paper. Indeed, the opening sentences of the two papers were almost verbatim the same. The point of the discussion was that it is only in an unchanging universe that we can safely assume the laws of physics to be constant in time. The cosmological principle justifies such constancy with respect to changes of place, but there is no rationally justified basis for the assumption that the laws in the denser state of the past are the same as those today. According to Mach's principle, one would expect that drastic changes in the density of the universe would also produce changes in the laws and constants of nature. Therefore, argued Bondi and Gold, if Mach's principle is admitted and if the laws of physics have to be constant—and this is necessary if experiments are to be repeatable—the universe must be unchanging. "We regard the reasons for pursuing this possibility as very compelling, for it is only in such a universe that there is any basis for the assumption that the laws of physics are constant; and without such an assumption our knowledge, derived virtually at one instant of time, must be quite inadequate for an interpretation of the universe and the dependence of its laws on its structure, and hence inadequate for any extrapolation into the future or the past."[142] Combining the "stationary postulate" with the restricted cosmological principle, they proposed what they called the perfect cosmological principle, namely, that "the universe . . . [is] not only homogeneous but also unchanging on the large scale." The term "perfect cosmological principle" was Gold's invention. Bondi at first found it presumptous to call it perfect, but was persuaded by Gold that it was a good name after all.[143]

It is to be emphasized that Bondi and Gold arrived at the perfect cosmological principle by purely philosophical arguments, and that they posed it as a postulate or fundamental hypothesis. On the other hand, although they would soon be accused of having introduced it as an a priori principle, they did not consider it as true by necessity: "We do not claim that this principle must be true, but we say that if it does not hold, one's choice of the variability of the physical laws becomes so wide that cosmology is no longer a science. One can then no longer use laboratory physics without relying on some arbitrary principle for their extrapolation. . . . this is the only assumption on the basis of which progress is possible without further hypothesis." In other words, relativistic cosmology is not scientific—or not as scientific as the alternative—because it rests on arbitrary assumptions with regard to the extrapolation of laboratory physics. Bondi and Gold evidently had complete confidence in the power and necessity of their perfect cosmological principle: "We regard the principle as of such fundamental importance that we shall be willing if necessary to reject theoretical extrapolations from experimental results if they conflict with the perfect cosmological principle even if the theories concerned are generally accepted. Of course we shall never disregard any direct observational or experimental evidence and we shall see that we can easily satisfy all such requirements."

The foundation of Bondi and Gold's steady-state cosmology was the perfect cosmological principle, but this principle alone was recognized to be insufficient to construct a cosmological model. This is shown by the fact that a static universe such as the Einstein model of 1917 also satisfies the perfect cosmological principle. However, not only was a static universe ruled out by observations of the redshifts, it would also be in a state of thermodynamic heat death with no evolution and direction of time. And, as Bondi and Gold argued, it is clear "from local scale physics and indeed from our very existence" that this is not the case. They therefore concluded that the perfect cosmological principle yields a stationary *and* expanding universe. Such a solution is only possible if matter is continually created. Matter creation was thus introduced by Bondi and Gold as a consequence of the perfect cosmological principle, which also provided an estimate of the rate of creation, namely, about 10^{-43} g·s^{-1}·cm^{-3}. The value follows from considering a volume of linear dimension R filled with matter of constant density ρ. Because of the expansion, we have, in comoving coordinates, that $R^3(t) = R^3(0) \exp(3Ht)$, where $t = 0$ is an arbitrary time before t. If the mass density is conserved, the mass in the universe must follow its expansion, i.e., $m(t) = m(0) \exp(3Ht)$, and matter is thus being created at the

$$\text{creation rate} = \frac{dm}{dt} \Big/ R^3(t) = 3H\frac{m(0)}{V(0)} = 3\rho H.$$

Inserting the values of ρ and H accepted at the time, the numerical value of the creation rate comes out. This is, of course, an exceedingly small value, which makes it impossible to detect the process in any direct way. It corresponds to the formation of three new atoms of hydrogen per cubic meter per million years; or, as Hoyle put it more picturesquely, it is "no more than the creation of one atom in the course of about a year in a volume equal to that of a moderate-sized skyscraper."[144] With the smaller value of the Hubble constant that came to be accepted in the 1950s, the predicted creation rate changed too, but its exact value was of little importance. Between 1950 and 1955, the Hubble time increased from about 1.8 to 7 billion years, corresponding to a decrease in the creation rate by a factor of about 4. This, and a changed estimate of the density, made Hoyle modify his illustration of the smallness of the creation rate to be "about one atom every century in a volume equal to the Empire State Building."[145]

The new matter was supposed to be created randomly throughout space and not mainly in the stars (additively and not multiplicatively, to use a later terminology). This hypothesis was also included in Hoyle's theory and was accepted as the only natural one during most of the history of steady-state cosmology. Only in 1964 did McCrea suggest the possibility of multiplicative creation of matter, an idea which was also developed by Hoyle and Narlikar at the same time (see section 7.3). In what form did the new matter appear? Bondi and Gold suggested that it would be as hydrogen atoms although they, like Hoyle, admitted that this was merely "for simplicity and definiteness." None of the steady-state versions predicted the nature of the new matter. In principle, it could just as well be large radioactive elements as hydrogen atoms. However, Bondi, Gold, and Hoyle argued that the newborn matter would have to be in a primordial form out of which heavier elements could be synthesized in stellar processes. Whether it was neutrons, protons and electrons, or hydrogen atoms, was less important at this stage of the theory.

What was important was the kinematic state of the created particles. The initial velocities of the particles imply a preferred direction everywhere in space-time and thus cannot be formulated in a Lorentz-invariant way.[146] This feature disagreed with orthodox relativity theory, but Bondi and Gold did not see the disagreement as damaging at all. After all, they did not share the belief that laboratory laws, such as expressed in the theory of relativity, might justifiably be extrapolated to the whole universe. In 1923, Herman Weyl had introduced the fundamental hypothesis that all world lines of galaxies are diverging, i.e., that they form a bundle of nonintersecting geodesics.[147] According to this postulate, generally accepted as a necessary ingredient of cosmology, all galaxies have a common origin in the distant past, and the system of world lines makes it possible to define a common cosmic time. The velocity field discussed qualitatively by Bondi and Gold was the same as the vector (4.2) in Hoyle's version, which satisfies the Weyl postulate.

Referring to Weyl's postulate and the existence of a cosmic time, Bondi and Gold wrote: "General relativity demands that the laws of cosmology should be invariant while admitting that the one and only application is not invariant. We can see no reason why the laws of nature determining the structure of our universe should be invariant, although the universe is unique and does not bear an invariant aspect." The same willingness to question orthodoxy turned up in their discussion of matter conservation (or energy-matter conservation, as it really is). The steady-state universe did not satisfy this fundamental conservation law in its ordinary interpretation, the theorem of continuity, but why should it do so strictly? Do we know empirically that matter is conserved to any degree of precision? Of course not, was the reply of Bondi and Gold: "Hydrodynamic continuity is no doubt approximately true but this does not compel us to assume that it holds without any deviation whatever. In the conflict with another principle [the perfect cosmological principle] which is much more far-reaching and capable of making more statements about the nature of the universe and the applicability of physical laws, there is no reason for upholding the principle of continuity to an indefinite accuracy, far beyond experimental evidence."

The same argument was used by Gold in 1949, when he may have shocked quite a few physicists by claiming that if the mass conservation law clashed with steady-state theory, "this is regarded as a matter of complete indifference."[148] What mattered, wrote Bondi in 1952, was which principle led to the simplest description of nature. Adoption of metascientific criteria such as simplicity is usual in scientific arguments, but seldom so explicitly as in this case: "The principle resulting in greatest overall simplicity is then seen to be not the principle of conservation of matter but the perfect cosmological principle with its consequence of continual creation. From this point of view continual creation is the simplest and hence the most scientific extrapolation from the observations."[149]

Contrary to Hoyle, Bondi and Gold were very critical of the use of general relativity in cosmology, and unequivocally distanced themselves from all attempts to harmonize the steady-state model with the general theory of relativity. This does not mean that they were antirelativists in the sense that they did not accept Einstein's theory; they just questioned its unrestricted application to the universe at large. Their main objection was methodological, namely, the extrapolatory nature of relativistic cosmology, and the fact that the relativistic field equations are much too wide in the sense that they cover many more possibilities than actually exist. Relativistic cosmology was not really a theory, they objected, but a supermarket of theories which could only be applied to the one and only universe if supplied with more or less arbitrary assumptions and parameters adjusted by observations. All this made relativistic cosmology seem "utterly unsatisfactory" to Bondi and Gold, who, from a methodological point of view, rather preferred Milne's deductive approach.

Given the fact that Bondi and Gold deliberately formulated their theory in a broad and qualitative manner—their paper contained almost no equations— it is remarkable how many definite consequences they were able to derive from it. First, they concluded that the metric of the steady-state universe had to be of the de Sitter type, i.e., a flat space expanding exponentially. This followed from simple, semiqualitative reasoning. According to the perfect cosmological principle, the rate of creation of particles must be constant in time. The rate is affected by the spatial curvature, k/R^2, which then must also be constant; since $R = R(t)$, we must have $k = 0$, i.e., a flat, Euclidean three-space. (The four-space, or space-time, on the other hand, is curved.) Also, the Hubble parameter, $H = R'/R$, must be a constant according to the perfect cosmological principle. From $R' = HR$ it follows directly that $R = \exp(Ht)$. It is to be recalled that the Robertson-Walker metric and the meaning of parameters such as k and R do not rely specifically on assumptions of relativistic cosmology. Bondi, Gold, and Hoyle could therefore use the general metrical theory for their new model.

As a second deduction, Bondi and Gold considered the distribution and formation of galaxies. New galaxies are formed at all times at such a rate that the number of galaxies per unit proper volume remains constant in spite of the expansion of the universe. Although some galaxies are older than others, there will be no systematic difference in age between galaxies nearby and far away. In any large volume of space there will be old and young galaxies, the ages being distributed according to a certain statistical law which was obtained in 1950 (see below). Also, the average luminosity of galaxies is a constant, contrary to the situation in relativistic cosmology, where distant galaxies are assumed to have been in an earlier stage of development at the time they emitted the light we observe now. This difference was soon to receive much attention in connection with the so-called Stebbins-Whitford effect, which will be discussed in section 6.1. Bondi and Gold derived for the number of galaxies located between r and $r + dr$ the expression $4\pi r^2 n dr(1 + r/T)^{-3}$, where n is the constant number of galaxies per unit proper volume and T is the Hubble time. The expression differs from the corresponding one in relativistic cosmology, where the last factor does not occur. The difference reflects the fact that in the steady-state theory the galaxies are distributed uniformly with respect to proper volume and not, as in relativistic cosmology, with respect to coordinate volume.

Bondi and Gold further studied the distribution of galaxies with magnitude, using as a parameter the quantity $0.6m - \log N(m)$, where $N(m)$ is the number of galaxies with brightness greater than the apparent magnitude, m. They found that their model fitted nicely with observations, although not markedly better than relativistic cosmology. Since there were no observations which clearly favored the steady-state model, Bondi and Gold stressed again the methodological differences between the two models: "In general relativity a very wide range of models is available and the comparisons [between theory and observation] merely attempt to find which of these models fits the facts best. The number of free parameters is so much larger than the number of observational points that a fit certainly exists and not even all the parameters can be fixed. In our theory, on the other hand, the parameters are determined by observations of near nebulae. The counts of distant nebulae serve merely as a check on the theoretical forecast."

Bondi and Gold ended their paper with a comment on Hoyle's forthcoming field-theoretic formulation, which they rejected for much the same reasons that they rejected relativistic cosmology. "We feel that this formulation is unsatisfactory and unacceptable," they wrote. One of the reasons for their dissatisfaction was that neither standard relativistic theory nor Hoyle's theory accommodated Mach's principle, without which Bondi and Gold thought that no scientific cosmology could be established. But Bondi and Gold also recognized the force of a field-theoretic formulation; Hoyle had been able to derive the mean density of matter of the universe, and this important result could not be reproduced in the Bondi-Gold version. In fact, Hoyle's expression for the matter density can easily be derived from the perfect cosmological principle, but only by means of arguments based on Newtonian cosmology. This was shown by McCrea in 1950.[150]

4.5 ELABORATION AND INITIAL RESPONSE

Early Opposition

At the end of their paper, Bondi and Gold promised an improved formulation of their theory, but it never appeared. They tried to produce a field-theoretic version, based on the vector field defined by the velocities of newly created matter, but failed to do so.[151] The basic features of the steady-state theory, in its two versions, were established with the two papers of 1948 and did not change significantly over the next decade. During the early phase of the theory, a few new results were derived, but most of the development that took place consisted in more elaborate versions, alternative explanations of already derived results, attempts to clarify the status and meaning of the theory, and numerous discussions in order to defend it against criticism. Most of the technical development that did take place was within Hoyle's theory; because of its field-theoretic formulation, this theory was much better adapted for further development than the closed, deductive system of Bondi and Gold. Whereas the Bondi-Gold version and the Hoyle version were originally presented as two very different theories leading to the same results, the disagreement between the two versions was eventually seen as less important. Faced with increasing opposition, the two versions were often seen as merely different representations of the same theory, *the* steady-state theory. As Bondi said in March 1949, when faced with objections against the theory, "I think that the three of us should present a united front."[152] And so they did.

It was one of the great advantages of the steady-state theory that it predicted a definite age distribution of galaxies without needing a detailed theory of galaxy formation. The 1948 papers argued that such a distribution existed, but the specific distribution function was only supplied in 1950. The following derivation, based on work by McCrea, is a nice illustration of how important observational results follow in a simple way from steady-state assumptions.[153] Let n be the constant number of galaxies per unit volume and $nf(\alpha)$ the number of galaxies of age equal to or larger than a certain age α. At distance r from the earth the galaxies have a recessional speed Hr according to Hubble's law. If the time corresponding to r is $t + \tau$, at the earlier time t the galaxies will be closer to the earth; it follows from the exponential steady-state expansion that the galaxies will be at distance $r \exp(-H\tau)$. The number of galaxies of age $\geq \alpha$ within a sphere of radius r at time t is $4r^3 nf(\alpha)/3$ and must be the same at time $t + \tau$. At this later time the number consists of those galaxies which at time τ had an age $\geq \alpha - \tau$ and were within the smaller sphere of radius r $\exp(-H\tau)$. Therefore $4/3 \; r^3 \; nf(\alpha) = 4/3 \; r^3 nf(\alpha - \tau) \exp(-3H\tau)$, from which follows $f(\alpha) = f(\alpha - \tau) \exp(-3H\tau)$. Since $f(0) = 1$, the distribution function is simply $f(\alpha) = \exp(-3H\alpha)$. The number of galaxies per unit volume with an age in the interval $(\alpha | \alpha + d\alpha)$ is then $| ndf(\alpha) | = 3nH \exp(-3H\alpha) \; d\alpha$. From this it follows that the fraction of galaxies that are older than age t is

$$\frac{1}{n} \int_t^\infty 3nH \exp(-3H\alpha) d\alpha = \exp(-3Ht)$$

and that the average age is

$$\frac{1}{n} \int_0^\infty 3nH \exp(-3H\alpha) d\alpha = \frac{1}{3H} = T/3.$$

Thus, according to steady-state theory anno 1950, the average age of galaxies in any large region of the universe was predicted to be about 6×10^8 years.

In the fall of 1948, shortly after the appearance of the steady-state theory, Hoyle was invited to give a series of three lectures on the new theory before the Royal Institution of London. To most nonastronomers, the summary account of the lectures appearing in *Nature* in February 1949 was their first meeting with steady-state cosmology.[154] Among the difficulties that faced relativistic models with an origin in a point source, Hoyle mentioned not only the familiar time scale difficulty but also the problems of accounting for the formation of galaxies. In Lemaître's model the time scale problem does not arise, but Hoyle pointed out that in all point source models, Lemaître's included, the density of matter was once very high and thus would be expected to have resulted in condensations during the early stage of expansion. If big-bang models of the Lemaître type were correct, why were all galaxies not formed at the beginning of the expansion, when the density of matter was much higher than the mean density of galaxies observed today? The problem of formation of galaxies would soon move to the forefront of cosmology. It turned out to be a particularly difficult one within both types of cosmology.

In connection with his discussion of Lemaître's big-bang model, Hoyle made a comment which, in retrospect, is highly interesting. He argued that even a large amount of radiation energy in a hypothetical early universe would be unable to prevent the condensations of matter. Basing his estimate on the previously mentioned work done by Adams and McKellar, he found the present radiation energy density to correspond to a blackbody temperature about or slightly smaller than 1 K. How he got this temperature, rather than the 2.3 K reported by McKellar, is unclear. At any rate, the temperature of 1 K corresponds to an energy density much smaller than the present energy density of matter. Hoyle's argument is not of particular interest, but his reference to Adams and McKellar and his estimate of the present background temperature is. Recall that the two Australian astrophysicists had suggested a certain spectral line to be associated with a thermal molecular energy corresponding to an excitation temperature of about 2.3 K. Whereas Alpher and the other American big-bang physicists were not aware of the work of McKellar and Adams, Hoyle had been familiar with it since the time of its publication. Shortly before the appearance of McKellar's paper, Hoyle and Lyttleton had suggested that H_2 molecules might exist in cosmic clouds, which was the first prediction of interstellar molecules.[155] Hoyle was therefore much interested in the Australian result. He even used it to calculate the present amount of radiation energy and its corresponding temperature; but naturally he did not connect it with any background radiation caused by the early expansion of the universe. Hoyle was, in fact, aware of Alpher and Herman's prediction and might therefore have drawn the correct conclusion, that the temperature inferred from molecular spectroscopy was the same as that calculated by Alpher and Herman. But although Hoyle had read Alpher's and Herman's paper of

1948, he did not pay much attention to the brief and slightly obscure section in which the background radiation was mentioned.[156] It was, after all, based on a theory he thought was wrong. Moreover, he realized that the later published predictions of a higher temperature were inconsistent with the result obtained by McKellar and Adams. Yet Hoyle was not far off the mark, for he discussed the McKellar-Adams result in connection with Lemaître's theory, "according ... [to which] the ratio of thermal energy to the rest energy of matter was greater in the past." Hoyle seems to have been the only scientist who related the work of McKellar and Adams to cosmology, but in assuming a steady-state universe he failed to link it to the big-bang alternative of Gamow, Alpher, and Herman.

In early 1949 Hoyle had ready a generalization of his cosmological theory and an answer to the severe criticism that Bondi and Gold had launched in their paper. Hoyle agreed with his two friends that "the discrepancies [between astrophysical data and relativistic cosmology] can only be resolved by a modification of [Einstein's field equations], or by an entirely new theory."[157] He clearly preferred the first choice, to keep as closely as possible to existing theory and try to incorporate the idea of matter creation in it. This choice was also justified for pragmatic reasons; as he pointed out, the Bondi-Gold approach would be very difficult to develop into a quantitative theory. Hoyle added an argument of a more metaphysical nature. He was not convinced of Bondi's and Gold's sense of aesthetics according to which an axiomatic-deductive theory was superior to one based on extrapolations of experiments. "It is believed," Hoyle wrote (referring to himself), "that the wide cosmological principle should follow as a consequence of primary axioms of the field form, ... and should not appear itself as a primary axiom." In general, Hoyle had little respect for the kind of aesthetically based considerations which traditionally were part of cosmological reasoning and which played such an important role in the Bondi-Gold theory.

The steady-state theory was first discussed publicly among astronomers in December 1948, when McCrea arranged that Hoyle, Bondi, and Gold gave a brief presentation of their works at a meeting of the Royal Astronomical Society held in Edinburgh.[158] Present at the meeting were, among others, W. H. M. Greaves (president), Milne, Edmund Whittaker, Erwin Finlay-Freundlich, and the physicist Max Born. Hoyle stressed again the difference between his view and that of Gold and Bondi, namely, that according to his version the perfect cosmological principle was introduced as a consequence of the creation-tensor "rather than a criterion that must be imposed for aesthetic reasons." As to the wider scope of the new cosmological theory, he mentioned that "The possibilities of physical evolution, and perhaps even of life, may well be without limit." This was more than just a casual remark, and was a topic which Hoyle would return to later. The initial response to the new theory, as judged from the reported responses at the meeting, was reluctant, but not unambiguously hostile. Born expressed briefly his general skepticism,[159] whereas Whittaker declared his "warm admiration for the work as a whole." Milne did not see the point of introducing "one dogma more" into cosmological theory when there already existed a satisfactory theory—his own theory of kinematic relativity, of course.

The next time the steady-state theory was brought up in the Royal Astronomical Society was on 11 March 1949, when Hoyle presented his new paper just referred to.[160] In the discussion following the presentation, both McVittie and Milne expressed their feelings that the hypothesis of continual creation violated the venerated principle of Occam's razor, i.e., that it was an unnecessary hypothesis. "I do not believe," objected Milne, "that the hypothesis of continual creation of matter is necessary, nor do I consider that it is on the same footing as the assumption that the universe as a whole was created at a particular epoch."[161] In a similar vein, Herbert Dingle argued that it was hardly justified to violate an established principle of science (energy conservation) just in order to remove a particular difficulty in existing cosmological theory (the time scale difficulty). Hoyle denied, of course, that his version of continual creation was an additional hypothesis, and claimed that it simply replaced one hypothesis by another.

One might believe that the deductive nature of the theory of Bondi and Gold, as well as their critical attitude to relativistic cosmology, would be features that appealed to Milne. But if so, they were not sufficient to make him like the theory. There was more to Milne's opposition to the new steady-state theory than dissatisfaction with its logical structure. This he made clear in his text for the Edward Cadbury Lectures, finished shortly before his sudden death on 21 September 1950 and published posthumously by the Clarendon Press.[162] Milne's project was to develop a rational theory of the universe in accordance with his conception of the Christian God and Creator. He believed that kinematic relativity was such a theory, and that a cosmology consistent with Christian belief had to begin with a point singularity created by God. He therefore took issue with "the latest fashion of continuous creation" which assumed spontaneous creation of matter without explaining the origin of the laws of nature governing the creation. Milne's rejection of steady-state cosmology was emotional and openly personal:

> [In steady-state theory] creation is limited to the routine production, with penny-in-the-slot regularity and monotony, of hydrogen atoms. And this has been going on from eternity, and will continue to eternity. All the wonder, all the marvel, of the created universe is a consequence of the sporadic formation of hydrogen atoms out of nothing in just the local frames of rest. . . . By concentrating on the creation of *matter* alone, and leaving out the mode of creation of *frames of motion*, the Providence of Bondi, Gold, and Hoyle does only half its job. This is not a Providence that I for one could worship as God; and to do the authors of this theory justice, they do not believe or assert that any transcendental omnipotence is behind the simple acts of creation at all.[163]

Up to the summer of 1949, the response to the Bondi-Gold-Hoyle theory had been fairly moderate, neither enthusiastic support nor stiff opposition. In fact, the steady-state theory was scarcely recognized as a particularly radical departure from existing cosmological thought. Outside England, the theory made almost no impact. This changed soon, first of all as a result of Hoyle's popular exposition of cosmology of 1950.

The Nature of the Universe

In the spring of 1949, Hoyle was asked to give a series of talks for the British Broadcasting Corporation, which he agreed to do with very short preparation. The series of five forty-five-minute talks was broadcast on successive Saturday evenings and was very successful. The talks were first broadcast on the BBC's Third Programme, aimed at some three hundred thousand well educated listeners, but were so successful that the lectures had to be repeated on the more popular Home Service program for about three million listeners. In an article in the *Listener*, the BBC's journal, Hoyle introduced the series with a survey of what he saw as the insurmountable difficulties facing earlier cosmological theories.[164] The new "creation theory," on the other hand, was pictured in rosy colors. Hoyle's broadcasts were clearly a partisan view and not a neutral account of the state of cosmology. A year later the talks were transformed into a book with very little editing.[165] According to Hoyle's recollections, his motive for making the broadcast series was entirely financial. The Hoyle family was short of money and the fee offered by the BBC (a total of £300) was substantial. Yet, having accepted the offer Hoyle was given the rare opportunity of having the entire nation know about his view of cosmology, including the relationship between big-bang theories and the new steady-state theory. This was an opportunity he would not miss, and it must have acted as an additional motivation. The steady-state theory proposed by him and his friends had not attracted the interest they thought it deserved, and so it was natural and tempting to Hoyle to present his views to a larger audience through the BBC, indisputably the most powerful institution in British cultural life.

The Nature of the Universe, as the title of the small book was, followed closely the original scripts. Being in the great tradition of British popular science books, which included famous astronomy authors such as Jeans and Eddington, it was a fine example of science popularization and sold very well. After less than half a year the book had sold about sixty thousand copies, a phenomenal number for a science book at that time (and even today a more than respectable number). Until the two last chapters the book was fairly uncontroversial, although Hoyle left no doubt that his aim was not to give an objective account of how the majority of astronomers thought about the universe, but to present his personal view. Whereas standard relativistic models of the expanding universe were given only a cursory treatment, he expounded the steady-state theory in considerable detail. The theory was presented as the only reasonable picture of the universe, whereas theories with a point source origin were criticized. For example: "On scientific grounds this big bang assumption is much the less palatable of the two. For it is an irrational process that cannot be described in scientific terms. . . . On philosophical grounds too I cannot see any good reason for preferring the big bang idea. Indeed it seems to me in the philosophical sense to be a distinctly unsatisfactory notion, since it puts the basic assumption out of sight where it can never be challenged by a direct appeal to observation."[166] Incidentally, this passage seems to be the one in which the term big bang was first used. Hoyle invented the name as a kind of nickname, with a pejorative connotation, but it was only much later that the name became generally used and then without the connotation. In the last chapter, entitled "A Personal View," Hoyle expanded his discussion to include ethical, religious, and political perspectives also. This was entirely within the tradition of British popular science books, but the content was not. Hoyle attacked not only "the marxists" and the "out-and-out materialists" but also, and far more controversially, Christian belief. "It seems to me," Hoyle wrote (repeating a favorite argument of marxists and materialists!), "that religion is but a desperate attempt to find an escape from the truly dreadful situation in which we find ourselves."[167] Hoyle's steady-state view made eternal life a possibility, but he had only scorn for the eternity offered by Christian belief and made no secret that he did not share the dogmas of the Christians. Arguing in a rather materialistic way, he claimed that Christian belief in an immortal soul existing without physical connections was a nonsensical notion; soul, mind, or spirit, in order to have any meaning, "must be capable of physical detection."

Hoyle's attack on Christianity undoubtedly aroused antagonistic feelings in many people and helped to make Hoyle a controversial figure in the broader public context also. The BBC broadcasts and the book made Hoyle a public figure and improved his financial situation. According to his recollections, they also impeded his career and made it difficult for him to have his papers published. He felt he was being tagged as a broadcaster, and that the BBC talks produced professional difficulties with his colleagues, who resented his candidness and rallied against the steady-state theory. It is difficult to say how much truth there is in Hoyle's claim of being persecuted by academic astronomers, but at least he felt that he was treated unfairly: "I found it difficult to get my papers published during the first two or three years of the 1950s. There were endless infuriating arguments with referees, and ultimately I got so worn down with it all that for at least a couple of years I published nothing at all."[168] Hoyle felt that there almost was a conspiracy against him in particular and against steady-state cosmology in general. Referring to the reception of papers on the steady state, he wrote in 1980: "Journals accepted papers from observers, giving them only the most cursory refereeing, whereas our own papers [his, Bondi's, and Gold's] always had a stiff passage, to a point where one became quite worn out with explaining points of mathematics, physics, fact, and logic to the obtuse minds who constitute the mysterious anonymous class of referees, doing their work, like owls, in the darkness of the night."[169] Yet, in spite of the difficulties, real or imagined, Hoyle's publications did come out in a steady flow and his productivity was scarcely matched by any other theoretical astronomer of the period.

It can hardly have surprised Hoyle that some of the reviews of his book were very critical. Herbert Dingle, the astrophysicist and professor in history and philosophy of science at London University, found the book an excellent example of popular science but criticized Hoyle's unreserved confidence in the correctness of the steady-state theory. Hoyle had stated that the cosmology of the future would probably not differ substantially from the picture outlined by the present steady-state theory, a confidence that Dingle found unjustified and to represent "the triumph of faith over reason and experience."[170] Dingle would soon launch a much sharper attack on the steady-state theory, but in his review of 1950 he was only moderately critical. To Dingle, the steady-state theory was objectionable because it rested on a universal principle, but so did big-bang cosmology and also Milne's cosmological system. Dingle did not find the latter theories more scientific than its new rival. Whether the cosmological creation process was conceived to take place continually or at one time in the past, it still was a deus ex machina foreign to science. In an essay from the fall of 1949, Dingle compared Hoyle's theory with Jordan's in the following way: "Jordan offends against the proprieties outrageously, but the law cannot touch him because of a loophole in it that escaped the notice of the legislators. Hoyle, on the other hand, knows that he is breaking the law but does it so surreptitiously that it is impossible to get the evidence on which to convict him."[171]

Many of the reviewers of *The Nature of the Universe* praised Hoyle's success as a popular writer, but also warned the readers that the book should be read with great caution. George P. Thomson, the British Nobel Prize–winning physicist who back in 1927 had confirmed the existence of matter waves predicted by quantum mechanics, preferred Gamow's big bang to Hoyle's continuous creation. He did so in part because the creation process violated a well-established physical law, "but also more on esthetic grounds." Alluding to Hoyle's atheism he wrote in his review: "Probably every physicist would believe in a creation [of the universe] if the Bible had not unfortunately said something about it many years ago and made it seem old-fashioned."[172] Another theme was taken up by the astronomer Frank Edmondson of Indiana University in his review in *Sky and Telescope*. Edmondson accused Hoyle of provincialism on the ground that he did not pay attention to the data and conclusions of recognized experts in astronomy outside Cambridge—that is, to American astronomers. "Hoyle's next book might be a better one if he were to drink tea at some other places in addition to Cambridge," he snapped. In Edmondson's opinion, *The Nature of the Universe* was a philosophical work written by a skilled manipulator, not by a true scientist. "The 'New Cosmology,'" he concluded, is "a large dose of philosophy spiced with a little science."[173]

An even less temperate reply came from Canada, where Ralph E. Williamson on the Canadian Broadcasting Corporation in June 1951 expressed the irritation that many astronomers felt against Hoyle's presentation. According to Williamson there was "a general dissatisfaction of astronomers with what Hoyle has said," a feeling that he had gone far beyond the limits of decent presentations of astronomy, and a fear that his immodesty and one-sidedness had harmed the profession. It was therefore important to stress that Hoyle in no way should be considered a representative of the astronomical community. In fact, according to Williamson, Hoyle was barely a real astronomer at all, for "he has had no real experience with handling the large telescopes which make modern astronomy possible."[174] Like Dingle, Williamson had a firm faith in "honest facts" and he accused Hoyle of replacing such facts with untested theoretical views. He even went to the extreme of comparing Hoyle with the notorious Velikovsky,[175] that is, intimating that Hoyle was a demagogue and pseudoscientist. This was the kind of criticism that enhanced Hoyle's dislike of the establishment of observational astronomers and confirmed his feeling of being persecuted.

Hoyle did not respond in public to Williamson's rather insulting review, but Bondi ridiculed the Canadian astronomer for his simplistic belief in "facts" and his "preposterous statement" that only people who had done observational work are entitled to discuss astronomy. "It is on the same plane," wrote Bondi, "as the statement that only plumbers and milkmen have the right to pronounce on questions of hydrodynamics." And with regard to facts: "But what is an astronomical fact? At most it is a smudge on a photographic plate! Does he expect Mr. Hoyle to give a broadcast talk on smudges?"[176]

Although Hoyle preferred to ignore Williamson's accusation of disregarding honest facts and basing his view on unfounded hypotheses, he had in fact given a kind of reply a few years earlier. Among the Cambridge trio, Hoyle was probably the least philosophical, and he rarely engaged in methodological or epistemic discussions of other than a trivial kind. Compared with Bondi, in most cases his philosophically oriented comments were brief and superficial. But when it came to the nature of scientific work, Hoyle's spontaneous philosophy of science could contain penetrating insights. In June 1948, just at the time when he had completed his steady-state theory, Hoyle wrote a survey of stellar physics together with Lyttleton. In this work the two astrophysicists developed in some detail their view concerning the nature of hypothesis. It was as follows:

> It is often held that scientific hypotheses are constructed, and are to be constructed, only after a detailed weighing of all possible evidence bearing on the matter, and that then and only then may one consider, and still only tentatively, any hypotheses. This traditional view, however, is largely incorrect, for not only is it aburdly impossible of application, but it is contradicted by the history of the development of any scientific theory. What happens in practice is that by intuitive insight, or any inexplicable inspiration, the theorist decides that certain features seem to him

more important than others and capable of explanation by certain hypotheses. Then basing his study on these hypotheses the attempt is made to deduce their consequences. The successful pioneer of theoretical science is he whose intuitions yield hypotheses on which satisfactory theories can be built, and conversely for the unsuccessful (as judged from a purely scientific standpoint).[177]

Although written in a different context this methodological morale covers well the situation in steady-state cosmology. Given the time of publication, it is not unreasonable to assume that Hoyle had his new theory in mind in particular. Hoyle and Lyttleton also commented on the nature of testing and the status of observational facts. They emphasized what later generations of philosophers of science would call the theory dependence of facts: "As far as the testing of hypotheses is concerned, this also is a matter of some difficulty, particularly when new ground is being covered, because the significance of observational data cannot adequately be assessed until the data begin to be theoretically understood. It is therefore often again necessary to make use of intuition in deciding what may be a fatal objection, or what may be only a temporary discrepancy. The theory and interpretation of the observations should proceed hand in hand as it were, no observational 'fact' being regarded as having definite significance until its theoretical counterpart is at least fairly well perceived and understood."[178] Hoyle, Bondi, and Gold were keenly aware that observational facts should not be taken for granted without critical scrutiny and that the transformation from "fact" to fact required some measure of theoretical understanding.

Another broadcasting response to Hoyle's book came from the Australian Broadcasting Commission, which dealt with it over three nights. The reviewer was Daniel O'Connell, an Irish astronomer who was also a Jesuit priest, newly appointed director of the Vatican Observatory, and a science adviser to the pope. In 1938 he had become director of the Riverview College Observatory in Sydney, Australia, where he stayed until 1952. O'Connell believed, as did most astronomers, that it was premature to ask questions about the universe at large before many more facts had been gathered. On the other hand, he clearly had sympathy for Lemaître's big-bang theory, "which so clearly implies a Creator." O'Connell could see no reason at all to support the steady-state theory, which he, as an astronomer and scientist, found hopelessly speculative and artificial. As a man of the church he used much of his review to criticize Hoyle's "naive" and "remarkably foolish" views on philosophy and religion. Whatever the scientific objections against Hoyle's universe, it is difficult not to recognize the force in O'Connell's angry criticism of the last chapter of *The Nature of the Universe*. Hoyle's cursory dismissal of Christian religion showed neither knowledge nor willingness to understand the subject. As O'Connell complained: "The fact is that, though Hoyle sets out expressly to teach philosophers and theologians their business, he makes no serious attempt to find what they hold, or what reasons they give for their beliefs."[179]

Mixed Receptions

Max Born, the German-born quantum pioneer who since 1936 had been a professor of physics in Edinburgh, introduced Jordan's cosmological theory to English-speaking physicists in the fall of 1949. Jordan's paper, which was written on Born's request, was a general survey of his work on a cosmology with creation of matter and a gravitational constant decreasing in time. Born's interest in the matter was occasioned by Hoyle's recent article on the steady-state theory, which Born found difficult to accept because of its violation of energy conservation. "For if there is any law which has withstood all changes and revolutions in physics, it is the law of conservation of energy."[180] Although matter was also created in Jordan's theory, there was no violation of the conservation law, and this made Jordan's theory more acceptable to Born. The slight degree of similarity between his own theory and that of Hoyle made Jordan judge the steady-state theory in a favorable light; he hoped that his old view of matter creation might now receive the serious attention it deserved. Both Born and Jordan referred only to Hoyle's work, which also was the one mostly discussed by other early commentators. Understandably, Bondi and Gold felt a bit offended at being ignored and having steady-state theory identified with Hoyle's version. The sharpness of Gold's reply to Jordan reflected dissatisfaction with regard to priority, and also Gold's wish to demarcate his and Bondi's theory from Hoyle's and, especially, Jordan's.[181] In many of the early comments on steady-state theory, the Bondi-Gold version was ignored, and since the continuous creation of matter was seen as the central feature of the theory, it was often associated with the theories of Dirac and Jordan.[182]

Another early response, also addressed only to Hoyle and spurred by his article in *Nature*, came from Reginald Kapp, who back in 1940 had suggested a kind of steady-state universe with continual disappearance as well as creation of matter. After becoming aware of "Mr. Hoyle's theory of continuous appearance of matter" he restated his view, which he considered as associated with, and a precursor of, the new steady-state cosmology. Kapp believed that his view had the advantage over Hoyle's that it avoided the consequence that some galaxies would have become infinitely old and hence infinitely massive.[183] In his reply to Kapp, Hoyle used the opportunity to chastise Born and Jordan also for their comments; as little as Gold did he want to have the steady-state theory mixed up with Jordan's theory.[184] The steady-state theoreticians cannot have been happy about Kapp's claim that their work consisted in a mathematical formulation of the *same* hypothesis that Kapp had suggested in 1940. There was a world of difference between the philosophical speculations of Kapp and the theory of the steady-state universe, and any association with Kapp's amateurish views would only harm the reputation of the theory.

Although Kapp's view did not qualify as a serious cosmological candidate, it was discussed at some length in connection with the steady-state theory.[185] He had one nice argument in favor of his own idea of continual disappearance, namely, that according to the steady-state theory, without this assumption at least some galaxies would have become infinitely massive. These would have receded beyond the horizon, and thus be unobservable (as Hoyle had pointed out), but they would still exist and thereby cause a conceptual problem (for this problem, see further in section 5.2). Some of Kapp's arguments against what he in 1940 had called the "once-upon-a-time theory" were similar to those of the steady-state physicists. For example, Kapp claimed that the hypothesis of a universe with a finite past was alien to the spirit of science because it postulated a mythical past in which "all sorts of things happened that could not happen now."[186] He believed that only models in which matter originated continuously, or originated and disappeared continuously, conformed to the principle of Occam's razor. Kapp never developed his view mathematically, but felt that it received a kind of legitimation from the steady-state theory. As late as 1960 he published an elaborate version of his unorthodox philosophical speculations, which was much more in the prewar tradition of Nernst and MacMillan than in the tradition of postwar scientific cosmology.[187]

Most of the astronomers who responded to the early version of steady-state cosmology were critical. Many dismissed the theory without examining it thoroughly, and perhaps without considering it a serious candidate for the structure of the universe. Cosmological models flourished and wasn't the new one just another short-lived fancy? It was tempting to dismiss it as speculative and to group it with other heterodox theories, such as those of Dirac, Milne, Jordan, and Eddington. This was what the French astronomer Paul Couderc did in a book of 1952, where he discarded the steady-state theory as "risky and over-imaginative."[188] Couderc spoke for the majority of astronomers when he warned against theoretical excesses in cosmology, and argued that now it was the time to let observations take over: "At the present time, new ideas—some mere readjustments, others quite fantastic in their novelty—are experiencing an extraordinary efflorescence. This is significant: we lack the support of data, and it is time that observation took over again. Everything possible has been extracted from existing observations; until the appearance of new data we would be wise to give our imaginations a rest. Let the giant telescopes be set to work at the limits of the observable region, while the humbler instruments (the weapons of the majority of astronomers) patiently investigate the nearer horizons, where there is still plenty of unexplored ground."[189] A defender of relativistic orthodoxy, Couderc was in favor of evolutionary cosmology and preferred Lemaître's model with a superdense early universe. Another mainstream theoretical astronomer, the German Otto Heckmann, judged that the steady-state theory was already proved wrong by the results recently obtained by Stebbins and Whitford; these, he claimed, indicated a galactic age effect incompatible with the steady-state universe.[190]

Einstein, the founder of modern scientific cosmology, never commented in public on either the steady-state theory or Gamow's simultaneous big-bang theory. Since his pioneering work of 1917, he had dealt only little with cosmology, but it was a branch of science he nonetheless followed with interest.

This is evident from an appendix attached to the second edition of his *Meaning of Relativity*, published in 1945, where he gave a concise and thoughtful survey of the field. Einstein did not mention alternative cosmological theories, such as Milne's and Dirac's, and there is no doubt that he considered the general theory of relativity the only reasonable framework for cosmology. This would have been reason enough to disregard the steady-state theory. When Hoyle lectured in Princeton in 1952, Einstein was asked what he thought of the new cosmology. He is said to have dismissed the theory as a "romantic speculation."[191] Another pioneer cosmologist, Georges Lemaître, also refrained from commenting on the new cosmology, which evidently challenged his own view. The steady-state theory differed too radically from Lemaître's own view to make him take it seriously. The difficulty, if not impossibility, of reconciling the steady-state view with Christian dogmas may have added to his lack of interest.[192]

Bondi, Gold, and Hoyle can hardly have been satisfied with the way in which their theory was initially received by astronomers and physicists. A theory as ambitious and radical as the steady-state cosmology was meant to attract attention from, and create a serious debate with, the established science it challenged, but at first this was not what happened. The challenge seemed to go unanswered and the theory did not receive the expected scientific attention. The popular interest caused by Hoyle's radio broadcasts could not make up for the cool response in scientific circles, in most cases consisting of either silence or brief dismissal. For example, when Erwin Freundlich wrote his review of cosmology for the International Encyclopedia of Unified Science, steady-state theory was not mentioned (but then neither was Gamow's big-bang theory).[193] Similarly, ter Haar's review of stellar and cosmological energy processes ignored the theory.[194] Alpher and Herman did mention the steady-state alternative in their review, but only briefly, and alongside the theories of Dirac and Jordan. This does not imply that the American big-bang cosmologists were not interested in or well informed about the steady-state theory. In reply to a question from the author, Alpher and Herman stated: "The three of us [Gamow, Alpher, Herman] did discuss the steady state approach in depth and often. We took it seriously, but were concerned of the fact that, as then formulated, the theory required continuous creation at a rate which would probably never be observable. . . . Rather than not take the steady state model seriously, we were vigorously pursuing the context and consequences of our own evolutionary approach and not writing comparisons periodically of the two approaches."[195]

This is not to say that there was no interest in the steady-state theory, but at first much of this interest came from quarters that Bondi, Gold, and Hoyle could not profit from and might even have wished to be without. It did not advance the theory's status to have it associated with the speculative amateur philosophical views of an emeritus engineering professor (Kapp) or with the unorthodox views of a positivistic German quantum theoretician with a Nazi past (Jordan). In view of the fact that the controversy which soon followed is usually presented as a controversy between steady-state theory on the one side and big-bang theory on the other, it is worth pointing out that until 1951 Gamow and his coworkers remained silent about the steady-state theory; and, with one exception, Bondi, Gold, and Hoyle did not refer to Gamow's big-bang theory, but only in a general way to relativistic models with a superdense state in the past. In his book of 1952 (prefaced October 1950), Bondi did not refer to the Gamow-Alpher-Herman theory at all, and in a survey article of 1951 Gamow was equally silent about the steady-state theory.[196] The limited interest devoted to the steady-state alternative by mainstream relativists soon changed, probably as a result of the support the new theory was able to attract, which proved that it was more than a short-lived fashion.

The mentioned exception was a review Hoyle wrote of Gamow's and Critchfield's book on nuclear theory in October 1950.[197] Hoyle found the book excellent except for its chapter on astrophysics. This part of the book made him wonder if this field "is not on a far lower plane of argument and experimental procedure than other branches of physical science." He was particularly upset about the authors' appendix on nucleosynthesis because it relied on the new hot big-bang model of Gamow, Alpher, and Herman, "a cosmological model in direct conflict with more widely accepted results." Among Hoyle's objections were the time scale difficulty and the problem of accounting for galaxy formation. Interestingly, Hoyle referred to the predicted temperature of the cosmic background radiation as an argument *against* the big-bang theory. Gamow had stated that the variation of the radiation temperature with time followed the expression $T = 2.14 \times 10^{10} \, t^{-1/2}$ K, which was not only numerically wrong but also a relationship only valid at early times in the expansion. With $t = 1.8$ billion years, the result is $T = 90$ K, which certainly was "much greater than McKellar's determination for some regions within the galaxy," as Hoyle remarked. Two points deserve mention. First, Hoyle was led astray by Gamow's miscalculation and apparently did not think more about the question. Had he looked a little more closely into the matter, he might have paid more attention to Alpher's and Herman's value of 5 K, and this might have resulted in second thoughts. But it did not happen this way. Second, assuming that Gamow read Hoyle's review (which he supposedly did) he also must have been aware of the possible connection to the result of Adams and McKellar. On the other hand, given Gamow's understanding of the background calculations (see section 3.4), he may not have considered this very relevant. Alpher and Herman did not know about Hoyle's review and so they missed the opportunity to connect the Adams-McKellar result with the big-bang theory. They only learned about the work of Adams and McKellar in 1965, after the discovery of the background radiation.[198]

By far the most valuable support the theory received came from William Hunter McCrea, professor of mathematics at the Royal Holloway College of the University of London. Born in Dublin in 1904 of Irish parents, he received his education at Chesterfield Grammar School in England and with a scholarship he went on to Trinity College, Cambridge University, to cultivate his deep interest in the mathematical sciences. McCrea graduated in 1926 under the physicist Ralph Fowler and did research in quantum mechanics until his interests shifted to relativity and theoretical astronomy in about 1930. In 1932 he went to London to take up a position as reader in mathematics at Imperial College and later in the decade he became professor at Queens College in Belfast. Soon emerging as a leading authority in relativity, cosmology, and theoretical astrophysics, McCrea did fundamental work in these areas during the 1930s, when he collaborated with Whittaker, Milne, and Eddington, among others.

Throughout his distinguished career, McCrea's approach to cosmology was characterized by philosophical reflections and a desire to obtain a theoretically coherent understanding of the field. McCrea had a deep interest in philosophy going back to his time as a student at Cambridge University, where he went to lectures by John McTaggart and Bertrand Russell. His interest and competence in foundational questions were expressed, for example, in a penetrating philosophical analysis of space-time theories in 1939.[199] This philosophically oriented approach, which was a general feature among British cosmologists and physicists in the period, contrasted sharply with the much more pragmatic and even antiphilosophical approach followed by American researchers. "McCrea is a 'postulating cosmologist' and have [*sic*] no connection with reality," Gamow wrote in 1961, expressing how many Americans looked upon the British tradition.[200] Indeed, in spite of all its novelty and radicality the steady-state theory was very much part of the British cosmophysical tradition with roots in the 1930s. The young cosmological turks of the late 1940s had more in common with this tradition that they wanted to admit. The common ground consisted in a shared belief in the legitimacy of cosmological principles from which models could be deduced, an interest in the conceptual foundations of cosmology, and in general an acceptance of some measure of rationalism, deductivism, and speculation. Had it not been for this common ground, there may not have been much of a controversy over steady state in the first place. After all, in order for a controversy to arise the involved scientists must at least agree that the subject under discussion is worthwhile enough to be controversial, rather than merely dismiss it as irrelevant or of marginal interest.

As mentioned, McCrea had become interested in the steady-state theory as soon as he knew about it, and he soon decided that it was worth further exploration. With a preference for deductive models with a minimum of hypotheses, such as Milne's, his metaphysical commitments were not far from those of Bondi and Gold. For example, McCrea shared their dissatisfaction with the state of relativistic cosmology, which, he wrote, "probably impresses the general theoretical physicist as a highly unsatisfactory subject."[201] His excellent 1950 review of the steady-state theory was positive throughout, and presented the theory as a promising new approach to cosmology. He found the theory particularly interesting because it promised a better understanding of the relationship between cosmology and subatomic physics, a revival of Eddington's ill-fated but never forgotten program. McCrea based his optimism on the fact that in the steady-state theory, the large-scale structure of the universe is determined by the rate of creation of matter, which presumably was an atomic parameter explainable by some future development of quantum mechanics.[202]

When Gamow commented on the steady-state theory, in the popular *Creation of the Universe* of 1951, he rejected it completely. In no case did he examine the theory carefully or enter a serious debate. It seems that he simply did not take the steady-state alternative seriously. It might be used for entertainment, he seems to have thought, but nothing else. Gamow's attitude to steady-state cosmology was reflected in one of his later popular science books, where he had Hoyle featuring as the author of and actor in a cosmic opera. Hoyle (in the dream of Mr. Tompkins) "suddenly materialized from nothing in the space between the brightly shining galaxies," bursting majestically into a song:

> The universe, by Heaven's decree,
>   Was never formed in time gone by,
> But is, has been, shall ever be—
>   For so say Bondi, Gold and I.
>   Stay, O Cosmos, O Cosmos, stay the same!
>   We the Steady State proclaim!
>
> The aging galaxies disperse,
>   Burn out, and exit from the scene.
> But all the while, the universe
>   Is, was, shall ever be, has been.
>   Stay, O Cosmos, O Cosmos, stay the same!
>   We the Steady State proclaim!
>
> And still new galaxies condense
>   From nothing, as they did before.
> (Lemaître and Gamow, no offence!)
>   All was, will be for evermore.
>   Stay, O Cosmos, O Cosmos, stay the same!
>   We the Steady State proclaim![203]

CHAPTER 5

Creation and Controversy

LIKE MILNE'S THEORY in the 1930s, the steady-state theory had a great effect on the development of cosmology by challenging the orthodox relativistic theory. Much of the development that took place during the 1950s was related to the controversy spurred by the steady-state challenge. However, it would be wrong to believe that cosmology in the 1950s was solely occupied with the question of whether the universe is in a steady state or evolves in accordance with the laws of general relativity. Much research did not deal with this issue, but consisted in either observational astronomy or mathematically oriented investigations based on the framework of general relativity in one of its several versions. In this tradition, which was a continuation of the tradition that had dominated cosmology in earlier decades, physical concepts such as the big bang and matter creation did not enter. For example, in 1957 Schrödinger published a small book on *Expanding Universes*, based on a lecture course held in Dublin in 1954. One looks in vain for any reference to either the big-bang or the steady-state theory. What interested Schrödinger and many other mathematical physicists was not the actual structure of the universe, but the mathematical and conceptual problems of classical world models such as first studied by Einstein, de Sitter, Lemaître, Tolman, and others.

By 1950 cosmology was still considered an immature science and there was no paradigmatic theory of the structure and evolution of the universe. Yet it was generally admitted that the universe was best described by the relativistic field equations. When steady-state theory arrived as a strong competitor it could neither be ignored nor shot down observationally. The new cosmology challenged the foundation of the evolutionary-relativistic world picture and resulted in predictions which, although not immediately testable, were within the practical possibility of testing, and for this reason the situation evolved into a controversy. Much of the controversy took place by means of arguments of a very general kind, theoretical or philosophical, whereas the observational results were still too uncertain to be of any decisive importance during the 1950s.

5.1 DEVELOPMENTS AND MODIFICATIONS OF STEADY-STATE THEORY

The cosmological theory invented by Bondi, Gold, and Hoyle in 1948 remained essentially unchanged during the 1950s. This is not to say that there were no attempts in the period to develop the theory. Like any living scientific theory, the steady-state model was explored and discussed, and several proposals for modifications were made. These were in part responses to observational challenges and in part attempts to reframe the theory into a more orthodox language or otherwise improve its logical structure.

The impression that one may get from later textbooks, retrospective accounts, and most historical writings on cosmology is that the steady-state theory, ever faithful to its original formulation of 1948, stubbornly remained the same until it was proven false by the 1965 discovery of the microwave background radiation. This impression is quite wrong. For one thing, the steady-state theory was not simply proved wrong in 1965; and, for another, the theory was far from a petrified corpus of doctrines. In fact, there was more development within steady-state cosmology during the 1950s than there was within the competing class of relativistic evolution theories. Indeed, from about 1960 proponents of the steady-state theory were criticized for making ad hoc changes in order to keep the theory in agreement with new observational evidence.

McCrea's Field Interpretation

In 1955 D. E. Littlewood discussed Hoyle's model in relation to the two time scales introduced by Milne in the 1930s. He argued that in order to reconcile cosmologies based on the perfect cosmological principle with general relativity, an anisotropic cosmological term would have to be introduced.[1] An earlier attempt to elucidate the relationship between the steady-state theory and other cosmological theories by means of Milne's idea was made by Martin Johnson, a physicist and astronomer at Birmingham University. Johnson speculated that perhaps big-bang and steady-state theories might then appear as equivalent descriptions of the universe. "Continuous creation has hitherto seemed an escape from the intractable problem of a zero in spatial expansion," he wrote, "but it may turn out to be one of the alternatives expressing the same facts as an adjustable time-scale." Nothing useful came out of applying the idea of two time scales to the steady-state theory.[2]

The technical contributions to the steady-state theory that took place during the decade were mostly attempts to modify it in such a way that it became more acceptable to the majority of astronomers and physicists. The concept of continual creation of matter and the relationship to the general theory of relativity were the focal points of the discussions among those scientists who basically were sympathetically inclined to the new cosmology. If the steady-state theory could be made to harmonize with general relativity and the mysterious creation of matter accounted for within the framework of existing physical theory, then the idea of a steady-state universe would become much more acceptable.

Almost all the technical contributions to the steady-state theory related to Hoyle's theory, which was not only better suited for development because of its field-theoretic formulation, but in its spirit was also much closer to general relativity than the Bondi-Gold version. Understandably, it proved difficult to develop the semiqualitative and methodologically radical Bondi-Gold theory, which was taken up by only very few scientists. One of those few was J. W. Dungey, a physicist at the Cavendish Laboratory, who in 1955 investigated the consequences of the perfect cosmological principle.[3] Dungey showed that many of the results obtained by Bondi in his book of 1952 could be deduced directly from the perfect cosmological principle, without relying on arguments based on Newtonian theory or kinematic relativity.

Although the steady-state theory usually appeared as one theory, and the original differences between the two versions were rarely considered important, Bondi and Gold continued to defend their original view that a field-theoretic formulation was unnecessary. As late as 1958, at the meeting of the Solvay Congress, Bondi defended his and Gold's attitude against Hoyle's.[4] As Bondi pointed out, the procedure adopted in the Bondi-Gold theory was phenomenological, closely analogous to that in thermodynamics: just as thermodynamics yields a number of observable consequences without being based on field theory or microscopic considerations, so did their cosmological theory. However, as Bondi explained three years later, by then he was no longer hostile to a field-theoretic formulation à la Hoyle. He just felt that it was premature at the current stage since the Bondi-Gold version made the same predictions. If observational tests clearly favored big-bang cosmology, then Bondi believed it would be worthwhile to try constructing a field theory; but until the observational situation was clarified he saw no point in engaging in the Herculean mathematical efforts such a formulation would presumably demand.[5] It is remarkable that apparently Bondi did not admit that there already existed a satisfactory field theory in the Hoyle (or rather Hoyle-Narlikar) version. In fact, when the situation envisaged by Bondi became a reality—with observations favoring big-bang theory—he showed no interest in the attempts by Hoyle and Narlikar to revise the steady-state theory.

Almost all the attempts to develop the steady-state theory took place in England, most of them made by people working in or associated with Cambridge (where Hoyle and Bondi were) or London (where McCrea was and to where Bondi came). To some extent the theory was discussed in a general way abroad also, but with one exception there seem to have been no scientists outside England who contributed to its technical development. The exception was a Japanese physicist, Hidekazu Nariai, who in 1952 suggested an alternative formulation of Hoyle's field equations.[6] Nariai argued for a different form of the creation tensor, but his formulation led to the same empirical results as Hoyle's. As was the case for many other scientists, it was the simplicity and uniqueness of the steady-state theory that appealed to the Japanese physicist. Nariai later wrote that he found the theory interesting because of "its clear and simple predictions for various phenomena to be tested in the near future."[7] However, although Nariai found the steady-state theory interesting, he did not support it. His own work was in the tradition of general-relativistic mathematical cosmology, and he criticized the steady-state theory for its rationalism and reliance on the perfect cosmological principle.[8]

The most important contribution to the steady-state theory in the 1950s was not due to Bondi, Gold, or Hoyle, but to McCrea, who presented a new version of the theory in 1951.[9] McCrea wanted to develop Hoyle's theory so as to make it accord better with general relativity and, in particular, to satisfy some kind of energy conservation. In Hoyle's work Einstein's field equations were modified by adding the C tensor in much the same way as the cosmological term was added by Einstein in 1917. The procedure was widely seen as ad hoc and as destroying much of the aesthetic appeal of the original field equations. Furthermore, the addition of a term (whether C_{mn} or Λg_{mn}) is not unique and thus opens the way for a variety of additions and a corresponding variety of cosmological models—which destroys the unique feature of steady-state cosmology that its advocates found so appealing. For these reasons McCrea argued that if modifications had to be made, they should not be modifications in the equations, but in the physical interpretation of the energy-stress tensor T_{mn}.

Ever since Einstein started his search for a unified field theory he had thought that his cosmological field equations (1.2) were "somehow in bad taste."[10] This was in part because of the (to a monist thinker) unaesthetical mixing of a geometrical and a physical tensor, and in part because of the arbitrariness in the energy-stress tensor. It was this freedom in the interpretation of the energy-stress tensor which McCrea exploited. As he noted, Hoyle's field equations would in a formal sense become identical with those of general relativity if the substitution

$$T_{mn} = S_{mn} + \kappa^{-1} C_{mn} \tag{5.1}$$

was made, i.e., if the creation term and the ordinary energy-stress tensor, S_{mn}, appearing separately in Hoyle's equation, were subsumed under a new tensor with a different physical interpretation.

McCrea made use of a result obtained by Edmund Whittaker in 1935, namely, that the quantity $\sigma = T^0_0 - T^1_1 - T^2_2 - T^3_3$ can be interpreted as the density of gravitational mass.[11] According to standard general relativity the values of the components are $T^0_0 = \rho$ and $T^i_i = -p/c^2$, and the gravitational mass density thus becomes

$$\sigma = \rho + \frac{3p}{c^2}. \tag{5.2}$$

McCrea employed a new kind of Newtonian analogy in which terms of order $1/c^2$ were included as significant in classical theory, contrary to what was the case in the 1934 McCrea-Milne theory. The strict Newtonian analogy was only valid for $p = 0$, but with the new method McCrea obtained formulas which corresponded completely to those of general relativity for $p \neq 0$ also. The extension amounted to deriving the continuity equation

$$\frac{d}{dt}(\rho c^3) + \frac{3p}{c^2} R^2 R' = 0,$$

in which $\rho > 0$ and p is assumed to be ≥ 0, which for $R' > 0$ (expanding universe) leads to a decreasing density, $d\rho/dt < 0$.[12]

Whereas in Hoyle's theory the pressure was zero, McCrea found that

$$\rho = \frac{3}{\kappa a^2} \text{ and } p = -\frac{3c^2}{\kappa a^2}, \tag{5.3}$$

where $a = c/H$. That is, $p = -\rho c^2$. By insertion into equation (5.2) one obtains $\sigma = -6/\kappa a^2$. The result can be obtained more directly from the Friedmann-Lemaître equations (2.3) and (2.4) with $\Lambda = k = 0$ and using the steady-state solution $R = \exp(Ht)$. Up to this point, McCrea had merely rederived results already known from the standard relativistic analysis of the de Sitter universe. In this and other classical analyses the pressure was assumed to consist of a matter and a radiation term and was thus positive or zero. McCrea now noticed that this interpretation of T_{mn} was not compulsory. What would happen, he asked, if a uniform *negative* pressure were admitted in accordance with equation (5.3)?

The suggested reinterpretation of T_{mn} corresponded to the introduction of a zero-point stress as an alternative to Hoyle's C tensor, viz., a change in the zero level in accordance with equation (5.1). In Hoyle's model the expansion was caused by an outward pressure produced by the created matter, whereas McCrea explained the expansion as a result of negative pressure, which corresponds to a negative gravitational mass between the observer and a galaxy. McCrea found the work done by the negative pressure (per unit volume and time) to be $3\rho c^3/a$, and it was this work which, if transformed by means of $E = mc^2$, reappeared in a rate of increase in mass density of $3\rho c/a$. Consider any sphere of fixed radius centered around an observer. The observer will assert that matter is continually receding across the surface of the sphere and being replaced by newly created matter. In 1953 McCrea explained the mechanism as follows: "By virtue of the negative pressure in the moving medium, the medium outside the sphere is doing work upon the medium inside. . . . *The net flow of energy across the surface is zero.* We can think of the medium as expanding under such a tension that it continually replenishes itself by the work it does upon itself. Therefore the process is energetically self-supporting. The only novel feature in the creation process is the supposition that the energy represented by the work done makes itself apparent in the form of, or is continually being converted into, matter of the same sort as the matter already present."[13] The pressure appearing in McCrea's theory, but not in Hoyle's, was of no direct physical significance because its gradient vanishes everywhere and thus will have no mechanical effects. Although not noticed by McCrea in 1951, his innovation was mathematically equivalent to adding a cosmological constant interpreted as a kind of vacuum energy, as first proposed by Lemaître in 1934 (see section 2.2).

McCrea's reinterpretation remained within the steady-state theory and led to the very same results as Hoyle's theory. But instead of making use of the creation process itself as the primary postulate, McCrea's introduction of a zero-point stress in space shifted the focus of the theory away from the mysterious creation of matter, which was no longer seen as a genuine *creatio ex nihilo* process, but rather as a kind of transmutation. McCrea later characterized the theory as describing "a universe that is forever renewing itself from its own resources," which brings to mind the earlier conceptions of Nernst, MacMillan, and Millikan.[14] Indeed, from a qualitative point of view, McCrea's notion of a cosmic stress was strikingly similar to Nernst's idea of an ether supplied with zero-point energy.

Methodologically, McCrea's work was closely related to Hoyle's, which it developed further in a direction away from the radical deductivism favored by Bondi and Gold, and toward conformity with the general theory of relativity. For example, McCrea did not consider the perfect cosmological principle a fundamental principle of nature, because it was not applicable to small-scale physics. One of his ultimate aims was to bridge cosmology and quantum theory, and for this purpose field theory, and not cosmological principles, would have to be used. Similarly, in McCrea's steady-state theory Mach's principle did not have the important function that it had in the Bondi-Gold theory. He emphasized as an advantage that in his theory the behavior of any large part of the universe was ascribed to conditions in that part. The element of localism was contrary to the holistic features in the Bondi-Gold theory. "It is only a matter of mathematical convenience to treat a model of the whole universe," McCrea wrote in 1951. "The essential features of the physical interpretation do not depend upon assumptions about the whole universe."

McCrea's alternative interpretation of the steady-state model inspired other British scientists to take up similar investigations. George C. McVittie, professor of mathematics at Queen Mary College, the University of London, and a leading cosmologist, was one of them. McVittie studied physics ("Mathematics and Natural Philosophy") at the University of Edinburgh and attended Whittaker's lectures on relativity during 1926–27. He recalled that he was once told by a fellow student that Whittaker "had suggested that he should look into the possibility of interpreting through general relativity the redshifts in the spectra of galaxies by 'someone called Hubble.'"[15] At that time the gulf between mathematical physics and observational astronomy was deep indeed. McVittie tended to favor mainstream relativistic cosmology, but his general attitude to science was also influenced by the more deductivist methods characterizing Milne's kinematic relativity and later the steady-state theory. As a former student of Whittaker and Eddington he was not immune to the rationalistic romance of steady-state cosmology.

McVittie had originally had some sympathy for Milne's approach, but in 1940 he criticized it on methodological and epistemic grounds. Although he never endorsed the steady-state theory, he seems for a short time to have found some appealing features in its version as worked out by McCrea. After he went to the United States in 1952, where he became director of the University of Illinois Observatory, McVittie turned increasingly toward relativistic evolution cosmology and soon became one of the main antagonists of the steady-state theory.[16] Much later, he indicated that a contributing motive for his emigration to the United States was the intellectual climate in British cosmology. He was glad to get away from England and "from the atmosphere of the steady state in this country. There was such a hullabaloo about the new revelation! Everybody—McCrea was trying to climb onto the bandwagon and did for a bit."[17] This is an interesting recollection, but it exaggerates the importance of the steady-state theory, which at no time held a dominating position in England (and even less so outside England).

In 1952 McVittie suggested a new world model that incorporated a revised form of McCrea's theory in a framework entirely within the realm of general relativity. "It will be shown," he wrote, "that a 'continuous' creation process can exist in a suitably chosen general relativity model of the universe. The method employed is that of apriori cosmology in which the model universe is established, not from observational data, but by postulating certain principles to which the universe is supposed to conform. These principles are, in effect, restrictions on the model universe imposed by the investigator because he believes them to be reasonable, or because they are in agreement with his epistemological or philosophical views."[18] This was a view entirely in agreement with the metaphysical commitments of the steady-state theoreticians. However, as we shall see, McVittie's attitude to the steady-state theory and other rationalistic cosmologies changed markedly during the 1950s.

McVittie's model was what he called a "gravitationally steady state" in which the gravitational mass density σ is constant and the acceleration is given by $R''/R = \Lambda/3 - 4/3\,\pi\kappa\sigma$. Such a universe can be shown to have an open past and an open future (i.e., it exists for all times), but with a scale factor R that varies hyperbolically and has a minimum at $t = 0$. For $t < 0$ the universe is contracting and for $t > 0$ it is expanding. The stress is time dependent, but negative for all t and with the same meaning as in McCrea's theory, i.e., a zero-point stress. Following McCrea, McVittie showed that in the expanding phase stress would be converted into matter, and in the contracting phase the reverse process would take place. The conversion follows from $\sigma = $ const, which is the same as

$$\frac{d\rho}{dt} = -\frac{3}{c^2}\frac{dp}{dt}.$$

McVittie showed that under certain assumptions the model would lead to a present mean density of matter of $\rho = 2.4 \times 10^{-29}$ g·cm^{-3}. This he considered more satisfactory that the much higher value of the Hoyle-McCrea theory (5×10^{-28} g·cm^{-3}) because it deviated less drastically from the observed value. The important feature in McVittie's work was that it presented, for the first time, a world model with matter creation in complete conformity with general relativity. In many respects it was a hybrid between steady-state and relativistic cosmology, but in spite of having matter creation and no beginning in time, it was not really a steady-state model, at least not in the classical sense. With a creation rate depending on the epoch, it did not satisfy the perfect cosmological principle. This feature would later reappear in the versions of steady-state theory developed by Hoyle and Narlikar, but then in a different context.

The works of McCrea and McVittie helped to demystify the concept of continuous creation of matter in the sense that it could now be argued that the process was compatible with general relativity and had its source in a universal stress of negative energy. However, the demystification had curiously little impact on the controversy. Most opponents of the steady-state theory ignored McCrea's idea of a negative intergalactic pressure and maintained that steady-state cosmology and general relativity were irreconcilable quantities. The British physicist William Bonnor was one of the few relativist cosmologists who publicly admitted the validity of the idea. As he wrote in 1955, granted the existence of McCrea's negative pressure, then "indeed, the steady-state model does satisfy Einstein's field equations."[19]

An idea somewhat related to McVittie's, but not building on standard relativity, was proposed by António Gião, a Portuguese mathematical physicist, in 1963.[20] Based on a generalization of the relativistic field equations, Gião claimed that none of the cosmological models usually discussed, the steady-state model included, were admissible. His alternative was what he called a generalized steady-state model in which the constancy of the density of matter was replaced by the constancy of the density of proper energy. From this starting point (which contradicts the perfect cosmological principle) Gião was led to an oscillating cosmological model with a period of about 16×10^9 years and including both creation and destruction of matter. During the phases of contraction, creation of matter would take place, and during those of expansion, a corresponding destruction. In Gião's model the Hubble parameter varied in time, attaining its maximum during the expansion and its minimum during the contraction. Although the model was ignored by other researchers, certain of its features reappeared in the theories developed by Hoyle and Narlikar a couple of years later.

Mechanisms of Matter Creation

Negative pressure or not, the creation process was completely obscure from a microphysical point of view, and the lack of a more definite mechanism made it impossible to discuss the hypothetical process within the framework of quantum physics. The lack of connection between quantum mechanics and cosmology was not peculiar to the steady-state theory, but because of the creation of matter it was felt to be more of a problem in this theory. In ordinary relativistic cosmology, it was just an aspect of the generally admitted dissociation between quantum mechanics and the theory of general relativity. In 1950, when Hoyle happened to be in Switzerland, he gave a talk at the Technische Hochshule (ETH) in Zurich, with Pauli in the audience. Afterwards

Pauli commented, "If you could understand the physics of how creation happens, it would be much better."[21] It would, but neither Hoyle nor others could come up with the kind of explanation wanted by Pauli.

Perhaps the only suggestion of a mechanism in terms of ordinary elementary particles was made by Claude de Turville, who proposed the creation of matter to arise from a postulated process in which a hydrogen atom was excited by an antineutrino and and a neutron was created:[22] $H + \bar{\nu} \rightarrow H + \bar{\nu} + n$. With the subsequent decay of the neutron, and the recombination of the produced proton and electron, a net process of $H + \bar{\nu} \rightarrow 2H + 2\bar{\nu}$ was claimed to occur. However, since this hypothetical process was purely speculative and with no foundation at all in particle physics—it violates baryon conservation—it did not throw any light at all on the creation mechanism. Even less credible was the suggestion of a Japanese astrophysicist, Toshima Araki, who thought that the violation of energy conservation in the steady-state theory might be avoided by considering the new matter as flowing in from a hypothetical higher-dimensional space in which our universe with its four-dimensional space-time was a subspace.[23] Araki's science fiction–like suggestion was not without historical precedents. For example, in his ether squirt hypothesis of the 1880s, Karl Pearson had suggested the idea of matter creation, or transfer, from other dimensions hidden to us. The revival of Pearson's idea in a mid-twentieth-century cosmological context did not help the steady-state theory to gain respectability.[24] Needless to say, perhaps, neither de Turville's nor Arakis's suggestions were taken very seriously.

Another suggestion, more in line with the works of Hoyle and McCrea, was earlier made by Felix Pirani, a young British-Canadian mathematical physicist who at the time was working as a postdoctoral assistant in Cambridge, where he discussed his work with Bondi and Hoyle. Pirani suggested regarding the creation of particles in steady-state theory as a kind of collision process in which new entities were produced in addition to the material particles (such as neutrons). The new entities, which he named gravitinos, would have strange properties.[25] Like the neutrino, a gravitino would have zero rest mass and be uncharged, but it would also have negative energy and momentum. Indeed, Pirani speculated that gravitinos might in fact be negative-energy neutrinos (not antineutrinos). He suggested that the proposed mechanism "may be susceptible to description in quantum-mechanical terms," possibly within the framework of the theory of weak interactions. If gravitinos were a kind of neutrinos this was a reasonable suggestion, but it remained undeveloped. Also, it is not clear whether Pirani thought of his particles as detectable or not. In any case, unlike the neutrino, the gravitino was met with silence by experimental physicists.

The advantage of the hypothesis was that energy and momentum conservation could then be upheld even in the process of creation itself. This could be done if it was assumed that gravitinos were produced in such a way that the total four-momentum in the creation process is zero. If gravitinos are produced together with material particles, the reverse process, annihilation of gravitinos and particles, must presumably also take place. However, Pirani argued that in an expanding universe, the annihilation rate would be much smaller than the creation rate and hence could be ignored. On the basis of the gravitino hypothesis, Pirani managed to build up a cosmological theory largely identical with the ordinary steady-state theory. There was the difference, though, that in Pirani's theory the mean density of matter would be four times larger than in Hoyle's theory (namely, given by $\rho = 12H^2/\kappa c^2$).

Pirani's work was intended as a contribution to the steady-state theory and, like McCrea's, was an attempt to find a mechanism which would bring continual creation within the ambit of more conventional physics. When Pirani arrived in Cambridge, he developed a strong interest in the steady-state cosmology, which appealed to him because it resolved the time scale difficulty and also because it minimized the scope for divine intervention in the universe.[26] When Bondi moved to King's College in London, Pirani followed him. However, with nobody following up the gravitino idea, Pirani turned to other aspects of mathematical physics and gradually lost interest in cosmology.

It was generally agreed among steady-state theoreticians that if the creation process were ever to be understood in more than a phenomenological way, it would have to be related to modern quantum and elementary-particle physics. However, it was also realized that such an understanding would have to wait for a quantum theory of gravitation and thus belonged to the future. In the Kaluza-Klein theories, first developed by the German Theodor Kaluza and the Swede Oskar Klein in the 1920s, an attempt was made to unify gravitation and electromagnetism with quantum theory by introducing a new, fifth dimension in addition to the four space-time dimensions. This kind of theory did not lead to a proper quantization of the gravitational field, however, and it was almost forgotten in the 1950s. But to an Australian astronomer, V. A. Bailey, the Kaluza-Klein approach offered the advantage that continuous creation of matter could be conceived as being supplied from the hypothetical fifth dimension into the visible four-dimensional world. It was thus an attempt at explanation of the same kind as the one offered previously by Araki. Bailey, who declared himself a supporter of the steady-state theory, hoped in this way to reconcile creation of matter with (five-dimensional) energy conservation, and thereby "cause relief in the minds of many persons who would otherwise be unable to accept the steady-state theory."[27] This was not the result that followed from Bailey's speculation, which was not further developed. In the 1960s, following new developments in quantum field theory, some attempts were made to give a quantum foundation to continual creation and McCrea's negative-pressure substratum.[28] However, the results were not encouraging, and the attempts had at any rate no effect on the cosmological controversy.

In the absence of a quantum theory of gravitational fields, macroscopic theorizing or crude analogies were all that could be offered to explain the continual creation of matter. In his presentation of the steady-state theory at the Solvay Congress in 1958, Hoyle suggested a simple way in which "the inherent plausibility of the creation of matter" could be demonstrated.[29] Hoyle started his plausibility argument by pointing out that the observable part of a steady-state universe is given by the horizon associated with the de Sitter line element, which is c/H. Although the universe is infinite, no signal emitted by a source at a distance larger than c/H will ever be received by the observer. The mass of this observable universe is thus 4/3 $\pi\rho$ $(c/H)^3 = c^3/2GH)$, where Hoyle's value of the density ρ has been used. Hoyle now noted that the gravitational potential of a mass m at distance r is Gm/r. By dividing the mass of the observable universe by c/H and multiplying by G, a quantity corresponding to the gravitational potential will appear. The result is $c^2/2$ or, what is the same in this context, about c^2. According to Hoyle, this meant that every particle in the universe exists in a potential well of a depth which is comparable to its rest energy, mc^2. "The process of creation can accordingly be thought of as involving no energy expenditure—a particle is created at a negative potential that compensates for its rest mass. Accordingly [*sic*] to quantum theory, particle creation might well be expected under these circumstances." However, as Hoyle realized, his picture was no more than a crude plausibility argument that skeptics might not, and in fact did not, find convincing.

Without using the mechanisms suggested by Hoyle or Pirani, P. Roman at the University of Manchester also took up the steady-state theory in McCrea's version. Roman argued, as other authors had done previously, that matter creation in an expanding universe could be accommodated within a slightly modified law of energy conservation. The mechanism of matter production, adapted from that suggested by McCrea and McVittie, was this: "The work done by the gravitational force is . . . the locally created energy. On the other hand, gravitation is necessary to keep the universe in a steady state . . . In a sense, we may say that gravitation works against a change in rate of expansion and keeps balance, thereby creating mass. However, the expansion is not brought about by the pressure directly, because p is constant and its gradient does not appear in the equation of motion. The pressure, through its negative contribution to the gravitational mass and hence to the gravitational force would tend to bring the system to an accelerated flow."[30] According to Roman, the mass density of the universe was $\rho = 24H^2/\kappa c^2 = 8\rho_c$ and the rate of matter creation about 7×10^{-46} g·cm^{-3}·s^{-1}, both values being empirically reasonable, although the density was rather too high.

At about the same time William Davidson wrote his Ph.D. thesis under the supervision of McCrea, whose steady-state theory he investigated within the framework of general relativity.[31] Davidson's analysis suggested that the state given by $\rho c^2 = -p = 3c^2/\kappa a^2$ constituted a natural state of relative stability of the intergalactic medium. By examining the effects of slight disturbances of this state, he found that the expansion tended to restore the steady-state conditions. A somewhat similar result was obtained in 1957 by William Bonnor, who, however, used it to argue against the steady-state theory.[32] Bonnor concluded that condensations could not form in a steady-state universe from small perturbations, and that galaxies were therefore unlikely to be formed. This was not the conclusion of Davidson, who pointed out the possibility of condensations forming from large disturbances, and hence the possibility of the formation of galaxies in the steady-state theory also.

The steady-state model of McCrea operated with Newtonian or relativistic dynamics in a way which was not altogether clear. The confidence in this model, whether in its original 1951 formulation or in its later versions, was diminished by an examination made in 1960 by Bonnor, who as an orthodox relativist was unsympathetic to the steady-state theory.[33] According to his analysis, the model led to results that made it physically unrealistic or, at least, drastically reduced its physical credibility. Bonnor showed that within McCrea's framework the motion of matter is indeterminate, so that, for example, there is no reason to assume that matter moves on a geodesic. The velocity of matter would be wholly indeterminate, that is, unpredictable. "This lack of determination weakens those versions [Newtonian or relativistic steady-state theories] to such an extent that they seem to be of little value as instruments of prediction in cosmology," he concluded.[34] At the same time Bonnor noted that since the steady-state theory (in its Bondi-Gold version) did not rely in itself upon either the general theory of relativity or Newtonian mechanics, then his criticism, in principle, left the theory unaffected. Nonetheless, Bonnor's work weakened its appeal: "What now seems clear, however, if my arguments are correct, is that those who wish to work with the steady-state theory must use a dynamics specifically designed for it, since they are not free to use the existing models of general relativity or Newtonian theory with any degree of confidence." Of course, no such special dynamics existed. Bonnor's theoretical analysis of 1960 thus resulted in conclusions very different from those of Davidson, and left the impression that McCrea's model was not, after all, acceptable.

Bonnor, of whom more below, was an important figure in British cosmology. He had an unusual career, which illustrates (as does, for instance, Gold's) the open disciplinary structure of the area in the 1950s. William Bonnor was born in 1920 in London, where he was brought up by his mother, a violinist. After studies at South Essex Technical College he worked during the war for the admiralty and also did research in applied chemistry for Shell. He received his Ph.D. in chemistry from the University of London in 1946 and started his scientific career with works on the rheology of greases and various aspects of physical chemistry. Only then did he seriously begin to develop an interest in mathematics, which resulted in a position as lecturer in this field at the University of Liverpool in 1949. His first contribution to general relativity came two years later and was quickly followed by a series of theoretical works in cosmology and general relativity. During most of the 1950s Bonnor was at Queen Elizabeth College, London, where he was appointed professor in 1962.[35]

An Electrical Universe

The cosmological theories of McCrea, McVittie, and others included the existence of a negative-energy cosmic stress which was not directly observable because of its uniformity. The existence of the stress could be taken as a primary postulate, or some physical source for the stress could be sought. The latter course was taken by Bondi and Lyttleton, who in 1959 suggested a new electrical cosmology based on the hypothesis of a universal charge excess.[36] The idea that gravitation may ultimately be understood in terms of electrical action was old, going back to the Italian physicist Ottaviano Mossotti in 1836, and later developed by Wilhelm Weber, H. A. Lorentz, and others. These early workers investigated the consequences of assuming that the electrostatic attraction between two elementary charges differed numerically from their electrostatic repulsion, and were thus led to speculations about "antigravitation"—repulsion between lumps of matter.[37] However, the idea lost its appeal with the demise of the electromagnetic world view. In 1959 it had been taken for granted for more than four decades that the numerical charges of the electron and the proton are exactly equal, but in fact this was an assumption for which there was no compelling theoretical reason.

As early as in his 1948 paper with Gold, Bondi had briefly considered the possibility that the electrical state of the created matter might not be strictly neutral.[38] Based on arguments from Newtonian cosmology, he concluded that an excess charge larger than about $10^{-19}e$ (where e denotes the elementary charge) would be inconsistent with the principle of homogeneity. In 1948 Bondi and Gold therefore assumed that the excess charge was zero, but in his work with Lyttleton eleven years later no such assumption was made.

Experiments proved that the two charges could differ at most insignificantly, but, as Bondi and Lyttleton pointed out, even a tiny difference might have significant cosmological consequences; and at any rate experiments could never establish strict equality, only equality within a certain limit of error. They therefore explored the consequences of assuming a charge excess so slight that it did not conflict with experiments. This was the same kind of reasoning which in 1948 made Bondi, Gold, and Hoyle assume a slight violation of energy conservation. It is to be noticed that the charge-excess hypothesis would have no drastic consequences for ordinary pair production, where pairs of particles and antiparticles are formed. Bondi and Lyttleton did not suggest that the positive elementary charge differed from the negative one in the sense that the electron's charge differed from the positron's, only that the numerical charges of protons and electrons were different.

If applied in conjunction with the assumption of continual creation of matter, the suggestion of charge excess is a very drastic hypothesis because it implies that the electrical charge is not a conserved quantity. If, for example, a hydrogen atom is created, it will consist of a proton of charge e and an electron of charge $(y - 1)e$, where $y \ll 1$. An excess charge of ye is created, contradicting Maxwellian electrodynamics which requires strict charge conservation. Consequently, Bondi and Lyttleton constructed a modification of the Maxwell equations which accommodated the slight violation of charge conservation. From the new equations applied to a homogeneous universe, they were able to derive the correct expansion law, i.e., a recessional velocity proportional to the distance between any two hydrogen atoms. It was found that this law had the form $v = q/3\eta \, r$, where q is the rate of creation of charge density and η is the charge density of the hydrogen atoms in intergalactic space.

Bondi and Lyttleton pictured the expansion process as an effect of the electrostatic repulsion and thus, remarkably, explained Hubble's law purely electrically, without recourse to gravitation theory. The electrodynamics of both steady and unsteady universes with charge excess was examined by L. G. Chambers in 1961; he presented a simplified analysis of Lyttleton's and Bondi's theory.[39] In order to secure numerical agreement with observation, Lyttleton and Bondi found that the excess charge would be given by $y \approx 2 \times 10^{-18}$, which was at the verge of what could then be established experimentally. A simplified argument is the following. Consider a spherical volume of hydrogen atoms of uniform mass density ρ, mass M, and radius r, so that the mass $M = (4/3)\pi\rho R^3$ contains M/m atoms each with an excess charge ye. A hydrogen atom situated at the surface of the sphere will experience an electrostatic repulsion of $F_{el} = (M/m)\,(y^2 e^2/r^2)$ as well as the ordinary gravitational attraction of $F_{grav} = GMm/r^2$. Let the ratio F_{el}/F_{grav} be denoted μ. Then

$$\mu = (ye/m\sqrt{G}\,)^2. \tag{5.4}$$

By insertion of the values of e, m, and G it then follows that the hydrogen atom will be repelled if the charge-excess parameter y is larger than 9×10^{-19}. The actual repulsive force will be $F_{el} - F_{grav} = F_{grav}\,(\mu - 1)$, which can be written $F_{rep} = 4/3\,\pi(\mu - 1)\rho Gmr$. That is, there is an outward acceleration corresponding to a cosmic expansion. From Hubble's law, $v = Hr$, it follows that the acceleration can be written $a = H^2 r$. By comparison, Lyttleton and Bondi were in this way led to the identification $H^2 = 4/3\,\pi(\mu - 1)\rho G$. With the accepted value of the Hubble constant, this led to $\mu = 5$. Finally, from equation (5.4) the mentioned value of $y = 2 \times 10^{-18}$ is obtained, which is well above the equilibrium value. Bondi and Lyttleton also found the average mass density of the universe, if in a steady state, and a relation between the density and the rate of creation. For the mass density they obtained $\rho = 3H^2/4\pi G$ and for the creation rate $3H\rho \approx 10^{-46}$ g·cm^{-3}·s^{-1}, both quantities of the same order of magnitude as in the original steady-state theory.

The Bondi-Lyttleton electrical theory was not restricted to world models with continual creation of matter, but the two authors clearly thought of it as a contribution to the steady-state theory. "Hitherto," the two authors claimed, "the steady state has been an assumption of cosmology, but the present analysis supplies an almost rigorous proof of it."[40] The assertion was questioned by William Swann, a seventy-seven-year-old British-born American physicist who in the 1920s and 1930s had done important work in cosmic-ray physics. According to Swann, the Bondi-Lyttleton theory did not justify a steady-state universe in particular.[41]

Lyttleton was, as he said in 1960, "certainly inclined to favour the steady state theory of the universe," and indicated that an experimental confirmation that $y \neq 0$ would vindicate the theory.[42] On the other hand, should experiment prove that $y = 0$ (or just much smaller than 10^{-18}) the steady-state theory would not be disproved. The association between the electrical theory and steady-state cosmology was made clear by Bondi and Lyttleton, who saw their theory as supplying the source for the negative pressure out of which, according to McCrea, matter was created: "The essential result of the electrical hypothesis . . . is that it supplies a reason for the stress, namely, the electromagnetic field due to the charge inequality."[43] Although Lyttleton had held a low profile in the cosmological debate, he made no secret of the fact that he belonged to the steady-state camp, together with his old friends. He made his preference clear in 1956, in a popular book in which he compared the steady-state theory favorably with big-bang cosmologies.[44] Lyttleton argued that the former theory had a much higher degree of aesthetic attraction, and that continual creation of matter was a more fruitful and scientific hypothesis than the theory with once-and-for-all creation of matter.

The electrical theory was quickly taken up by Hoyle, who modified and extended it as well as correcting some mistakes made by Bondi and Lyttleton.[45] Hoyle argued that the theory might throw light on one of the often discussed questions of cosmology, viz., why the world is made up of matter and almost no antimatter at all. If the matter creation process was governed by quantum theory, as it presumably was, one would expect particles of ordinary matter to be created symmetrically together with antiparticles; however, since the observable world consists of ordinary particles, some mechanism for separating the two kinds of matter would then be required. Hoyle sketched an explanation of the separation as due to the electrical forces arising from the assumed charge excess, and argued that intergalactic magnetic fields would be formed in the wake of the separation. He judged the electrical steady-state theory to have "impressive advantages," among which was that it allowed matter to be created in pairs, in accordance with quantum theory. It also indicated a mechanism for the generation of cosmic rays of ultrahigh energies, the origin of which was not otherwise understood. Another advantage, which was highligthed also by Bondi and Lyttleton, was seen in the smallness of the charge excess—estimated by Hoyle to be $y = 5 \times 10^{-19}$ rather than 2×10^{-18}—which made it almost the inverse square root of the magical number of Eddington and Dirac, 10^{39}. Since this number is also the ratio of the radius of the observable universe to the radius of the (classical) electron, it suggested that the size of the universe might be determined by the value of the charge-excess parameter y. A larger charge excess would mean a smaller universe! In general, wrote Lyttleton, the electrical steady-state cosmology "raises fresh hopes in several directions, for it seems to promise just those links between cosmology, quantum theory, and relativity that have for so long been dimly perceived peeping over the horizon."[46] Eddington's dream was still much alive.

Of course the electrical hypothesis also had its problems. For one thing, gravitational condensations of matter would seem unable to be formed in a smoothed-out universe where the electrical repulsion overcomes gravitation; and with no condensations, there could be no formation of galaxies. The problem was solved by Bondi, Lyttleton, and Hoyle by arguing that the intergalactic matter did not consist of neutral atoms, but was ionized. In this case, the charge excess would disappear in local regions of space and condensations would be able to occur. The weak point of the ingenious electrical theory was that it rested entirely on the assumption of a charge excess of an order of magnitude that might be tested in the laboratory any day. This was in contrast to the steady-state creation of matter, which, because of the extreme smallness of its rate, was directly testable in principle, but not in practice.

Millikan's famous oil-drop experiments, made during the second decade of the twentieth century, included positive as well as negative charges, and allowed the American physicist to conclude that if there was a charge excess it was smaller than 3×10^{-16} of the electron's charge.[47] In 1925 two Swiss physicists, A. Piccard and E. Kessler, performed another kind of experiment, from which they concluded that the charge excess was smaller than would correspond to $y = 10^{-20}$. The experiment was instigated by a hypothesis which Einstein had suggested at a meeting of the Swiss Physical Society in Lucerne in 1924, namely, that a charge excess of $y \approx 3 \times 10^{-19}$ might suffice to explain the magnetic fields of the earth and the sun. In favor of the hypothesis, Einstein pointed to the numerical relationship $ye \approx m\sqrt{G}$, which is essentially the same relationship discussed by Lyttleton and Bondi thirty-five years later. [It follows from Bondi's and Lyttleton's $\mu = (ye/m\sqrt{G})^2$ with μ of the order unity.] Apparently Einstein abandoned the idea, but his conversations with Piccard and Kessler led them to design their experiment. The result unambiguously refuted Einstein's hypothesis.[48]

These early experiments were not well known in the late 1950s. Lyttleton and Bondi were unaware of the Piccard-Kessler result, which clearly disagreed with their theory. So did new experiments, made as tests of the electrical theory. As early as September 1959, before Bondi's and Lyttleton's paper had appeared, two British physicists, Anthony Hillas and Thomas Cranshaw, claimed to have refuted the electrical hypothesis by confirming the results of Piccard and Kessler. They found in an experiment the charge of an argon atom to be smaller than $12 \times 10^{-20}e$, from which they concluded that the deviation of the proton's charge from e was at most $4 \times 10^{-20}e$, or that $y \leq (1 \pm 3) \times 10^{-20}$. If confirmed, this would mean the end of the electrical theory.[49]

On the other hand, this and most other experiments presupposed in their design the exact validity of Maxwell's equations and therefore could be suspected of systematic errors. To design an experiment with macroscopic equipment able to test the electrical hypothesis would be very difficult, because the effect of the modified Maxwell equations on the behavior of the equipment would have to be taken into account. (That is, if the Maxwell equations need modification; the hypothesis of charge excess without matter creation accords completely with Maxwellian theory.) Confronted with the claim of the two experimentalists, Bondi and Lyttleton at first refused to accept the validity of the conclusion.[50] Following a strategy that had worked well in other areas of the cosmological controversy, they suggested various sources of error. In their view, these made the "extremely obscure" experiment unable to decide the fate of the charge-excess hypothesis. In a situation where an experiment contradicted what they considered a beautiful theory, Bondi and Lyttleton preferred to place the guilt on the experiment.

Yet it was experiment which quickly made the electrical theory of Bondi, Lyttleton, and Hoyle disappear from the scene of cosmology. Hillas and Cranshaw showed that the experimental objections raised by Lyttleton and Bondi were unfounded and that no loophole seemed to exist.[51] The Hillas-Cranshaw result was confirmed by John King, an American physicist, who in the fall of 1960 reported experiments of a type similar to those made by Piccard and Kessler in the 1920s. King found hydrogen molecules to exhibit a charge approximately forty times less than that required by the theory of Bondi and Lyttleton.[52] Finally, experiments made in the early 1960s by Vernon W. Hughes and coworkers proved that the limits found earlier were reliable, and that the charge excess was definitely smaller than $y = 10^{-18}$. This was the inescapable conclusion based on experiments of different types.[53] "All these results," wrote Hughes in a review article of 1964, "provide strong evidence against the form of the Lyttleton-Bondi proposal which requires [$y = 2 \times 10^{-18}$]; they do not test the alternative, although less attractive, form of the Lyttleton-Bondi proposal, which requires a greater number of protons than electrons in the universe."[54]

For all practical purposes electrical cosmology was killed by high-precision laboratory experiments, and after 1960 the hypothesis was not revived. Although Bondi, Lyttleton, and Hoyle apparently accepted the experimental verdict, they did not publicly retract their theory. It just disappeared, as if it had never been proposed. It would have been possible to maintain a modified form of the theory, for example, by assuming the densities of the positively and negatively charged particles in the universe to be slightly different. However, this was not a strategy that was followed. In any case, the experimental repudiation of the charge-excess hypothesis made no great difference to the steady-state cosmology, which did not rely upon the electrical hypothesis. On the other hand, the failure probably contributed to the general feeling among many scientists that the steady-state theory was in decline.

5.2 Is Cosmology a Science?

In the beginning of the 1950s steady-state theory had established itself as one cosmological candidate among several others. It had not, however, succeeded in attracting much positive interest from astronomers and physicists, and by 1951 only four cosmologists clearly defended the steady-state theory—Bondi, Gold, Hoyle, and McCrea. To most astronomers, cosmology was scarcely a branch of science, and certainly not one that they felt any obligation to deal with. Most of the few astronomers and physicists who did have an interest in cosmoloy either dismissed or ignored the steady-state alternative, and favored relativistic cosmology in a broad sense; only rarely did this imply direct support of the big-bang theory, in either Lemaître's or Gamow's version. Cosmology was in an unsettled state, with no paradigm ruling the field, although the relativistic expanding universe was generally considered the best candidate for the structure of the world. In this situation most scientists looked for new observations in order to decide among the various alternatives, but in the early 1950s it was realized that observational data were still too restricted and not reliable enough to act as crucial experiments. The lack of unambiguous results of cosmological relevance furnished a fertile soil for discussions of a more general kind which can be loosely called extrascientific.

In the present context, extrascientific arguments are taken to be arguments that do not primarily deal with the content of the theory itself or with the relationship between theory and experiment. If an argument goes beyond the domain of the theory or extrapolates it to areas where its applicability is doubtful, it is extrascientific. Of course, there is no strict way of demarcating a theory's scientific domain, and all kinds of extrascientific arguments may influence the evaluation of a theory's scientific validity. All the same, scientists usually have no difficulty in deciding whether an argument belongs to the scientific discourse or not. Indeed, it is part of the socialization of scientists to learn the distinction, which is generally, if tacitly, agreed upon.

One indicator is to look at the place where the argument appears. If it appears in a scientific paper published in a professional journal, the argument will almost always be scientific, whereas extrascientific arguments normally appear elsewhere, for example, in books, public addresses, or journals outside the scientific specialty. In the present case, the situation is a bit more complex, and we cannot rely either on any simple methodological dichotomy between science and extrascience or on social indicators such as the site of publication. The reason is that the separation between what belongs to the scientific discourse and what does not to some extent presupposes a science in a so-called normal state, that is, a state characterized by consensus about the foundation and the demarcation of the scientific discipline. In periods of a potential revolution in science, where there is uncertainty about the meaning and scope of the science in question, it is much harder to determine which arguments are scientific and which are not. Indeed, such questions are largely decided by the outcome of the discussion between competing research programs.

I shall not discuss this complication here, and for the moment I assume that one can reasonably divide arguments into scientific and extrascientific categories in the present context also. I believe this is the case, and it suffices here to note that it is a belief shared by virtually all scientists. For the sake of clarity I further divide the extrascientific arguments into two classes, one which primarily deals with philosophical questions in a rather narrow meaning, and one which deals with implications of the theory that relate to ideological questions. The term ideology is here taken in its standard meaning, referring to political and religious views in particular. It should be obvious that this division is based more on considerations of convenience than on principle.

There is nothing surprising in the fact that cosmology in this period was discussed from a philosophical perspective, with the twofold aim of deciding the logical and conceptual structure of cosmology and also the wider implications for man's place in the universe. These topics have always been intimately connected with cosmology, indeed, they have been inseparable from it during most of its history. The new aspect was that with the emerging controversy between relativistic cosmology and steady-state theory, the topics also became important in the intertheoretic relationship between the two world views. In the absence of crucial observational data, much interest was directed toward the philosophical foundations and implications of the theories. The one which showed the stronger philosophical basis would attract new support and appeal to more scientists. On the other hand, if one of the theories turned out to be less scientific than its rival—judged by criteria which necessarily must be philosophical—it would be likely to lose support.

The Appeal of Steady-State Theory

Much of the philosophically oriented discussion was occupied with the question of which of the theories was the more appealing in a sense that was rarely defined. This may seem unsatisfactory, but it is a historical fact that cannot be denied and which has to be taken seriously. Some British astronomers clearly had a sympathy for the steady-state theory, which appealed to them on grounds that may be called philosophical, but were basically aesthetic. This was the kind of reasoning that made young Dennis Sciama feel attracted to the steady-state theory. Sciama was born in Manchester in 1926 of Jewish (nonreligious) parents with roots in the Middle East; his mother was from Egypt and his great-grandfather came from Aleppo in Syria. Dennis's father, who was in the cotton trade, wanted him to take over the business, but to his dismay Dennis was determined to become a scientist—he hoped for a career in mathematics.[55] He entered Trinity College, Cambridge, in 1944 and after an interruption of two years' military service, which he mostly spent working on solid state physics, he obtained his M.A. in 1949. In his graduate studies he had Dirac as his supervisor for a period, who influenced his general attitude to physics but was of little help in his Ph.D. work. At the time he obtained his Ph.D. degree, in 1953, he was well aware of the steady-state theory and knew Bondi and Gold well. The examiners for his Ph.D. work were Bondi and McCrea.

Young Sciama was impressed by the steady-state theory's simplicity and predictive power. He also had an instinctive sympathy for the theory because he felt it represented a generational revolt against established views. Referring to Bondi, Gold, and Hoyle, he later said that "They were the young rebels, and they were an exciting influence for a younger person like myself."[56] One of Sciama's first scientific papers, an ambitious attempt derived from his Ph.D. work to explain the origin of inertia in accordance with Mach's principle, demonstrated the influence of the Bondi-Gold philosophy. Although not relying specifically on the steady-state cosmology, Sciama's theory was in general agreement with the holistic views of Bondi and Gold. In accordance with the views of these scientists, he emphasized that local phenomena are strongly coupled to the universe as a whole, and that the outcome of local experiments can therefore yield information about the structure of the universe.[57] At the latest by 1954, Sciama had decided that, as far as he was concerned, the steady-state theory was a desirable picture of the universe.

Sciama's fascination with the steady-state theory was in part rooted in what he considered its empirical success, but also, and more importantly, in its philosophical appeal. Only the steady-state theory promised a cosmology in which the physical laws governing the universe were not mere contingencies but in some way were unique and necessary, reflecting requirements of self-consistency. As he wrote in 1955: "The steady-state theory opens up the exciting possibility that the laws of physics may indeed determine the contents of the universe through the requirement that all features of the universe be self-propagating. . . . The requirement of self-propagation is thus a powerful new principle with whose aid we see for the first time the possibility of answering the question why things are as they are without merely saying: it is because they were as they were."[58] Variations on this theme often appeared in Sciama's works, which clearly show his indebtedness to the original criticism of relativistic cosmology launched by Bondi and Gold. The methodological appeal of the Bondi-Gold theory was a decisive factor in Sciama's continuing support of the steady-state theory up to 1966. To mention just one example, in 1961 he reaffirmed his commitment to the steady-state theory, not because it fitted better with observations, but because of its appealing theoretical structure. He wrote: "A theory which, whilst it can be tailored to fit this unique universe, nevertheless has to present certain aspects of it as arbitrary, as though they could have been different, is therefore less satisfactory than a theory in which these aspects are essential."[59] The first kind of theory was, of course, the one based on general relativity, the second the steady-state theory.

Whereas the self-consistent kind of cosmology, with its clear rationalistic flavor, appealed to Sciama and some other steady-state advocates, other scientists and philosophers were repelled by the vision. It was a vision which continued to be discussed, although not necessarily as an argument for or against steady-state cosmology specifically. In 1973 McCrea argued that the ultimate goal of physics was to show that "things are as they are because they couldn't be otherwise" instead of merely explaining the state of things by reference to the fact that "they were as they were."[60] This kind of argument was an important source of the interest in the anthropic principle that occurred in the 1970s.

The ambiguity in appeals to metascientific principles such as unity and simplicity is clearly illustrated in the cosmological debate. Whereas Sciama saw the steady-state model as providing a unified world picture—a *Unity of the Universe*, as his book of 1959 was entitled—McVittie used similar arguments to dismiss the theory. According to McVittie, the steady-state theory was "unscientific" because it violated a unified description of the universe: "If [the steady-state theory is] accepted, it is necessary to suppose that the mechanics of the nebular systems are in some way different from the mechanics of all other astronomical systems. If the object of science is to unify phenomena into theoretical systems with as wide an amplitude as possible, the general relativity interpretation may be accepted until it leads to some prediction seriously contrary to observation."[61]

In 1951 Martin Johnson expressed his general sympathy for Hoyle's new cosmology which he found to be a way out of the conceptual difficulties to which big-bang cosmology led. The choice between cosmological models was, according to Johnson, "an aesthetic or imaginative choice" rather than a rational one. In his opinion cosmology had "more in common with the poetic or artistic attitude towards experience than with the solely logical," and he apparently rated the steady-state theory highly by such standards.[62] Ernst Öpik agreed that the choice between cosmological models would remain "a matter for esthetic judgment" and that the steady-state theory "can at present be adressed only from the standpoint of [its] esthetic value."[63] However, Öpik had a preference for cyclical models and he found the steady-state theory to be both wrong and ugly. Öpik was at the time director of the Armagh Observatory in Northern Ireland, and he often criticized the steady-state theory in the pages of the *Irish Astronomical Journal* and elsewhere.

The same kind of general, but rather uncommitted sympathy that Johnson had aired in favor of the view of Bondi, Gold, and Hoyle was also expressed by Sir Harold Spencer Jones, royal astronomer and former president of the Royal Astronomical Society. He did so in particular in a talk to the Royal Institution in the spring of 1952.[64] In his review of the steady-state theory, Spencer Jones emphasized as particularly appealing the fact that the theory was subject to direct experimental testing, a feature which was often stressed also by Bondi, Gold, and Hoyle. He also found the age distribution of galaxies to be evidence in favor of the steady-state theory, and noted with satisfaction its contrast to "the older view that the universe is running down and will come to an end after a finite time." Neither Johnson nor Spencer Jones supported the steady-state theory in an unqualified manner, but the royal astronomer's address was widely seen as official approval, and made its way to the newspaper headlines.[65]

Outside England, responses to the steady-state theory were fewer and less appreciative. The Swedish physicist Oskar Klein, professor at the Technical High School in Stockholm, had done important work in quantum mechanics in the 1920s and 1930s, when he was a close associate of Niels Bohr. His name is known from the Klein-Gordon equation, the first relativistic extension of the Schrödinger equation, which Klein found in 1926, and also from the five-dimensional Kaluza-Klein theory of gravitation and electromagnetism. During the 1940s Klein turned to astrophysics and also developed an interest in cosmology. In 1953, at the Liège conference on nuclear astrophysics, he admitted that cosmology was a field where "personal taste will greatly influence the choice of basic hypotheses." Klein's taste was for neither big-bang nor steady-state models. As to the first kind of theory, he pointed out that he and "many [other] physicists would . . . be very reluctant in accepting a theory on which we have to assume a literal creation of the world at a given, calculable time." This reluctance did not make Klein an adherent of the steady-state theory, which he objected to in methodological terms. Concerning the perfect cosmological principle, he wrote: "This postulate, which may very well be valid in a world of infinite possibilities, seems hardly suited, however, to form the basis for any special restriction of such possibilities demanding specific modifications of the laws of nature derived from experience. But this is again a matter of taste, and it would certainly be unwise to raise objections against the further development and testing of the mentioned attempts of applying the postulate in question under the assumption that the observed art of the supergalactic system is representative of the whole world."[66]

Klein's taste led him to sketch a cosmological model very different from both the big-bang and the steady-state theories. As a starting point he assumed a dilute cloud of cold hydrogen à la Laplace, which condensed under the forces of gravitation. With increasing density the radiation (arising from ionization caused by atomic collisions) would be enclosed in the cloud, and Klein argued that at some stage the radiation pressure would balance the gravitational contraction. After the equilibrium state had been reached—Klein estimated that the radius of the cloud would then be about 10^{24} m—the radiation pressure was assumed to result in an expansion corresponding to the one observed. Klein's model was of no particular importance in the period, but his general attitude to the cosmological problem was undoubtedly shared by many astronomers, who regarded both the big-bang model and the steady-state theory with suspicion. We shall return to Klein's cosmological ideas in chapter 7. What matters here is that they so clearly depended on taste, and thus exemplify the importance of subjective factors in cosmology around 1950.

Dingle's Attack

Between 1951 and 1953 Herbert Dingle served as president of the Royal Astronomical Society. Annoyed with the recognition that Spencer Jones had accorded to the steady-state theory, he used his presidential address of 13 February 1953 to launch a full-scale attack on the new cosmology.[67] Present at the meeting was Lemaître, who on the occasion received the first Eddington medal and was also admitted as a full member of the society. Dingle was well prepared for the attack on the steady-state theory, which to some extent was a repetition of earlier criticism against what he saw as rationalistic theories intruding into natural science. As mentioned, in 1937–41 he had led the opposition against the cosmophysics of Milne, Eddington, and Dirac, whom he accused of perverting the true meaning of science.

Trained in astronomy and physics, Dingle was a brilliant writer and a sharp debater with a strong interest in the history and philosophy of science. This was, in fact, the subject of his chair at University College, London, where he had been head of the Department of History and Philosophy of Science since its foundation in 1946. Dingle, who served as president of the historical commission of the International Astronomical Union and was among the founders of the *British Journal for the Philosophy of Science*, was a man of great reputation in the historical and philosophical aspects of science. But he was also a controversial figure because of his frequent engagements in polemics. In spite of considering himself a guardian of orthodox science against pseudotheories, many of his own contributions to science were in fact unorthodox and not well received. He was opposed not only to the steady-state theory, but to relativistic cosmology as well, and indeed to any theory which rested on a universal principle of some sort. Dingle was even led to dismiss Einstein's special theory of relativity and to propose his own alternative, a sure sign of unorthodoxy and estrangement from mainstream physics.[68] During the 1950s, Dingle became involved in a controversy over the nature of special relativity in general, and the meaning of the clock paradox in particular. He concluded that Einstein's version of special relativity could not be true, and his troubles in having his arguments published led him to believe that he was the victim of a conspiracy led by mathematical physicists within the Royal Society.[69]

Dingle became an ardent critic of mathematical (or theoretical) physicists, a species of scientists he accused of arrogant confidence in the power of mathematical reasoning. In 1961 he declared that the mathematical physicists "have simply lost the power of understanding of what they are doing. They have substituted mathematics for reasoning, and now automatically believe that its categorical character absolves them from considering any unorthodoxy."[70] While Eddington, Milne, and Dirac had previously been Dingle's favorite targets, in the 1950s he replaced them with Bondi, Hoyle, and McCrea, who were not only theoretical physicists of a rationalistic bend but also—after all—defenders of orthodoxy when it came to the theory of relativity. The fact that, in this particular respect, McCrea and Bondi repre-sented established physics and objected to Dingle's interpretation of special relativity presumably made his feelings against the steady-state theory even more hostile.

The presidential address of 1953 was a general attack on cosmologies based on a priori reasoning; but it was particularly directed against the steady-state theory, which Dingle dismissed as quackery, and as nothing but the emperor's new clothes. When divested of its mathematical-symbolic clothing, Dingle found that "the substance underneath appears ridiculous because it is ridicu-lous." He went on:

> It is hard for those unacquainted with the mathematics of the subject, and trained in the scientific tradition, to credit that the elementary principles of science are being so openly outraged as they are here. One naturally inclines to think that the idea of the continual creation of matter has somehow emerged from mathematical discussion based on scientific observation, and that right or wrong, it is a legitimate inference from what we know. It is nothing of the kind, and it is necessary that that should be clearly understood. It has no other basis than the fancy of a few mathematicians who think how nice it would be if the world were made that way. The mathematics *follows* the fancy, not precedes it; the fancy is credited because it gives scope for mathematical exercise, not because there is any reason to believe it true.[71]

One thing which upset Dingle in particular was the philosophically based arguments of Bondi, Gold, and Hoyle for the steady-state theory. Dingle had no objections to philosophical arguments, but in the case of the three Cam-bridge physicists he concluded that it only led to "unscientific romanticizing." This sad result he ascribed to their assumed total lack of knowledge of the history of science. Had they only taken the time to read the works of Galileo, Newton, and Faraday, they would have been able to recognize their follies. Dingle was so upset that he did not refrain from distorting the views of his targets, and in general presented a caricature of the steady-state theory. For example, the accusation against the steady-state theory for being merely a mathematical game certainly did not apply to the Bondi-Gold version, which was remarkably nonmathematical. As another example, Dingle claimed that Hoyle's philosophical criticism of big-bang cosmology amounted to the doctrine that "a confession of ignorance is unscientific," and that the steady-state theoreticians lacked that supreme virtue of the scientific mind, humility. Although humility was never among Hoyle's virtues, he had never claimed what Dingle ascribed to him. All what he argued was that the concept of the big bang was unscientific because it would forever be beyond scientific analy-sis. This is very different from claiming a confession of ignorance to be unscientific. Dingle further claimed that the steady-state theory asserted con-tinual creation of matter to be a primary axiom, whereas Bondi, Gold, and Hoyle had in fact carefully explained that matter creation followed from their theory. One might believe that Dingle's aggressiveness was rooted in his commitment to an alternative cosmological system, but this was not the case. He offered no alternative except a cautious empiricist wait-and-see attitude, which effectively would deny any role to theoretical models of the universe.

Dingle disliked cosmological principles of any kind and saw in them the main cause why cosmology had degraded into a state of pseudoscience. He proposed "calling a spade a spade and not a perfect agricultural principle," and that the cosmological principle should be renamed "the cosmological assumption"and the perfect cosmological principle "the cosmological presumption." Both were essentially unscientific, "the cosmological presumption" only doubly so because of its even more sweeping nature. This principle, he claimed in 1956, has "precisely the same nature as perfectly circular orbits and immutable heavens," that is, it is in principle inviolable.[72] In 1956 as in 1937, Dingle's favorite antiscientist was Aristotle, his favorite scientist Galileo. Yet Bondi stressed in the same year that the perfect cosmological principle should be considered a falsifiable working hypothesis, not an a priori assumption.

Dingle's address caused considerable public attention but seems not to have been very effective in scientific circles. It was reported in the newspapers and considered important enough to merit full republication in *Science*, the leading American science journal.[73] Remarkably, the accused parties—mainly Bondi, Gold, and Hoyle—refrained from replying to Dingle's vitriolic attack. They may have thought that the unbalanced nature of the address spoke for itself, and that it would be pointless to enter a debate with the angry scientist-philosopher. Gold recalled that "I couldn't bring myself ever to take Dingle seriously . . . We ridiculed him. I mean, we just couldn't take him seriously in the least. I vaguely remember that he attacked it [the steady-state theory], but it was only gibberish so far as I was concerned."[74]

In America, Dingle's accusations aroused the criticism of Cornell physicist Phil Morrison and the philosopher Adolph Grünbaum, who both objected to Dingle's claim that the steady-state theory was based on a continuous series of miracles in the form of creation of matter. As Grünbaum argued: "It is only a dogmatic insistence on the claim that the conservation of mass (energy) is cosmically the 'natural' state of affairs which can lead to Herbert Dingle's characterization of the Bondi-Gold hypothesis as requiring 'a continuous series of micracles.'"[75] Both Morrison and Grünbaum felt that Dingle's wisecrack was unfair to the steady-state theory, which certainly did not rely more on miracles than did the big-bang theory. In a letter to Grünbaum, Bondi acknowledged the intervention of the American philosopher. "Naturally I found it particularly enjoyable that you discussed the matter with reference to an unsound criticism of our theory," Bondi wrote, referring to his and Gold's theory.[76]

Dingle's accusation of continual creation being miraculous was not the last of such claims, however. In 1959 the Argentine physicist and philosopher Mario Bunge, whose position can roughly be described as a marxist-oriented positivism, claimed that the steady-state theory was nonscientific because it violated what he called the genetic principle, that is, that nothing comes out of nothing or passes into nothing. Approvingly quoting Dingle's address, Bunge stated that "the concept of emergence out of nothing is characteristically theo-logical or magical—even if clothed in mathematical form."[77] A similar objec-tion was later raised by the Swiss-American astronomer Fritz Zwicky, who found that steady-state cosmology was "not a theory at all, scientifically speaking" because of the sudden creation of matter out of nothing. Zwicky ironically suggested admitting from the start that stars and galaxies are created from nothing, all at the proper time and proper rate, "since by this bold hy-pothesis we can easily overcome all the difficult theoretical problems con-fronting us."[78]

At the time he wrote the letter to *Scientific American*, Grünbaum was deeply interested in cosmology and its philosophical implications. The previ-ous year, 1952, he had given a comprehensive review of the subject in which he included not only the steady-state theory, but also the new cosmology of Gamow, Alpher, and Herman.[79] Anticipating Dingle's objection and support-ing Hoyle's, Grünbaum pointed out that as far as the *ignorabimus* attitude was concerned, Gamow was even more guilty than the steady-state theoreticians: in Gamow's big-bang theory, the beginning of the universe was a process which in principle evaded scientific analysis. Many years later, Grünbaum again examined the question of creation in cosmology, both in the big-bang theory and in the then defunct steady-state theory. He now concluded that the cosmologists in the 1950s as well as their successors had confused the ques-tion of the temporal origin of the universe with the question of the creation of matter in the universe. Whatever the kind of cosmology, the latter is a pseudo-problem, Grünbaum argued.[80]

Debate over Creation

Whereas Grünbaum considered the steady-state theory to be perfectly scien-tific, the theory was severely criticized by another American philosopher, Milton Munitz, who in 1954 offered a more detailed and sober version of Dingle's criticism.[81] Munitz had earlier attacked the philosophical basis of rationalistic cosmologies, namely, those of Eddington and Milne. In accor-dance with the view of Dingle, he found these cosmological theories to be modern versions of Platonic idealism.[82] Munitz now targeted the concept of continual creation of matter, which he claimed was meaningless and part of a mystifying supernaturalism. It was not even a necessary consequence of the steady-state theory, he suggested. The steady-state creation of matter caused a heated debate among physicists, astronomers, and—in particular—philoso-phers, most of whom tended to see this element of the theory as an unforgiv-able sin, almost a crime against science.

The passions aroused over this question are a bit surprising, not least in the light of the fact that it was far from the first time that scientists had seriously proposed small-scale violation of the law of energy conservation. This law, also known as the first law of thermodynamics, was established in the

1840s and soon acquired a paradigmatic status. However, the impression that the law remained unquestioned—that it became either a dogma or a tautology—is wrong. On the contrary, its absolute validity was constantly discussed and often questioned. For example, in late Victorian physics it was sometimes argued that the law only had validity if it incorporated both the realm of material substances and the ether, and that it would therefore presumably fail in ordinary experiments confined to the materialist cosmos.[83] In the early part of the twentieth century, the law was subjected to a major reinterpretation, now becoming a law of conservation of combined energy and matter according to Einsteinian mass-energy equivalence. Later on, in an intellectual climate very different from that of Victorian physics, the idea of energy nonconservation continued to be defended. Within the context of quantum theory it was suggested several times in the 1930s by authorities such as Bohr, Dirac, Jordan, Landau, and Schrödinger. The steady-state physicists were thus in quite good company. Yet the discussion in the 1950s ignored the history of the subject and went on as if it were the first time in the history of modern science that someone had dared to question the law of energy conservation. Since Bondi, Gold, and Hoyle never mentioned their historical precursors or argued for energy nonconservation by means of historical arguments, one may assume that they were not aware of this part of the history of modern physics.

Munitz's careful analysis, being the first and still one of the best philosophical examinations of the steady-state theory, merits close attention. He divided his criticism into epistemological and methodological parts, the first primarily directed against Hoyle and the latter against Bondi. As in the case of Dingle and most other philosophers entering the debate, Munitz felt provoked to intervene because of the appeal to philosophy and the scientific method so openly made by the steady-state trio. (In a sociological perspective, this may be seen as an attempt to demarcate the territory of philosophy of science from that of science.) William Bonnor, who was himself a steady-state adversary, noticed in a review of Munitz's book, "it is highly unusual now for philosophers to criticize the professional writings of physical scientists, yet this is what Professor Munitz most certainly does." Bonnor believed that this indicated the lack of maturity of cosmology as a science: "Since most of the [cosmological] theories are well able to explain all the facts, each author resorts to philosophical principles to show the superiority of his own theory. Cosmologists thus become fair game for Professor Munitz."[84] In fact, it was the Cambridge trio's use of philosophical arguments that made the steady-state theory a target for some philosophers, not the theory as such. It is remarkable that Dingle and Munitz used almost no space to comment on the big-bang theories of Gamow or Lemaître, which was undoubtedly because the proponents of this class of theories did not make their philosophical premises explicit. By refraining from philosophical arguments and sticking to physics, Gamow and his coworkers avoided the philosophers' critical analysis. Had big-bang theory been treated by the same critical standards as the philosophers used in their analysis of the steady-state theory, it would not have fared well. But in this respect Munitz's analysis was not impartial.

Munitz exploited skilfully the inconsistencies, different formulations, and philosophical weaknesses of the steady-state theory. Whereas Dingle had accused Hoyle of claiming that "a confession of ignorance is unscientific," Munitz argued that Hoyle's alleged claim that continual creation of matter was an ultimate, irreducible act was dogmatic and unscientific because it made an explanation (in terms of deductions from higher-level laws) impossible. Now it is questionable if a statement, in order to be scientific, needs to be derivable from some other statement (which then explains it), but Munitz at least held that this was the case. He interpreted steady-state continual creation of matter—because it was introduced as an ex nihilo process—as being unexplainable in principle and hence unscientific. His main argument was throughout against ex nihilo creation, a concept which he found invited "dogmatism and the surrender of the search for intelligibility" (almost a quotation of Hoyle's criticism of big-bang theory!). Hoyle, as well as Bondi and Gold, had intimated that creation of matter might be unexplainable in the sense that it was the natural state of affairs of the world, the background situation upon which all phenomena would have to be judged. For example, on one occasion Hoyle answered the question of the origin of the new matter by the following argument: "This is also [like the question of the origin of the universe] a meaningless question. It is only asked because in everyday life people have got used to the idea that matter must be conserved. When a conjuror pulls a rabbit out of his hat we know that the rabbit did not suddenly come into existence at the moment we see it and therefore it makes sense to ask 'Where did the rabbit come from?' But if the rabbit were indeed created by the conjuror, it might make no sense to ask this question."[85] Such a view Munitz held to be not only unscientific, but also meaningless. Yet it is difficult not to recognize the elementary force in Hoyle's argument. After all, which state of nature is the natural and unperturbed one, and hence not in need of explanation, is a question that cannot be answered by reference to tradition. About 1600, Aristotelians maintained that the natural state of sublunar bodies was rest or motion toward the center of the earth, and that any other motion needed explanation; but according to Galileo's principle of inertia (or rather to Newton's later version of the principle), uniform motion is the natural state of affairs that requires no explanation in terms of external causes. The only explanation of uniform motion the new physics could offer was that it followed from Newton's second law in the absence of an external force. It was in much the same way that Bondi and Gold suggested density conservation to be more natural than conservation of matter and energy.

Whereas Dingle considered Aristotle the spiritual ancestor of steady-state theory, Munitz believed the theory had its roots in Plato, who for him symbolized the nonscientific attitude. "It is from Plato that metaphysical idealism and supernaturalism derive their inspiration in constructing a cosmology," Munitz wrote, and continued: "As we turn to the present day, the manner in which we find creation appealed to in the steady-state theory is one which, in effect, carries this progressive mystification to its last stage. . . . Scientific cosmology, of course, now not only makes no claims about the designful character of the universe; it also stops short of making any reference to the Creator or the process of His making. It is not even claimed that these are mysteries whose existence is to be believed in even though not understood. All that it would retain is the fact that matter in an elemental form is created continuously. But if the Maker, the process of making, and the purpose are gone, *what is there left of the concept of creation?*[86] Exactly the same argument could be raised against big-bang creation but, as mentioned, Munitz did not use much of his ammunition against the philosophical basis of this theory. He did point out, however, that evolutionary cosmologies with a sudden origin of the universe also made the methodological error of indulging in an *ignorabimus*. Munitz was no more attracted by the big-bang theory than was Dingle, and like his British colleague he refrained from advocating a particular cosmological model or view. Another American philosopher and early commentator on the theories of Gamow and Hoyle, Oliver Reiser, had no such reservations. He followed Munitz in criticizing the big-bang as well as the steady-state theory for presenting matter creation as inexplicable, but contrary to other philosophers he suggested his own, highly speculative cosmological theory, a pantheistic "cyclic-creative universe."[87]

The issue of whether there are legitimate questions of cosmology to which science will never be able to provide an answer was an important component of the philosophical debate associated with the cosmological controversy. In a general way, this discussion was a continuation of an old debate over the boundaries between science and faith that had taken place for almost a century. It was highlighted by the German physiologist Emil Du Bois-Reymond, who in 1872 gave a famous lecture in which he argued that there were indeed such mysteries of the universe (*Welträthsel*) that could never be decided by scientific means. Hence Du Bois-Reymond's motto: *Ignorabimus*—we shall remain ignorant. Du Bois-Reymond had in mind such mysteries as the essence of matter, the origin of life, and free will, and not the origin of the universe; but otherwise the discussion between cosmologists and philosophers in the 1950s had much in common with the discussion in the late nineteenth century.

According to Munitz, continual creation was a meaningless concept. Meaningless or not, although the creation of matter in steady-state theory could not be verified directly, it was in principle a verifiable concept. Future experiments might prove that hydrogen atoms are in fact created with the rate predicted by the theory, but even then Munitz would claim that Hoyle's *concept* of continual creation was meaningless because it was an ex nihilo process with no mechanism or purpose. Yet it was not at all essential to the steady-state theory that the creation of matter took place in a way that made it meaningless to ask why and how the new matter was born. All that Hoyle, Bondi, and Gold claimed was that matter is continually created. Munitz's objection was that creation was not merely the appearance of elemental matter where none had been before, but the unexplainable appearance of such matter out of nothing. "The authors of the theory insist on taking the second expression as the correct one," Munitz wrote; "but whatever strength there is in the content of their proposals and methodological soundness in their procedure lie in actually using the first expression. To say that matter is found in the universe leaves open the possibility of explaining its appearance, whereas to say that it is created not only denies such a possibility but also *employs a term without any significant content*."[88]

Munitz pointed to what may appear to be a weakness in the arguments of the steady-state advocates, namely, their insistence on ex nihilo creation. When steady-state theory appeared they may have felt that this was the best strategy to follow in order to avoid questions of how and why the new matter came into existence. But it was surely unnecessary to insist on this point. Hoyle, Bondi, Gold, and McCrea could have said that for the time being it was not possible to account for the creation of matter, but that this might well be possible in some future development of the theory. Such an answer, which would have left the entire steady-state theory intact, would presumably have satisfied Munitz. If a mechanism for the creation could be found it would not change the theory a bit. As we have seen, this was what happened when McCrea, McVittie, and Pirani introduced the idea of transformation of negative-energy stress as a kind of cause of the continual creation of matter. Munitz admitted this as a methodological improvement, but believed that with this change in the foundation of the theory it was no longer justified to speak of creation of matter at all.

It should be noticed that Munitz's demand that creation of matter be explained, or at least explainable, did not extend to a corresponding demand for conservation of matter to be explainable. Is there an explanation of matter conservation? If so, the explanation lies in the general theory of relativity, from the equations of which matter conservation follows. But Hoyle's modified equations had the very same structure as Einstein's, only including the C tensor, and so these equations explain creation of matter in the same abstract sense that the field equations of general relativity explain matter conservation. As Bonnor remarked: "Does Professor Munitz demand that creation of matter be given *more* explanation than the conservation of matter?"[89] Hoyle made a similar point when he argued that physicists do not try to explain electricity, gravitation, and magnetism causally, by asking why there are such forces or where they come from. They "explain" such phenomena by working out testable consequences based on the hypotheses that they exist, by asking *how* the phenomena operate and not *why* they exist. "Exactly the same situation applies to the creation of matter. We cannot say why matter is created or where it comes from, but we can say, 'If matter is created continuously, then it is created in such and such a way'."[90]

Many of the objections raised by Munitz were repeated in a stronger form and wording by Bunge, who believed that cosmology—not only the steady-state theory—was in a state of crisis and hardly deserved to be called scientific. The kind of cosmology that unfortunately flourished was "science-fiction cosmology," the worst example of which was the steady-state theory, which Bunge grouped together with "Eddingtonian neo-Pythagoreanism, ESP, psychoanalysis, and philosophical psychology."[91] Once again it was the continuous creation of matter that so provoked the physicist-philosopher. "The creation fantasy," he argued, was pure magic and an ad hoc hypothesis of the worst kind. The only reason for introducing continuous creation was to save the perfect cosmological principle, another unjustified dogma. Bunge had no sympathy for big-bang cosmologies either and indeed disliked relativistic cosmological models. In fact, he found the only unobjectionable element in the steady-state theory to be its insistence on the world being temporally infinite. Bunge's claim that cosmologies which operated with a finite age of the world were necessarily nonscientific implied that all the major cosmological models of the period were "magic," although some more than others. In general, the philosophical criticism was unconstructive and felt to be of little value by the scientists who were actively engaged in cosmological research.

The criticisms of Dingle, Munitz, and Bunge were not concerned only with the steady-state model, but with the entire project of scientifically studying the universe at large. Is cosmology a science like physics and chemistry? What are the criteria of truth and how do they differ from those adopted by other sciences? Which of the several competing theories is the most scientific? How sound is the conceptual basis of cosmology? These were questions discussed by astronomers, physicists, and philosophers of science. Undoubtedly inspired by the challenge from the steady-state theory, there was in the 1950s a great deal of philosophical interest in cosmology. One of the subjects which fascinated philosophers and philosophically minded scientists was the concept of the age of the universe, a central question in the cosmological controversy. In 1953 the *British Journal for the Philosophy of Science* announced a prize essay on this topic to be judged by a committee consisting of Harold Jeffreys, Fritz Paneth, Karl Popper, and Lancelot Whyte. Twenty-six essays were received, of which six were published the following year. In one of the essays, written by the American philosopher Michael Scriven, it was concluded that the notion of a beginning of the universe is unprovable in principle: "No verifiable claim can be made either that the universe has a finite age or that is has not. We may still believe that there is a difference between these claims: but the difference is one that is not within the power of science to determine, nor will it ever be."[92] It is symptomatic of the state and reputation of cosmology at the time that Scriven's essay, which denied the scientific validity of the central question of the cosmological controversy, was awarded the first prize in the competition.

Of course philosophers did not agree on this question any more than astronomers did. At about the same time, a British philosopher, Brian Ellis, argued that there are no compelling philosophical reasons against a cosmic beginning. Although the conclusion that the universe has a beginning in time cannot be the result of a direct observational report, Ellis argued that it could well be obtained observationally and hence was testable.[93]

The Bondi-Whitrow Dispute

The steady-state theoreticians did not respond to Munitz's criticism, but carried on the philosophical debate with the purpose of showing the methodological superiority of steady-state theory. It was mostly Bondi who explicitly engaged in philosophical discussion, as was the case in 1954 when he and Gerald Whitrow discussed the philosophical basis of cosmology in the pages of the *British Journal for the Philosophy of Science*.[94] Whitrow was a highly reputed British astronomer and mathematician who since the 1930s had worked with cosmological models, in part together with Milne. He had a strong interest also in the history and philosophy of science and, contrary to Dingle and Munitz, he held no prejudice against the kind of rationalistic cosmological theory represented by Milne and Bondi. Although critical of the steady-state theory, Whitrow shared in a general way some of the attitudes and values which characterized the steady-state theoreticians. For example, in a book of 1949 he wrote about cosmology that it is a subject in which "a clear understanding of the problem will always depend as much on fruitful theoretical concepts, freely constructed by the intellect, as on 'stubborn and irreducible' observational data"—a view which Bondi, Gold, and Hoyle would gladly have subscribed to.[95] However, although Whitrow's role in the cosmological debate was primarily that of a neutral referee, he came to view the steady-state theory with suspicion. He warned against cosmological arguments that presupposed that the universe is necessarily unique and self-contained, a feature characteristic of the steady-state theory.[96] Rather than limiting the boundaries of cosmological inquiry, Whitrow preferred to keep the possibilities open with regard to the scope and nature of cosmology. On the other hand, he was also critical of relativistic orthodoxy and believed that it was wrong just to dismiss alternative views.[97]

When Bondi and Whitrow engaged in their dialogue in 1954, it was not the first time they had crossed swords over cosmological models. The year before, Whitrow had challenged the steady-state protagonists with what he believed was a paradoxical situation specifically connected with their world model.[98] If the paradox was real it would imply an inconsistency in the logic of the steady-state theory, and hence make it a less attractive model compared with the evolutionary models. Whitrow pointed out that the steady-state model had what he called a world horizon, a distance from each fundamental observer (galaxy) at which matter recedes with the velocity of light, so that signals originating outside the sphere can never reach the observer. The horizon problem involved in a de Sitter universe was discussed long before the discovery of the universal expansion, first in the debate between Einstein and de Sitter in 1917. It was later illustrated by Eddington, according to whom "light is like a runner on an expanding track with the winning-post receding faster than he can run."[99] Various kinds of horizons also turn up in other cosmological models, but it was a topic that had never received serious and systematic attention. Hoyle had noticed the existence of a horizon in the steady-state theory, but without making its meaning clear or considering it a problematic feature.[100]

Whitrow noted that in the steady-state theory the horizons of two fundamental observers A and B in causal contact will be different. From this it follows that if monochromatic light is emitted in the direction CBA from a galaxy C situated just at the horizon of A, then it will arrive at A with an infinite wavelength because it is Doppler shifted according to the recessional velocity $v = c$ of C. However, one can imagine that when the beam of light passes B with some wavelength λ, it, or part of it, is absorbed and instantly a new beam of light with the same wavelength is sent toward A. The two cases should be indistinguishable, and yet they are not so, for in the second case the light received at A will only be Doppler shifted by a finite amount. This was Whitrow's paradox.

A brief controversy followed after a rather brash reply from Bondi and Gold, who apparently saw Whitrow's query as an attempt to discredit the steady-state theory. They denied that there was any difficulty at all and claimed that Whitrow's paradox was purely illusory: "We do not know what hurricanes will be directed against our cosmological edifice; but we are a little aggrieved to think that it is being credited with so little structural strength that Dr. Whitrow's puff could make it shudder."[101] Bondi and Gold did not, in fact, answer the puzzle posed by Whitrow, and it was only in August 1954 that a satisfactory explanation was offered. Pirani showed that the misunderandings between Whitrow and the three steady-state physicists were due to different meanings being ascribed to the concept of the observable region. A closer analysis showed that the steady-state horizon, in Whitrow's meaning, was a spherical shell not of finite, but of infinite radius.[102] As Whitrow admitted, this leads to a resolution of the paradox, for then there will be no sudden disappearance of galaxies. The total number of galaxies visible to an ideal telescope of unlimited power will be infinite; however, the number of galaxies visible to any real telescope, however powerful, will be finite because the intensity of the received radiation from highly redshifted galaxies will tend to zero.

It turned out, then, that Whitrow's horizon paradox, although more than a "puff," posed no real threat to steady-state cosmology. In this sense the controversy was a victory for Bondi, Gold, and Hoyle. As a result of the discussion a closer study of the horizon problems was taken up, as they appear both in the steady-state theory and in relativistic models. The horizon of the steady-state theory was clarified by Gold and Hoyle in a discussion with an anonymous author, presumably Whitrow.[103] Following Pirani's work, the subject was studied by the Indian physicist Sen Gupta, and more carefully and systematically by Wolfgang Rindler, a student of Whitrow's.[104] Coming from pure mathematics, Rindler had taught himself relativity and cosmology from the books by McVittie, Eddington, and Tolman. At the time he clarified the horizon problem, he had a clear sympathy for the steady-state theory.[105]

Although the paradox was illusory in steady-state theory, this theory behaved epistemically so strangely in other respects that Whitrow felt that it could not be the correct description of the universe. It can be shown that the steady-state model will at all times contain regions of galaxies that cannot be observed, even in principle, and that the universe thus contains infinitely many subuniverses which are causally unconnected. As Whitrow remarked, this is indeed a puzzling situation "from the point of view of the theory of knowledge" since the model then postulates the existence of galaxies that are, always have been, and always will be unknowable by empirical means.[106] How seriously can one take such a model as a candidate for the real universe? Can a model with this degree of built-in *ignorabimus* claim to be truly scientific?

Related to the conceptual problems pointed out by Whitrow is the consequence that not only are there some galaxies that are unobservable, but there are also galaxies that are of gigantic mass (and, in both cases, an infinite number of them). Since extremely old galaxies continue to grow in mass by accreting the material continuously formed throughout space, there will be a large number of extremely massive galaxies, for example, billions of times more massive than the Milky Way. That such objects are not observed may be explained by the fact that they have long ago passed beyond the horizon, but even so there will be problems. As McVittie pointed out, an observer placed near such a supermassive galaxy would have a very different world picture than observers in our observable universe.[107] Such an observer would not see, even at a very large scale, a homogeneous universe and this contradicts the cosmological principle on which the steady-state theory builds.

The conceptual objections of Whitrow and McVittie were attempts to show that the steady-state theory led to consequences that were contradictory, or at least highly bizarre. An objection in the same category was raised by Richard Schlegel in 1962, namely, that the number of atoms (or galaxies) in the steady-state universe is too large.[108] There is, of course, an infinite number of particles in the steady-state universe, but mathematically an infinity is not just an infinity. There are different kinds of infinities, and Schlegel argued that for physical spaces—where atoms occupy a volume different from zero—the number of particles cannot possibly be larger than a so-called denumerable infinity. Since he believed he had demonstrated that this was the the case in the steady-state theory, he considered the theory contradictory. None of the steady-state physicists responded to the charge, which in any case turned out to be based on shaky mathematical ground.[109]

Now return to the 1954 discussion between Whitrow and Bondi, the key question of which was whether modern cosmology is a science or not. That this question could be seriously discussed at all showed, if nothing else, the lack of professional maturity of the field. The opinion of philosophers was one thing, but this was a discussion between two eminent scientists. According to Whitrow, cosmology anno 1954 was not truly scientific; moreover, it was unlikely ever to become a science in the ordinary sense. His main reason for this pessimistic view was the very nature of cosmology, the study of an absolutely unique object, the universe. This uniqueness had made cosmology "a borderline subject between the special sciences and philosophy," and Whitrow believed that it would remain such a borderline semiscience. Cosmology's lack of scientific character was socially reflected in the lack of unamimity among the practitioners. According to Whitrow, agreement about fundamental standards and methods was a hallmark of science, and who could claim that such agreement existed between steady-state theory and relativistic evolution cosmology?

Bondi did not deny the disagreement between cosmologists about fundamental issues, but viewed the situation more optimistically. According to him, cosmology was quickly on its way to becoming a truly scientific endeavor and was indeed already a science. As to lack of unanimity, this only showed that physical cosmology was a young science which had not yet entered a phase of maturity. Bondi found this a natural state of young sciences, and likened the state of cosmology to those of other young sciences such as psychology and virology. The disagreement between Whitrow and Bondi with regard to this particular question depended on their different conceptions of the historical development of cosmology. In Whitrow's view, cosmology was a very old subject which had been cultivated since antiquity; hence it was unlikely that unamimity would be achieved within a short span of time, and more reasonable to conclude that lack of unanimity was an integral part of the field. Bondi, on the other hand, would only admit physical cosmology a history of about one hundred years, say, from the recognition of the galaxies as the building blocks of the universe. From this perspective it made sense to compare the scientific state of cosmology with those of psychology and virology. In addition to "lack of unamimity" as an indicator of lack of scientific character, both Whitrow and Bondi mentioned "importance of philosophical considerations." Philosophy's role was reduced to the heuristic level in mature sciences such as physics and chemistry, but was still of great importance in cosmology. Again, whereas Bondi believed that observation and physical theory had now largely replaced philosophy also in cosmology, Whitrow argued that philosophical arguments would always remain an essential part of the study of the universe.

Theory and Observational Facts

The following year, 1955, featured a celebration of the fiftieth anniversary of the theory of relativity in Berne, the Swiss city where Einstein worked as a patent clerk in 1905. At a conference there in July, Bondi gave a talk on the steady-state theory in which he spelled out the same point he had discussed with Whitrow. "I should like to draw attention to how scientific cosmology has become," Bondi said. He illustrated his claim as follows: "The cosmological papers today have all dealt with empirical tests of cosmological theories and nobody has referred to how satisfying or how beautiful or how logical this or that theory is. A few years ago, the emphasis would probably have been the other way round."[110] Yet aesthetic and philosophical considerations remained important ingredients in the continuing cosmological debate, including Bondi's own arguments.

One reason for Bondi's optimistic belief in the scientific level of cosmology was methodological, namely, that in spite of other disagreements, there was, he claimed, broad agreement with regard to what criteria should be used to settle controversies. Personal taste and metaphysical commitments such as preference for simplicity would always play a role, but a firm basis for demarcating science from nonscience could only be found in the possibility of experimental disproof: "Although the adherents of some theories are particularly concerned with pointing out the logical beauty and simplicity of their arguments, every advocate of any theory will always be found to stress especially the supposedly excellent agreement between the forecasts of this theory and the sparse observational results. The acceptance of the possibility of experimental and observational disproof of any theory is as universal and undisputed in cosmology as in any other science, and, though the possibility of logical disproof is not denied in cosmology, it is not denied in any other science either. By this test, the cardinal test of any science, modern cosmology must be regarded a science."[111] Whitrow largely agreed with Bondi's Popperian view, but warned against considering falsifiability a final and absolute criterion. For one thing, Whitrow pointed out, testing is a difficult procedure because we tend to read our theories into our experimental results. The warning against relying on "facts pure and simple" was readily accepted by Bondi, who admitted that there were situations where fact should be subordinated to theory. For example, "when an extremely difficult and uncertain experiment contradicts a hypothesis otherwise exceedingly well confirmed by experiments, the results of this uncertain experiment are disregarded." In general, although Whitrow and Bondi conceived cosmology differently, they had no difficulty in communicating fruitfully, and they both adopted the kind of methodology which would later be called sophisticated falsificationism.

Bondi's view on the matter is further illustrated by an article which, after several years delay, was published in the first volume of the annual review *Vistas in Astronomy* in 1955. Observational astronomers—by far the largest fraction of the astronomical community—were almost unanimously against the steady-state theory, which they felt was a theoretical construction without a solid foundation in observational fact. Possibly inspired by Williamson's hostile review of Hoyle's broadcast series, Bondi decided to problematize the notion of observational fact in astronomy. Mainly by relying on examples from the modern history of astronomy, he argued that observational knowledge is no more reliable than theoretical knowledge; and, furthermore, that errors in observation are likely to be more frequent and persist for longer than errors in theory. When presenting his paper to the Royal Astronomical Society in April 1952, he contrasted the uncritical atmosphere of observational astronomy with "the healthier atmosphere of scientific scepticism and frank discussion common in theoretical astronomy," a claim which naturally provoked Dingle (the society's president) and the observational astronomers present at the meeting.[112] Bondi's provocation aroused considerable hostility among astronomers, with the result that his paper was rejected for publication in the Royal Astronomical Society's *Monthly Notices*.[113]

An observational result is not just a fact and certainly not a hard and incontrovertible one, Bondi pointed out, but the end of a long chain of inferences in which the raw data are only a small part. Simply to talk about observational results as facts was "an insult to the labours of the observer, a mistaken attempt to discredit theorists, . . . the curse of modern astronomy." No, the situation was that "in observational work long chains of inferences are based frequently on somewhat uncertain data, whereas in physical theories of astronomy, though long chains of inferences are also used, they are generally based on much more reliable experimental data."[114] Furthermore, while theoretical work was always checked and received with a critical attitude, there was a widespread prejudice that observational results were by their very nature reliable and not in need of critical checking or repetition. To Bondi it was particularly important to destroy the myth of the supremacy of observational knowledge, because of the conflicts between observational and theoretical results that so often turned up in modern astronomy. In such conflicts one should not accept the observational claims without critical examination, and it would often be advisable to give theoretical results priority over observations. Bondi did not mention the current controversy over cosmology among his examples, but there can be little doubt that he had it in mind. Many astronomers believed that the steady-state theory was already refuted by observations, and it was an important part of the strategy of Bondi and his allies to weaken the objections by arguing the relative lack of reliability of observational inferences. As we shall see in chapter 6, the steady-state advocates were quite successful in this line of defense.

Although Bondi's view later received a good deal of philosophical support and today is uncontroversial, in the 1950s it was widely seen as heretical. Observational astronomers, in particular, disliked it, and sometimes associated it with what they felt was steady-state cosmology's preference for theory and disdain of observations. Allan Sandage, the American astronomer, later reported that Bondi's position contributed to the bad reputation of the steady-state theory in the United States: "[Bondi's] first theorem is that whenever there is a conflict between a well-established theory and observations, it's always the observations that are wrong. And when he announced the first theorem in England and it was read by the Mount Wilson astronomers, they just dismissed all the steady state boys. That was the beginning of the rejection. Bondi was associated with Gold and Hoyle, and they made such outrageous statements that they just couldn't be believed."[115] Bondi's view was in broad agreement with the view expounded by Hoyle and Lyttleton in 1948, when they stressed that "the significance of observational data cannot adequately be assessed until the data begin to be theoretically understood."[116] Intuition, according to Bondi, Gold, Hoyle, and Lyttleton, has a legitimate role to play in the testing of theories, and this is of course problematical. When it comes to major changes in the world picture there are no standard intuitions common to scientists. Intuitive reasoning becomes subjective and arbitrary. Sandage's intuition with regard to the structure of the world differed from that of the steady-state physicists.

That the relationship between theory and observation was more complex than imagined by many scientists had been known to philosophers for a long time. The dependency of data on the theoretical framework had been discussed by the French philosopher and chemist Pierre Duhem at the beginning of the century, but did not become generally accepted until the 1960s. In 1962 Ernest Hutten, a British philosopher of science, gave indirect support to Bondi by criticizing astronomers for having been too hasty in accepting data as evidence for a theory which is not logically acceptable. He had in mind the big-bang theory and the recent radioastronomical measurements of Martin Ryle, which were widely interpreted as a disproof of steady-state cosmology.[117]

If Bondi's paper was seen as methodologically provocative by orthodox scientists, so, undoubtedly, was a paper published by Gold one year later. Like Bondi, he discussed astronomy and cosmology in a general, philosophical perspective, concluding that the (one and only) "scientific method" so highly rated by Dingle and others was an illusion. Nineteen years before Paul Feyerabend started preaching his "anything goes" antiphilosophy of science, Gold had this to say about the dignified scientific method:

> In no subject is there a rule, compliance with which will lead to new knowledge or better understanding. Skilful observations, ingenious ideas, cunning tricks, daring suggestions, laborious calculations, all these may be required to advance a subject. Occasionally the conventional approach in a subject has to be studiously followed; on other occasions it has to be ruthlessly disregarded. Which of these methods, or in what order they should be employed is generally unpredictable. Analogies drawn from the history of science are frequently claimed to be a guide; but, as with forecasting the next game of roulette, the existence of the best analogy to the present

is no guide whatever to the future. The most valuable lesson to be learnt from the history of scientific progress is how misleading and strangling such analogies have been, and how success has come to those who ignored them.[118]

This is quite a remarkable statement, anticipating much of what happened in the philosophy of science during the following two decades.

McCrea's Uncertainties

The discussion about cosmology's scientific status was given a particular twist some years later in a brief exchange of opinions between McCrea and the British theoretical astronomer William Davidson, a former student of McCrea's. The years which had passed since the Whitrow-Bondi discussion had witnessed the controversy between the evolutionary models of the universe and the steady-state model, but also an increased feeling that observations might, in fact, decide between the rival theories in a way similar to how observations act in the more ordinary branches of science. In 1960, the steady-state theory was under pressure from radioastronomical measurements, which may have been the background for McCrea's suggestion that the distinction between the two theories might be meaningless. Such a suggestion would imply that neither of the theories could meaningfully be said to be the correct one, and that the steady-state theory was therefore saved from being refuted.

From very general, philosophical arguments, McCrea reasoned that in a universe with an infinite number of objects we cannot by observing the universe for any finite time get sufficient information to make exact predictions. McCrea's argument was based on two uncontroversial facts, the speed of light as the maximum speed of any signal, and the expansion of the universe. From this he concluded that "even in principle, we cannot predict the behaviour of a remote part of the universe with as much assurance as we can predict that of a nearer part."[119] The unpredictability he had in mind was unavoidable, a kind of cosmological uncertainty principle similar to Heisenberg's uncertainty principle in the subatomic domain. McCrea argued that this uncertainty would be proportional to the cosmological redshift $(1 + z)$ and thus would be serious for very large distances only. But since it was exactly at these distances that the various cosmological theories differed in their predictions, it followed that "the difference between these theories loses any meaning." McCrea's conclusion was this: "In fact, all problems concerning the creation of matter as hitherto formulated appear to lose significance. It is suggested that this is basically why all efforts to reach a decision between the two types of theory by observational tests have failed." He thus supplied arguments for the view that Whitrow had defended some years earlier, namely, that it was in the very nature of cosmology that it would never become a science on a par with ordinary physics.

McCrea's disillusionment was related to the criticism of cosmological theory which originally had been put forward by Bondi and Gold in their paper of 1948. The lack of repeatability of the universe as a whole implied, he argued, a limitation in our theoretical knowledge about the universe. Science is, at least in principle, able to predict the behavior of small numbers of galaxies, for we can assume that arbitrarily many realizations of such systems exist, and in this sense they are repeatable. "But, as we extend the scope of our enquiries to include more and more of the universe, we should expect predicting to merge into recording, until for the most general features of the universe we can do no more than record."[120] The attractiveness of steady-state theory, as McCrea had experienced it in 1948, was that it promised a new world picture by bringing creation processes into physics. Now, twelve years later, he had become disillusioned, and tended to believe that a precise formulation of steady-state theory, valid for the universe as a whole, could not only not be obtained, but would be meaningless. "Much of the attractiveness [of the steady-state concept] disappears, however, when we realize that what looks to be such a revolution of ideas results in observable differences that, even on the interpretation supplied with the theory, are almost impossible to detect."[121]

McCrea developed his radical view in a paper of 1962 and also in his 1963 presidential address to the Royal Astronomical Society. He now argued that the knowledge of the universe has to have epistemic priority over physical laws, and that this makes it "absurd" to use laws of physics to predict or explain the behavior of the universe. As mentioned, the problem area with which he groped was a sophisticated version of the one dealt with by Bondi and Gold in their work of 1948. Their argument, that the cosmic applicability of physical laws was only justified in a nonchanging universe, was wholly supported by McCrea: "I take the view that in some sense the universe determines its laws. So, if the universe changes, its laws must change. The way in which they do so must be unpredictable. For, were it predictable, we should be using a 'meta-law' not depending on the state of the universe, to predict the change in the law."[122] Following Bondi and Gold, McCrea argued that ordinary physical laws, formulated as differential equations, could not exist for the universe; such laws must be independent of the system, and a physical law existing apart from the universe is inconceivable. Similarly, in order to make predictions from a law, boundary conditions are needed, and no such conditions can be formulated for the universe as a whole. However, McCrea did not find the existing steady-state model to provide a way out of the dilemma, because the domain of applicability of the perfect cosmological principle was too ill defined. He was therefore driven to repeat his conclusion of 1960, that "the universe is fundamentally unpredictable."

The pessimistic conclusion was in part based on an investigation made by Phil Morrison and his student A. Metzner, who applied information theory to calculate the flow of information in various cosmological models.[123] The two Americans found that while complete information could be gained from nearby sources, information would vanish as the boundary of the observable universe was approached. Furthermore, for big-bang models, Morrison and Metzner concluded that no information at all could be obtained about the creation event postulated by these models. Their work reinforced McCrea's belief in the unavoidable restriction in knowledge that could be obtained in cosmology.

Davidson objected that McCrea's position was prematurely pessimistic and that there was no reason to believe that there were any serious inherent limitations in the observational knowledge that could be gained from the universe.[124] According to Davidson, the game of choosing between different cosmological theories did not differ fundamentally from what goes on in local physics. It was a matter of fitting the facts to the most likely interpretation, whereas McCrea's suggestion would amount to giving up cosmology as a science. Davidson granted that there were all kinds of difficulties, but also noticed that the steady-state theory nonetheless led to quite definite predictions, which not only could be tested, but were in fact being tested. However, all tests, and indeed the entire foundation of scientific cosmology, rested on the cosmological principle, in either its restricted or perfect form. Most astronomers, including Davidson, felt that if the cosmological principle was rejected there would be no firm basis on which to found cosmology. But the cosmological principle is itself based on an extrapolation of finite observations, which would not be justified if McCrea's suggestion was accepted. As McCrea wrote: "Following these ideas, the cosmological principle would now assert that the universe as seen by a distant observer is like the universe as seen by us to within a factor of uncertainty $(1 + z)$ in the description. Thus we should be asserting almost nothing about what the universe is like at great distances (in space or time). This provides a view of cosmology that essentially leaves room for endless observational surprises. It seems more satisfactory than the recent trend towards a belief that the nature of the whole universe has already been discovered."[125]

However, at the time when McCrea made his radical suggestion, observational cosmology was already on its way to making it obsolete. It was Davidson's optimistic view that turned out to be justified by the developments taking place over the next few years. In 1970, when the situation in cosmology had largely been clarified and the big-bang model established as the standard theory of the universe, McCrea recanted his pessimism. Impressed by the progress in cosmology, he now suggested what he called a principle of compensation according to which "The less information we can get, the less we need in order to make predictions that are testable by observation."[126] This principle, which was largely a rationalization of the last decade's history of cosmology, was, as McCrea noted, complementary to his earlier principle of uncertainty. However, he interpreted it to mean that it would in principle be impossible to have precise knowledge of the earliest stages of the state of the universe. Even this remnant of pessimism has proved untenable.

The question of the scientific nature of cosmology was discussed in 1961 from a logical-philosophical perspective by the Oxford philosopher Rom Harré.[127] As a starting point, Harré adopted a modified version of Bondi's and Gold's old classification of two different approaches to cosmology. What

Bondi and Gold in 1948 had called the extrapolatory approach—the extrapolation of locally valid concepts of physics to the entire universe—was called type elevation by Harré; and Bondi and Gold's alternative, the deductive method based on assuming a homogeneity postulate such as the perfect cosmological principle, was called the method of indefinite generalization. After having defined certain criteria for a theory to be scientific, Harré purportedly demonstrated that cosmogony (the study of the origin of the universe) is *not* a science. Theories of cosmogony, he argued, are either false, namely, if there was no first event; or, if there was a first event, then it is not identifiable by any law of nature. The study of the physical processes of the universe, called cosmophysics, was argued to be scientific, but only insofar as it is based on indefinite generalizations.

Although Harré's exercise in philosophical analysis dealt with cosmology only in an abstract sense, indirectly it provided support for the steady-state theory by denying the scientific nature of cosmogony, and also of cosmophysics if based on the extrapolatory approach characteristic of relativistic expansion theories. With regard to the continual creation of matter, Harré contradicted earlier critics by arguing that "it is wholly irrational to complain that the process by which this occurs has not been explained." According to Harré, the steady-state theory, like any cosmological theory, expresses a particular conceptual system, and the one adopted by Bondi and his allies leaves no room for other conceptual systems in which matter creation can be explained.[128] So at least one distinguished philosopher was willing to defend continual creation of matter as a rational idea. Another philosopher of distinction, the American Norwood Russell Hanson, also dealt with the situation in cosmology and the concept of matter creation, which he considered to be equally serious in the two competing theories. According to Russell Hanson much of the controversy between the steady-state view and the "Disneyoid picture" of a big bang was semantic in nature, that is, rooted in different meanings given to words such as "creation," "universe," and "energy conservation."[129]

In a very different way, the philosophical aspects of steady-state cosmology were taken up by Bernard Lovell, the leading radio astronomer and director of the Jodrell Bank Observatory. In a series of lectures broadcast on the BBC in 1958, he reflected on the two rival theories of cosmology without clearly preferring either of them. As far as creation of matter was concerned, Lovell agreed with the steady-state advocates that this theory was more scientific—or rather "materialistic," as he preferred it—than the big-bang theory. Repeating earlier steady-state arguments, he claimed that the big bang would forever be beyond scientific understanding, whereas continual creation of matter was at least verifiable in principle. "Philosophically," he said, "space and time had a natural beginning when the conditions of multiplicity occurred, but the beginning itself is quite inaccessible." This was a view in complete accord with, indeed a repetition of, the one which Lemaître had stated in 1931 and more recently repeated at the 1958 Solvay Congress. Lovell claimed that even if continual creation of matter were to be proved a fact, the formation of hydrogen atoms would still be a series of miracles because the theory provided no "information about the nature of the energy input which gave rise to the created atom."[130] This was clearly a misconception of the steady-state theory, in which, as its founders had often stressed, creation of matter was not an energy-to-matter conversion but a creation out of nothing. At the time when Lovell read his lecture, there were even possible explanations of the creation mechanism, such as McCrea's suggestion that a cosmic negative pressure provided the energy input. However, this was not really relevant to Lovell, who was convinced that creation of cosmic matter, whether abruptly or continuously, had to be a divine act. He freely admitted an *ignorabimus* attitude, having metaphysics and religion coming in where science had to give up. As remarked by Grünbaum, he thus shared with Dingle the mistaken claim that creation of matter would be miraculous, although Lovell did not use the claim in rejecting the steady-state theory in particular.[131]

The reluctance of philosophers to accept cosmology as a science is further illustrated by Stephen Toulmin in his book *Discovery of Time*, written with June Goodfield shortly before the path-breaking events of 1965. "Cosmological theory is still basically philosophical," Toulmin and Goodfield declared.[132] They based their judgment on the disagreement about fundamentals concepts and methods, and also argued that cosmology still faced the same objections that were raised by Kant in his *Critique of Pure Reason* in 1781. Arguing that cosmology, in order to become a science, must abandon its a priori principles and concentrate on the accumulation of empirical data, Toulmin and Goodfield approvingly quoted the great philosopher from Königsberg: "At the present stage in physical cosmology, it is no good our being in a hurry. As in history and archaeology, the only hope of solid results lies in a step-by-step advance. There will always be some point in our map of the past history of the universe beyond which we have nothing particular to show. The same goes for the future—and how could it be otherwise? If we force astrophysics to serve us with a revised version of Genesis and Revelation we dig a pit for ourselves."[133] That is, the skeptical attitude of Toulmin and Goodfield was essentially the same as that held by Dingle and Munitz. It was not an unreasonable attitude in the mid-1950s, but to repeat it ten years later was a sign of ignorance about the progress which, after all, had taken place during the decade.

Popper and Cosmology

Popper's philosophy of science played an important role in the methodological discussions that were part of the cosmological controversy. The Austrian Karl Popper wrote his major work in the philosophy of science, entitled *Logik der Forschung*, in 1934. This work, later so famous, was discussed by continental philosophers associated with the positivistic Vienna circle, but it was not well known in the Anglo-Saxon world. In the 1950s, when Popper had settled in England as professor at the London School of Economics, he was not primarily known as a philosopher of science, but rather as a political and social philosopher. His critical rationalism, the name given to his philosophy of science, had received no systematic treatment in English and was presumably unknown among many scientists. However, to scientists with a philosophical interest, Popper's principal ideas were well known and appreciated. The general awareness of Popper's system had to await the publication of *The Logic of Scientific Discovery* in 1959, an enlarged English edition of his main work, which quickly brought him into the center of Anglo-Saxon philosophy of science. The fact that Popper's views were endorsed by Bondi and Gold years before the appearance of *The Logic of Scientific Discovery* presumably means that they knew of them through the early German edition.[134]

The element in Popper's philosophy that appealed in particular to the steady-state theoreticians was his insistence that a theory's scientific status and value are given by its degree of falsifiability. A good theory, according to Popper, is not one which is shielded from refutation, but on the contrary one which constantly faces refutation from experiments. Indeed, a theory which is so structured that it cannot be refuted empirically does not belong to science at all. The famous falsification criterion not only functioned as a demarcation between science and nonscience, it also established a criterion for which of two or more rival theories should be (temporarily) preferred, namely, the one with most "potential falsificators," or possibilities of refutation. It is easy to see why steady-state protagonists welcomed Popper's program. Since the very beginning of the steady-state theory, they had complained that the evolutionary theories did not allow precise predictions of a kind which, if contradicted by observations, would lead to their refutation. Contrariwise, steady-state cosmology led directly to a number of such crucial predictions, such as the age variation of galaxies and a definite space-time metric. Although they could not use Popper in defeating the evolutionary models, they could use him in arguing that, from a philosophical view, steady-state theory was superior to, that is, more scientific than, its rivals.

In his obituary of Popper, Bondi recalled that "to me his thoughts came as a flash of brilliant light."[135] Popperian themes, sometimes referring to Popper's name and sometimes not, were recurrent in the rhetoric of Bondi in particular. For example, when Bondi gave the Joule Memorial Lecture in Manchester in 1958, he included a strong support of Popper's views and stressed the vulnerability of the steady-state theory as agreeing with these. If just one prediction of the steady-state theory, such as the independence of distance of the density of galaxies, turned out to be wrong, he would not hesitate in declaring the theory "stone dead."[136] Two years earlier he had made it clear that he considered the foundation of the steady-state theory, the perfect cosmological principle, a scientific working hypothesis exactly because of its refutability. "This possibility of a clear-cut disproof establishes the scientific status of the P.C.P.," Bondi stated.[137] The high opinion of Popper's philosophy was evident in the enthusiastic review of *The Logic of Scientific Discovery* that Bondi wrote in 1959 together with Clive Kilmister, a mathematical physicist and Bondi's colleague at King's College, London. The two reviewers found that "Popper speaks as a working scientist to the working scientist in a language that time and again comes straight out of one's heart," and they pointed out how relevant Popper's analysis was to the situation in cosmology: "For here the correct argument has always been that the steady state model was the one that could be disproved most easily by observation. Therefore, it should take precedence over other less disprovable ones until it has been disproved."[138] On several later occasions, Bondi described himself as a follower of Karl Popper and praised his theory of science in similar words.[139]

In the late 1950s Bondi came to know Popper well, and they sometimes discussed the situation in current cosmology. However, Popper never intervened in the philosophical debate over cosmology, nor did he as much as mention modern cosmology in his works. He did have an interest in the subject, however, as is indicated by his role as a judge in the 1953 prize essay on the age of the universe. Popper had followed the development of relativistic cosmology since the early 1930s. On one occasion, in 1940, he entered the discussion in cosmology with an analysis of non-Doppler explanations of the redshift.[140] In this little-known paper, which was inspired by Milne's system of different time scales, he preferred a tired-light theory, but argued (in accordance with Milne) that such a theory was equivalent to the Doppler interpretation: "To ask whether 'in reality' the universe expands, or c [velocity of light] decreases, or the frequencies speed up, is not more legitimate than, when prices of goods fall throughout the economic system, to ask whether 'in reality' the value of money has increased or the value of the goods has decreased." Originally, when he learned about Friedmann's and Lemaître's ideas, Popper admired the big-bang theory. However, the admiration seems to have been short-lived. Popper came to dislike the big-bang theory, in part because of the inexplicability of a beginning of time, and in part because he felt the theory to be complicated and almost irrefutable because of its many auxiliary hypotheses. According to his view, it was scarcely a scientific theory at all. But although he found the steady-state theory much more acceptable than the big-bang theory, he did not like its foundation in the perfect cosmological principle. Popper shared with some other philosophers a distrust in principles of this sort, and apparently saw the merit of the steady-state theory in the continual creation of matter. When it was claimed that the original version of the steady-state theory had been refuted in the early 1960s, Popper remained opposed to big-bang theory. He believed that although the observed part of the universe was evolving, the universe might still be in a steady state over very large distances.[141] Such an idea was developed by Hoyle and Narlikar in the mid-1960s.

Continual Discussion

Many of the philosophical issues related to the controversy over steady-state theory turned up in a small book of 1960, based on a symposium arranged by the BBC and broadcast in late 1959.[142] The discussion, chaired by Whitrow and involving Bondi, Bonnor, and Lyttleton as speakers, summarizes much of the debate as it had taken place until the end of the 1950s. Although Bonnor represented relativistic orthodoxy, his views on cosmology were not completely antagonistic to those held by Bondi and the other steady-state theoreticians, which makes the discussion between the two even more illuminating. In a popular book of 1964, Bonnor summarized the controversy and his own position as follows: "The steady-state theory has aroused great controversy, some of it forthright to the point of rudeness, though the authors have defended their position with patience and restraint. . . . I must warn the reader at once that, from the scientific and mathematical points of view, I do not think very much of the steady-state theory, and I expect its rivalry to relativistic cosmology to be as short-lived as that of Milne's theory."[143] Bonnor was not far wrong in his prediction. From an epistemic point of view, however, he was much less opposed to the steady-state theory.

Bonnor was strongly against theories with a sudden creation of cosmic matter, in part for the same reasons as Bondi, Gold, and Hoyle, namely, that such theories introduced a divide between what is accessible to scientific analysis and what is not. According to Bonnor, it was plainly unscientific to call on metaphysics or theology when it came to the origin of expansion, or, for that matter, in some other area: "It is the business of science to offer rational explanations for all the events in the real world, and any scientist who calls on God to explain something is falling down on his job. This applies as much to the start of the expansion as to any other event. If the explanation is not forthcoming at once, the scientist must suspend judgment: but if he is worth his salt he will always maintain that a rational explanation will essentially be found. This is the one piece of dogmatism that a scientist can allow himself—and without it science would be in danger of giving way to superstition every time that a problem defied solution for a few years."[144]

In view of the earlier-mentioned criticism of steady-state cosmology for relying on miracles, it is interesting to notice that Bonnor—on this point in line with Hoyle, Bondi, and Gold—dismissed the big-bang theory because it, in his view, assumed a miracle. The following quotation from 1957 sounds very much like the criticism of a steady-state cosmologist: "To say that there is a singular state is an euphemestic way of confessing our ignorance. As far as the theory is concerned, the beginning of the expansion at time $t = 0$ is a *miracle*. . . . It seems to me that it is against the spirit of science to suppose, without a good reason, that there was a period in the history of the universe of which we cannot have any knowledge."[145] Rather than accepting a primeval singularity with all its pitfalls and associations with miraculous intervention, Bonnor wanted a singularity-free universe with an unlimited past and future— but one consistent with the general theory of relativity. His favored candidate was an oscillating or cyclical model in which the universe oscillated smoothly, i.e., with no big squeezes and no big bangs. This model had some of the conceptual advantages of steady-state theory, such as avoiding the question of the creation of the universe, while at the same time keeping on the firm ground of general relativity.[146]

Several years later, in 1965, a somewhat similar model was proposed by Jaroslav Pachner, a physicist at Prague's Technical University.[147] In 1960–61 Pachner proposed a new relativistic theory in which the cosmological constant was related to either creation of matter from nothing (if $\Lambda > 0$) or annihilation into nothing (if $\Lambda < 0$). However, because "the hypothesis of creation or annihilation of matter is strange to any scientific theory," he concluded that $\Lambda = 0$ and with this choice he ended up with an oscillating universe of a period of 44×10^9 years.[148] Pachner wavered on the subject of matter creation, and in 1965 he developed his model into a new version. He now built on McCrea's idea of a negative, universal stress and examined a model in which the stress was proportional to the fourth power of the space curvature ($-p/c^2 = aR^{-4}$). In this case he also found the energy density to depend on the curvature. Pachner was in this way led to an oscillating, singularity-free model which, in the present epoch of cosmic evolution, was practically indistinguishable from the finite-age Friedmann-Lemaître universe. Since the 1965 model could be ascribed an infinite age, and since it operated with matter continually created by the negative stress, it was a kind of steady-state universe. However, appearing at a time when steady-state cosmology was no longer taken very seriously, the model was not much noticed.

While Bonnor agreed with Bondi in his rationalism and in some of his criticism of relativistic evolution theories, he nonetheless rejected the steady-state alternative. He did so partly for observational reasons and partly for reasons of a philosophical nature. As to the latter, Bonnor argued that the steady-state theory was not really a fundamental theory in the ordinary sense, but at most a theory sketch or a phenomenological model. What was missing, he claimed, was a good explanation of why the universe is expanding. Without a proper mathematical foundation in terms of field equations, the scientific status of the steady-state theory was on the same level as Boyle's gas law before it was explained by the kinetic theory of gases. Bonnor was, of course, aware of the several attempts to supply steady-state theory with field equations, or otherwise to explain the expansion, but none of these impressed him as comparable to the beautiful mathematical structure of relativistic cosmologies. Like most other steady-state critics, he also saw continual creation of matter as highly objectionable because it violated the law of energy conservation. Although admitting that the law might turn out not to be strictly true, he considered continual creation to be such a revolutionary proposal that it would turn physics upside down, and hence it could only be accepted if all other alternatives failed. He claimed that energy nonconservation, even on the tiny level suggested by the steady-state theory, would "sweep away many of the most fundamental parts of physics," but did not mention which.[149] In this connection it may be relevant to recall that another fundamental law of physics, the conservation of parity (left-right symmetry), had been overthrown a few years earlier. Until 1956 it was generally believed that parity is universally conserved, yet it turned out that the law is not absolute, and fails to be satisfied in weak interactions. The discovery of parity nonconservation did not "sweep away many of the most fundamental parts of physics," but merely restricted the generality of the law.

Bonnor was not alone in arguing in this way. McVittie also felt that matter creation was a violation of the basic rules of scientific reasoning. "It's like breaking the rules when you are playing a game," he said in 1978. "If you allow yourself in the game of American football to take knives on board with you and stab your opponent, now and again, of course the results will be very remarkable, particularly if one side only has the knives and the other is merely the recipient."[150] Weizsäcker, who was not himself engaged in the debate, pointed out that because of the close connection of energy conservation with the general structure of physics, very strong empirical evidence would be needed in order to give up the conservation law. He could see no such empirical evidence, and, as he noticed in his Gifford Lecture of 1959, "the idea of an infinite duration of time is certainly not an empirical one."[151] Although the large majority of physicists (as well as astronomers and philosophers) found continual creation of matter to be inadmissible, there were exceptions. Interestingly, the founder of modern big-bang theory was one of them. Gamow rejected the steady-state theory, of course, but not for this reason. Recalling his own fascination with Bohr's idea of energy nonconservation in the early 1930s, he later said about the steady-state theory: "It's perfectly logical. Matter is not conserved—so what?"[152]

The criticism of Bonnor and others that steady-state theory was phenomenological, describing the universe rather than explaining it, was directed against the Bondi-Gold version in particular. It was a criticism that had followed this theory from its very start. In their work of 1948, Bondi and Gold made the methodological remark that cosmology differed from the rest of physics by not obeying laws in the ordinary sense. They traced the difference to the object of cosmology, the universe, being a unique entity, whereas laws are abstract descriptions of a large number of idealized experiments. It was a point to which Bondi often returned in his later discussions of the philosophical basis of cosmology. He would readily admit that cosmology was phenomenological, but saw this as a necessity, not a weakness of a particular model. Bondi flatly disagreed with Bonnor and other critics as to how fundamental a cosmological theory could be. As he wrote in 1956, "The most complete 'explanation' that a theory of cosmology can give is therefore a description."[153]

Simplicity was another of the metascientific concepts that often turned up in the cosmological controversy. For example, Bonnor argued that exact energy conservation was preferable on the grounds of simplicity, to which Bondi replied that the result of introducing continual creation of matter, the steady-state universe, was simpler than the evolving types of universe. As one would rather expect, the simplicity criterion was too elastic and ill defined to be of any real use in the discussion. The difference in the conceptions of science held by Bonnor and Bondi came up in relation to the role played by the perfect cosmological principle. Bonnor found it uneconomical to adopt such a wide hypothesis, but Bondi strongly disagreed: "It seems to me that the essence of scientific work is that we *should* make assumptions. . . . the bigger the assumption we make, the more testable it is likely to be by experiment and observation. It is the purpose of a scientific hypothesis to stick out its neck, that is to *be* vulnerable. It is because the perfect cosmological principle is so extremely vulnerable that I regard it as a useful principle. . . . your views on what constitutes science must differ markedly from mine. I certainly regard vulnerability to observation as the chief purpose of any theory."[154] Bonnor was not against Popperian standards, but thought that something more constructive than mere refutation had to come out of a hypothesis. To accept a wide-ranging assumption such as the perfect cosmological principle would require a substantial reward, and he could see none. Moreover, he turned Popper's philosophy against steady-state theory by noticing that, contrary to the relativistic cosmologies, the steady-state theory was from the very beginning constructed specifically as a theory of the universe. That is, he intimated a certain ad hoc quality in the foundation of steady-state theory. "Now both Milne's theory and the steady-state theory were designed especially to solve the cosmological problem, so it is not remarkable that they both yield unique models."[155]

Together with Williamson and Bonnor, McVittie developed during the 1950s into a staunch opponent of steady-state cosmology. As mentioned, McVittie had in the 1930s and 1940s worked in the rationalistic tradition of Milne, and had still in the early 1950s expressed views in general agreement with the methodology behind steady-state cosmology and kinematic relativity. But he converted to mainstream relativistic cosmology and became increasingly critical of what he now saw as rationalistic fancies without a proper empirical foundation. McVittie came to see himself as belonging to the "observational school," which was the approach to cosmology that "does not aspire to finality but only to discovering what model best fits the data."[156] Ten years later, in an essay review of 1961, he complained about the steady-state scientists' a priori assumptions and substitution of logic for observation, thus repeating the charges of earlier critics. He had little respect for Bondi's appeal to falsificationism, which he saw as a vulgarization of Popper's views, "as if the excellence of a theory were measurable by the rapidity and ease with which it can be disproved." As McVittie noticed, if (what he took to be) Bondi's version was right, then "we should be justified in inventing a theory of gravitation which would prove that the orbit of every planet was necessarily a circle. The theory would be most vulnerable to observation and could, indeed, be immediately shot down."[157] He underscored his lack of respect for the steady-state theory by considering it, from a philosophical point of view, as belonging to the same class as Kapp's amateurish pseudocosmology.

From a long career in cosmology, McVittie knew that "the temptation to substitute logic for observation is peculiarly hard to resist in astronomy and especially so in cosmology."[158] Given the scarcity of relevant and reliable data, what should the astronomer do? One alternative, which McVittie associated with the steady-state theory, would be to "supplement what is directly observed by additional items of information based on the *absence* of detectable phenomena." This was an approach that McVittie found totally unacceptable. Instead, the proper approach would be to emphasize positive, inductive knowledge, to "discover how much can be found out about the universe through measurements that yield non-null results rather than by the consideration of logical possibilities which might conceivably be the case." McVittie discussed the same themes at a conference in Padova in the fall of 1964, celebrating the quatercentenary of Galileo's birth.[159] Several of McVittie's arguments were similar to those put forward by Dingle in the debates of 1937 and 1953. He even associated the rationalist school of cosmology with Aristotle's system and the empirical approach with Galileo's physics. And as Whitrow had objected to Dingle's historical analogy in 1937—pointing out that Aristotle was in fact more empirically oriented than Galileo—so he now, twenty-seven years later, made the same point in reply to McVittie's analogy.

Needless to say, the difference in attitude between McVittie and the steady-state theoreticians was philosophical in nature, and hence could not be resolved by scientific means.

5.3 RELIGION, POLITICS, AND THE UNIVERSE

The cosmological controversy was not only fueled by philosophical arguments, it was also tied up with views of a religious, ethical, and political nature, that is, with the prevailing ideological climate of the period. That there was such influence is beyond doubt, but it is not obvious in what way, if any, it affected the controversy. I shall argue that in the end it was, by and large, much ado about nothing. Although interesting in its own right, and certainly illuminating as a case in the history of ideas, the discussions about cosmology's religious and political implications had virtually no impact on the path followed by scientific cosmology.

Theology and Cosmological Models

For most of its history, cosmology has been part of mankind's religious rather than scientific world view. With the progress of astronomy and the advent of cosmological models based on the laws of physics, the association between cosmology and religion loosened, but it never disappeared. It probably never will. In previous chapters we have seen how religious views played an essential part in Millikan's cosmology, and how the relationship between cosmology and religion was constantly discussed by both theologians and astronomers, including de Sitter, Jeans, Eddington, Milne, Barnes, and Lemaître. The expanding universe and Lemaître's big-bang theory were sometimes interpreted as support of the Christian view of creation, although most astronomers refrained from drawing such conclusions. But, as Bertrand Russell, a sharp critic of any such interpretation, observed, "Theologians have grown grateful for small mercies, and they do not much care what sort of God the man of science gives them so long as he gives them one at all." On the whole, the attempts to turn modern physical science into support of religion were less than convincing and often contradictory. Russell teasingly remarked: "Eddington deduces religion from the fact that atoms do not obey the laws of mathematics, Jeans deduces it from the fact that they do."[160]

The temptation to use cosmology as a scientific argument for Christianity, in one of its many versions, was evident in England at the time when the steady-state theory emerged. Milne explicitly interpreted his cosmological theory in religious terms, and about the same time a similar view was defended by another British mathematical physicist, Edmund Whittaker. In books from the 1940s, Whittaker suggested giving a modernized Thomist proof of God's existence based on the new cosmology. He argued that the knowledge of a temporal beginning of the universe proved the existence of God as the ultimate cause of the world. Moreover, modern science, according to Whittaker, led to a single and transcendent God in accordance with the Christian view. "Recent researches have led to the conclusion that the universe cannot have existed for an infinite time in the past, at any rate under the operation of the laws of nature as we know them: there must have been a beginning of the present cosmic order, a creation as we may call it, and we are even in a position to calculate approximately when it happened."[161] Whittaker believed that not only did science support deism, it also implied a rejection of pantheism; for if God was identified with creative evolution then it would be necessary for God to be born, and such a notion he considered meaningless. As to the problem of the creation of the world, his view was the conventional one, namely, that "the Creation itself being a unique event is of course outside science altogether."[162]

Milne agreed with Whittaker on this point, but believed that he could give reasons why the world was created as a transcendental singularity. In his *Modern Cosmology and the Christian Idea of God*, he wrote: "We can make no propositions about the state of affairs *at* $t = 0$; in the divine act of creation, God is unobserved and unwitnessed, even in principle. . . . We can form no idea of an actual event occurring at $t = 0$; we can make propositions in principle only *after* the event $t = 0$. As for why the event happened, we can only say that had no such event happened, we should not be here to discuss it."[163] Apart from the creation event, Milne believed that everything else about the universe is rational. "To say that the universe is rational is to say that its Creator is rational," he declared.[164] However, Milne's rational ideas of a rational God creating a rational universe were wanting in clarity. As pointed out by a theologian, if the universe is rational in the sense that it is logically necessary, it is difficult to see how any theistic implications can follow.[165]

There can be little doubt that the discussions among Hoyle, Gold, and Bondi, which led to a tentative formulation of the steady-state theory in 1947, were colored negatively by the views expounded by Whittaker, Milne, and other religious scientists. The three steady-state pioneers were atheists and either hostile or indifferent to organized religion; the same was the case with Sciama, the most important of the younger theoreticians. Although the motives behind the steady-state model were not religious (or, rather, not antireligious), it must surely have added to their satisfaction that it was possible to design a universe in which there allegedly was no room for a Creator.[166] At any rate, Hoyle made a point of associating the steady-state theory with atheism, and, conversely, the big-bang theory with religion in general, and theism in particular. He did so on many occasions, first and most provocatively in his *Nature of the Universe* of 1950. These utterances—one can hardly call them arguments—appeared in his popular works and never turned up in his scientific articles and addresses, where he followed the unwritten rule of avoiding explicit references to religious and political matters. As we have seen, Hoyle reacted strongly to the notion of a temporal beginning of the universe, a concept he found intolerable from both a scientific and a philosophical view. It was a notion "quite characteristic of the outlook of primitive peoples" who postulate the existence of gods to explain the physical world. When Hoyle was in his early teens he concluded that religious ideas were just fairy tales with no foundation in reality, and he never changed this simplistic atheistic view. Bondi and Gold largely shared his views,[167] but they left the overt association with religion to their colleague, who had no hesitations in spelling it out. He asserted that "It is not a point in support of this [steady-state] theory that it contains conclusions for which we might happen to have an emotional preference," and yet it is all too evident that Hoyle did have such emotional preferences.[168] Clearly, Hoyle was not only antireligious but also, and especially, anticlerical. His dislike of organized religion popped up in the most unexpected places, such as in a popular book on the universe in which he explained that the conflict in Northern Ireland could be settled simply by arresting "every priest and clergyman in Ireland and to commit every man jack of them to long jail sentences on the charge of causing civil war."[169] Somehow Hoyle believed that the steady-state theory was preferable not only from a scientific point of view, but also from a political and an ethical point of view. He intimated that this cosmological model might help in realizing higher human values, such as "exposing the futility of nationalistic strife. It is in just such a way that the New Cosmology may come to affect the whole organization of society."[170] How this should come about he did not explain.

Hoyle's rather offhand remarks about religion can hardly be considered a serious attempt to discuss the theological implications of cosmology, and they were probably not intended as such. But they had the effect of polarizing the public debate and causing concern in a large part of the religious community that some astronomers were now trying to undermine Christian faith. Theologians maintained that the steady-state theory held no authority in science, and that faith in God had anyway nothing to do with what cosmological view happened to be accepted at the time being. Yet the concern was widespread. At the Modern Churchmen's Conference in Cambridge in 1950, several mem-

bers were disturbed by Hoyle's recently published book and its effect on people's attitude to Christianity.[171]

In the case of Wallace Sargent, a fifteen-year-old English schoolboy,
Hoyle's lectures did indeed have the feared effect. Sargent, who later made a
distinguished career in astronomy, recalled that listening to Hoyle's broadcast
series and reading his book made him "violently antireligious," and brought
him into trouble with the school authorities. When Sargent did graduate work
in astrophysics in the late 1950s, he clearly favored the steady-state model.
Among his reasons was that he considered the theory to be associated with an
atheistic world view.[172]

Now there is no one-to-one relationship between cosmological and religious views, if there is any at all, and sympathy for the steady-state theory did
not necessarily imply atheism, any more than sympathy for the big-bang
theory implied theism. Recall that Dean W. R. Inge defended a kind of steady-
state universe as being "more in accordance with what we may imagine to be
the will of God" than the big-bang universe.[173] The attitude of Lovell provides
another example of the lack of one-to-one correlation between steady-state
theory and atheism.

Bernard Lovell was, like Whittaker, a devoted Christian, but he did not
regard the steady-state theory a threat against theism; creation of matter,
whether continuous or sudden, was for him a sign of divine activity. Lovell's
view concerning the relationship between science and religion was influenced
by the organic-metaphysical system of the mathematician and philosopher
Alfred North Whitehead's. According to this system, God is immanent rather
than transcendent: "He is not the beginning in the sense of being in the past of
all members. He is the presupposed actuality of conceptual operation, in uni-
son of becoming with every other creative act." Whitehead's and Lovell's
God was a constantly intervening and interacting universal being, who "is not
before all creation but *with* all creation."[174] From such a perspective there is no
fundamental separation of science and religion. As far as cosmology is concerned, it is a religious view which can easily accommodate, indeed, is in
harmony with, an eternal universe with continual creation of matter.

It may furthermore be argued that the negation of the heat death in the
steady-state theory, and the possibility of endless life that this model implies,
are appealing features to the religious mind. Such considerations may have
had an effect in some astronomers' sympathy for steady-state cosmology. But
then, of course, one doesn't have to be religious to appreciate the prospect of
eternal life. Sciama, who was an atheist, once referred to his devotion to
steady-state theory in about 1960 as follows: "Partly, I think, because it's the
only model in which it seems evident that life will continue somewhere. . . .
even if the galaxy ages and dies out, there will always be new, young galaxies
where life will presumably develop. And therefore the torch keeps being carried forward. I think that was probably the most important item for me."[175]
Contrary to the other steady-state theoreticians, McCrea was a practising
Christian (Anglican). He had strong religious convictions and believed that

ultimately cosmology requires the concept of God. In 1974 he emphasized that, as far as he was concerned, the universe is purposeful and "purpose is inseparable from person, and the Person of the Creator is revealed in the person of Christ."[176] Whether this view also influenced McCrea's scientific work is not clear, but Gold suspected that McCrea had religious motivations for supporting the steady-state theory. He recalled that the Cambridge trio felt embarrassed at the thought that McCrea used the theory as part of what they believed was religious propaganda.[177]

The debate about continual creation of matter in cosmology replayed to some extent themes discussed in the nineteenth century's controversy over the spontaneous generation of life. In both cases, it was the *spontaneous* creation, of either matter or life, that caused so much heat. Darwin thought that "it is mere rubbish, thinking at present of the origin of life; one might as well think of the origin of matter."[178] Yet spontaneous generation was historically associated with Darwinism and other evolutionary theories, and in some quarters spontaneous generation was seen as a threat against the Christian Church. In the 1870s religiously motivated attacks on spontaneous generation were common as part of the attempts to counter the harmful influence of Darwinism. A professor of medicine at Paris, Paul Chauffard, linked spontaneous generation with atheism, materialism, and modernism. He warned that the new biology would lead to "general doctrines whose subversive application will lead to the ruin of a civilization imbued with spiritualism."[179] But just as twentieth-century continual creation of matter could be reconciled with religion, so could nineteenth-century spontaneous generation of life. At least one French scientist argued that God's creative faculty did not stop with the first origin of life; if life were perpetually and spontaneously generated, this would only demonstrate the continual presence of divine activity. The biological debate in the 1870s was again a replay of themes discussed earlier in geology. For example, according to the 1789 steady-state vision of George Toulmin, the eternity of the world and the slow, continual changes in its structure "enforce the excellence of moral rectitude; and the existence of a Supreme Being—infinite in wisdom, goodness, and intelligence."[180]

The atheist Hoyle feared that big-bang cosmology might lead to religious propaganda of the sort he knew from Whittaker; indeed, he claimed that the big-bang picture included a first miracle which could only make sense in a religious context. Although he rarely referred to Gamow, it was his version of the big bang-theory he often had in mind. Gamow's theory was, after all, the main rival of the steady-state theory, and it was only Gamow who matched Hoyle in presenting cosmology in best-selling popular works with a wide audience. But Gamow was not the best target for Hoyle's arrows. He was not a religious man, and like most other cosmologists he was careful not to have his science drawn into the fuzzy realm of theology. In fact, the model he favored about 1950 was not a big-bang theory of the ordinary type, but an oscillating model in which each big bang was preceded by a big squeeze of an earlier universe. Such an oscillatory model is as far from theistic interpretations as is the steady-state theory. Still, the very title of his popular book of 1952, *The Creation of the Universe*, could not help invoking religious ideas in many people. Apart from a casual reference to St. Augustine, Gamow did not mention religious questions in the book, but in the second printing he nonetheless felt it necessary to include in the preface this note: "In view of the objections raised by some reviewers concerning the use of the word 'creation,' it should be explained that the author understands this term, not in the sense of 'making something out of nothing,' but rather as 'making something shapely out of shapelessness,' as, for example, in the phrase 'the latest creation of Parisian fashion.' "[181] This, together with a few other joyful remarks, was about all what Gamow had to say about the religious implications.

A Papal Intervention

Papers in the *Physical Review*, the world's leading journal for physics research, rarely contain references to God or the church. It may have disturbed some readers of the eighty-sixth volume of 1952 to find a paper by Gamow introduced by a lengthy quotation of an address by Pope Pius XII in which the pope in no uncertain terms endorsed the big-bang theory.[182] The quotation in a paper which otherwise had nothing to do with religion was presumably meant as just an eye opener, an unconventional joke of the sort Gamow excelled in. The background was an unusual intervention of the Catholic Church in the cosmological controversy which took place in 1951, and which fueled the discussion over the relationship between cosmology and religion. Had it not been for this rather clumsy intervention there might have been little more to say on the subject.

Pope Pius XII was a learned and enlightened man. He had an interest in astronomy and the other sciences, the latest results of which he wanted to utilize as rational support for the doctrines of the church. He was fascinated by the theory of the expanding universe and influenced by the writings of Jeans, Milne, and Whittaker, in particular. In 1950 the pope issued an encyclical letter on evolutionary biology in which he admitted the theory as a legitimate subject of scientific study and one which did not necessarily conflict with the teachings of the church. The following year, on 22 November 1951, he delivered an address to the Pontifical Academy of Sciences in the presence of several cardinals and the Italian minister for education. In his address, the pope dealt in considerable detail with the support to the notion of a Creator which he thought had recently come from cosmology. The basic argument of the pope was not only that is there no disagreement between the astronomers and the church, but that the results of modern science actually give ample evidence for the existence of a transcendental Creator. He endorsed the big-bang picture unreservedly from the start of his address: "Everything seems to indicate that the material content of the universe had a mighty beginning in time, being endowed at birth with vast reserves of energy, in virtue of which, at first rapidly, and then ever more slowly, it evolved into its present state."[183] After having cited various methods of determing the size and age of the universe, he pointed out that these figures "involve no new idea even for the simplest of the faithful. They introduce nothing different from the opening words of Genesis, '*In the beginning* God created heaven and earth . . .' —that is to say, at the beginnings of things in time." The essence of the pope's message is contained in the following excerpt:

> What was the nature and condition of the first matter of the universe? The answers given differ considerably from one another according to the theories on which they are based. Yet, there is a certain amount of agreement. It is agreed that the density, pressure and temperature of primitive matter must each have touched prodigious values.* . . .
>
> Clearly and critically, as when it [the enlightened mind] examines facts and passes judgment on them, it perceives the work of creative omnipotence and recognizes that its power, set in motion by the mighty *Fiat* of the Creating Spirit billions of years ago, called into existence with a gesture of generous love and spread over the universe matter bursting with energy. Indeed, it would seem that present-day science, with one sweep back across the centuries, has succeeded in bearing witness to the august instant of the primordial *Fiat Lux*, when, along with matter, there burst forth from nothing a sea of light and radiation, and the elements split and churned and formed into millions of galaxies. . . .
>
> What, then, is the importance of modern science in the argument for the existence of God based on change in the universe? By means of exact and detailed research into the large-scale and small-scale worlds it has considerably broadened and deepened the empirical foundation on which the argument rests, and from which it concludes to the existence of an *Ens a se*, immutable by His very nature. . . . Thus, with that concreteness which is characteristic of physical proofs, it has confirmed the contingency of the universe and also the well-founded deduction as to the epoch when the world came forth from the hands of the Creator. Hence, creation took place. We say: therefore, there is a Creator. Therefore, God exists![184]

The pope's presentation of the position of contemporary cosmology was biased in that he gave a harmonious picture of the field which had no justification except that it served his purpose. After all, the big-bang theory was far from unanimously accepted in 1951. "It is worthy to note," said the pope, "that modern scholars in these fields [astronomy and physics] regard the idea of creation as quite compatible with scientific conceptions, and that they are even led naturally to such a conclusion by their researches." This might sound like a reference to the creation cosmology of Bondi, Gold, and Hoyle, the only theory in which creation of matter was claimed to be "compatible with scientific conceptions." In fact, it was a reference to the big-bang theories of Lemaître and Gamow, and the Pope only indirectly alluded to the fact that there were rival cosmologies such as the steady-state theory. He did so in discussing the law of entropy, the consequence of which—the ultimate heat death—he accepted as agreeing with Christian belief. The "unduly gratuitous" hypothesis of "continued supplementary creation" was briefly dismissed. It is remarkable that physicists in the 1920s—Millikan and MacMillan—introduced a steady-state universe in order to save the world from the heat death and thereby to argue a Christian world view, whereas thirty years later the pope sanctioned a diametrically opposite view. And recall that Dean Inge found the stationary, recurrent universe to be in better accord with Christian belief than the big-bang universe.

More important than the pope's partisan view of the cosmological scene was the very essence of his message, the claimed concordance between science and religion. The rationalistic message, that big-bang cosmology's notion of the beginning of the universe justified or supported the religious concept of a created world, was hard to swallow for many theologians, both within and without the Catholic Church. According to many theologians, both then and now, the cosmological creation is something very different from creation in the religious sense. Science does not, and cannot, support religion in any direct way. If this was the case, religion itself would seem to have to follow the changes which necessarily take place in science, to be subservient to science instead of being a supreme truth of revelation. As a theologian concluded in 1956, "The whole question whether the world had a beginning or not is, in the last resort, profoundly unimportant for theology."[185] All the same, there is no doubt that the pope's intervention left among many people the impression that the biblical Genesis had literally been proved by big-bang cosmology; and, conversely, that a good Christian, or at least a good Catholic, could not possibly accept the steady-state theory.

Lemaître, whose theory of the primeval universe formed the backbone of the pope's argument, was not at all happy with the address. Himself a Catholic priest and high-ranking member of the Pontifical Academy, Lemaître believed that astronomy and theology were two separate contexts which should not be mixed. He also felt that the big-bang theory was still a hypothesis, and that the pope had presented it in a much too authoritative way. He was therefore quite upset and found it necessary to intervene, together with Daniel O'Connell, the director of the Vatican Observatory and science adviser to the pope.[186] Apparently they succeeded in persuading the pope that the close association between science and theology that he had argued was helpful neither to science nor to the church. Less than a year after the speech at the Vatican Academy, the pope delivered an address to six hundred and fifty astronomers gathered in Rome for the Eighth General Assembly of the International Astronomical Union. It may have been tempting for the pope to proceed along the same track he had followed in 1951, not least because the Rome conference witnessed Baade's announcement of a time scale much smaller than had previously been accepted; in the view of many astronomers, this discovery made the big-bang theory appear even more likely, and the steady-state alternative even less likely. But this was not what happened. The Rome discourse, delivered in Castel Gandolfo, differed markedly from the previous address, being much more moderate and avoiding specific references to the metaphysical and religious implications of the big-bang theory.[187] Never again did Pope Pius XII try to make cosmology support Christian dogma. Incidentally, the Rome meeting illustrated the politically and ideologically sensitive aspects of cosmology. The meeting was originally planned to take place in 1951 in Leningrad (St. Petersburg), but was canceled because of the Korean War. It caused much resentment among the Soviet delegates that the meeting was transferred to a North Atlantic Treaty Organization country and that it included a papal discourse. Recalling the pope's propaganda the previous year, the four Soviet delegates stayed away from the discourse and the subsequent audience.

The pope's 1951 address was exactly the kind of religious interpretation and use of cosmology that Hoyle detested. It must have confirmed him in his suspicion of an unholy alliance between big-bang cosmology and organized religion. Although the Catholic Church did not proceed along the path taken in November 1951, of course the discussion of the relationship between religion and the two rival cosmologies did not end. I shall not deal systematically with this relationship during the 1950s; it suffices to recall the positions of Lovell and Bonnor in this respect. The atheist Bonnor rejected big-bang theory for largely the same reasons as Hoyle did; among these, that it lent support to divine creation. "The underlying motive is, of course, to bring in God as creator," Bonnor stated. "It seems like the opportunity Christian theology has been waiting for ever since science began to depose religion from the minds of rational men in the seventeenth century."[188] This motive, which Bonnor rejected emphatically, was clearly exhibited in the pope's address of 1951. But, as mentioned, Bonnor nonetheless rejected steady-state cosmology. Lovell, the theist, avoided choosing between the two models. He found both equally unsatisfactory because they did not provide an explanation of creation of matter, a common lacuna which he—as earlier the pope—reserved for divine intervention. Lovell's argument for religion differed slightly from the pope's, though: whereas the pope saw big-bang cosmology as positively leading to a transcendental Creator, Lovell chose to see divine signs in those border areas of cosmology which seemed to defy science.

Naturally, the question of the relationship between religion and cosmology, or science in general, continued to be a concern of the Catholic Church. In a message of 1 June 1988 Pope John Paul II stressed that "Christianity possesses the source of its justification within itself and does not expect science to constitute its primary apologetic." Although arguing for some kind of consonance, he warned specifically against "making uncritical and overhasty use for apologetic purposes of such recent theories as that of the 'Big Bang' in cosmology."[189] The pope's remark undoubtedly referred to the dominant big-bang theory in the late 1980s, but it could as well have referred to his predecessor in Rome, Pius XII.

Cosmology in the Soviet Union

The role of religion in the cosmological controversy in the 1950s was intimately connected with the political and ideological situation in the period. In order to understand the controversy in a wider sense it is necessary also to look at these factors, the zeitgeist, as prevalent in the Western world in the postwar years. Generally speaking, the decade after the emergence of the steady-state theory in 1948 was characterized by the cold war and its associated values, including a reaction against materialism and, of course, communism. If a view, scientific or not, could somehow be associated with marxist values, it would be more easily discredited than a view which reflected opposite values. This sort of mechanism can also be witnessed in the cosmological controversy as it took place outside the scientific journals and meetings. In this covert struggle, cosmological models were sometimes intimated to be associated with either positive or negative values, the latter typically being materialism, marxism, atheism, and totalitarianism. The negative values were derived from Soviet communism—where they were considered positive, of course.

Since the late 1920s there had begun in the Soviet Union an attempt to make the country's astronomy more congruent with the official ideological line of the communist party. Astronomers should serve the party by providing anticlerical propaganda and exposing the idealistic cosmological views of the West, in particular those which implied a creation of the world.[190] According to the party's ideology, as it was formulated in the late 1930s, cosmological models with a finite time scale had to be rejected because of their theistic implications. In 1947, Andrei Zhdanov, the notorious chief ideologue, expressed it in this way: "The reactionary scientists Lemaître, Milne and others made use of the 'red shift' in order to strengthen religious views on the structure of the universe . . . Falsifiers of science want to revive the fairy tale of the origin of the world from nothing . . . Another failure of the 'theory' in question consists in the fact that it brings us to the idealistic attitude of assuming the world to be finite."[191] As in physics and other areas of science, in astronomy Stalinism led to a sycophantic tradition of hailing Lenin and Stalin, if not as great scientists then as great philosophers of science. A Soviet astronomer, P. P. Parenago, ended a book on astronomy with a tribute to "the greatest genius of all mankind, comrade Stalin."[192]

Although relativistic cosmology was not necessarily seen as bourgeois idealism, the very application of physical theories to the universe as a whole was regarded as suspect. Soviet authorities claimed that it was unscientific and against the spirit of dialectical materialism to extrapolate local laws of physics, such as the theory of relativity, to the entire universe. In accordance with this view, cosmology as such was often seen as unmarxist. Incidentally, the criticism of relativistic cosmology's extrapolatory approach was, apart from its basis in Marxist-Leninist ideology, largely the same as the one later argued by the steady-state theoreticians. In 1948, party officials renewed their efforts to clean Soviet astronomy of bourgeois attitudes, the most dangerous of which was the relativistic theory of a closed expanding universe. This theory was, the astronomer V. E. Llov warned in 1953, a "cancerous tumor that corrodes modern astronomical theory and is the main ideological enemy of materialist science."[193] A conference of the U.S.S.R. Society of Astronomy and Geodesy taking place in Leningrad in December 1948 illustrates the heavy politicization of Soviet astronomers, but also their different opinions concerning ideological questions.[194] Some of them argued that the expanding universe was a capitalist myth and that the redshift had to be explained otherwise; other participants disagreed, and pointed out, as did Dmitri Iwanenko, that the theory of the expanding universe originated with Friedmann and thus, as a *Soviet* theory, should not be dismissed as bourgeois idealism.

Naturally, the pope's address of 1951 was taken as the final proof, if such was needed, that big-bang cosmologies were religious and not scientific views. The left-wing intellectual and procommunist French magazine *La Pensée* introduced in 1951 an article on Soviet cosmogony in this way: "For the first time, in fact, in the history of the Catholic Church, a sovereign Pontiff, abandoning all attempts for conciliation with Genesis, throws his support behind certain cosmogonic hypotheses, which, even they, too, postulate a creation of kinds ('expanding universe' of Lemaître, Milne, and others), are far removed from the creation story of the Scriptures. . . . This new position of the Pope . . . seems to announce a closer united front of diverse idealist philosophies, against materialism and true science."[195] Whereas cosmology was widely seen as a suspect and quasi-religious field, cosmogony in the narrower meaning—the formation of the earth, the moon, the stars, and the galaxies— was an accepted branch of astronomy which was pursued by several Soviet scientists. The theory of terrestrial and lunar cosmogony which *La Pensée* contrasted with Western cosmology was the one proposed in 1949 by Otto I. Schmidt, a leading Soviet mathematician, geographer, and arctic explorer. Neither Schmidt nor other Soviet cosmogonists dealt with the entire universe, so the comparison may seem somewhat flawed.

Soviet views on cosmology in the 1950s may be exemplified by the work of Victor Ambarzumian, probably the most important Soviet astrophysicist of the period. Belonging to the same generation as Gamow and Bronstein, Ambarzumian had a distinguished career in national and international astrophysics and was considered an authority on stellar evolution. Born in 1908, he graduated in mathematics and astronomy from Leningrad State University in 1928. Six years later he organized a department of astrophysics at the University, the first in the Soviet Union, and in 1946 he founded the Biurakan Observatory in his native Armenia. Ambarzumian was also an academician (i.e., a member of the U.S.S.R. Academy of Sciences) and a convinced marxist who was deeply influenced by the doctrines of dialectical materialism. According to Ambarzumian, cosmology proper was a myth, an unjustified extrapolation of observations and theories based on the empirically accessible part of the universe (the "metagalaxy") to the hypothetical construct of the entire world.[196] Ambarzumian was not opposed to relativistic cosmology (the only scientific form of cosmology he admitted), if cosmology was taken in a more restricted sense; but he found it premature to discuss world models of the kind suggested by Western cosmologists. In a lecture in 1963 he affirmed this position, which was also held by many astronomers in the West. "I personally think that at the current stage it does not even make sense to compare these models with observations in a detailed fashion."[197]

The Soviet communist party was much less interested in cosmology than it was in genetics. There were no purges and no Lysenko in Soviet astronomy, but then there was no Vavilov either. Yet the official ideology had a serious effect on Soviet astronomy, which responded by simply avoiding ideologically sensitive areas, including cosmology in the Western sense of the word. There was no official ban on cosmology, but the few studies which were published all avoided model construction as the field was pursued in Great Britain, in particular. The official view continued to be that cosmology cannot be treated scientifically; that the universe is infinite in space and time; and that matter is conserved. From 1934 to 1958 there appeared no cosmological models from Soviet astronomers or physicists which corresponded to the kind of models that were discussed in the West.[198] On the other hand, there were no attempts to formulate an independent, dialectical-materialist cosmology either. That is, Soviet astronomers conformed to the dogma of the communist party by giving up the study of the universe as a whole.

It is important to realize that the official attitude to Western cosmological models did not discriminate very much among the various versions. Basically, they were all unscientific and bore the imprint of bourgeois idealism. It goes without saying that theories with a creation in the past, such as Lemaître's and Gamow's, were categorically rejected. At a meeting in Leningrad in December 1948, Soviet astronomers confirmed in a resolution the necessity to fight against the "reactionary-idealistic 'theory' of a finite widening of the universe . . . [and] to expose tirelessly this astronomical idealism, which helps clericalism."[199] The same year one of the most distinguished astronomers in the Soviet Union, Boris Vorontzoff-Velyaminov, attacked George Gamow's big-bang theory. This theory was not only unscientific, he claimed, it was also invented by a former Soviet citizen who had betrayed his country. So: "The Americanized apostate Gamow . . . advances new theories only for the sake of sensation [and] with amazing ease, sometimes even after a few months, discards them in order to propose a new, equally sensational theory."[200] From the undeniable fact that official Soviet astronomy rejected evolutionary cosmological theories it does not follow, however, that they endorsed steady-state theories. Still, this was what Gamow intimated when he quoted Vorontzoff-Velyaminov's attack in the preface of his *Creation of the Universe*. The forced logic seems to have been that since big-bang cosmology was so repulsive to Soviet communism, then the main rival of this cosmology, the steady-state theory, must be regarded with sympathy among communists. Gamow did not actually say so, but he claimed that Vorontzoff-Velyaminov adopted a steady-state cosmology of the same kind as that of Hoyle, Bondi, and Gold. Although he added that Vorontzoff-Velyaminov's reasons were "entirely different" from those of the British astronomers, and that the Russian advocated the view "in the field of stellar evolution" (and not cosmology), the mere association between marxist orthodoxy and steady-state theory would suggest to many readers that the latter theory somehow was politically suspect.

In fact, Soviet astronomers and ideologues seem to have considered the steady-state theory no less reactionary and bourgeois than the big-bang theory. According to the American, Russian-born astronomer Otto Struve, Llov accused not only Gamow, Lemaître, and Weizsäcker of cultivating a reactionary bourgeois ideology, but also Hoyle, Bondi, and Gold.[201] The infinity of the steady-state universe in both space and time might have been an appealing feature, but if so it was all destroyed by the continual *creation* of matter, which was strictly intolerable to true Marxist-Leninists. Creation of matter out of nothing, whether taking part all at once or continuously, just smacked too much of religion and idealism. Together with the steady-state theory, the theories of Dirac and Jordan were therefore categorically rejected. Hoyle visited Moscow in 1958 to participate in the meeting of the International Astronomical Union. He recalled: "Judge my astonishment on my first visit to the Soviet Union when I was told in all seriousness by Russian scientists that my ideas would have been more acceptable in Russia if a different form of words had been used. The words 'origin' or 'matter-forming' would be O.K., but creation in Soviet Union was definitely out."[202]

The homogeneity principles were a stumbling block, too, and the perfect cosmological principle was seen as even more suspect than the narrow principle. In 1953 two Soviet astronomers, B. V. Kukarkin and A. G. Masevich, who had both attended the meeting of the International Astronomical Union in Rome the previous year, criticized Jordan's theory for being open to religious exploitation. They also explicitly denounced the steady-state theory as "the thoroughly idealistic and absurd theory of the creation of matter."[203] It must be concluded that the various intimations of some sort of association, however indirect, between Soviet communism and either of the two rival cosmologies are unfounded. The only association was negative, and it held equally for both kinds of cosmologies.

The weak position of cosmology in Soviet astronomy was noted at a meeting of the Commission for Cosmogony of the U.S.S.R. Astronomical Council in December 1956. Most of the speakers agreed with Ambarzumian, who admitted that "cosmological problems are somewhat neglected in the USSR" and called for more work in cosmology and extragalactic astronomy.[204] It was decided to stimulate such work, but it took some years before cosmology became visible in Soviet astronomical journals. For example, the first three volumes of *Soviet Astronomy*, first published in 1957, included no articles under the section "Cosmology and Cosmogony"; during the 1960s the number increased to an average of seven papers per volume. Moreover, when cosmology was dealt with it was in a different way than in the West. There seems to have been no interest in the controversy between big-bang and steady-state models, which was barely referred to at all. Almost all scientific papers avoided mentioning the steady-state theory, and when it was alluded to it was under other names, such as "the concept of a nonevolving universe remaining in a stationary state developed in recent years by F. Hoyle, D. Bondi [*sic*] and others."[205]

The neglect of cosmology in the Soviet Union in the 1950s does not mean that the subject was completely absent from the country's science. For example, in 1959 twenty-two-year-old Igor Novikov took a course in cosmology under A. L. Zelmanov, a Moscow theoretician who since the late 1930s had written surveys of relativistic cosmology and studied possible generalizations of the Friedmann-Lemaître theory. Zelmanov's course was basically mathematical, rather than physical, but it included a thorough discussion of all relativistic world models, including Lemaître's big-bang model. The course taught Novikov mathematical cosmology but did not arouse an interest in the physical aspects of the subject.[206] It was only a couple of years later, when Novikov came under the influence of Zel'dovich, that he discovered that cosmology might be studied also from a physical and astronomical point of view.

Yakov Zel'dovich was a rising star in, and an energetic promotor of, cosmology in the Soviet Union. Born in Minsk in 1914, Zel'dovich studied chemistry at the Physical-Technical Institute in Leningrad. He was soon drawn into chemical physics and nuclear physics, and became a leading member of the Soviet nuclear bomb program. After his long-time occupation with nuclear physics, the chemist-turned-physicist drifted into space science and astrophysics, and from there to cosmology. Zel'dovich made it clear in 1962 that the steady-state theory was unacceptable. His reasons for rejecting it were that it was unnecessary, in conflict with the theory of relativity, and rested on the illegitimate concept of spontaneous creation of matter.[207] These were scientific reasons shared by many of his Western colleagues, and Zel'dovich did not include political or ideological arguments in his criticism. He was convinced that the general theory of relativity was complete and of universal applicability, that is, that all cosmology had to be based on Einstein's field equations. The question was to use existing theories correctly, not to introduce new ones: "In the past chemistry and astronomy have made great contributions to physics: the Mendeleev [periodic] table, the doctrine of molecules, the laws of electrolysis, formed the basis of the ideas about the structure of matter, and astronomy provided the law of universal gravitation and the first measurement of the speed of light. Now, however, in the second half of the twentieth century, it is the deep conviction of the author (not shared, by the way, by many of his colleagues) that it would be naive to expect from astronomy new data about nuclear reactions, the creation of elementary particles, and the laws of the general theory of relativity."[208]

From about that time Soviet cosmology experienced a shift, with more astronomers and physicists being involved in the field, and with a much reduced importance of ideological considerations. Remarkably, in 1962—at a time when the big-bang theory attracted little interest among Western scientists—Zel'dovich concluded that "it is deemed probable that in the earlier stages of the evolution of the universe there existed a homogeneous isotropic Friedmann nonstationary solution with the density of matter decreasing from an infinite value at the initial instant."[209] The kind of cosmology that Zel'dovich and his younger associate Igor Novikov favored was scarcely distinguishable from the one cultivated by Western mainstream big-bang relativists. Unorthodox theories, such as those of Hoyle or Dirac, were dismissed as quasi-scientific: "We do not agree with the 'theories' appearing from time to time with features that violate the fundamental laws of physics. Such theories are, for example, those in which there is constant creation of matter 'out of nothing' far from the singularity (the theory of the steady-state universe) or those which involve a decrease of the gravitational constant with time. . . . We adopt the viewpoint that the homogeneous and isotropic Universe can be examined within the realm of GTR [general theory of relativity]."[210]

Zel'dovich did not differ from his Western colleagues in his relativist orthodoxy, but he presented his view more candidly, as a conviction or belief which first and foremost was theoretically based. In a book-length review of cosmology completed in the beginning of 1965, which "lays no claim of impartiality," Zel'dovich made his stand clear: "We assume that there was a moment of infinite density, at $t = 0$, which is a singular solution. The existence of the infinite density can be regarded as essentially a question of belief. If so, *I do believe* and shall attempt to draw all possible conclusions from this belief, comparing these conclusions with the facts observed until some irresistible contradiction dissuades me. So far, I do not know of any such contradiction."[211] The same attitude characterized Zel'dovich's view of the steady-state theory, which he admitted was not decisively refuted by observation (this was shortly before the discovery of the cosmic microwave background). "The negative attitude toward the theory of creation is based on theoretical principles of a general character," among which energy conservation and general relativity counted heavily.[212]

When it came to the creation itself, Zel'dovich and Novikov were, like most of their colleagues, whether in the East or the West, vague and cautious. In a review article of 1967 they claimed that the fact that $R \rightarrow 0$ for $t \rightarrow 0$ "does not imply the creation of the universe 10^{10} years ago (i.e. at $t = 0$)." Their argument was that one can imagine a previously existing, contracting universe out of which our universe was born at $t = 0$, including conservation of baryons and entropy. "But the jump itself at $t = 0$ from one [contracting] solution to the other [expanding] is outside the limits of application of the Friedmann solution and the whole modern physics," they wrote.[213] Incidentally, this sounds very much like a repetition of the views previously stated by Milne, Whittaker, and Lemaître—the very scientists who a few years earlier were so scorned for importing religion into cosmology.

Not all Soviet cosmologists agreed with Zel'dovich's preference for homogeneous models and his acceptance of a cosmic singularity at the initial moment of time. M. F. Shirokov and I. Z. Fisher studied in 1962 a type of inhomogeneous world model which admitted discrete masses in the universe.[214] Their modified field equations had the same structure as Hoyle's steady-state equations (4.2) but, as Shirokov and Fisher emphasized, the analogy was purely formal. Their theory remained on relativistic ground, but had the advantage that it avoided a space-time singularity at $t = 0$. Instead of this "unnatural property" the Shirokov-Fisher theory led for $t \to 0$ to a minimal universe, the density of which the two Russians estimated to be only about 10^{-4} g·cm^{-3}.

The dramatic change in Soviet cosmology is illustrated by the contents of the decennial jubilee volumes on astronomy of 1947, 1957, and 1967 (celebrating the thirty-, forty-, and fifty-year anniversaries of the 1917 revolution).[215] In the 1947 volume, published in 1948, forty-six publications on cosmology were included. Of these, twenty-five dealt with nonstandard explanations of the redshift, fourteen were popular or philosophical works, and only eight dealt with cosmology in the usual meaning of the term. The next volume, published in 1960, contained no papers on cosmology, although it included a bibliography. A drastic change in the field appeared with the 1967 volume. It now included a comprehensive review article on cosmology (by the veteran A. L. Zelmanov) as well as a list of references to a long series of works by Soviet astronomers and cosmologists, including Zel'dovich, Novikov, Fock, Lifshitz, Sakharov, Markov, Shklovsky, V. L. Ginzburg, and N. S. Kardashev. More important than the increase in numbers of Soviet articles on cosmology was the spirit of the post-1960 contributions: they were freed from ideological content and were in general favorable to the new big-bang theory.

As another example of the revolution in Soviet cosmology we may mention a paper by Andrei Sakharov, one of the fathers of the Soviet hydrogen bomb (together with Zel'dovich), and later a famous political dissident. In 1966 Sakharov suggested that the observed charge asymmetry—the nonexistence of cosmic antimatter—was a result of the violation of fundamental conservation laws in the primeval universe. He had in mind the violation of so-called *CP* invariance, which is the combined particle-antiparticle and left-right symmetry. The existence of *CP* nonconservation had been proved experimentally by American physicists in 1964, but only as a tiny effect in the decay of a particular kind of meson (the neutral kaon). According to Sakharov, *CP* nonconservation was important in the earliest phase of the big bang. Moreover, drawing on a proposal of his compatriot Moisei Markov, Sakharov speculated that "neutral spinless maximons (or photons) are produced at $t < 0$ from contracting matter having an excess of antiquarks, that they pass 'one through the other' at the instant $t = 0$ when the density is infinite, and decay with an excess of quarks when $t > 0$, realizing total *CPT* symmetry of the universe."[216] The hypothetical maximons had recently been suggested by Markov as primordial particles of the Planck mass $m = (\hbar c/\kappa)^{1/2} \approx 10^{-5}$ g. Markov's conclusion is worth quoting: "Since an energy of $\sim 10^{28}$ eV is necessary for the production of maximons, the possibility of producing such particles even in accelerators of the remote future is excluded. But one may assume that in its initial stage of development the matter in the Universe was composed predominantly of maximons. Assuming with the passage of time the initially present maximons are partially converted into forms of matter which we know, via the collapse mechanism of small masses, it is still possible to assume that part of the initially present maximons could have been preserved up to the present time."[217]

What I want to call to attention is not the specific content of either Sakharov's or Markov's works, but rather their spirit. For one thing, the authors unhesitatingly made use of the early big-bang universe and wrote freely about infinite densities and the time before the big bang; and, for another, they perceived the early universe as a testing ground for hypothetical particle physics. Such daring views—scarcely in the spirit of dogmatic dialectical materialism—would have been quite unheard of in the Soviet Union just a few years earlier. But by 1966 the tide had changed, and apparently irreversibly.

In the strange extrascientific debate of cosmology it was important, it seems, to present the preferred model as antimaterialistic, "materialism" being a naughty word in the period, associated as it was with communism. The opponents of the steady-state theory often labeled it materialistic, but Hoyle claimed that it was just the opposite. In a book of 1956, dealing amateurishly with almost everything from world politics to social problems, Hoyle described himself as a scientist "who fight[s] under the banner of Anti-Communism." The target of his cold war crusade was not only communism, but also, somewhat surprisingly, Catholicism. What these two ideologies had in common was, according to Hoyle, that they were both totalitarian belief systems: "Both Catholics and Communists argue by dogma. An argument is judged 'right' by these people because they judge it to be based on 'right' premises, not because it leads to results that accord with the facts. Indeed, if the facts should disagree with the dogma then so much the worse for the facts."[218] Hoyle did not refer to cosmology in this context, but it is revealing that he sometimes described the "official" cosmology with the same choice of words. He felt that the response of the astronomical community to the steady-state theory was dogmatic, and, bearing in mind the pope's address of 1951, he might have felt that big-bang mainstream cosmology and Catholicism shared at least their dogmatism; since communism certainly was dogmatic, too, he managed to tie together big-bang cosmology and communism. To be fair to Hoyle, this is my interpretation; Hoyle did not say so. It is worth noticing, if only to increase the confusion, that if the term "Catholics and Communists" in the quotation is replaced by "steady-state theory" we have almost exactly the criticism that Dingle leveled against the new cosmology of Bondi, Gold, and Hoyle!

The sometimes heated extrascientific debate about the two cosmological theories proceeded in an irregular and covert manner. It is difficult to judge its effects on the wider audience, but as far as the astronomical community is concerned it seems to have been ineffective. Scientists may have emotional preferences for a theory for all kinds of reasons, and it is possible that some of the astronomers and physicists who joined the cosmological controversy in the 1950s did so motivated, consciously or not, by political or religious reasons. But if they did, it did not turn up in their scientific arguments and work. Even those scientists who admitted ideological and religious factors to be relevant, such as Hoyle, Sciama, and Bonnor, kept them strictly apart from their scientific work. All things considered, the extrascientific debate had almost no influence on the scientific developments in cosmology. Of much more importance were the observations and experiments to which we shall now turn.

The Universe Observed

NEITHER PHILOSOPHICAL discussions nor religious and other metaphysical considerations made the steady-state theory appear a much less likely candidate for the structure of the universe by about 1960. The outcome of the controversy was decided by observations and experiments, pretty much in the same way that the fate of more ordinary physical theories is settled. But although observational testing was on the program ever since 1948, it took more than fifteen years before observations clearly indicated that evolutionary theories fitted better with the universe than the rival steady-state theory. And even then there was no undisputable proof of a big-bang universe, only increased evidence.

The difficulty in testing cosmological theories should not be surprising. Because of the subject matter of such theories, testing is necessarily based on indirect observational claims and long chains of inferences with ample room for discussion of each step. The cosmological tests in the 1950s and 1960s demonstrate how complicated and delicate testing is, and how intimately it is bound up with theory. But they also bear witness to the ingenuity of experimentalists and observational astronomers, and show that even theories so grand and fundamental as those in cosmology can be subjected to observational tests in essentially the same way as other theories in the more mundane parts of physics. Indeed, in special cases a cosmological theory can be tested by ordinary laboratory experiments, namely, if it relies crucially on a hypothesis which can be so tested. This was the case with the electrical universe model of Lyttleton, Bondi, and Hoyle. Dirac's cosmology with a varying gravitational constant was also refuted by experiments, although in this case they involved measurements from the Viking landers on Mars.

The kind of empirical work used in astronomy is usually referred to as observation, and not experiment. Ordinary experiments take place in the laboratory, where the scientist is able to maintain a certain amount of control of the objects or phenomena studied; they can be prepared and manipulated in the way required, and they can often also be produced. This active intervention of the experimentalist is not possible in astronomy, where, for obvious reasons, the objects are out of the astronomers' control. What the astronomer can do is basically to observe, register, and classify signals from celestial bodies. Yet there is no clear-cut distinction between observation and experiment, and in astronomy also ordinary experimental work is of the utmost importance in detecting and analyzing the signals from the heavens. The view, held by some philosophers of science, that astronomical knowledge has an epistemic nature different from that of terrestrial laboratory physics seems difficult to justify and is in disagreement with the history of modern astronomy.

The most direct way of testing a theory consists in confronting the theory's predictions with observations, either directly (or rather "directly") or by means of experiments. If observation unambiguously shows the predicted result to be wrong, the theory will, *in principle*, fall. There are, however, several reasons why this kind of simple refutation rarely takes place in the physical sciences. What is tested is not really the theory as such, but one of its predictions, and it is often possible to modify a theory in such a way that the falsified prediction becomes unimportant and the falsification leaves the modified theory intact. Furthermore, a good theory usually yields more than one prediction. It may happen, as indeed it often does, that while most of the predictions are verified, one turns out to be wrong. In such a case scientists will be wary of just dropping the theory, especially if there is no other theory to take its place. Again, even the most careful experiments and observations may be wrong—a trivial point, perhaps, but one of considerable importance. In other words, scientists may in certain cases legitimately deny "facts" and stick to theory. In general, there are a variety of reasons why testing of a theory is far from a straightforward procedure. Needless to say, the farther away from the observer the information comes, the more possibilities there are for questioning the data. Theories relating to extragalactic astronomy, and cosmology in particular, are by their very nature extremely difficult to subject to simple observational testing. Yet the difference in this area between cosmology and the other sciences is one of degree, not of principle.

Testing may be considered in a broader perspective than merely comparison between experimental results and the predictions of a theory. Apart from proper predictions, a theory also leads to forecasts of a much less precise nature, expectations of what new observations will be like. These are usually not of a quantitative nature and they do not follow logically from the theory, but are expectations derived from the theory's general character. They agree with the theory, but do not follow unambiguously from it. Such expectations are also subject to testing, in the sense that they may turn out to be right or wrong. If observations contradict an expectation, the theory will lose support, but it will not be falsified. Finally, one may speak of testing in a rather different way, which does not necessarily involve observations at all, but which relates to the structure of the theory and the consequences derived from it. It may be that the theory is incoherent, includes ad hoc hypotheses, or is otherwise methodologically flawed; or the theory may lead to paradoxes or other consequences that are judged fantastic or impossible. Examples are the objections raised against the steady-state theory by Whitrow, McVittie, and Schlegel. Nonempirical testing of this sort sometimes take place via thought experiments. The different forms of testing have in common the fact that they are all used in evaluating a theory's soundness, to decide whether it shall survive or not. Testing of a fundamental theory, as in cosmology, is rarely a matter of one crucial observation, but is a process in which the confidence in the theory is judged by accumulated results of all kinds relevant to the theory.

In the present case the following tests will be discussed:

1. The time scale problem
2. The Stebbins-Whitford effect
3. Redshift-magnitude relationship
4. Redshift-diameter relationship
5. Nucleosynthesis, formation of elements
6. Nucleosynthesis, formation of helium
7. Radio astronomical source counts
8. Gamma- and x-ray backgrounds
9. Quasars, redshifts and distribution
10. Cosmic microwave background
11. Formation of galaxies
12. Cosmological arrow of time

Of these test methods, we shall deal with items 1–5, 7, and 9 in the present chapter. The other methods will be discussed in chapter 7.

6.1 OBSERVATIONAL CHALLENGES

A Longer Time Scale

The age of the world was one of the observations that played a considerable role in the early phase of the controversy. As mentioned previously, the time scale problem was that according to most evolutionary models with a finite past, the age of the universe, as inferred from the Hubble constant, came out too low, namely, smaller than the age of the earth and the stars. Since the age of the world must necessarily be larger than that of its constituents, cosmological models that do not satisfy this criterion are in trouble. To put it more directly, models that definitely disagree with the criterion must be wrong.

Although there were various ways to avoid the problem, it was considered especially serious in closed world models with a zero cosmological constant. The fact that the time scale problem did not exist in the steady-state theory was one of the advantages of this theory, and one which steady-state advocates used against the evolutionary models. According to some antagonists of the theory, it was also the only reason why it could demand attention. In his criticism of steady-state cosmology, Dingle claimed that the theory was simply a speculative attempt to overcome the time scale difficulty. This was quite wrong, as regards both the emergence and the justification of the theory, but it was a view that continued to be brought up in more polemical contexts. For example, Bonnor claimed in 1964 that the steady-state theory became unnecessary in 1952 with the revised distance scale, but that "its originators showed a sturdy reluctance to recognize the fact."[1] In reality, the time scale problem was of relatively little importance in the arguments of the steady-state protagonists, except perhaps in the earliest phase of the development of the theory.

The accepted value of the Hubble time in the period from 1948 to 1952 was 1.8 billion years, with an estimated uncertainty of only 10% or so; being smaller than even the age of the earth, this was definitely too low a value. It was obtained from Hubble's relation $T = r/v$, where v, the recessional velocity of the galaxies, was measured by means of their redshifts. The distance r was obtained by the Cepheid method, that is, by identifying variable Cepheid stars in the galaxies and observing their period. The time scale of 1.8 billion years therefore relied on the Cepheid method and the calibration of the period-luminosity curve. Up to about 1952, most astronomers had complete confidence in the results derived by this method, perfected as it had been through three decades. During the 1940s, the Swedish astronomer Knut Lundmark had found evidence that the galactic distances were too small, but the discrepancies among different methods of distance measurements did not cast serious doubts on Hubble's results.

Such doubts first arose in 1951, when the German astronomer Albert Behr concluded that all intergalactic distances had to be increased by a factor of 2 or more. According to Behr's estimate, the age of the universe was close to, or perhaps larger than, 3.8×10^9 years.[2] A similar, but better argued, conclusion was announced by Walter Baade at the meeting of the International Astronomical Union in Rome in 1952. According to Baade, existing galactic distances were seriously wrong.[3] Baade, a German-born astronomer who since the late 1920s had done outstanding astronomical observations with the large American telescopes, worked at the time at the recently completed 200-inch Hale telescope on Mount Palomar, by far the most powerful instrument for astronomical observations at the time. Baade was born in 1893, received his doctorate from Göttingen University in 1919, and worked at Hamburg University's Bergedorf Observatory until 1931, when he emigrated to the United States. He returned to West Germany in 1958 and died two years later, in Göttingen.

Guided by the hypothesis that the Andromeda galaxy is comparable in size with our own Milky Way—and not much smaller, as believed by Hubble and others—Baade reexamined the Cepheid method. He found that the Cepheids on which the calibration of the curve had been based had an absolute luminosity considerably larger than previously believed. This discovery of a zero-point error led to a recalibration of the period-luminosity curve, and then also to new values for the distances to the galaxies. (Distances within the Milky Way were not affected.) In 1952 Baade announced that the distance scale had to be doubled, and subsequent analysis showed that it might even have to be tripled. In the words of the *Transactions of the International Astronomical Union* of 1952: "Above all, Hubble's characteristic time scale for the Universe must now be increased from about 1.8×10^9 years to about 3.6×10^9

years." The wording of the report (which avoided interpreting the Hubble time as related to the age of the universe) was in fact due to Hoyle, who at the meeting acted as secretary of the Commission on Extragalactic Nebulae and therefore transcribed Baade's talk.

The consequences for the time scale were evident: The velocities of the galaxies were unaffected, and with r doubled or tripled in the $T = r/v$ formula the Hubble time also suddenly became two to three times as large as previously accepted. The results presented by Baade in 1952 were only preliminary and he did not publish his full data until several years later. Although support for his estimate followed quickly,[4] there was some uncertainty with respect to how authoritative it was. Most astronomers accepted it because of Baade's reputation, but it was agreed that control and further measurements were needed. In McCrea's review of cosmology of 1953, he referred in an appendix to the new development and noticed that it would imply a time scale of about four billion years. "It is certainly disconcerting to learn," he wrote, "that such a drastic revision may be required at almost a single stroke."[5]

Yet Baade's estimate was conservative, and new measurements with the Hale telescope soon showed that the Hubble time was even larger. In 1956, Milton Humason, Nicholas Mayall, and Allan Sandage determined the best value to be $T = 5.4 \times 10^9$ years,[6] and it was recognized that the true value would probably be even greater. In fact, two years later Sandage estimated the Hubble constant to lie between 100 and 50 km·s^{-1}·Mpc^{-1}, corresponding to an age of the universe between 6.5×10^9 and 13×10^9 years if a Euclidean space was assumed (the Einstein–de Sitter model). Sandage found the values consistent with astrophysically based ages, and concluded that "there is no reason to discard exploding world models on the evidence of inadequate time scale alone, because the possible values of H are within the necessary range."[7]

The American astronomer Allan Sandage was at that time on his way to becoming the leading observational cosmologist of the second half of the twentieth century, a worthy follower of his mentor Hubble. He had studied at Caltech under Robertson and Baade, and became Hubble's assistant at Mount Wilson in 1950, at the age of twenty-four. Some years later he took over the observational program at Mount Palomar, which he gave a clearer cosmological orientation. Sandage taught himself theoretical cosmology so that he "finally understood what the papers in the earlier journals were saying," and he presented the subject in a form that also appealed to, and could be understood by, astronomers without a solid education in mathematics and general relativity.[8] In the 1970s Sandage became engaged in a controversy with Gérard de Vaucouleurs and others concerning the value of the Hubble constant, with Sandage advocating a long constant (50 km·s^{-1}·Mpc^{-1} or $T =$ ca. 19 billion years) and Vaucouleurs a short one (100 km·s^{-1}·Mpc^{-1} or $T =$ ca. 9.5 billion years).

What was the effect of the revised time scale upon the cosmological controversy? Seen in isolation, it was largely irrelevant with regard to the steady-state theory, in which the Hubble time has nothing to do with the age of the universe, being infinite in that model. It changed the numerical value of those parameters in which the Hubble constant entered, such as the average lifetime of the galaxies and the rate of creation of matter, but not to an extent that made them either more or less plausible. However, the revised Hubble time also affected the average density of cosmic matter, which according to Hoyle's version was given by $\rho \sim T^{-2}$. This solved a problem which existed with the older time scale. Together with the Andromeda galaxy, the Magellanic Clouds, and some smaller galaxies, the Milky Way is a member of the Local Group of galaxies, which is a cluster held together only very weakly. If the density of the group were only slightly less, it would not remain a bound system, but each galaxy would independently be subject to the universal expansion. Now the steady-state cosmic density based on $T = 1.8 \times 10^9$ years gives a value larger than the observed (or inferred) density of the Local Group; this implies, as Hoyle admitted in 1955, a "flat disagreement" because the Local Group would then expand apart. With $T = 7 \times 10^9$ years, on the other hand, the predicted cosmic density would be reduced by a factor of about 15, just enough to keep the group together.[9]

Although the revised time scale thus solved at least one problem within the steady-state theory, indirectly it weakened the position of the theory because it removed one of the few solid objections against the evolutionary models. The longer time scale did not suddenly make the time scale problem disappear, but psychologically it had the effect that it was no longer considered a real problem. If Baade's estimate of 3.6×10^9 years was accepted, the time scale difficulty would still be there. After all, the Hubble time is not identical with the age of the universe, and in most of the relativistic models it is shorter. For example, in the Einstein–de Sitter world the age is two-thirds the Hubble time, which should be compared with the age of earth, in 1952 held to be (2.9 ± 0.3) × 10^9 years. This was the result suggested independently by Arthur Holmes and Fritz Houtermans in 1946, based on measurements of the abundances of lead isotopes.[10] The result is still an age of the universe smaller than the age of the earth, and even with the 1956 value the age would be uncomfortably close to that of the earth. At about the same time as the Hubble time was revised, a major revision of the age of the earth took place—in the same direction. In 1953 Houtermans and Clair Patterson independently concluded that the best estimate for the age of the earth is $(4.5 \pm 0.3) \times 10^9$ years. That is, it was larger than the age of the universe according to the Einstein–de Sitter model. This did not really matter, however. With Baade's revision, almost all astronomers realized that the value of the Hubble time was rather uncertain and might well be considerably larger than two billion years.

However, the age criterion holds for any component of the universe, not only for the earth, which was recognized to be a relatively young object. During the 1920s and most of the 1930s the long time scale of James Jeans, based on his theory of stellar dynamics, was generally accepted. Although this time scale, according to which the ages of the oldest stars were as much as 10^{12} years, was abandoned in the late 1930s, it was agreed that most stars were considerably older than the earth. Theories of stellar evolution from the late 1950s resulted in ages of the oldest stars of between fifteen and twenty billion years, which thus would reinstate the time scale difficulty in spite of the revised estimates of the Hubble times. Hoyle, himself one of the main contributors to the new theory of stellar evolution, concluded in his 1960 Guthrie Lecture: "The ages of the oldest stars in our own galaxy appear to be at least 1.5×10^{10} years, and may indeed be as high as 2×10^{10} years. If we accept the present-day estimates of the galaxies, the certain requirement that our galaxy be younger than the whole universe rules out normal cosmologies with $\lambda = 0$. . . . Only Lemaître's cosmology with $\lambda \neq 0$, and the steady-state theory, survive the test."[11] However, this conclusion evidently presupposed that the estimations of stellar ages were reliable. It could thus easily be escaped, using the same strategic arguments that steady-state proponents used when their model was in trouble. For example, at about the same time Sandage admitted that observational data favored evolutionary models with a Hubble time of 7.4 billion years, clearly below the age of the oldest stars. But he chose to attribute the inconsistency to inadequate stellar models and to conclude that "for the moment it is better to assume that the observational data, . . . are not well known rather than to claim that current [relativistic] cosmological theory is inadequate."[12]

Even the discrepancies between stellar and universal ages did not matter much. Both steady-state and big-bang advocates seemed to have agreed, at least tacitly, that Sandage's judgment was reasonable, and that observational uncertainties prevented any firm conclusions with regard to world models. The decision to declare the time scale dificulty dead was as much psychological as scientific during most of the 1950s, but with ever-increasing observational values the astronomers' intuition proved right. At any rate, since the mid-1950s the time scale problem stopped being an important part of the cosmological controversy. At least, so it was thought. Much later, in the fall of 1994, newspapers all over the world carried headlines such as "Big Bang: Just So Much Noise?," "Big Bang's Defenders Confronted by Crisis," and "Theory of Universe's Age Poses New Cosmic Puzzle." The reason for the unexpected reappearance of the time scale problem was new measurements of galactic distances with the Hubble space telescope, indicating a Hubble time of 7.3×10^9 years, which is much smaller than the 16.5×10^9 years estimated for certain clusters of galaxies.[13] But in 1994 the puzzle was no longer interpreted in favor of the steady-state theory, which at that time had long ceased to be a viable alternative to the big-bang world view.

It is interesting to notice that the time scale problem, traditionally used as an argument against evolutionary theories with a finite age, also became a difficulty for the steady-state model. In this model the characteristic reproduction time is $T/3$, which is the average age of galaxies. Although there are galaxies of any age, it follows from the age distribution function that only 5% will have ages larger than T. This means that if most galaxies have an age about or larger than T, steady-state cosmology will be in trouble. For example, with the accepted Hubble time before Baade's revision, 1.8×10^9 years, and with the age of the Milky Way being about 16×10^9 years, as indicated by astrophysical evidence, our galaxy would be older than all but one in about 10^{10} of those around us! This might be possible, but would seem highly unlikely. In this sense, the longer time scale accepted during the 1950s helped steady-state cosmology as much as it helped the big-bang models. Even the increased value of the Hubble time seemed not sufficient to make the predicted mean age of the galaxies convincing. Gamow pointed out that with $T = 5 \times 10^9$ years the nearest galaxy of similar age to the Milky Way should be about twenty times as far from us as the average distance between the galaxies. This he considered implausible: "Since the steady-state theory of the universe does not deny to individual galaxies the right to evolve in time, we should find ourselves surrounded by a bunch of youngsters, as the galactic ages go!"[14]

In fact, with better estimates of galactic ages the time scale problem could be considered a problem for the steady-state model in particular. About 1960, the estimated ages of galaxies became more reliably determined by basing the estimates on calculations of stellar evolution. Most specialists agreed that the majority of galaxies had ages of the order of $2T$, which disagreed with most evolutionary models, but even more with the steady-state model. The discrepancy was pointed out by the American astronomer Ivan King: "The steady-state model of the universe is inconsistent with a view of stellar populations and evolution that has gained wide acceptance among astronomers who have no prejudice toward any particular cosmological hypothesis. . . . Instead of resolving the time scale dilemma, the steady-state hypothesis has considerably worsened it."[15] The steady-state reply to this challenge was, as expected, to question King's argument and the estimates of galactic ages. Thus Hoyle argued that, according to the steady-state theory, the mean age of stars in any galaxy might well be smaller than $T/3$ if the galaxies evolved from one structural form to another (e.g., from spirals to ellipticals).[16]

The Stebbins-Whitford Effect

In order to disprove the steady-state theory, one of its central predictions would have to be refuted, that is, a prediction that derived directly and unambiguously from the core of the theory. One such prediction was that the age, or some other intrinsic property, of the galaxies must on average be the same in any sufficiently large region of the universe. This was required by the perfect cosmological principle, on which the steady-state theory built. In particular, there must not be any systematic dependence of the ages of the galaxies on their distance. This criterion was admitted at an early stage by the steady-state theoreticians, according to whom it exemplified the greater falsifiability of their theory compared with the evolutionary theories. When Bondi, Gold, and Hoyle proposed their model in the summer of 1948, they did not know of any evidence to the contrary, but it so happened that in the same year two American astronomers published data which soon became used as a weapon against the steady-state theory.

Joel Stebbins and Albert Whitford were pioneers of astronomical photo-electrical photometry and had a high reputation in the astronomical community. Stebbins, who had developed photoelectrical methods since about 1910, was considered the founder of this important method in astronomy, i.e., the use of photoelectric detectors to measure the intensity of starlight. When Stebbins retired as director of Washburn Observatory in 1948, Whitford succeeded him. After the Second World War, the two astronomers applied the new scintillation counters and photomultipiers to amplify the feeble light signals of faraway galaxies into measurable electrical currents.[17] In 1948, Stebbins and Whitford published the results of their measurements of two-color indices of various elliptic galaxies, that is, the intensity of continuous spectra at two different wavelengths.[18] The measurements were made in order to provide more reliable data for the redshifts, but they found that the spectra were shifted much more toward the red than the redshifts could account for. Furthermore, this excess reddening increased with the distance of the galaxies. Stebbins and Whitford did not relate their finding to cosmological models, but pointed out that there seemed to be only two possibilities with regard to the explanation of the phenomenon. It might be due to absorption and scattering of the light on dust particles during its journey through intergalactic space, or it might indicate an intrinsic difference in color between galaxies at different distances. Subsequent measurements seemed to confirm the existence of an excess reddening depending on the distance. The effect was only found for elliptical and not for spiral galaxies. Rather as a byproduct of their analysis, Stebbins and Whitford argued for a minor correction of the Hubble constant. Taking into account the new values of the color indices they suggested that the value of H was 580 km·s^{-1}·Mpc^{-1}, corresponding to the reduced Hubble time of 1.7×10^9 years.

Stebbins gave an account of his and Whitford's work at the meeting of the International Astronomical Union in Zürich in August 1948, where he tentatively interpreted the effect as the result of intrinsic differences among the galaxies. The dust scattering hypothesis was considered unlikely, because it would require an amount of intergalactic material much larger than usually assumed to exist. The significance for cosmology of the Stebbins-Whitford effect, as it soon came to be known, became clear at an early stage. In the fall of 1950, when Stebbins delivered the George Darwin Lecture before the Royal Astronomical Society in London, he repeated his suggestion that the effect showed that more distant galaxies were redder than those closer by, possibly as a result of a larger number of red giant stars in the former.[19] Listening to Stebbin's lecture were Hoyle and Gold, who realized that if this was the case the new steady-state model would be in serious trouble. The Stebbins-Whitford effect could be accounted for by the big-bang theory, in which the more distant galaxies are also the younger ones; it would be an evolutionary effect, the younger galaxies containing more red giants than the older ones.

This kind of evolution could not exist according to the steady-state theory. Here galaxies had different ages, but with a distribution that was the same at any distance. So the situation was, in the words of Bondi, that "any systematic change of colour with distance, over and above that due to red-shift, would disprove the steady-state theory."[20]

Steady-state opponents were quick in announcing that the Stebbins-Whitford effect refuted the new cosmological theory. In 1951, Gamow claimed that the effect "strongly contradicts" steady-state theory, a conclusion which was repeated by Heckmann the same year.[21] Although there was some uncertainty with regard to the method used by Stebbins and Whitford, the extra reddening was generally accepted and interpreted as an age effect. The steady-state theory could not be modified to accommodate the Stebbins-Whitford effect, and so the steady-state theoreticians critically reexamined the data, in order, if possible, to question the existence of the effect. That the data did not necessarily warrant the conclusion of Stebbins and Whitford had already been argued by the French astronomer Gérard de Vaucouleurs in 1948, who found that there was no major discrepancy between the calculated values of the color indexes and those observed. His criticism seems to have been ignored. De Vaucouleurs suggested that the discrepancy might be purely an instrument effect due to inadequate spectral resolution, a suggestion which later turned out to be largely correct.[22] Then, in a paper of 1954, Bondi, Gold, and Sciama subjected the data to critical analysis and showed that they were inadequate to establish the effect claimed by Stebbins and Whitford.[23] In order to calculate the effect of the redshift on the color index, the Americans had used observations of M 32, assuming that this was a typical galaxy; the uncertainty in this assumption spoiled the justification of the Stebbins-Whitford effect.

The criticism did not kill the Stebbins-Whitford effect at once, and the same year Whitford reaffirmed its existence.[24] But further work by Whitford and his colleague A. Code showed that the effect was indeed questionable. Photo-electric measurements of M 32 revealed that it had an unusual spectrum, and when allowance was made for this the apparent color excess virtually disappeared.[25] However, Whitford did not withdraw the claim so unambiguously as the steady-state proponents might have wanted. The withdrawal was hidden in one sentence in the annual report of the observatory of which Whitford was director, the Wisconsin Washburn Observatory,[26] and was also included in Whitford's address to the 96th Meeting of the American Astronomical Society, held in New York in December 1956. According to *Sky and Telescope*'s abstract, "Dr. Whitford has made a new study of representative elliptical galaxies in four clusters at greatly different distances. . . . [He] found that the energy curves for all the galaxies, near and remote, were the same within observational error. Any age effect in elliptical galaxies therefore appears to be too small to be detected by his measures. . . . The new observational result, voiding the Stebbins-Whitford effect, is consistent with the Hoyle-Schwarzschild theory of the evolution of stellar systems."[27] Bondi and his steady-state followers were dissatisfied with the withdrawal appearing in a popular journal and wanted the mistake clearly and publicly admitted in one of the major journals of astrophysics.[28] For some years the Stebbins-Whitford effect continued to be cited as a serious argument against steady-state theory.[29]

After about 1957 the Stebbins-Whitford effect ceased to be a challenge to the steady-state theory. The lack of an age effect agreed with this theory, but could also be accounted for on the evolutionary theories. So the Stebbins-Whitford effect turned out to have no lasting influence on the controversy. Since it had first been claimed to contradict the steady-state theory, the final result was a considerable psychological victory for the steady-state supporters. They had successfully defended their theory by criticizing what was generally believed to be an observational fact. With this success, they were confident that they could also handle other challenges, and that the steady-state theory offered a better picture of the universe than the evolutionary models.

The Stebbins-Whitford effect was the most discussed attempt to refute the steady-state theory by means of an observable cosmological quantity that varied in time, but it was not the only one. In 1959 a German astrophysicist, Kurt Just, pointed to an effect of a similar nature which he believed contradicted the steady-state theory. He considered the "richness" of galaxy clusters, a measure of the distribution of magnitudes among the members of a cluster; if a cluster included a high frequency of high-luminosity (or low-magnitude) members, it was said to be rich. From a statistical analysis of data from almost seventeen hundred clusters, Just concluded that there was a clear overrepresentation of luminous galaxies in old clusters, i.e., that they were richer than young clusters. Therefore, "the steady state theory is disproved by the demonstration of a temporal development of the galactic clusters."[30] However, Just's alleged disproof did not arouse the same kind of interest as the Stebbins-Whitford effect had, and steady-state advocates did not even reply to it.

Redshift-Magnitude Measurements

The classical method of determining observationally the geometry of the universe, and then discriminating between competing world models, was by extending Hubble's redshift-distance relation into very distant parts of the universe. This program was already started by Hubble and his collaborators before the war. Even earlier, in 1933, the German physicist Max Kohler had provided a detailed theoretical discussion of the relation between redshifts and apparent luminosities according to different world models.[31] Kohler's work was the first systematic attempt to discriminate between relativistic world models by means of observations, but astronomers seem not to have taken notice of it. At any rate, during the 1930s the range of observations was too limited to serve as reliable tests of the geometry of the universe. It took two decades until the pioneering works of Kohler, Tolman, Hubble, and others could be used effectively.

Let us briefly summarize the essence of the redshift-magnitude method as it was known about 1955.[32] For relatively small distances, Hubble's law was established to be a linear relationship

$$cz = H_0 r \qquad (6.1)$$

between the redshifts $z = \Delta\lambda/\lambda$ and the distances of galaxies, r. Whereas the redshift is a relatively unproblematic quantity to measure, the distance is not. As noticed in chapter 1, the Hubble law can also be expressed in the alternative form

$$m = M - 5 \log H_0 + 5 \log (cz) + \text{const}, \qquad (6.2)$$

where m and M are the apparent and absolute magnitudes, respectively, and the base of the logarithm is ten. However, equation (6.2) assumes a Euclidean geometry in which the inverse-square law for illumination is valid. It does not take into account either the expansion of the universe or the curvature of space, which will modify the inverse-square law. Qualitatively, this modification may be understood as a result of the expansion of the universe and the finiteness of the velocity of light. When astronomers measure the redshift of a distant galaxy, they do not find it as it is today, but as it was, say, a billion years ago; namely, at the time in the past when the light was emitted by the galaxy. If the expansion velocity was higher in the past, this implies a larger redshift than the one following from the simple Hubble law (6.1). There would then be a deviation from the redshift-magnitude relation (6.2), the plot curving slightly upwards instead of being linear. If the expansion was slower in the past, the result would, of course, be an opposite deviation.

If this is taken into account and subjected to quantitative analysis, the result turns out to be

$$m = M - 5 \log H_0 + 5 \log (cz) + 1.086(1 - q_0)z + \text{const},$$

or, since H_0 is a constant,

$$m = M + 5 \log (cz) + 1.086(1 - q_0)z + \text{const}. \qquad (6.3)$$

The introduction of a new term in z corresponds to a corrected version of the Hubble law. Instead of the linear form it now becomes $cz = H_0 r + ar^2$, where a is a constant. A relation of the form (6.3) was first derived in 1942 by Heckmann.

An important quantity is the so-called deceleration parameter, q_0, which is a measure of the rate of slowing down of the expansion. It is defined as $q_0 = -(R''/RH^2)_0$ and is a dimensionless quantity. The deceleration parameter was introduced in cosmology in the mid-1950s, first by Robertson and then by Hoyle and Sandage.[33] Earlier works, such as Heckmann's, made use of the second derivative of R. The expression (6.3) is valid if the world model satisfies a Robertson-Walker metric, and it does not matter whether the model is relativistic or not. Moreover, it is only approximate, involving the first term of a series expansion. A similar relationship between redshift and apparent magnitude can be obtained in closed form, valid for any z, as was first done in 1958 by Walter Mattig, a student of Heckmann's.[34] For our purpose the less exact form (6.3) will do. The point is that different world models have different deceleration parameters, so that this is a quantity which can be used to differentiate between them. For very small z this is not possible, for then (6.4) reduces to (6.2), in which information on q_0 has disappeared, corresponding to the fact that local space is Euclidean.

It follows from the metric of the steady-state universe, where the expansion takes place exponentially, that $q_0 = -1$. In relativistic evolution models with a zero cosmological constant, q_0 is related to the space curvature by the relationship

$$\frac{kc^2}{R^2} = H_0^2 (2q_0 - 1).$$ (6.4)

This implies that

if $q_0 > 1/2$, then $k = +1$, and space is closed (elliptic),

if $q_0 = 1/2$, then $k = 0$, and space is flat (Euclidean),

if $q_0 < 1/2$, then $k = -1$, and space is open (hyperbolic).

In other words, if the value of the deceleration parameter can be found from (6.3), it yields a determination of the geometry of space, and hence indicates which world models are ruled out and which are not. In principle, the method is simple. Select a suffcient number of galaxies with large redshifts and the same absolute brightness. Then M in (6.3) is a constant, and a plot of $\log(cz)$ against m yields a curve which for small z will be a straight line, in accordance with Hubble's original law (6.2). If the curve continues along a straight line also for higher z values, $q_0 = 1$; if it bends downward, $q_0 < 1$; and if its bends upward, $q_0 > 1$. With sufficiently precise data, the magnitude-redshift curve should give the "right universe." A lucid discussion of this program, including the significance of equation (6.4), was given by Hoyle and Sandage in their 1956 paper. The Hoyle-Sandage formulation soon became the standard way of discussing the subject and was incorporated in the new generation of textbooks on cosmology that began to appear in the late 1960s.

Of course there are many problems involved in working out the program, which relies on magnitude observations in the high-redshift region where data are few and uncertain. For example, in order to produce a redshift-magnitude diagram, the included galaxies must have the same absolute magnitude so that M becomes constant in (6.3). For this reason observers used the brightest elliptical galaxies of clusters, which were believed to be almost equal in absolute magnitude. This was a hypothesis that involved some degree of uncertainty. Furthermore, since galaxies are systems of stars in evolution, the luminosity L of a galaxy will also evolve in time, but in a way which is not well known. In most measurements from the 1950s it was simply assumed that $dL/dt = 0$, although this was known not to be the case. It was only in the late 1950s that theories of evolution of galaxies appeared, giving functions $L(t)$ from which evolutionary effects on the observed magnitude could be calculated. Another problem, which had played a role since the mid-1930s, related to the fact that the light received at wavelength λ from an object with redshift z is emitted by the galaxy at wavelength $\lambda/(1 + z)$. Comparison of two galaxies of different redshifts therefore introduces an uncertainty. This is taken into account by introducing a correction factor known as the K term in the expression for m. The form of the K term had been a major problem for astronomers ever since Hubble's observations in the 1930s. There are several other uncertainties involved in magnitude-redshift measurements, and in the 1950s it was realized that the net result of these might prevent the magnitude-redshift test from discriminating effectively between at least some of the world models.

It is often stated that the steady-state theory is unique among all cosmological models in predicting a deceleration parameter as low as $q_0 = -1$. This is not quite the case, however. In the cosmological model introduced by Lemaître in 1931, the universe starts in a big bang and expands into a quasi-stationary world, where it stays for some time until it starts to expand again. The final expansion tends toward a de Sitter universe, that is, an exponential expansion just as in the steady-state theory. For this reason, the Lemaître model geometrically approaches the steady-state model. Although no definite value of q_0 can be ascribed the Lemaître universe—it depends on the value of the cosmological constant—it can well be -1. This means that even if $q_0 = -1$ were favored by observations, it would not necessarily mean a confirmation of the steady-state theory. Because the Lemaître model is evolutionary, however, there are other ways to decide between this model and the steady-state model.

The first estimate of what later became known as the deceleration parameter is sometimes ascribed to William A. Baum in 1953,[35] whose result is quoted as $q_0 = 1 \pm 0.5$. In fact, Baum did not report this value in 1953, but only four years later. In 1956, Humason, Mayall, and Sandage analyzed the accumulated data for 474 galaxies with redshifts up to $z = 0.2$. For eighteen clusters, they chose the brightest galaxies, assumed to have the same absolute luminosity, and found in this way the expected departure from linearity in the $\log cz$-$\log m$ plot (figures 6.1 and 6.2). As a best value for the deceleration parameter they found $q_0 = 2.5 \pm 1$, but due to the uncertainties and the considerable scattering of the points the result was admitted to be highly provisional. "Although it would be appropriate to end this paper with a definite statement of the possible cosmological models consistent with the present data, such a statement cannot be given at the present," they concluded.[36] All the same, the result seemed incompatible with the $q_0 = -1$ predicted by the steady-state

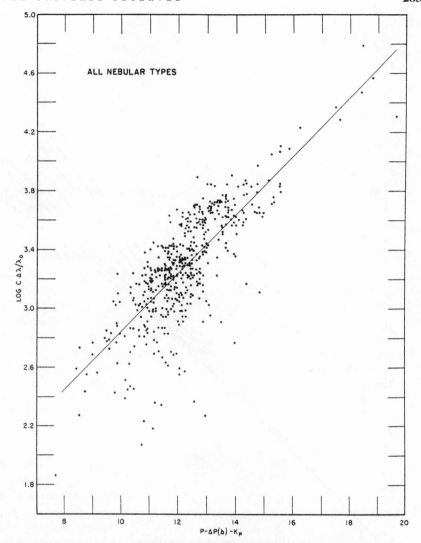

Figure 6.1. The overall result of the redshift-magnitude observations reported by Sandage and his collaborators in 1956, including data for 474 galaxies of all types. *Source:* Humason, Mayall, and Sandage (1956), p. 141.

model. One year later Sandage felt justified in concluding from the same data that this model "does not fit the real world."[37]

As expected, this was a conclusion that the steady-state cosmologists did not accept. Hoyle realized that the data from the 200-inch telescope were important and might constitute "the most serious potential contradiction of the steady-state theory," but he emphasized the uncertainties which, in his view, made the data inconclusive as a cosmological test.[38] As far as he was con-

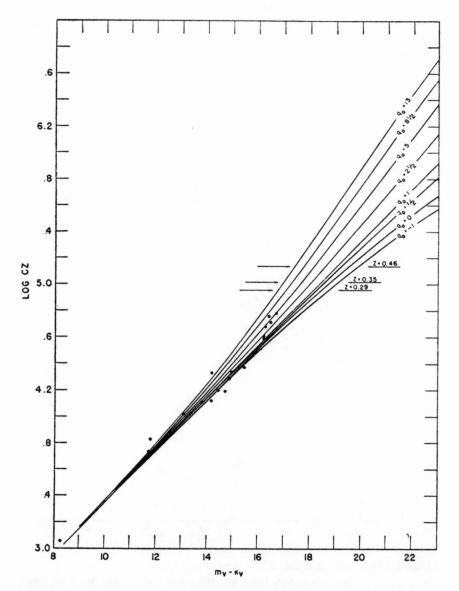

Figure 6.2. The theoretical relationship between redshifts and magnitudes, showing lines for different values of the deceleration parameter q_0. Data for eighteen clusters of galaxies are plotted in order to estimate the best value for q_0. From the comparison Sandage concluded that $q_0 = 2.5 \pm 1$, although the value was viewed as "highly provisional." *Source:* Sandage (1961), p. 367.

cerned, the contradiction was only "potential." As previously noticed, Sandage was opposed to the steady-state theory. It was not because he had any strong feelings about it, he just found it hopelessly speculative and disagreeing with sound observational astronomy. Sandage found it something of a mystery that the theory could be taken seriously at all. In 1978 he stated: "I think really the steady-state theory was dead as soon as it was formulated. And almost every observational astronomer felt that way. . . . I think it's true to say that no one in Southern California ever took steady-state seriously."[39]

The caution advised by the steady-state theoreticians with regard to the data of Humason, Mayall, and Sandage seemed justified, for the following year Baum published a new determination of q_0 based on an improved photometric technique. The result differed markedly from the one found by the Mount Palomar astronomers.[40] Baum found that q_0 might vary between 0.5 and 1.5, with both limits being quite possible. He concluded: "[T]he universe either is Euclidean or is mildly curved inward, that is closed and finite. . . . Strongly curved models, either inward or outward, would not be compatible with the present results, nor would the steady-state model of Hoyle, Bondi, and Gold." Although Baum's result thus still contradicted the steady-state prediction, it also indicated a considerable lack of certainty in the magnitude-redshift method. At this stage, scientists could choose to emphasize either the uncertainty or the qualitative agreement of the results of Baum and that of Humason, Mayall, and Sandage. From the perspective of Hoyle and his followers, the data did not constitute a refutation of the steady-state theory, but were a sign of the inconclusive nature of the test. It was still only a potential contradiction.

Astronomers opposed to the steady-state theory evaluated the data differently than Hoyle did. At the Paris Symposium on Radio Astronomy in 1958, McVittie concluded that although the current value of q_0 would undoubtedly change with future measurements, it was most unlikely that the accepted value might become negative. This, he held, "rules out a large class of relativistic models, and unfortunately for the supporters of the steady-state theory, it rules out that model too."[41] Sandage agreed. He considered Baum's value of q_0 to be in qualitative agreement with the earlier result obtained at Mount Palomar and emphasized its disagreement with the steady-state prediction. In his review of 1961, he concluded: "Baum's results undoubtedly have higher weight than those of HMS [Humason, Mayall, and Sandage], and it seems quite clear that q_0 cannot be greater than 3 and likely lies between 2 and 0."[42] The attempts to determine more accurate values of q_0 from the redshift-magnitude diagram continued during the 1960s, but in spite of much work, this direct approach did not give the unambiguous results hoped for. Disappointingly few galaxies were identified that had redshifts larger than 0.2 and were of the kind that allowed reliable distance measurements. Most astronomers believed with Sandage and McVittie that the results ruled out the steady-state theory, but it was to some extent a matter of belief and by no means a clear-cut observational refutation. The accumulated redshift-magnitude observations of galaxies provided some evidence against the steady-state theory, but it was ineffective in deciding the matter.

In the late 1950s another method related to the redshift-magnitude method was suggested as a geometric test of cosmological models.[43] The idea of the diameter-redshift test is simple: the angular diameter of distant objects of the same absolute size, as observed from the earth, will depend on their distances, the relationship being inversely proportional if a Euclidean geometry is assumed. In that case a double-logarithmic plot of angular diameter versus redshift will give a straight line with negative slope. If space is non-Euclidean— i.e., $k = 1$ or $k = -1$—one would expect deviations from linearity for high values of the redshift, just as in the redshift-magnitude test. In order to apply the test, objects of roughly similar size and with sufficiently high redshifts must be identified. Galaxies will not do, but spherical clusters of galaxies turned out to be good candidates. Such clusters seem to be about the same absolute size and their angular diameter can be determined in a relatively well-defined way. The advantage is that the evolutionary effects which contributed to the uncertainty of the redshift-magnitude method are not serious for clusters of many hundred of galaxies, since it is unlikely that the size of the whole cluster changes significantly during the time it takes light to travel between the cluster and the earth. The method was thus to observe the redshifts and angular diameters of clusters of galaxies, and to plot these in a double-logarithmic diagram. Theory predicted what the high-redshift part should look like according to cosmological models with different geometries or deceleration parameters.

The diameter-redshift test was discussed in particular by Hoyle, who pointed out a qualitative difference between the curves predicted by the steady-state theory and the evolutionary models.[44] According to the steady-state theory, the angular diameter will vary with the redshift as $\Delta\theta = \text{const} \times (1 + 1/z)$, where "const" denotes the constant quantity $DH/2c$, D being the absolute diameter of the object. This means that there will be a limiting value $\Delta\theta = DH/c$ for very high redshifts. According to almost all evolutionary models (with $q_0 \geq 0$), on the other hand, clusters of galaxies will exhibit a minimum apparent diameter at a certain redshift and then, for even larger redshifts, begin to look bigger (figure 6.3). For $z \to \infty$ the object will ultimately fill the whole sky! In the case of the Einstein–de Sitter model, Hoyle showed that the corresponding formula is

$$\Delta\theta = \text{const.} \times \frac{(1+z)^{3/2}}{(1+z)^{1/2} - 1}.$$

It follows that the minimum diameter will be about $3.4DH/c$ and will appear at $z = 1.25$, much larger than any redshift observed at the time. The difference in apparent angular diameters would seem to make a nice test, but unfortunately the predicted differences only appeared at such large distances that optical telescopes were of little help. It was therefore hoped that radio astro-

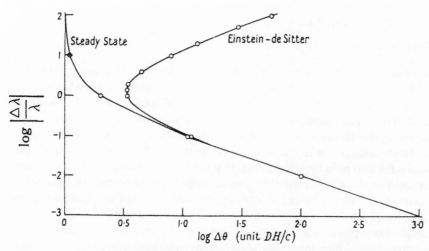

Figure 6.3. The apparent angular diameter of a source of absolute diameter D plotted against the redshift. The theoretical results are shown for two cosmological models, the relativistic Einstein–de Sitter model and the steady-state model. *Source:* Hoyle (1961), p. 14.

nomical measurements might be able to provide data for the test, but this was not the case. Hoyle used data for the radio sources Cygnus A and 3C 295, from which he tentatively concluded that the situation was favorable to the steady-state theory, but the data were all too uncertain to constitute a proper test.

By 1961 the diameters of almost four hundred radio sources had been measured at Jodrell Bank, primarily with the purpose of testing the predictions of cosmological models. However, the interpretation of the data remained uncertain. Davidson analyzed the Jodrell Bank observations and found the results "not inconsistent" with steady-state predictions.[45] Henry Palmer, in an analysis of 133 Jodrell Bank data, found that there was "no obvious relationship" between the data and either the steady-state or the Einstein–de Sitter model.[46] Conclusions more decisive than this were not possible. The diameter-redshift data were just not precise enough and did not cover sufficiently large distances to enable astronomers to choose clearly between the steady-state theory and the relativistic evolution theories. Once again, the situation seemed a stalemate.

It was only much later, after the steady-state theory had disappeared as a viable alternative, that the diameter-redshift test became effective. The method was VLBI (very long baseline interferometry), and the sources were quasars and active galaxies of high redshifts. In 1993 new data were reported that clearly ruled out the steady-state prediction. The results strongly indicate a deceleration parameter of 0.5, that is, an Einstein–de Sitter universe.[47] This is highly interesting, but of course it is of no historical relevance.

6.2 GALAXIES AND ATOMIC NUCLEI

Formation of Galaxies

Astronomy and cosmology are evolutionary sciences which are supposed to be able to account not only for the structure of the universe but also for how the celestial objects were or are formed. Since galaxies make up the main part of the bulk matter in the visible universe, it was generally agreed that a good cosmological theory should be able to explain the formation of these objects and their distribution in clusters. This requirement thus constituted a test between the two rival theories, although a kind of test that did not involve new measurements, but just the ability to account for the existence of galaxies: a cosmological theory unable to offer a good explanation of how galaxies are or were formed would not necessarily be rejected; but it would stand in a weak position compared to a rival theory in which the problem was solved. Both relativistic evolution theories and the steady-state theory built on the cosmological principle, that is, they accepted the assumption of the universe being essentially homogeneous. From this shared starting point the problem was to explain the lack of homogeneity, which after all is so characteristic a feature of the universe even in large regions. How did local irregularities in the distribution of matter come about, and how did they evolve into the localized high-density regions called galaxies?[48]

The problem of galaxy formation was part of the earliest theories of the expanding universe, and in the evolutionary theories it remained a problem that seemed to defy solution. It was studied by Lemaître, among others, in the 1930s; and, even earlier, investigations of the influence of matter condensations on the stability of a static universe led to some of the first theoretical works on the expanding universe (by Eddington, McCrea, and McVittie). The formation and evolution of galaxies were also an important part of Milne's kinematic-relativistic program, but his account of this matter played almost no role in post-1945 cosmology.

In very general and qualitative terms, the problem within the evolutionary cosmologies was to establish a balance between the gravitational forces, which will keep matter together, and the thermal motion and the expansion, both of which effects will tend to disperse matter and hence prevent the formation of galaxies. Although the concentration of matter would, in itself, be favorable for galaxy formation in the earliest times of the universe, the temperature was at this stage so high that matter would easily evaporate from the condensations that were accidentally formed. What was needed was a sufficiently low temperature combined with a sufficiently high density and a sufficiently low expansion rate.

Lemaître attempted to explain the formation of galaxies within his theory of 1931 in which the universe passes through a long stagnation phase where it is in a quasi-static Einstein state. By assuming the young universe to have a fluctuating density, he sought to develop schemes in which small high-density regions would gather mass and eventually develop into protogalaxies, and from these into galaxies. The right condition for such a process was found to exist in the stagnation phase. This phase, and thus Lemaître's theory of galaxy formation, depended on the assumption of a cosmological constant different from zero. For the majority of cosmologists, this was reason enough to reject Lemaître's theory and to look for other ways of explaining the formation of galaxies.

The starting point of almost all theories of galaxy formation was a famous piece of mathematical physics due to James Jeans in 1928, but which went back to calculations made as early as 1902.[49] Conjecturing that matter was originally distributed uniformly throughout space, Jeans showed that in such an unbounded, primitive medium, small density fluctuations would result in gravitational instabilities that in some cases would cause the gas to break up into individual gas clouds—protogalaxies. He deduced the minimum size and mass of such structureless condensations in terms of the temperature and density, formulas which are known as the Jeans conditions. According to Jeans, the formation of galaxies thus happened purely accidentally, as the result of random fluctuations. Jeans's results were based on Newtonian physics and the assumption of a static universe, but it turned out that they could be adapted also to the new expanding universe that entered in 1930.

In Gamow's big-bang theory also it was important to account for the formation of galaxies. He had already sketched an explanation in 1948, which he developed during the following years.[50] In the early, radiation-dominated universe there would be no possibility of condensations, the highly diluted gas being completely uniform because of the violent impact of the photon gas. Only when matter began to dominate, after some thirty million years, would condensations take place in accordance with the mechanism proposed by Jeans. By inserting into the Jeans conditions the numerical values for the density and temperature of the universe at this stage of development, Gamow found a minimum mass of galaxies (before star formation) of about 10^{37} kg, in reasonable agreement with observations. However, Gamow's world picture was a rapidly expanding universe, and since Jeans had assumed a stationary universe, his formula had to be revised. As mentioned in section 3.2, Gamow and Teller had studied the conditions for gravitational condensations in a uniformly expanding universe already before the war, concluding that they were possible when the distance scale factor was six hundred times less than it is today. In collaboration with Ulam and Metropolis, the two physicists reconsidered the problem in the light of the big-bang theory in 1949, but the result was disappointing. Calculations showed that galaxies could not be formed by a Jeans mechanism within the life span of the universe if condensations did not already exist. They therefore looked for some new agent which could be made responsible for the beginning of the condensation process. At first they believed that interaction between gas particles and photons might be such an agent, but calculations showed this idea to fail also.

The solution that Gamow settled on was the action of turbulent motions in the early universe. The inspiration came from Weizsäcker's theory of the formation of the solar system, first proposed in 1943 but only generally known after the war, in part through a detailed summary given by Gamow and the astrophysicist Joseph Hynek.[51] The basis of this theory was the postulate of an originally flattened galaxy with systems of convection currents in it. Because of turbulence, small regions of condensed heavy elements would form, and out of these the planets would form by further gravitational condensations. Weizsäcker also applied his hydrodynamic scheme to cover the formation of stars and galaxies and in 1951 he published an English summary of his ideas. Although he stated that the theory "excludes all discussions of models of the universe," he also mentioned that "no permanent creation of matter [is] assumed."[52] In spite of his wish to eschew cosmology, Weizsäcker's work fitted better with a big-bang type of universe than with a steady-state universe.

Gamow found Weizsäcker's theory very attractive and sought to develop it into a theory of the formation of galaxies in a big-bang universe. He reasoned that in order to overcome the dispersive effects of the expansion at a stage when the primordial gas had cooled off, there must temporarily exist regions of space with more atoms than in their surroundings. Following Weizsäcker, this was thought to be accomplished as a result of the random turbulent motion of the rarefied cosmic gas. In this way there would be formed concentrations of matter with gravitational attractions sufficiently strong to prevent dissipation. Gamow was able to estimate the masses of those permanent structures arising from the concentrations by arguing that they could neither be too big nor too small; in the first case, they would never have been formed as a result of the turbulence, and in the second case, the turbulent motion would have swirled the structure apart again.

As usual for Gamow, when he came up with a new idea, he had much confidence in his cosmological version of Weiszäcker's theory. In 1952, a paper on this subject was presented to the Royal Danish Academy of Sciences (of which Gamow was a member). Gamow wrote in a letter that he was eager to know if Bohr liked his theory, and added: "Now it seems that things can be fixed by assuming turbulence . . . Anyway I am quite persuaded that it is correct, at least, in principle."[53] A year later he was still optimistic and wrote to Bohr: "A student of mine is now trying to analise the observed distribution of galaxies in space, and to see if it realy satisfies the law of turbulent motion. If it does, it will be the solution of the truble . . . I hope it will be so!"[54] However, it was evident that the very foundation of Gamow's theory, the existence of strong initial turbulence, was an assumption that was grafted upon the big-bang theory and was only justified because it seemed to work. There was no explanation of how the turbulent motions originated in an otherwise homogeneous universe. Since such an explanation seemed scarcely possible, "it may be well to introduce the primordial turbulence on a postulatory basis along with the original density of matter and the rate of expansion," Gamow wrote in 1952.[55] To many astronomers, and of course to the steady-state supporters in particular, it was a methodological blemish to introduce violent turbulence as one more initial condition. Methodological objections apart, further investigations showed that Gamow's theory was also technically in trouble. In a series of works from 1956 to 1957, Bonnor carefully reviewed the problem of galaxy formation within relativistic cosmologies, concluding that the ordinary density fluctuations allowed by statistical theory are much too small to account for the present number and sizes of galaxies; or, put differently, the time scale of roughly four billion years is much too small to allow fluctuations to give rise to condensations from which galaxies can form.[56] Among the standard expanding models, only Lemaître's might avoid this problem, and even in this case, Bonnor concluded, it seemed difficult to account for the formation of condensations from small perturbations.

Bonnor could not disprove Gamow's turbulence theory, but he pointed out that it was unsatisfactory both on methodological grounds—no explanation of the turbulence was given—and because there was no reason to assume that the postulated primordial turbulences would persist during the initial phase of expansion. As an alternative, he argued for an oscillating universe in which the inhomogenities in the early universe were fossils from the preceding contraction. This kind of model appealed to Bonnor also for metaphysical reasons. Another alternative would be to use a model with a longer time scale, which meant accepting a nonzero cosmological constant. Bonnor mentioned that the old Lemaître-Eddington model, which is not of the big-bang type, might be reconsidered now that the main objection to this model was removed with the recognition that the formation of the heavy elements did not need a primordial high-density and high-temperature universe. However, no revival of the Lemaître-Eddington model took place.

By 1957 big-bang cosmology was in a rather messy state as far as galaxy formation was concerned. The theory of galaxy formation from cosmic turbulences had not been refuted, but neither was it further developed. Until the mid-1960s the subject shared the same fate as big-bang cosmology in general: it attracted little interest. If the big-bang–inspired turbulence theory stagnated, what about galaxy formation according to steady-state cosmology? Clearly, the entire problem is very different within this theory, where there is no need to explain how the first galaxy was formed—for the good reason that galaxies have always existed. This is a great advantage, for then the gravitational perturbations of existing galaxies on intergalactic dust can be used as a source of condensation instead of the dubious turbulences needed in the big-bang theories.

The formation of galaxies was one of Hoyle's many interests in the late 1940s. In August 1949, one year after the steady-state theory had been proposed, he attended a symposium in Paris dealing with cosmic aerodynamics. Weizsäcker lectured on his theory of turbulence, and in the subsequent discussion Hoyle criticized it for presupposing turbulence in the intergalactic medium without explaining the origin of the turbulence.[57] According to Hoyle, turbulent motion was a secondary phenomenon and therefore could not be used as an explanation of galaxy formation. At that time Hoyle was unable to come up with a better explanation, and neither he nor any other of the participants discussed the problem within a cosmological context.

A steady-state theory of galaxy formation was first proposed by Sciama in 1955. Like other steady-state advocates, he laid great stress on the methodological merits of the theory, namely, that it determined the formation and distribution of galaxies solely by requiring a self-propagating system. Contrary to the arbitrariness of Gamow's theory, Sciama claimed to have accounted for the distribution of matter in the universe entirely in terms of the general laws and constants of nature: "No adjustable parameters are involved, since the state of the system is uniquely determined by the condition that it be self-propagating and stable. This condition enables one to calculate the average properties of the system of galaxies that exists in a steady state universe."[58]

Sciama considered a galaxy moving through the intergalactic gas, and by using the theory of accretion which had earlier been applied to stars by Lyttleton, Hoyle, and Bondi, he found that as a result of the galaxy's gravitational attraction a concentration of material would take place in the wake. The density of this region was found to be large enough for it to collapse under its own gravitation, creating a "child galaxy," the mass of which was determined by the original galaxy. Using some results earlier derived by Hoyle,[59] Sciama showed that the child galaxy would have about the the same mass as its parent. Without further assumptions, representative values for the mass of a galaxy followed from Sciama's arguments, namely, $10^{41} - 10^{42}$ kg, in good agreement with what was estimated from observations.

Apart from single birth processes, where a child galaxy escapes from its parent, Sciama also studied gravitational capture, in which case a double galaxy would be formed, and developed this into a theory of formation of clusters. Among other results, he deduced the average distance between galaxies, the maximum and average numbers of galaxies in a cluster, and the percentage of galaxies that are single. Most of the results obtained by Sciama, for both galaxies and clusters, were in remarkable agreement with the rather uncertain observations, and none were clearly contradicted. Given the fact that these results followed from rather simple arguments and relied uniquely on the steady-state assumption, it is understandable that Sciama considered his theory a triumph for the steady-state theory. "It appears," he concluded, "that the steady state theory of the universe provides an adequate framework for a quantitative discussion of the formation, distribution and properties of galaxies, which is free from adjustable parameters and which is in reasonable agreement with observation." Sciama later extended his theory to include magnetic and thermal forces also, where the latter were assumed to arise as a result of heating caused by cosmic rays.[60]

Sciama's steady-state explanation of galaxy formation was undoubtedly superior to the corresponding theories within relativistic cosmology, both in a methodological sense and as far as predictive and explanatory power was concerned. Whereas the problem was unsolved in the relativistic evolution theory, a good start had been made with the more direct and much simpler steady-state theory. Sciama's theory of 1955, in which existing galaxies gave birth to new ones through gravitational effects, was not the only steady-state explanation of the formation of galaxies. A different theory, but not necessarily an alternative to Sciama's, was proposed in 1958 by Hoyle and Gold, in what constituted a major change in the physical picture of the steady-state universe. The "hot steady state universe" was argued by Hoyle at the Solvay conference in June 1958 and two months later in a joint paper by Gold and Hoyle at a symposium on radioastronomy held in Paris.[61] The basic assumption of this theory was that the present intergalactic gas was hot, having a (kinetic) temperature of about 10^9 K, which allowed Gold and Hoyle to consider galaxy formation a thermal process and not primarily a gravitational process. Such high temperatures were possible in the steady-state theory, but not in relativistic cosmologies, where the adiabatic expansion from the initial high-density universe would reduce the temperature drastically. It may seem strange that the almost empty universe can be ascribed a temperature of one billion degrees, but the temperature in question is the kinetic temperature, which is a measure of the average kinetic energy of the particles. Because of the high rarefaction, the high-temperature intergalactic space will still be extremely cold. In the steady-state theory, there was no high-density phase of the universe and no resulting coupling between matter and radiation. Contrary to the situation in Gamow's big-bang theory, it was therefore possible to have a high kinetic temperature of matter and, at the same time, a low radiation temperature.

According to the Hoyle-Gold theory, condensations might occur in a hot intergalactic gas by local radiative cooling in regions where the cooler parts were compressed by the surrounding hotter gas. By requiring the cooling to take place in a time scale of $1/3H$—the characteristic period of steady-state processes—the temperature of the hotter gas was estimated to be 10^7 K and its density 10^{-27} g·cm^{-3}; this would result in cool protogalaxies with $T \approx 10^7$ K and $\rho \approx 10^{-24}$ g·cm^{-3}. The temperature of the hotter gas was obtained by a primary cooling process from the cosmological gas of $T \approx 10^9$ K. In this way Hoyle and Gold could explain the condensation of a very large cloud into a number of galaxies, rather than into a single giant galaxy. Furthermore, by following the steps through the primary and secondary cooling processes, they were able to get the correct order of magnitude of the size and mass of the formed galaxies, about 0.1 Mpc and 10^{11} sun masses.

As to the secondary cooling process, they noticed that the thermal energy density of the very hot cosmological gas could be reproduced if it was assumed that the matter created continually according to the steady-state theory consisted of neutrons. These are radioactive, decaying into protons, electrons, and antineutrinos, and the kinetic energy of the decay products (except the neutrinos) gave just the right energy density. Hoyle and Gold thus believed that they had solved the problem left untouched in earlier versions of the steady-state theory, i.e., the nature of the continuously created matter. It consisted of neutrons, neither protons and electrons nor neutral hydrogen atoms. Although the Hoyle-Gold theory did not lead to the same kind of detailed

predictions as Sciama's theory, they considered it confirmed by its explanatory power in areas outside galaxy formation. In particular, they used the hot-
universe model to offer an explanation of one of the unsolved problems of
astrophysics, the origin and energy spectrum of the primary cosmic radiation.
According to Hoyle and Gold, cosmic rays were not accelerated by purely
local galactic processes, but by large-scale magnetic fields arising from cosmological thermal processes. In addition, they were able to offer an interpretation of the newly discovered Zwicky tails, faint extensions from many galaxies that were otherwise unexplained.

Hoyle also pointed out that a small fraction of the thermal energy of the
intergalactic matter would be expected to turn up as a detectable x-ray background radiation of energy about 50 keV. The absence of such a radiation
would not disprove the theory, however, since it would only appear if the
temperature was about one billion degrees. For lower temperatures, it would
not be expected. The weak point in the Hoyle-Gold theory was, of course, that
its basis, the assumption of a hot cosmic gas, was a postulate with only indirect support. Since such a hot gas emits electromagnetic radiation, one might
believe that its existence could easily be detected, but this was not the case.
The radiation from the hot, ionized hydrogen gas would be in the form of 1
keV quanta, and since these are absorbed by oxygen, carbon, and other atoms
in the interstellar gas, the radiation would not be observable from the earth.

With two independent theories at hand, Sciama's and the one of Hoyle and
Gold, steady-state theory seemed in a strong position with regard to the problem of galaxy formation. But further investigations showed that not everything was fine. Martin Harwit, a Czechoslovakian-born astronomer with a
fellowship at Cambridge University, in 1961 subjected the steady-state formation of galaxies to a careful analysis.[62] While a graduate student at MIT, he
noticed that Sciama had overlooked a feature in his accretion theory, and was
encouraged by Gold to pursue the problem. At that time Harwit appreciated
the steady-state theory for its clear predictions, but he had no particular preference for any cosmological model. Harwit concluded that if Hoyle's value of
the average density of the universe was accepted, gravitational forces alone
could not bring about galaxy formation; the only escape from this conclusion
would be to postulate new matter to be created in a localized process, contrary
to the steady-state assumption. The Bondi-Gold version of steady-state theory
did not fix the density, but Harwit showed that this theory would also require
an unrealistically high density to form galaxies—at least ten times the density
prescribed by Hoyle's theory. He therefore concluded that if the steady-state
theory was to explain the formation of galaxies, it had to be by means of forces
other than gravity. The critical analysis cast serious doubt on Sciama's gravitational accretion model, but left the Hoyle-Gold model untouched. It was
therefore not a serious blow to steady-state cosmology—after all, relativistic
cosmology did not have an acceptable gravitational mechanism either.

The situation with regard to galaxy formation in the early 1960s was, then,
that this type of test also failed to discriminate clearly between the two rival

cosmological theories. Steady-state theory could offer interesting and encouraging results, but nothing that was beyond criticism; most evolutionary models failed to account for the formation of galaxies, yet the entire problem was too complex to warrant the conclusion that the big-bang theory must be wrong. All the same, in a general way the problem of galaxy formation was a small victory for the steady-state supporters. In 1963, Hoyle and his two collaborators Margaret and Geoffrey Burbidge summarized the situation as follows: "Undoubtedly, the greatest shortcoming of all cosmological theories lies in their failure to provide a working model of the formation of galaxies. Evolutionary cosmology provides no model at all. Galaxies are supposed to arise from initial fluctuations, every necessary property being inserted into the theory as an initial condition. In short, evolutionary cosmology achieves nothing more than its hypotheses, its deductive successes are nil. The steady-state cosmology is faced by uncertainties at just the point where cosmology and astrophysics should properly be connected—the condensation of the intergalactic medium."[63] The sympathy of the authors of these lines was with the steady-state theory, but their description of the situation could presumably be accepted also by many astronomers with no interest in the cosmological controversy.

Nucleosynthesis

It was agreed by all cosmologists that a good theory of the universe should be able to account for the formation and cosmic distribution of the chemical elements, i.e., to explain how the various atomic nuclei heavier than hydrogen were or had been synthesized. The better the agreement between the predicted species of nuclei and those found empirically, the better the theory. As stressed in chapter 3, Gamow's big-bang theory rested on and was guided by such nuclear archaeological considerations from its very beginning. By 1953, with the appearance of the theory of Alpher, Herman, and Follin, the theory had become a sophisticated, quantitative theory which scored significant successes in the prediction of the abundances of the light elements. But the problem of bridging the mass gaps at atomic masses 5 and 8 remained a stumbling block for the theory as far as the heavier elements were concerned.

Compared with big-bang theory à la Gamow, steady-state theory was completely different with respect to the problem of nucleosynthesis. If nuclear physics was taken out of Gamow's theory, there would no longer be a theory; it would just be ordinary relativistic cosmology. If it was taken out of steady-state theory, it would make no difference at all, for the simple reason that this theory had no direct connection with nuclear physics. The Bondi-Gold version did not consider nucleosynthesis at all, and neither Bondi nor Gold felt at home with nuclear physics, a field in which they did not have much training. (However, Bondi entered nuclear astrophysics on one occasion; see below.) The Hoyle version of steady-state cosmology also remained silent for a long time on the matter, which played almost no role in the British discussion until about 1956. With very few exceptions, the cosmological discussion was seen as a controversy between steady-state and relativistic evolution theories, and not between the steady-state and the Gamow big-bang theories. For this reason questions of nucleosynthesis only turned up on rare occassions. Furthermore, most of the British astronomers and physicists engaged in the debate lacked interest and insight in advanced nuclear physics. In fact, Hoyle was one of the very few discussants who had the necessary background to follow the calculations of the American nuclear cosmologists. It is no wonder, then, that while by far the largest part of the cosmological controversy took place in Great Britain, the scene shifted toward the United States when it came to nucleosynthesis.

In spite of the methodological and disciplinary divide between big-bang and steady-state theories, it was of course realized that the formation of elements was an area that the steady-state theory also had to come to grips with. Whatever the differences between Hoyle's version and the version of Bondi and Gold, a cosmic origin was precluded by the very foundation of the steady-state theory, the perfect cosmological principle. In other words, the formation of elements out of hydrogen also has to take place today, i.e., in existing sources such as stars and novae. For this reason, the problem of nucleosynthesis was methodologically more acute for the steady-state theory. The big-bang alternative had at its disposal two kinds of nuclear ovens, both the stars and the early universe, and it therefore had a greater freedom in accounting for the distribution of elements. In this respect, as in most others, the steady-state theory was more easily falsifiable. Yet this was mostly in principle only, for until the mid-1950s the discussion took place as if it were an either-or: were the elements produced under the extreme conditions of the very early universe, or could they be accounted for under the rather different, if equally extreme, conditions in the interior of stars?

Gamow's research program was to demonstrate that *all* building-up nuclear processes were of cosmological origin. He saw failures to account in this way for the production of certain elements as a weakening of the big-bang theory, as much as he saw failures to account for the production of elements in stars as a weakening of the position of the steady-state alternative. In fact, the first mentioned failure could not amount to a disproving of the big-bang model. The essence of this model would not be contradicted even if stellar explanation proved perfectly adequate and cosmological explanation did not. Although Gamow preferred cosmological nucleosynthesis, he realized this situation, of course, and used it when he felt a need for it. For example, in 1954 he admitted the failure of big-bang theory in building up elements heavier than boron, but he took the position that this was not a very serious problem; after all, it is not incompatible with the big-bang theory to have elements generated in stars. Gamow simply retreated to the statement that "I would agree that the lion's share of the heavy elements may well have been formed later in the hot interior of stars."[64]

The question of element formation entered the cosmological debate in 1951–52. Bondi admitted it as a test in the sense that he recognized that if no stellar processes were found to synthesize heavier elements in the right proportions, then steady-state theory would be faced with "a very serious difficulty."[65] At about the same time, Gold reflected on the different status of nucleosynthesis in the big-bang and steady-state theories. Referring to the mentioned fact that the big-bang theory could make use of two kinds of nuclear ovens, both a primordial and a stellar one, he wrote: "Indeed the amount of freedom for speculation that one has in such a theory is embarrassingly great."[66] On the other hand, he also pointed out that the big-bang theory "loses an important advantage" if it could be established that the generation of heavy elements is a process still going on in the stars or elsewhere. As it turned out, this was just what happened during the following years: the big-bang theory lost an important advantage, but not more than that.

Now there were no attempts to find nuclear building-up processes within the steady-state theory specifically; there was no, and could not be any, steady-state nucleosynthesis as a counterpart to big-bang nucleosynthesis. All that steady-state theoreticians could offer were ordinary stellar processes, or perhaps modifications in supernovae explosions (such as Hoyle had suggested in 1946–47, before the steady-state cosmology). There was nothing corresponding to the ambitious and largely successful big-bang program of reproducing the cosmic abundances of elements. What appeared to be the inability of the steady-state theory to cope with nucleosynthesis was seized on by Gamow, who considered it a strong reason to dismiss this theory. In his popular book of 1951, he stated that the steady-state theory did not have "a satisfactory quantitative explanation of the abundances of chemical elements." Hoyle's preference for formation by stellar processes was judged "artificial and unreal."[67]

The problem of generating elements heavier than helium was placed in a new perspective in 1951–52, when Ernst Öpik and Edwin Salpeter independently proposed a way in which carbon could be produced in the cores of contracting stars at temperatures of a few hundred billion degrees.[68] Although Öpik was the first to suggest the mechanism, which was included in his 1951 theory of stellar evolution, his contribution was little noticed, and it was only with Salpeter's work that the mechanism made an impact. Salpeter was (like Bondi and Gold) born in Vienna of Jewish parents. As a result of the *Anschluss* the family fled to Sydney, Australia, where Edwin received his higher education. After the war he went to Birmingham, specializing in quantum field theory, and then to Cornell University, where he worked with Bethe on problems of quantum mechanics. Invited by Fowler, he spent the summer of 1951 at Caltech working in nuclear astrophysics, a field which was new to him. The Kellogg Laboratory at Caltech was headed by William Fowler, a pioneer of nuclear astrophysics and a future Nobel laureate for his contributions to the field. Since 1946 the focus of the laboratory had been the study of stellar nuclear reactions, making Kellogg the only physics institution in the world with such a focus. Until about 1955, neither Fowler nor his collaborators worried about cosmology or possible cosmological implications of their work. On the contrary, the Kellogg approach was very much in the pragmatic nuclear-engineering style that also characterized most other American specialists. It was during his stay with Fowler that Salpeter found a method to overcome the problem of mass gaps at atomic weights 5 and 8.

According to Salpeter, the triple-alpha burning process $3\,^4\text{He} \rightarrow\,^{12}\text{C} + 2\gamma +$ 7.3 MeV could take place at a reasonable rate under the temperature and pressure conditions in the interior of a star. Beryllium-8 has no stable existence because it disintegrates spontaneously into two alpha particles, but with the thermal energies at temperatures exceeding 10^8 K, small amounts of the isotope can be formed. Salpeter reasoned that although the highly unstable beryllium-8 would exist only for a fraction of a second, it would survive long enough to form carbon by the consecutive processes $2\,^4\text{He} \rightarrow\,^8\text{Be}$ and $^8\text{Be} +$ $^4\text{He} \rightarrow\,^{12}\text{C}^* \rightarrow\,^{12}\text{C} + 2\gamma$. Salpeter further suggested that, once carbon was formed, subsequent capture of additional alpha particles might result in heavier elements such as oxygen-16 and neon-20. A similar suggestion was made by Öpik in his work of 1951. The possibility of a triple-alpha process (but not through beryllium-8 as an intermediate) had already been considered by Bethe in his classic work on solar energy production.[69] However, Bethe's interest focused on the conditions of the sun, and he had concluded that under these conditions no carbon-12 could be produced permanently.

At the time when Salpeter devised his scheme, he collaborated to some extent with Bondi, who spent part of the year 1951 as a visiting research fellow at Cornell University, invited by Bethe. Bondi and Salpeter together published a brief review of nucleosynthesis, Bondi's one and only contribution to the field and the first time that Salpeter's triple-alpha process appeared in print.[70] Although cosmology did not explicitly enter the review, the two authors concluded that stellar reactions and supernovae explosions might account for all the elements in the universe; the work thus implicitly supported the steady-state view. Salpeter was not very interested in the cosmological question, which he preferred to keep in the background. His attitude was that of a practical engineer, meaning that the emphasis was on solving a technical problem in nuclear physics.[71] He did, however, have a slight preference for the steady-state theory because of its aesthetic and methodological virtues.

Two years later, Salpeter's mechanism was taken up by Hoyle during the first of his many stays at Caltech, where he spent the first three months of 1953. Hoyle predicted the existence of a resonance state of $^{12}\text{C}^*$ at 7.68 MeV (later found to be 7.655 MeV) and showed that the Salpeter bridge depended critically on this state. Only if the resonance state existed would it be possible to reproduce the cosmic abundance ratios of helium, carbon, and oxygen by means of the reaction. Without it, the carbon produced would transform into oxygen according to the scheme $^{12}\text{C} +\,^4\text{He} \rightarrow\,^{16}\text{O} + \gamma$. When Hoyle told the Caltech experimentalists about the predicted resonance, at first they responded skeptically to the self-confident Briton. Fowler recalled: "I was very sceptical that this steady state cosmologist, this theorist, should ask questions about the carbon-12 nucleus. . . . Here was this funny little man who thought that we should stop all this important work that we were doing otherwise and look for this [resonance] state, and we gave him the brushoff. Get away from us, young fellow, you bother us."[72]

The skepticism did not last long. Fowler and his collaborators recognized the force of Hoyle's theoretical argument and a young postdoctoral assistant, Ward Whaling, suggested an experiment to test it. Within two weeks the predicted resonance state was confirmed experimentally, even before Hoyle's paper had appeared.[73] The cosmological significance of Salpeter's mechanism was that it would take place at a sufficiently high reaction rate under the physical conditions corresponding to the late phase of a red giant star ($T \sim 10^8$ K, $\rho \sim 10^5$ g·cm^{-3}), but not under the conditions usually assumed to govern the early big-bang universe. So here was a mechanism that could be utilized by the steady-state physicists in order to build up a theory of formation of the heavier elements. In 1956, Chushiro Hayashi and Minoru Nishida, two Japanese astrophysicists at Kyoto University, found a way to bridge the mass gaps 5 and 8 within a big-bang context.[74] Their work was a continuation of the theory which Hayashi had proposed in 1950. The two Japanese managed to get the right conditions for the triple-alpha process by using a higher density of protons and neutrons in the primordial universe than the one ordinarily assumed. (Hayashi and Nishida assumed $T \sim 10^{10}$ K and $\rho \sim 10^7$ g·cm^{-3}.) However, in that case further nuclear processes would yield a concentration of the heavy metals in disagreement with observations, so the theory of Hayashi and Nishida did not succeed in providing a satisfactory big-bang explanation for the abundances of the heavier elements.

Burbidge, Burbidge, Fowler, and Hoyle

During Hoyle's stay at Caltech in 1953 he also developed a theory of nucleosynthesis in which collapsing stars at very high temperatures exploded and scattered their materials into space. The intergalactic material, or parts of it, would then be trapped into galaxies out of which new collapsing stars would be formed. In this way a self-perpetuating mechanism of element synthesis was sketched. As Hoyle pointed out, the steady-state theory was "the particular cosmological framework into which the writer would seek to fit the present theory."[75] A more comprehensive astrophysical theory of nucleosynthesis became a reality after Fowler, impressed by Hoyle's work, spent his sabbatical year 1953–54 in Cambridge in order to work with Hoyle. In England, he and Hoyle were joined by Margaret and Geoffrey Burbidge, a young married couple of British astrophysicists who had spent several years in the United States. Fowler and the Burbidges at once started a collaboration in a study of the nuclear processes taking place in the stars, Hoyle being unable to participate in the early phase because of a heavy teaching load.[76]

In 1956 Hoyle had another sabbatical leave at Caltech, where the Burbidges now stayed as research fellows. Together with Fowler they collaborated to develop an ambitious theory of element synthesis that would explain, at least in outline, the formation and distribution of all elements. In May 1956 the British-American team had its first paper ready, a suggestion that certain characteristics of supernovas might be explained by assuming that vast quantities of the transuranic element californium were synthesized in the explosion.[77] The isotope californium-254 decays by spontaneous fission, and the liberated energy, about 200 MeV per fission, was believed to explain the late phase of a supernova. But the main result of the collaboration was what subsequently was often referred to as the B^2HF work, published in a preliminary form in the fall of 1956, and with the full report appearing a year later.[78]

Earlier attempts to establish a theory of nucleosynthesis had presumed the reactions involved to belong to the same category, being either equilibrium or nonequilibrium, and either stellar or cosmological. Hoyle and his coworkers realized that the first constraint had to be given up, but they found that they could keep to stellar sites. Since what is of interest in the present context is the cosmological significance of theories of nucleosynthesis, it is sufficient for our purposes to mention only the main results obtained by the Burbidges, Fowler, and Hoyle. Many of these results were also obtained by Alastair Cameron, a physicist at the Chalk River atomic plant in Canada, who in 1957 reviewed the problem of stellar nucleogenesis.[79] Although Cameron covered largely the same ground as Hoyle and his coauthors, his paper was qualitative and less detailed than the B^2HF work, and for this reason it was overshadowed by the theory of Hoyle, Fowler, and the Burbidges. Cameron kept to the stars and did not mention cosmological questions.

First of all, the B^2HF paper was a theory of stellar element formation, considering only stars in their various phases of development and not more exotic nuclear ovens. Supernovae explosions were assumed for the very heavy elements, but cosmological genesis remained outside the theory in the sense that it was shown to be unnecessary. The authors mentioned that their conclusions would be equally valid for a big-bang synthesis "in which the initial and later evolving conditions of temperature and density are similar to those found in the interior of stars."[80] However, in fact the conditions of the primeval universe, as given in the Gamow-Alpher-Herman theory, differed greatly from those in the stars, so the admission had no significance.

According to the B^2HF team, in main-sequence stars hydrogen was burned to helium, following earlier theories, and out of the helium produced nuclei of the type $(A, Z) = (4n, 2n)$ were formed. In this way the formation of elements up to calcium-40 was explained. These processes took place at increasing temperatures, from about 10^7 to 10^9 K. The formation of the medium-weight elements with high binding energy was explained as a mixture of equilibrium processes (*e* processes) and neutron capture processes at different rates. In the slow *s* process neutrons were captured by iron-group and lighter elements, with up to 10^5 years between successive captures; in the rapid *r* process, supposed to take place in supernovae, the time interval was less than a second. The *s* process was shown to be able to synthesize elements up to bismuth, whereas the even heavier elements were the results of *r* processes. In some cases, proton-capture processes were also invoked. Finally, an unknown *x* process was assumed to explain the production of light nuclei such as lithium and beryllium. All these processes, taking place under different conditions of temperature and density, are summarized in figure 6.4.

The end result of the complicated network of nuclear reactions was impressive. The Burbidges, Hoyle, and Fowler were guided by new empirical abundance data compiled by Harold Urey and the German-American physicist Hans Suess, which gave a more detailed picture than the classical compilation of Goldschmidt.[81] Comparing their calculations with the Suess-Urey data, the authors concluded: "We have found it possible to explain, in a general way, the abundances of practically all the isotopes of the elements from hydrogen through uranium by synthesis in stars and supernovae." This was no exaggeration, as shown by a typical case, the calculated values of the (logarithmic) abundances of chromium isotopes relative to chromium-52. With the observed values given in parentheses, the Burbidges, Hoyle, and Fowler obtained the results

chromium-50	−1.89	(−1.27)
chromium-53	−0.85	(−0.94)
chromium-54	−1.78	(−1.50)

The subject of the B^2HF paper was not cosmology, but simply nuclear astrophysics. That it was also part of the cosmological controversy only appeared on close reading, the references to cosmology occupying less than one page of the 104-page article. Yet there is no doubt that the authors considered the "primeval theory" of Gamow and his coworkers to be wrong. Only on one occasion did they refer to the steady-state theory by name, emphasizing that "studies concerning the origin of the radioactive elements may lead to objective tests of the various cosmological models."[82] Assuming steady-state conditions, they found that the proposed synthesis of uranium led to the mass ratio 0.75 between uranium-235 and uranium-238 at the time when no further contribution to the material of the solar system occurred. The present-day ratio is 0.0072, a result of uranium-235 having a shorter half-life than the heavier isotope, and a simple calculation showed that this would require the age of the solar system to be about 5×10^9 years, in good agreement with the accepted value.

Hoyle, Fowler, and the Burbidges emphasized that their stellar theory of element synthesis had the methodological advantage that it built on nuclear processes currently taking place in the stars; because the stars change during their evolution, there were a variety of processes available for building up the different nuclear species. All this was different from the big-bang theory, where the nuclei were synthesized in a hypothetical state of the universe, and where "all the varying conditions occur in the first few minutes, and it appears

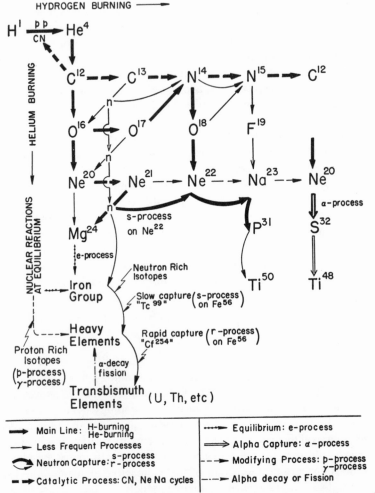

Figure 6.4. Schematic overview of stellar element formation according to the 1957 theory of the Burbidges, Hoyle, and Fowler. *Source:* Burbidge et al. (1957), p. 552.

highly improbable that it can reproduce the abundances of those isotopes which are built on a long time scale in a stellar synthetic theory."[83] Although the B^2HF theory was not explicitly associated with steady-state cosmology, it was definitely a non-big-bang theory; and since the only important alternative to this theory was steady-state cosmology, the B^2HF theory gave valuable support to the view of the universe held by Bondi, Gold, and Hoyle. At the Solvay Congress in 1958, where Hoyle gave an exposé of the B^2HF theory, Heckmann asked how the theory related to steady-state cosmology. Hoyle avoided taking the theory as evidence for a steady-state view and replied diplomatically that the B^2HF theory would also be consistent with a big-bang type of cosmology "provided any superdense state of matter that may occur in non-stationary cosmology satisfies the requirement that matter emerges from the superdense state essentially as hydrogen."[84] He knew very well that this requirement was not met in existing big-bang theories. Robert Oppenheimer used the occassion to criticize the steady-state theory, but also to praise what had come out of it: "Dr. Heckmann expressed misgivings about the arbitrary alterations of the basic equations of relativity involved in the 'steady-state theory.' In these I concur. But, by providing an incentive for understanding the present state of the cosmos in terms of processes that can now be in progress, this theory has led to the beautiful work reported yesterday by Hoyle on element synthesis. Even if the hypothesis is, as I believe, quite wrong, it has thus led to great progress in our understanding."[85]

The B^{2}HF theory successfully explained the abundances of the elements by making use of multiple and complex nuclear processes taking place in stars and supernovas under varying conditions. This feature was essential for the success of the theory, but it also resulted in a theory which was anything but simple and might seem to lack unity and aesthetic qualities. It was not the kind of theory that appealed to Gamow, who considered the big-bang explanation promising because it presented a relatively simple picture of element formation, taking place in a short interval of time and under well-defined and homogeneous conditions. Some years before the B^{2}HF theory, Gamow indicated that the stellar theories were unsatisfactory, and more or less ad hoc, because of their complexity and multifariousness. He expressed his dissatisfaction in his own, inimitable way: "What van Albada and Hoyle demand sounds like the request of an inexperienced housewife who wanted three electric ovens for cooking a dinner: one for the turkey, one for the potatoes, and one for the pie. Such an assumption of heterogeneous cooking conditions, adjusted to give the correct amounts of light, medium-weight, and heavy elements, would completely destroy the simple picture of atom-making by introducing a complicated array of specially designed 'cooking facilities'."[86]

Another humorous description of the situation in nucleogenesis was given by Barbara Gamow, George's wife, several years later:

NEW GENESIS

In the beginning God created radiation and ylem. And ylem was without shape or number, and the nucleons were rushing madly over the face of the deep.

And God said: "Let there be mass two." And there was mass two. And God saw that deuterium was good.

And God said: "Let there be mass three." And there was mass three. And God saw tritium and tralphium [hydrogen-3 and helium-3], and they were good. And God continued to call number after number until He came to transuranium elements. But when He looked back on his work He found that it was not good. In the excitement of counting, He missed calling for mass five and so, naturally, no heavier elements could have been formed.

God was very much disappointed, and wanted first to contract the Universe again, all to start all over from the beginning. But it would be much too simple. Thus, being almigthy, God decided to correct his mistake in a most impossible way.

And God said: "Let there be Hoyle." And there was Hoyle. And God looked at Hoyle . . . and told him to make heavy elements in any way he pleased.

And Hoyle decided to make heavy elements in stars, and to spread them around by supernovae explosions. But in doing so he had to obtain the same abundance curve which would have resulted from nucleosynthesis in ylem, if God would not have forgotten to call for mass five.

And so, with the help of God, Hoyle made heavy elements in this way, but it was so complicated that nowadays neither Hoyle, nor God, nor anybody else can figure out exactly how it was done.

Amen.[87]

The B^2HF theory provided a brilliant defense of steady-state cosmology, but it did not seriously weaken the big-bang alternative. As mentioned, this theory cannot be refuted by a stellar theory of element formation. All the same, the success of the B^2HF theory naturally reduced the motivation to develop the primordial theory. Why consider other sites if stars suffice as the ovens in which the elements are cooked? According to Robert Wagoner, himself a pioneer in nucleogenesis, big-bang theories of element synthesis lay fallow for almost a decade after the B^2HF paper.[88]

That the B^2HF theory was taken as support of steady-state cosmology is perhaps most clearly seen from Bondi's postscript of 1963 to his article on "Some Philosophical Problems in Cosmology." Bondi described the stellar theories of element synthesis as "a tremendous triumph" for the steady-state point of view.[89] In accordance with his Popperian views, he did not present the stellar theory of the origin of elements as a direct confirmation of steady-state theory, but as the passage of a test: "The fascinating point . . . is that a theory as uncertain as the steady-state theory should have inspired and directly caused one of the most important advances in physics during the last decade, . . . this theory of the origin of heavy elements means that the steady-state theory has effectively passed a severe test." Whether or not the steady-state theory was accepted as the correct picture of the universe, it was generally recognized that it had led to great progress in the understanding of nucleosynthesis. For example, this was what Oppenheimer stated in his comment at the 1958 Solvay Congress, quoted above.

We have examined four different kinds of observational tests that were discussed in the 1950s. On the whole, the situation was unsettled and confused, with some tests favoring a big-bang universe and others the steady-state model. By 1960, the time scale problem was no longer much discussed, but redshift-distance measurements indicated a universe in disagreement with the steady-state doctrine, although the evidence was uncertain and subject to discussion. The Stebbins-Whitford effect had graciously vanished from the scene, leaving the steady-state theory with a psychological victory, but not more than that. Similarly, the debate over the origin of the elements had, by and large, been won by the steady-state astrophysicists, but not completely, for the abundances of the very light elements were still best explained within a big-bang context. Finally, the formation of galaxies also tended to favor steady-state cosmology, but with no possibility of ruling out the rival big-bang theory. As far as these observational tests were concerned, the steady-state theory had performed remarkably well, indeed as well as, or even better than, the big-bang theory.

The theme of the Eleventh Solvay Congress, already mentioned several times, was the structure and evolution of the universe. If this prestigious meeting gives some indication of the strength of the steady-state theory at the time, it must be concluded that the theory was in a fairly strong position. Among the participants were all the leading steady-state theoreticians— Hoyle, Gold, Bondi, and McCrea—but also opponents of the steady-state theory, including Heckmann, Sandage, and Baade. The controversy between the steady-state theory and evolutionary cosmology was an important part of the congress, but, as was usual in Europe, Gamow's big-bang theory was absent from the discussions. The big-bang idea, in a different version from Gamow's, was defended only by Lemaître. Whatever the real degree of popularity of the steady-state theory, Gamow felt that it was dominating the scene in Europe. He had been refused an invitation to the Solvay meeting, which he interpreted as a result of his uncompromising opposition to steady-state cosmology.[90]

The most characteristic feature of the various observational tests was their lack of possibility of discriminating clearly between the two cosmological theories. Some of the tests favored one kind of theory, others the other kind of theory, but none of them were even close to discriminating crucially between them. We have not yet included radio astronomical measurements, although these had entered the cosmological debate already in the mid-1950s. As it turned out, radio astronomy was able to determine the question in a manner which was much more direct and unambiguous than the other attempts at testing. In a sense, it was radio astronomy that solved the cosmological controversy as far as observational evidence was concerned.

6.3 IMPLICATIONS OF RADIO ASTRONOMY

More than traditional optical astronomy, it was the new science of radio astronomy that helped create the conviction among a majority of astronomers that the steady-state theory of the universe was wrong. The radio astronomical measurements did not clearly and unambiguously *prove* that this was the case, but a gradual accumulation of data amounted to convincing evidence in the early 1960s. Attempts to defend the steady-state theory against the evidence were dutifully produced, but were not accepted except by the few who already believed in the correctness of that theory. Indeed, it was only about 1960, and then as a result of the radio astronomical measurements, that some supporters of steady-state cosmology began to lose faith in a universe with no global evolution. This was the first clear sign that the controversy was coming to an end, but it was a sign only, neither an armistice nor a complete victory of evolutionary cosmology.

Radio Astronomy Faces Cosmology

Radio astronomy is traditionally dated back to 1932, when Karl Jansky, a radio engineer employed by the Bell Telephone Company, published a paper in which he reported a static radio noise coming from the Milky Way.[91] Although the discovery excited popular interest and was widely discussed among radio engineers, it failed to attract much attention among physicists and astronomers until the late 1930s. The credit for making astronomers acquainted with celestial radio signals goes to another American radio engineer, Grote Reber, who during the mid-1930s built his own rotatable parabolic reflector. In 1940 his first measurements appeared in the *Astrophysical Journal*, indicating that the subject was now seen as also a branch of astronomy and not of radio engineering only. However, the emergent astronomical interest in the cosmic radio signals was temporarily halted when the war broke out.

The real beginning of radio astronomy took place in Great Britain and was initially a by-product of that country's preoccupation with radar research. The physicist James S. Hey, who spent the war years developing radar, discovered in 1942 that the sun emits radio waves associated with its sunspot regions. Because of security regulations, the paper in which this important discovery was announced only circulated among a limited number of researchers; it was only when it was republished in 1946 that it became generally known. Other work done by Hey and his group at the end of the war revealed radar signals reflected from the trails of meteors and from the moon. More important, Hey also discovered in 1946 that some extrasolar radio sources are localized, that is, that there exist discrete sources of radio emission, so-called radio stars. The first extrasolar localized radio signals came from an unknown source in the direction of Cygnus, which was therefore denoted the Cygnus A source. Three years later, Australian researchers at the Radiophysics Laboratory in Sydney identified the first radio star (Taurus A) with an optically known object, namely, the Crab Nebula. In 1950 British astronomers detected radio emission from the Andromeda galaxy, the first identified extragalactic radio source.

At that time, radio astronomy began to be organized as a research branch in Britain, first in Cambridge and at the new Jodrell Bank Observatory near Manchester. At about the same time, the Radiophysics Laboratory in Sydney started a research program in solar radio astronomy. These three centers—Cambridge, Manchester, and Sydney—dominated radio astronomy during the 1950s, sometimes in friendly competition and sometimes in not-so-friendly rivalry. Other early activity in radio astronomy took place in The Netherlands, the United States, and the Soviet Union, but for our purpose only the British and Australian works need consideration. The scientists working with radio astronomy in the late 1940s were mostly physicists and engineers with war-time radar experience, and it was only after some time that they changed their professional identities and became—i.e., conceived of themselves as and were recognized by others as—astronomers. In this respect, the early history of radio astronomy shows some similarity with the emergence of cosmology in general, and with steady-state cosmology in particular.

Originally, radio astronomy was not considered relevant to cosmology. Radio cosmology only started in 1954–55, but several years earlier the Dutch astrophysicist Hendrik Van de Hulst had called attention to the cosmo-logical implications of the scarcely nascent field of radio astronomy. In an important work of 1945, where Van de Hulst predicted the existence of a 21-cm microwave spectral line originating in a quantum transition of inter-stellar hydrogen, he also discussed the possible use of radio waves in distin-guishing between various world models. Van de Hulst considered an expand-ing and a static universe and concluded from the predicted radio background in these two models that the static model was wrong.[92] The predicted 21-cm spectral line was detected in 1951 and became at once an important tool for astronomers.

The main opponent of the steady-state theory from the field of radio astron-omy was Martin Ryle, later astronomer royal (1972) and a Nobel laureate (1974). Born in 1918, Ryle entered Oxford University in 1936 and graduated in physics three years later. As a devoted radio amateur he wanted to do iono-spheric radio physics but, like so many physics graduates of his generation, was drawn into radar development during the war. From 1940 to 1945 he worked on various problems of radar within the Air Ministry Research De-partment, specializing in methods to jam German radar. In this work he was very successful, and Ryle and his team made important contributions to the British war efforts. Among other things, he discovered how to jam the V2 rocket guidance system and developed a method of simulating electronically a spoof invasion shortly before D day.[93] After the war, he went to Cambridge University where he became absorbed in the newly started work in radio as-tronomy and soon emerged as a leader of the field. Ryle at first concentrated on improving instrumental techniques, which about 1954 led to the invention of the so-called aperture synthesis, a method to turn the interferometer into a radio telescope suited to map the sky for radio sources. The motivation behind the Cambridge radio astronomers' interest in making surveys of the distribu-tion of radio sources was not cosmological, at least not primarily, but it soon turned out that the surveys had important consequences for the study of the universe at large.

Although early radio astronomy did not bear on cosmological questions, certain events from the early 1950s came to have some significance for the later controversy. About 1951 the general belief among astronomers was that radio stars were peculiar stars located within the Milky Way, and that the few examples of radio-emitting nebulae (such as Virgo A, identified with the galaxy M87) were not representative of radio sources. This was also the opinion of Ryle, who at a meeting in London in April 1951 argued from the newly established isotropical distribution of radio sources that most of them were galactic. Gold, however, argued that the sources might well be other galaxies in which there was a relatively large population of dense, collapsed stars with strong magnetic fields.[94] In the discussion following Gold's suggestion, Hoyle supported it, whereas McVittie and Ryle maintained that radio stars were stellar objects within our galaxy. Although Gold stressed that his suggestion was merely a hypothesis—"a possible point of view, though I do not necessarily wholeheartedly accept it"—the discussion soon took on a sharp nature. "I think the theoreticians have misunderstood the experimental data," Ryle replied caustically, and Hoyle ended by accusing Ryle and McVittie of being dogmatic: "Professor McVittie and Mr. Ryle have dogmatically asserted that the discrete sources cannot be of the extragalactic origin, although of the half dozen or so discrete sources that have indeed been identified, five have been found to correspond with nearby extragalactic nebulae. Presumably a discrete source ceases to be a discrete source as soon as it is identifiable as a galaxy."

Later the same year, at a meeting of the Royal Astronomical Society on 12 October, the exchange of views between Gold and Ryle was reiterated, Ryle arguing against extragalactic radio sources and Gold repeating his suggestion of an extragalactic origin.[95] This may seem to have been only a minor incident and one unrelated to cosmology. However, the sharpness of the exchange of views and the fact that Hoyle and Gold had previously challenged Ryle at scientific meetings suggest that there was more to it than an ordinary scientific debate. Ryle, who preferred an engineering and observational approach to astronomy, had obviously little patience with theoreticians like Gold and Hoyle. Many years later, Hoyle described the Gold-Ryle debate as developing into a feud. In his opinion, "it was this disagreement that first led Ryle to seek a disproof of the steady state theory."[96] Although there is no documentary evidence that Ryle was interested in steady-state cosmology (or cosmology at all) at this date, it is quite possible that his disagreement with Gold and Hoyle, as well as his general lack of respect for theoreticians, colored his view when the cosmological question turned up in radio astronomy a couple of years later.

At any rate, from being a somewhat heretical suggestion, Gold's idea soon became the accepted view. From late 1952 Ryle began to consider the possibility he had earlier dismissed, that many of the radio sources might be extragalactic. There was no sudden conversion, but gradually he and many other radio astronomers realized that observational evidence favored Gold's view. In 1955, when this view was confirmed, Gold referred in a discussion to the earlier opposition of Ryle and other astronomers. Ryle was defended by Lovell, who claimed that "I do not think that we have ever suggested that many of the radio sources could not be extragalactic."[97] In view of Ryle's un-ambiguously stated arguments for the radio sources being of galactic origin, Lovell's reply was not convincing.

The recognition that radio sources were extragalactic was of great cosmological importance, for then there was the possibility that "we were in the cosmology game," as Ryle expressed it in an interview of 1976.[98] The realization that radio astronomy could contribute to cosmology, and perhaps even decide between rival cosmological models, had been hinted at by Bernard Mills in Sydney in 1952, and also by J. G. Ratcliffe, the head of the Radio Physics Group at Cambridge University.[99] It was explicitly stated by Ryle in the fall of 1953, when he said, in a talk given to the Cavendish Physical Society, that "if indeed it [the seemingly extragalactic component] is extragalactic, it offers the possibility of being able to distinguish between some of the cosmological theories. Whether the observations will ever be sufficiently accurate one cannot say, but it is nice to think that the cosmologists may one day not lose complete freedom of choice of the conditions beyond the optical limit."[100] The lack of respect for cosmologists, which Ryle here aired, was shared by many astronomers. It was more clearly spelled out in other notes from the same period. In one of these, Ryle wrote: "Cosmologists have always lived in a happy state of being able to postulate theories which had no chance of being disproved—all that was necessary was that they should work in the [optically] observable universe . . . Now we do seem to have some possibility of exploring these most distant regions. Even if we never actually succeed in measurements with sufficient accuracy to disprove any cosmological theory, the threat may discourage too great a sense of irresponsibility."[101]

In the summer of 1954, Ryle became convinced that most of the discrete radio sources were indeed extragalactic and that observations could in fact be used to test some cosmological models. Since the completion in 1950 of the first, limited survey (the 1C survey, with only fifty sources), Ryle and his group had constructed a much larger and more precise antenna system from which data were collected for almost two thousand sources, almost all of them without optical identification. Meanwhile, the Sydney group surveyed the southern sky and in 1952 Mills produced the first double-logarithmic plot of the number of sources versus their intensity ($\log N$-$\log S$). This method, which will be considered below, soon became a standard tool in radio cosmology. Observations for the new Cambridge survey, known as 2C, were completed in May 1954, and soon thereafter Ryle and his collaborators began to contemplate their cosmological significance.

John Shakeshaft, a student of Ryle's, published the first radio cosmological paper, in which he tentatively suggested that the steady-state theory was unable to account for the data. Shakeshaft concluded that "[it is] possible to account for the polar intensity with the extra-galactic sources alone on the relativistic theory but not the steady state theory."[102] But he also added that

"The present results cannot yet, however, be used to distinguish conclusively between the two theories." It was only in the spring of 1955, after months of hard work to reduce all the data and complete the 2C catalog, that the Cambridge radio astronomers were ready to address the cosmological debate at full scale. The final survey included 1936 radio sources of which 30 were found to be extended and hence presumably located within the Milky Way; the number of cosmologically relevant sources smaller than 15' was thus 1906.

The 2C Survey

From about 1954 the Cambridge radio astronomers' research program became increasingly oriented toward cosmology, and more specifically toward a refutation of the steady-state theory. It has been argued that the entire 2C survey program was motivated by a desire of the radio astronomers to justify the new science in the most effective way, namely, to demonstrate its power as a cosmological test. As noted by Benjamin Martin, the rather mundane business of cataloguing radio sources seems, at first sight, to be a somewhat dull kind of scientific work.[103] Why, for example, didn't Ryle and his group decide to design apparatus to extract information from some of the nearby sources with the aim of solving the mystery of the mechanism of radio sources? In Martin's view, the chosen strategy reflected the radio astronomers' wish to legitimize their science to the optical, "real" astronomers, many of whom still considered radio astronomy more radio engineering than astronomy proper. A successful survey program, the ultimate intention of which was to disprove the steady-state theory (profoundly disliked by most main-line astronomers), would not only prove that radio astronomy could succeed where optical astronomy had failed; it would also be seen as rendering great service to the astronomical community. However, although sociological motives of this kind may have played a role from about 1954, they cannot explain the survey program, which was initiated several years before Ryle realized its significance for cosmology. The observations of the 2C survey were independent of the cosmological controversy, but the conclusions and the following work were not. Before turning to the conclusions of Ryle and coworkers in 1955, we shall introduce the most important of the analytical tools used in radio cosmology, the log$N$-log$S$ plot.

Let the flux density of a celestial object, that is, the power (energy per time) emitted in a certain frequency range through a unit area, be denoted S. In radio astronomy, the usual unit for S is 10^{-26} W·m^{-2}·s, which is called a Jansky (Jy), named after the pioneer in radio astronomy. If it is assumed that the sources are distributed spherically in a static Euclidean space with a constant space density, it follows from simple considerations that the number of sources N with a flux density larger than or equal to S varies as S to the power of $-3/2$. That is, there exists a relationship of the form $N \sim S^{-1.5}$ or

$$\log N = \text{const} - 1.5 \log S.$$

Thus, a logarithmic plot of N versus S will exhibit a straight line with slope −1.5. If the space is not Euclidean, or the sources are not distributed uniformly, one cannot expect the same kind of relationship. In general, different cosmological models will lead to different $\log N$-$\log S$ predictions, which then can be compared with the $N(S)$ plot obtained empirically. The use of number counts as a cosmological test goes back to the mid-1930s, when Hubble attempted to apply a similar method for optical sources, the ordinary galaxies. Hubble used the apparent magnitude instead of the flux density and expressed the relationship, valid for galaxies uniformly distributed in flat space, as

$$\log N = 0.6m + \text{const.} \tag{6.5}$$

From his observational material he found a different expression, viz.,

$$\log N = 0.501m + \text{const} \tag{6.6}$$

and argued that the deviation from (6.5) was due to a redshift effect on the apparent magnitudes.[104] Since the relationship between the magnitude and the flux density is $m = -2.5 \log S + \text{const}$, the expressions (6.5) and (6.6) can also be written as relationships between $\log N$ and $\log S$. In the first case the slope is −1.5, in the second −1.25. Hubble hoped that the observationally established deviation from (6.5) could be used to gain information about the structure of space. However, technical difficulties associated with the myriads of visible galaxies forced Hubble to abandon the test. The number of radio sources known from the 2C survey was much smaller, yet large enough to be statistically useful, so that the $N(S)$ method could be used as a cosmological test. The method, as first applied in 1955, was to count the number of radio sources with flux density larger than S, and then make a $\log N$-$\log S$ plot, which can be compared with the predictions from the $S^{-1.5}$ law and those of various cosmological models.

Although it is possible to calculate different $N(S)$ relations from the different relativistic models, evolutionary effects make it almost impossible to separate the model dependence and compare it with observations. For example, although the Einstein–de Sitter model gives a curve disagreeing with observations, agreement can be obtained by requiring N to be larger in the past, as it must have been in this model, and by supposing a weakening of radio sources with age. This possibility does not exist in the steady-state theory, where the density and intensity of radio sources has on the average remained the same at all times. It was therefore only the steady-state model that could be conclusively tested by the radio number count method. Since the space of this model universe is Euclidean one might expect the predicted slope to be simply −1.5, but this is not the case because of the expansion of the universe. If the redshift is taken into account, by using results derived by Bondi and Gold as early as 1948, it turns out that $N(S)$ will be smaller than in the static case, i.e., will lie beneath the curve with slope −1.5. Although the precise form of the steady-state curve required knowledge of the spectra of the radio sources—which did not exist—the steady-state theory unambiguously predicted a curve which, for faint sources (the left part of the log$N$-log$S$ plot), is less steep than that given by the static, Euclidean case. For this reason, only the steady-state theory was the subject of the $N(S)$ test. The test boiled down to the question: are there more or fewer faint sources than predicted by the $S^{-1.5}$ law? If the answer was more, it seemed that the steady-state theory was refuted. The opposite answer would agree with the steady-state prediction, but without causing serious trouble for the evolution theories.

The log$N$-log$S$ diagram was a focal topic of the first published announcement of the 2C results, written by Ryle and his student Peter Scheuer. It appeared clearly that whereas the most intense sources were grouped around a line with slope −1.5, the major part of the curve was considerably steeper, with a slope of about −3 (figure 6.5). The flattening of the curve for small intensities (or flux densities) was considered an instrumental artifact due to the limited resolving power. Having discarded alternative explanations, Ryle and Scheuer concluded that the origin of the observed distribution was cosmological and that only evolutionary theories were able to account for the large slope. On the other hand, "Attempts to explain the observations according to steady-state theories offer little hope of succes."[105] This was an understatement of what Ryle really meant.

A more direct cosmological interpretation was offered by Ryle in Oxford on 6 May 1955, when he delivered the Halley Lecture: "If we accept the conclusion that most of the radio stars are external to the galaxy, and this conclusion seems hard to avoid, then there seems no way in which the observations can be explained in terms of a steady-state theory. . . . If future observations confirm the present conclusions, we have at last a powerful means of exploring those regions where the predictions of different cosmological theories diverge. Already it seems possible to make a distinction between the two main groups of cosmological theory; a detailed comparison of the observations with the predictions of different evolutionary models may not only narrow down the field still further, but it may also throw light on the problems of galactic evolution."[106] Naturally, the steady-state cosmologists were not happy with this conclusion from radio astronomy, which came as an unpleasant surprise to most of them. In the following debate over Ryle's interpretation, both radio astronomers and steady-state theoreticians participated, but whereas the Cambridge radio astronomers were forced to take the first group seriously they had little respect for the objections of the latter. They were primarily experimentalists and to the extent they took an interest in cosmology they were convinced that the steady-state theory was wrong. A member of the Cambridge group recalled many years later: "Some of the people who were against [2C]—you know, who were criticizing the observations—we dismissed as mere theoreticians. I mean [the steady-state cosmologists] had the greatest entrenched reason to criticize the observations—but they weren't taken as seriously."[107]

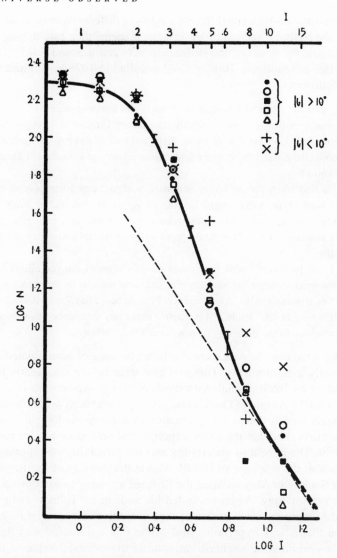

Figure 6.5. The main cosmological result of the second radioastronom-ical Cambridge survey (2C), as presented by Ryle in August 1955. Values of log$N$ are plotted against log$S$ (log$I$ in the figure). The dashed line corresponds to a slope of -1.5 as in the steady-state prediction of uniform spatial density. *Source:* Ryle (1955), p. 142.

McCrea, who had assisted in the preparation of the Ryle-Scheuer paper, received a preprint a few days before the paper was submitted for publication. He expressed his admiration for the work, and then went on: "In spite of not seeing anything obviously wrong, one has the irrational impression that it is all too easy! This may be simply because we have grown used to the idea that any discrimination between different models or different theories can be made only on the basis of 'second order' effects. It seems then astonishing if it can be made just by noticing the difference between 1.5 and 3.0."[108] Another steady-state cosmologist, Tommy Gold, recalled in 1978 his immediate reaction as follows:

> I went into his [Ryle's] lab after he'd made the announcement, in the first cata-
> logue, that he had shot down the steady state theory. I remember going to his lab and
> sitting myself down on a table there. Ryle and two or three other members of his
> team collected around there, and I said, "How secure are your data? I mean, do you
> really know?"
>
> Hoyle had taken the data very seriously, you see, and I kept saying to Hoyle,
> "Don't trust them, there might be lots of errors in this and it can't be taken
> seriously." Hoyle said, "You must take observational data seriously, otherwise
> you are nowhere." I said, "I will take them seriously when I know they're correct and
> not before."
>
> . . . I said [to Ryle], "Well, how much error in intensity are you allowed to make,
> and how much error in the intensity measurement are you in fact making." And he
> said, "Oh, it doesn't make any difference. Peter Scheuer has shown it wouldn't make
> any difference at all." I said, "If it doesn't make any difference what intensity you
> judge a source to be, then why do you have a radio telescope?"[109]

Gold suspected that the data were unreliable because of accumulated errors in the intensity measurements. This was also what he argued, clearly to Ryle's irritation, at an International Astronomical Union symposium in August at Jodrell Bank.[110] Although Gold could not explain exactly what the suspected errors were, he felt that it was premature to draw cosmological conclusions from the curve, and that the entire experimental procedure needed more careful control. The appeal to uncertainty and the possibility of experimental errors was also the response of Bondi, who at the meeting of the Royal Astronomical Society in May recalled the ill-fated attempts to use optical number counts in cosmology. As he reminded his audience: "Tolman deduced from them [the intensity measurements] that the universe was highly elliptical. McVittie that it was hyperbolic, and Milne that space-time was flat. Heckmann showed how the slightest variation in the assumed value of the (very uncertain) recession constant would throw the interpretation from one extreme to the other." The disinclination of Gold and Bondi to accept the 2C data reflected their general and common attitude toward the comparative validity of experimental and theoretical results. As mentioned in an earlier chapter, they did not see experimental data as intrinsically more reliable than theoretical knowledge, rather the contrary. Furthermore, together with Sciama they had recently succeeded in resolving another claimed refutation of the steady-state theory, namely, the Stebbins-Whitford effect. If this effect was spurious, wouldn't it be possible that neither could Ryle's results stand up to closer scrutiny?

The program in radio source counts not only accentuated the rivalry between the steady-state model and evolutionary models of the universe. It also increased the atmosphere of hostility between two of Cambridge's most eminent astronomers, Ryle and Hoyle. Although there is no doubt that Ryle was against the steady-state theory, in his publications he downplayed the cosmological controversy and presented his results as following soundly and neutrally from observations and technical analysis. In 1964, in an interview with *Newsweek*, he denied that he was a "Big Banger" and expressed his dislike of being involved in an "argy-bargy match."[111] Hoyle, however, felt that Ryle was driven not by a quest for truth, but rather by a desire to disprove the steady-state theory. Many years later he complained about Ryle's strategy in his crusade against steady-state cosmology in general and, Hoyle felt, himself in particular. "His programme, which he pursued relentlessly over the years, does not seem to have been directed towards any other end [than disproving the steady-state theory]. There was no question of establishing the correct cosmology, but only of disproving the views of a colleague in the same university, a situation which I have never felt to have deserved the plaudits which the scientific world showered upon Ryle."[112]

Since none of the steady-state theoreticians had competence in practical radio astronomy, they were unable to criticize the reported data from a technical point of view. Instead they tried to weaken Ryle's conclusion by checking with independent measurements. This meant at the time Mill's group in Australia, the only other place where surveys comparable with the Cambridge observations were made. It soon turned out that the Australian measurements disagreed with the Cambridge data, and thus could be used as ammunition in the defense of the steady-state theory. The disagreement was already indicated at the August 1955 International Astronomical Union symposium, and when the Sydney survey of the southern hemisphere appeared in 1957 the controversy with Cambridge became explicit. The data for the overlapping areas were considered to be "almost completely discordant" and the disagreement with the Cambridge data to be "striking." Mills and his group found that the major part of the $\log N$-$\log S$ curve had a slope of -1.8, as compared with the -3.0 found by their colleagues in Cambridge. Accordingly, "deductions of cosmological interest derived from its [the Cambridge group's] analysis are without foundation."[113]

The Sydney group did not, and could not, interpret their data in favor of steady-state cosmology, but simply as unsuited to infer any cosmological conclusion. All the same, since the target of the Cambridge claims was steady-state cosmology, indirectly the Sydney results acted as an argument in favor of the possible validity of this model of the universe. The immediate result of the Cambridge-Sydney disagreement was confusion and caution, just the situation the steady-state cosmologists could use in their defensive strategy. As Joseph Pawsey, one of the Sydney radio astronomers, wrote in 1958: "In the field of cosmology there has been a halt. Previously announced radio results which appeared to invalidate the steady-state continuous creation hypothesis concerning the universe are now believed to be invalid and the point is now non-proven. More certain knowledge must await the construction of more powerful tools."[114]

While waiting for this more certain knowledge to appear, the third relevant party in the debate, the Jodrell Bank group, largely remained neutral. With no data of their own to contribute, Lovell and his group were inclined to downplay the cosmological significance of radio astronomy, which in effect meant siding with Sydney against Cambridge. "At Jodrell we . . . thought that the whole business was unsound," one Manchester radio astronomer recalled. "We thought that to try to deduce the structure of the cosmos from the analysis of a great mass of flux densities of unknown origin was, to say the least, premature."[115] In a comparison between the divergent results obtained at Cambridge and Sydney, two Manchester radio astronomers concluded that the Sydney results were the more reliable. The Cambridge number counts were judged to be "extremely doubtful" and so unreliable that the observed increase in slope "cannot be taken to have any cosmological significance."[116] During the summer of 1958 the confusion was at its height, just at the time when radio astronomy was discussed at the International Astronomical Union conference in Paris and at the Solvay Congress in Brussels. In Paris, the Jodrell Bank astronomer Robert Hanbury Brown criticized the Cambridge group's conclusions and stated that "the radio astronomers must make considerable progress before they can offer the cosmologists anything of value." Mills argued from the new Australian 3.5-m measurements that the slope of the number-intensity curve was only -1.65; he assumed that this result, although "not conclusive," did not differ significantly from -1.5.[117]

In Brussels, Lovell gave a talk in which he denied that the radio astronomical surveys necessitated addressing special cosmological models. In his general conclusion, Lovell stated that "It seems that the only safe conclusion to be drawn from the work [at Cambridge and Sydney] is that, as yet, there are no radio astronomical observations which can influence significantly the existing views on the large scale structure of the universe."[118] Radio cosmology was predominantly a British-Australian affair, but in a few cases astronomers from other countries also entered the discussion. Iosif Shklovsky was an eminent Soviet astrophysicist from Moscow's Sternberg Astronomical Institute and an independent discoverer of the 21-cm radio waves usually attributed to Van de Hulst. In a book completed in the fall of 1958, he discussed the results of the Cambridge and Sydney groups, and his opinion did not differ from that of the majority of his Western colleagues. "Thus the Cambridge results, which had created a sensation, now appear to be in error," he wrote. Unusually for a Soviet astronomer at the time, Shklovsky referred explicitly to the cosmological controversy and the steady-state theory "as developed by certain theoreticians but by no means well established at the present time."[119]

All things considered, the steady-state cosmologists had reason to be moderately satisfied with the situation in radio astronomy at the time. By 1959 it had become evident to most astronomers that the results from Cambridge 2C contained large systematic errors—as claimed by Gold from the very start— and that more and better measurements were needed. This was not the opinion of all astronomers, however, and especially not of those who were already convinced that the steady-state theory was wrong. It is clear, and not particularly surprising, that the opinions of the involved parties depended on their previously held sympathies and antipathies. At a meeting on relativity theory in Royamont, France, in the summer of 1959, McVittie surveyed the observational situation in cosmology. At that time he tended toward a relativistic model universe with hyperbolic space and negative cosmological constant as the one in best agreement with the data. As far as the steady-state theory was concerned, he emphasized that it failed to reproduce the results obtained by the Australian radio astronomers. For McVittie there was no doubt: the steady-state theory was just wrong. Barely able to hide his irritation that the theory was still alive and well, he ended his paper as follows:

> Supporters of the steady-state theory suggest that new kinds of observations are needed in order to "test" their theory as against general relativity. The observers also appear to believe that the question is still open. In fact, it has been known since 1956 that the steady-state theory predicts the wrong sign for the acceleration parameter. We have also pointed out that the predicted average density of matter is rather too high. And lastly, one result of the present paper has been to show that the steady-state theory fails to reproduce the empirical law of distribution of Class II radio sources. In view of these considerations, it is not clear how further "tests" could validate the steady-state theory. Its model universe simply does not agree with observations whereas, as we have seen, certain general relativity model universes do.[120]

Ryle and his group also continued to believe in the basic validity of their conclusions and their cosmological significance. They were already collecting the first data from an extended survey (3C), which, they hoped, would settle the matter. An indication of what was to come was presented by Ryle at the Paris symposium, where he reconsidered the situation with the newest data available. He insisted that "the arguments presented here suggest that the discrepancy between observation and the predictions of the steady-state model are considerably greater than could be established hitherto [in 1955]."[121] Within a few years Ryle's claim would receive confirmation, with a drastic change in radio cosmology as a result.

From Controversy to Marginalization

AT THE END OF THE 1950s, after a decade of controversial existence, the steady-state theory was as alive as ever. At that time the more emotional and philosophical resistance to the theory had weakened and no longer played a significant role. The steady-state model was undoubtedly a view of the universe still rejected by most astronomers and physicists, but years of debate over conceptual issues had led to no real clarification, and it was as if scientists had grown accustomed to the basic ideas of steady-state cosmology. These ideas were resisted, but nonetheless they formed a legitimate part of the cosmological discourse. For example, whereas the idea of continual creation of matter had been hotly debated in the early and mid-1950s, this was an issue that no longer attracted strong emotional interest among most steady-state opponents. Interest shifted to observational and experimental issues, a change that to a large degree was induced by new observations that seemed to allow for new possibilities of discriminating between the two main rival models of the universe.

The general change in climate in favor of relativistic evolution models with a big bang that took place about 1960 was probably related to the simultaneous revival of interest in the general theory of relativity. Einstein had originally suggested three tests of his theory, the gravitational redshifts of spectral lines, the deflection of light by the sun, and the precession of the perihelion of mercury. By 1920 the theory had passed these classic tests, but the successes seemed to exhaust the theory's contact with experimental reality. Since the mid-1920s general relativity had had almost no connection with experiments and, so it seemed to many, was slowly degenerating into pure theory. The theory was appreciated by mathematical physicists because of its beauty and challenging theoretical problems, but their experimental colleagues at best found it irrelevant.[1] With the postwar development of new technologies, such as the laser, the possibility of testing the theory by means of ordinary laboratory experiments became a reality. "Einstein's theory of gravitation, his general theory of relativity of 1915, is moving from the realm of mathematics to that of physics," concluded the mathematical physicist Alfred Schild in 1960; "After 40 years of sparse meager astronomical checks, new terrestrial experiments are possible and are being planned."[2] The result of the new experiments was a complete confirmation of the theory, with the effect of increasing physicists' interest and confidence in general relativity.

Now the validity of the general theory of relativity was not, in fact, a main subject of the cosmological controversy. Most steady-state theoreticians were experts in the theory, which they valued highly. But they doubted if it was unrestrictedly valid, that is, if it could be applied without changes to describe the entire universe also. Of course, the new experiments could say nothing about this issue and so were, strictly speaking, of no relevance with respect to the cosmological controversy. But they nonetheless helped in creating an atmosphere among specialists that Einstein's theory was universally true and no longer a subject of discussion. By implication, cosmological theories deviating from general relativity, or based on alternative theories of gravitation, tended to be regarded with more suspicion than earlier. The renewed faith in the general theory of relativity was given enthusiastic expression by John Wheeler, who in 1962 identified two trends in recent gravitation physics: "(1) Interest has fallen in inventing new theories of gravitation, spacetime, and the expanding universe. (2) Increasing numbers of investigators share the conviction that Einstein's 1915–1916 analysis of the curvature of space by energy is a unique theory, of unrivalled scope and reasonableness against which no objection of principle has ever been sustained, and out of which one should now try to read the deeper meaning and consequences."[3]

At the same time that observations tipped the balance definitely in favor of relativistic big-bang theory, a few steady-state theoreticians continued to develop and modify the theory, in part as a response to the threatening observational situation. The result was a series of heterodox cosmological theories, which were the heirs of the original steady-state theory, but in many respects differed substantially from it. The original divide between the Bondi-Gold version and the Hoyle version, which during the 1950s had played a minor role only, now developed into a separation: Hoyle definitely abandoned the perfect cosmological principle, and the Bondi-Gold cosmology, building on that principle, disappeared from the scene, with the result that the steady-state theory was left to Hoyle and a few collaborators. At the end of the 1960s Hoyle and Narlikar were busy with such theoretical work, undeterred by the fact that the great majority of astronomers and astrophysicists had declared the steady-state theory—in whatever version—dead and buried.

Although the steady-state theory was no longer considered a serious alternative to the big-bang theory by 1970, the controversy continued in a different form, now as an attempt to discredit the victorious big-bang theory. The opposition against the new big-bang orthodoxy included many scientists with no association with the steady-state program, yet much of this phase had its roots in the former steady-state opposition. It is therefore natural to include a brief discussion of the modern criticism of big-bang cosmology insofar it is related to the older controversy.

7.1 NEW OBSERVATIONS, NEW DEBATES

A major reason for the low status of cosmology and the undecided controversy between the two rival systems of the universe was the paucity of observations of clear cosmological significance. For astronomers of an empiricist inclination (which means almost all astronomers) this had always been a valid objection against theoretical cosmology. In 1936 Hubble had written about scientists that "one of the few universal characteristics is healthy skepticism toward unverified speculations. These are regarded as topics for conversation until tests can be devised."[4] When Malcolm Longair began research in radio astronomy as a research student in 1963, some tests had been devised, but cosmology was still widely considered a topic of conversation and not a decent scientific area of research. Longair's supervisor, Peter Scheuer, warned him that "there are only 2 ½ facts in cosmology."[5] The two facts, according to Scheuer, were that the sky is dark at night (Olbers's paradox) and the Hubble recession of the galaxies; the half fact was that the universe is in a state of evolution. What was desperately needed, according to Scheuer and all other astronomers, were new facts.

During the first half of the 1960s the cosmological scene changed as a result of a number of spectacular discoveries that showed the existence of new objects and phenomena of cosmological relevance. By far the most important channel of cosmic information is electromagnetic radiation (photons), which for a long time had been restricted to the visual part of the spectrum or those in its immediate neighborhood, the infrared and ultraviolet parts. It was only in the 1930s that visible light as a source of information was extended with the much longer radio waves, resulting in the important radio astronomical programs of the 1950s. The other, high-frequency end of the electromagnetic spectrum began to be exploited about 1960, when cosmic gamma rays and x rays were first detected, and in 1965 came the surprising information gained from the microwave window.

Gamma and X-Ray Astronomy

On 31 January 1962, President Kennedy in his annual report to the Congress gave an account of his new multibillion-dollar space program. Reviewing the latest space accomplishments, the President mentioned that "a most important scientific result was the discrediting of one major theory of the universe, a version, but not a widely accepted one, of the socalled 'steady-state' theory."[6] The unusual reference to cosmology was to results obtained by the artificial satellite Explorer XI, which was launched into orbit on 27 April 1961. The satellite carried with it a gamma-ray detector designed to identify gamma photons from outer space with an energy larger than about 100 MeV. The scientists behind the experiment, William Kraushaar and George Clark from MIT, found fewer than one hundred such photons, but even this small sample provided the cosmological test mentioned by President Kennedy.[7] It was not, however, a test of steady-state cosmology, but only of particular models which assumed large amounts of antimatter in the universe, whether these were created continuously or not. As McVittie concluded in a careful analysis of the Kraushaar-Clark measurements, "the steady-state theory gives as much of an acceptable interpretation of the data as does general relativity."[8]

Ever since Dirac's prediction of antiparticles in 1931, a few scientists had speculated about the possible existence of cosmic antimatter made up of positrons, antiprotons, and antineutrons. The first such speculation was mentioned by Dirac in his Nobel Prize speech of 1933. The discovery, or rather manufacture, of the antiproton in 1955 encouraged physicists to take the question more seriously. Since particles and antiparticles are symmetrical—the antiproton is only "anti" by convention—how is it that our world seems to consist of particles only? In 1956, Hoyle and Geoffrey Burbidge suggested that the hypothesis of cosmic antimatter might have cosmological implications and be developed into "a modified form of the steady-state theory of the expanding universe."[9] If half of the particles created continually in space were antiparticles, some of these would annihilate with particles at a rate estimated to be 3×10^{-22} cm^{-3}·s^{-1}. Hoyle and Burbidge believed that cosmic proton-antiproton annihilation processes might well be important, but they also stressed that their considerations were highly tentative. Given the tentative nature of their suggestion, and the fact that neither they nor other steady-state theoreticians returned to it, the idea of a symmetrically continual creation of matter scarcely deserves to be called a theory and certainly not "a major theory of the universe." At any rate, it was proven false by the Explorer XI experiment, which showed that there were much fewer proton-antiproton annihilations than the mentioned 3×10^{-22} cm^{-3}·s^{-1}. Protons and antiprotons annihilate into pions, and since neutral pions decay into high-energy gamma photons, measurements of the cosmic gamma intensity provide an upper limit for the annihilation rate. It was by measuring the gamma intensity that Kraushaar and Clark were able to disprove the hypothesis of a symmetrical creation of matter.

The problem of the missing antimatter continued to worry some astrophysicists. It still does. Big-bang cosmologists in the 1960s either avoided the problem or claimed that there was no problem. As Gary Steigman explained in 1969, the lack of symmetry could simply be ascribed to an assumed asymmetry in the initial conditions of the universe.[10] This kind of explanation, relying on the formula that "things are as they are, because they were as they were," was, of course, found unpalatable by steady-state advocates. Hoyle speculated in 1969 that a symmetric steady-state universe may be possible if matter is not created continually through space, but in particle-antiparticle pairs in the highly condensed nuclei of galaxies.[11] If so, the annihilation gamma photons would be absorbed in the nuclei and never be detected, and only the particles would be expelled from the nuclei. The suggestion, difficult to test and admitted to be just a speculation, failed to attract interest.

Whereas the gamma-ray test was of little relevance in the cosmological controversy, the exploration of cosmic x-ray sources turned out to be more important, both to the controversy and to astronomy in general. X-ray astronomy was a new field which started in the early 1950s, when Herbert Friedman and his group at the U.S. Naval Research Laboratory explored solar x rays with rocket-borne Geiger counters. Inspired by the successes of radio astronomy, a search began for x rays emitted from sources outside the solar system. The first important result was obtained in June 1962, when Riccardo Giacconi and his collaborators in American Science and Engineering, a private research company associated with MIT, detected a surprisingly intense x-ray source that was later named Scorpius X-1. In addition to the discovery of this x star, soon followed by several other discrete sources, Giacconi also detected a diffuse x background radiation.[12] The background radiation was assumed to be of extragalactic origin and, as first pointed out by Hoyle in early 1963, it was therefore of possible cosmological interest. Hoyle argued that the x rays were produced by a bremsstrahlung process due to the slowing down of energetic electrons, and a quick calculation showed that the electrons had to have energies corresponding to a hot intergalactic medium. This was just the kind of medium required by the hot steady-state universe suggested by Gold and Hoyle in 1958–59 and which, Hoyle now claimed, predicted an x-ray flux in agreement with the one measured.[13]

However, what Hoyle had presented as a verification of the hot steady-state universe soon turned out to be just the opposite. Half a year later, Geoffrey Burbidge and Robert Gould reconsidered Hoyle's calculations, which they performed more accurately, including factors that Hoyle had left out. The result was very different, a flux prediction from the Gold-Hoyle theory that exceeded the observed flux by a factor of almost 100. The theory of Gold and Hoyle required an intergalactic medium at a temperature of about 10^9 K, but the calculations of Burbidge and Gould showed that the temperature could at most be about 10^7 K if agreement with observations was to be obtained. The conclusion was unambiguous: "Thus the hot universe model proposed by Gold and Hoyle must be ruled out."[14] A similar conclusion was obtained by George Field and Richard Henry at Princeton University, who found a maximum temperature of 4×10^6 K according to the steady-state model, again disagreeing with the hot version.[15]

According to Stanley Jaki, a Benedictine priest and historian of science, the steady-state theory is quasi-scientific and antireligious. He contends that Hoyle's unwillingness to give up the steady-state theory in the light of the x-ray measurements was a result of Hoyle's metaphysical commitment to the theory: "Orbital satellites in the mid-1960s failed to detect the radiation postulated by the steady-state theory. [This negative result] . . . did not impress a cosmologist as antimetaphysical as Hoyle."[16] However, it did not require any metaphysical commitment to keep to the steady-state theory in spite of the negative result. Although x-ray astronomical data were incompatible with the hot steady-state model, they did not touch the core of the model, the postulate of an unchanging universe. Whereas there was good reason to abandon the hot version as a result of the measurements and the calculations of Burbidge and Gould, they in no way contradicted the steady-state theory as such. Hoyle, Sciama, Narlikar, and the few other steady-state advocates could accept the results without giving up the essence of their model, an eternal universe with matter creation. A crucial cosmological test based on x rays was suggested in

1964 by Gould and Sciama, who considered the possible existence of characteristic x-ray lines superimposed on the continuous background.[17] Such lines will be spread out into a spectrum by the redshift in a way depending on the cosmological model, with the steady-state theory leading directly to a result that can be compared with experiment. However, no characteristic lines were observed and so the test remained ineffective.

If the x-ray measurements were not crucial, neither were they irrelevant to the fate of the steady-state theory. The hot-universe version was steady-state theory's best offer of a physically plausible cosmology, and to give it up implied a loss in reputation. There is no doubt that Hoyle's unfortunate attempt to present the x-ray background as support of the hot-universe model contributed to a general decline of confidence in steady-state theory. According to the MIT astrophysicist Philip Morrison, it was "the beginning of the rout of the steady-state theory."[18] There is little doubt that the beginning of decline took place at that time, but to ascribe it to the results of x-ray astronomy would probably be to overestimate the importance of the x-ray background. In this process, the results of radio astronomy were much more important.

Radio Astronomy: The Decisive Turn

Contrary to the tests that could be designed from gamma- and x-ray astronomy, the source count method of ordinary radio astronomy attacked the very heart of the steady-state theory. The problem with the 2C data was that they were not sufficiently reliable and were contradicted by results obtained in Sydney. When the first, preliminary results from the new 3C survey were published in early 1958, they did little to clarify the matter.[19] The slope of the $\log N$-$\log S$ curve was reported to be between -2.2 and -2.7, far away from steady state's -1.5 but also far away from the Sydney group's -1.65. As a result, there continued to be doubt as to what cosmological inferences could be drawn, if any. When Rudolph Minkowski summed up the results from the Paris symposium on radio astronomy in the summer of 1958, he concluded that "at this moment the available data are obviously not a sound basis for cosmological discussions . . . considerable time will elapse before the study of radio sources has reached a state in which the results may be used with confidence to attack cosmological problems."[20] It turned out that the considerable time was less than three years.

With new data and analyses coming in from both the northern and southern hemispheres, progress did take place. This happened in a slow but steady process, the result of which was a growing consensus and recognition of the reliability of the new Cambridge data. In 1960, at a conference of the International Scientific Radio Union, the Sydney radio astronomer John Bolton, then working in California, reported that work done with Caltech's Owens Valley interferometer gave results generally supporting the 3C survey. Although Bolton expressed some doubt with regard to the great slope proposed by Ryle, he concluded that the slope could not be forced down to the steady-state value of -1.5.[21] The following year marked the point of no return and the beginning of the end of dissension as far as the radio astronomical data were concerned. Based on extended counts of sources and an improved statistical method due to Scheuer, Ryle's group had ready new and definite conclusions in the beginning of 1961. These were published in papers by Ryle, in collaboration with Clarke and Scott, respectively, and presented on 10 February at a meeting of the Royal Astronomical Society. Two weeks later Ryle gave a Royal Institution lecture on the new Cambridge work, where he discussed the implications of the work for a broader audience.[22]

The essence of the new work was that in the energy flux range between 2×10^{-26} and 30×10^{-26} W·m^{-2}·Hz^{-1} (soon extended to 100×10^{-26}) the slope of the log$N$-log$S$ diagram was found to be clearly incompatible with -1.5. Ryle and Scott concluded that the best fit to the observations was a straight line of slope -1.80, and that the allowed variation due to errors and uncertainties was in the range -1.68 to -1.93. The steady-state theory gave a unique $N(S)$ relationship in which the only major uncertainty was the luminosity of the sources. By plotting $N(S)$ in terms of the number of sources according to a static and Euclidean universe, Ryle was able to present convincing evidence that even with the largest permissible variation in the source luminosity, the steady-state model failed to agree with observations (figure 7.1). The conclusion was this: "A comparison of the predicted and the observed curves shows a marked discrepancy, even when the smallest permissible source luminosity is adopted; the observed number of sources in the range $0.5 < S < 2 \times 10^{-26}$ watts (c/s)$^{-1}$m^{-2} is 3 ± 0.5 times that predicted by the steady-state model. If a luminosity function similar to that of the identified sources is assumed, the discrepancy is 11 ± 2. . . . These observations do . . . appear to provide conclusive evidence against the steady-state theory."[23] However, Ryle knew that even "conclusive evidence" might be debated, and in his Royal Institution lecture he added wisely that "there may be special cases of the Steady-State model which fit the present results." The report of 10 February was widely discussed in the press, both in England and abroad. According to McCrea, news about Ryle's conclusion had leaked out to the press, so that "astronomers on their way home after the meeting were able to read all about the final overthrow of steady-state theory!" Martin Harwit recalled similarly: "The afternoon editions of the newspapers being hawked on the streets outside Burlington House, as I was coming up the street to Ryle's talk at the RAS that afternoon, had banner headlines and were being hawked as 'The Bible Was Right.' "[24]

Contrary to the 2C survey, the new Cambridge results remained stable and they soon received support from other radio astronomers. At a meeting of the International Scientific Radio Union in Tokyo in 1963, Ryle reported an improved figure for the log$N$-log$S$ slope. It was now settled to -1.8 ± 0.1. Finally, the Sydney group announced the result of new measurements of frequency 408 MHz in 1964, ending up with a revised figure of -1.85 ± 0.1, evidently "in very good agreement" with the Cambridge value. It was

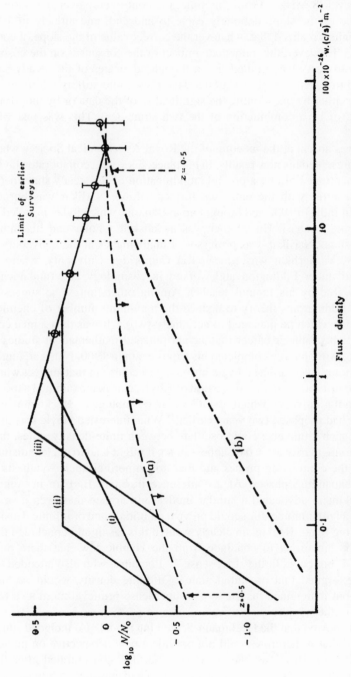

Figure 7.1. Ryle and Clarke's comparison of January 1961 of the observed N/N_0 relationship with that predicted by the steady-state model. The predicted distribution, depending on the assumed luminosity function, lies between the two dashed lines (a) and (b). *Source*: Ryle and Clarke (1961), p. 361.

further confirmed by results from the 4C survey, carried out during 1958–64 at frequency 178 MHz.[25] From this time, the controversy over the observational value of the slope definitely came to and end, and already in 1961 most specialists realized that, whatever the correct value of the slope, it could not be −1.5. However, the immediate impact of the consensus on the cosmological controversy was limited. Even though advocates of the steady-state theory had to accept the observational data, they could still avoid giving up the theory, either by questioning the significance of the data or by modifying the theory (or by a combination of the two strategies). This was just what happened.

Hoyle was absent at the meeting of the Royal Astronomical Society where Ryle first presented his new results. In his place, his young collaborator Jayant Narlikar mentioned briefly a possible modification of the steady state theory that would agree with the data (see further below). Narlikar was born in Kolhapur in India in 1938 and brought up in Banaras, which today is called by its old name Varanasi. His mother was a Sanskrit scholar and his father, Vishnu Vasudev Narlikar, was professor of mathematics at the University of Banaras. V. V. Narlikar was educated at Cambridge University, where he had studied under Eddington and worked in cosmology and fundamental theory, inspired by his famous teacher. Among other things, he suggested the use of Eddington's theory to deduce the maximum number of chemical elements (for which he managed to get ninety-two).[26] It was thus a nice continuation of his father's career that Jayant pursued mathematical studies in Cambridge. After having completed his tripos exam in 1960, Narlikar's interest in cosmology was aroused by an introductory course in that subject which Hoyle gave in 1960.[27] On the suggestion of Hoyle, Narlikar examined a model of a spinning universe which the German cosmologists Heckmann and Schücking had proposed two years earlier.[28] What interested Hoyle was their claim that such a universe could oscillate between finite-density states, thus making the singularity at $t = 0$ disappear; if so, it might lead to nuclear disintegration in the contracting phases and then to primordial nucleosynthesis in the early expanding phases. At a conference in Santa Barbara in August 1961, Heckmann advocated a similar model, and argued that even a small amount of angular momentum could prevent an initial state of infinite density, a singular big bang. Heckmann clearly wanted this result and concluded that it would be "incorrect to conclude from the Hubble law that there must have been a 'big bang' in the strict sense."[29] Lemaître, who also attended the conference, argued that an initial state of infinite density would be "too extreme," but maintained his view of a superdense primeval atom as a true beginning of the world.

Narlikar showed that the Heckmann-Schücking model also included singularities, and that it therefore could not provide a new perspective on nuclear synthesis.[30] With Hoyle as his supervisor, Narlikar concentrated after this work on steady-state cosmology, a theory which had appealed to him since he first became acquainted with it. His reasons for preferring the steady-state model were in part of a philosophical nature, largely of the same kind that originally motivated Bondi, Gold, and Hoyle, and which also held a strong appeal to Sciama. Then as later, Narlikar stressed the methodologically unsatisfactory features in the big-bang theory, such as the nonrepeatability of the big-bang scenario and its nonobservability even in principle. As we shall see, Narlikar fought vigorously to renew the steady-state alternative and discredit the big-bang view in the 1960s and onwards.

Bondi's immediate response to Ryle's announcement was to repeat his earlier doubt about the reliability of the Cambridge measurements. In the discussion following Ryle's presentation to the Royal Astronomical Society, Bondi opined: "With regard to the historical development of Ryle's work we should remember that six years ago he gave the slope of the $\log N$-$\log S$ relation as -3 and now it has been reduced to -1.8. Perhaps there is still a residual error which might account for the relatively small discrepancy between -1.8 and -1.5."[31] Ryle did not appreciate Bondi's sarcasm and, according to a witness of the debate, "flew into a rage, which resulted in the nastiest public display of tempers between scientists that I have seen in more than 30 years as a professional astrophysicist."[32]

Bondi's evaluation was in part the result of a spontaneous reconstruction of the latest history of the cosmological debate, which he, furthermore, rationalized as related to the inherent vulnerability of the steady-state theory. In 1959 he expressed the general relationship between the steady-state theory and observations as follows:

> In fact, since the steady-state forecasts a perfectly definite answer to each of the observations enumerated, while the others [evolutionary theories] forecast a whole range of answers, one would expect almost every observation to disagree with the steady-state theory at the limit of observation, where errors are always large, even if the theory were in fact correct. Since observational errors are frequently under-estimated, as experience shows, one would expect, if the theory were correct, to meet claims that the steady-state theory had been disproved just at the very limits of the observation each time a new type of observation has been made, and for the claim to be withdrawn after a more critical analysis of the errors involved. In fact this has happened more than once in the last ten years.[33]

With such a general view it is understandable that Bondi did not take Ryle's new claim too seriously. In Ithaca, New York, Gold responded to the news in much the same way as Bondi. "I certainly do not consider this a death to continuous creation," he told a newspaperman, "and I do not think that the kind of observations being referred to are capable of giving such a verdict. A similar statement has been made by Professor Ryle in 1955. But the observations on which it was based were later found to be incorrect."[34] In a letter to Gamow he wrote ironically about "the Ryle effect," leaving no doubt that he had no faith in the reliability of Ryle's data.[35] With a slope of -3 in 1955 and a gradual decrease to -1.8 in 1961, wouldn't one suspect further measurements to give about -1.5? However, this kind of loose argument no longer worked. The new Ryle value turned out to be stable and there was just no identifiable residual error to fall back on.

In spite of the solid foundation of Ryle's data, none of the leading steady-state cosmologists accepted that they amounted to a refutation of their theory. Neither McCrea, Sciama, nor Narlikar converted to evolutionary cosmology at the time, and the three originators of the steady-state theory believed that it was still possible to win the battle. In August 1961 the General Assembly of the Internatonal Astronomical Union gathered in Berkeley, where Ryle gave a talk about the cosmological consequences of the most recent radio astronomical observations.[36] Gold, Bondi, and Hoyle were all present, none of them convinced of Ryle's claim of having definitely killed the steady-state theory. The *New York Times* reported on the situation among the steady-state advocates, who were reported to be worried but in good spirits: "The 'steady state' team admits that it is fighting an up-hill battle. And, Dr. Bondi says, they are going to be fighting it for a long time to come."[37] Two years later, Bondi had stopped working in cosmology, but the reason was not that he had by then become convinced of the strength of the observational arguments of Ryle and other steady-state opponents. In 1963 Bondi reaffirmed that "if, indeed, the tests are shown to go against it [the steady state theory] conclusively, then, of course, it must be dropped;" but he also made it clear that this was still not the situation as he saw it: "Personally, I would like to reserve judgment on this because I am not all that convinced of the reliability of some of the observations."[38]

Neither did the new Cambridge data make much impression upon McCrea, who continued to defend the steady-state theory. In a summer school in 1962, devoted to the latest developments in radio astronomy, he gave a survey of cosmological theories without as much as mentioning the significance of Ryle's data. "It appears," McCrea said, "that no thoroughly satisfactory observational test has yet been possible"—a statement which must have surprised many of the radio astronomers present at the summer school, including Martin Ryle. McCrea's positive view of the steady-state theory had not been weakened over the years and was not shattered by the Cambridge claim. "To sum up," McCrea ended his review, "we may say that steady-state cosmology possesses practically all the satisfactory general features that we previously noted for Newtonian or relativistic cosmologies, while it avoids the unsatisfactory features associated with a definite age of the universe. Steady-state cosmology possesses also the philosophically agreeable feature of yielding an effectively unique model, instead of the infinite set yielded by other theories. *Its predictions appear not to have been contradicted by observation.* At present it has the disadvantage of not resting upon any accepted 'field-theory.'"[39] It is characteristic that McCrea ignored the Cambridge radio astronomical verdict, and also that his evaluation was expressed in terms of aesthetically loaded wordings such as "satisfactory general features" and "philosophically agreeable."

The whole situation in radio cosmology was that the observations seemed designed to refute the steady-state theory and only indirectly, by ruling out its main rival, gave support to evolutionary models of the big-bang type. In fact, the slope of −1.8 was difficult to reconcile with evolutionary models also, and did not help at all in determining which model was the most likely candidate for the real universe. This unsatisfactory state was noticed by the German astronomer Wolfgang Priester in a careful analysis of 1958.[40] Although later explanations of the source counts in terms of evolutionary theories brought some progress, for a while the situation did not change significantly. In 1961 William Davidson began a research program of fitting counts to evolutionary models. After much work he was able to show in 1964 that at least some of the relativistic models, such as the Einstein–de Sitter model, were compatible with the available radio data.[41] In his work of 1961, Davidson analyzed the Cambridge data independently and reached the same conclusion as Ryle, that the steady-state theory was invalidated "beyond all reasonable doubt." On the one hand, he considered this regrettable because of the appealing simplicity of the model; but, on the other hand, he found it an advantage that astronomers could now finally forget about the steady-state theory and concentrate on the many relativistic models. The detailed mathematical investigations of Davidson and others were of limited use as guides in the jungle of relativistic models, and mainly confirmed that the cosmological use of the radio source counts was restricted to the negative level, to discard the steady-state rival.

Before the situation clarified in 1964, this point was picked up by some steady-state advocates, who argued that if the radio source counts did not fit with plausible evolutionary models either, this would indicate some unknown effect influencing the observations; and then the claimed refutation of steady-state theory could not be taken for granted. The theme was taken up by McCrea in his presidential address to the Royal Astronomical Society in February 1963:

> Consider the Cambridge radio survey published two years ago. It is in strong disagreement with the steady-state model. I think W. Davidson has done the best that can be done to fit an evolutionary model to the results. . . . The result is very interesting but not compelling. For suppose the steady-state model to be clearly wrong and suppose evolutionary cosmology to be clearly acceptable and suppose also the observations to mean exactly what they appear to mean. There is a wide range of evolutionary models to choose from and we should then expect a considerable range of these to be easily compatible with the Cambridge observations, because we should still have to accommodate details of the red-shift and all the rest. With great respect to all who have worked so hard in this field, I have to say that the natural meaning of the present situation is that either some unknown effect is influencing the observations or no existing theory is likely to fit the observations.[42]

However, with the continuing work of Davidson and others, McCrea's interpretation of the situation was no longer tenable.

The steady-state theoreticians developed two independent and very different answers to the Cambridge claim. Hoyle and Narlikar produced a modified steady-state theory specifically designed to account for the data, and Sciama suggested a reinterpretation of the data while keeping to the simple steady-state model. The first Hoyle-Narlikar theory (several others were to follow) relied on the hot steady-state universe which Gold and Hoyle had proposed in 1958, and in which the intergalactic medium was assigned a kinetic temperature of about 10^9 K. Within this framework, Hoyle and Narlikar introduced expanding regions of "blobs"—clusters of galaxies—of an initial size about 30 Mpc that by now would have extended to the enormous size of about 300 Mpc.[43] This introduced a large-scale nonuniformity in space, which, Hoyle and Narlikar claimed, was nonetheless compatible with the steady-state postulate. In these regions all galaxies would be age correlated, i.e., they would not follow the ordinary age distribution of the steady-state theory. Whereas it had generally been assumed that the probability $k(\tau)$ of a galaxy being a radio source was independent of the age of the galaxy, Hoyle and Narlikar argued that this assumption was ungrounded, and that in their model k would increase with τ. By choosing a particular function for $k(\tau)$, corresponding to k increasing by a factor 100 as τ increases from T to $2.3T$ (T is the Hubble time), Hoyle and Narlikar could construct an example in which the log$N$-log$S$ slope was steeper than -1.5 and in rough agreement with Ryle's result. They may not have considered their model a candidate for the real universe, but rather a counter-example demonstrating that the value of -1.8 could be reconciled with a modified steady-state theory.

Whatever the reasons behind the Hoyle-Narlikar alternative, it was severely criticized and had to be abandoned after a short time. It was briefly discussed by Ryle at the Jodrell Bank summer school in 1962, but he seems not to have taken it seriously. Ryle repeated his conclusion that "the present results show so large a departure from those predicted by the steady-state model that we must recognize that we live in an evolving universe."[44] In the subsequent discussion, no one saw any reason to contradict him. Methodologically, the Hoyle-Narlikar alternative was clearly ad hoc, building on hypotheses rather than inferences, and some of the hypotheses appeared implausible. For example, the theory required the radio sources to emit energy for an interval of time of a couple of billions years, an energy output of an order that no known physical mechanism could explain. More importantly, the theory depended critically on the correctness of the hot-universe model; the expansion of the initial irregularity to a size of about 300 Mpc required a temperature of the order of magnitude 10^9 K. With the demonstration from x-ray astronomy that the temperature was much smaller, the Hoyle-Narlikar theory had to be abandoned together with the hot-universe model. The theory, ingenious but artificial, was not welcomed even by other steady-state sympathizers, and to most astronomers it merely signaled that Hoyle's defense of the steady-state dogma had become desperate. However, later research proved that the notion of superclusters and large-scale inhomogeneities, first proposed by Hoyle and Narlikar, was not so fantastic after all.

Sciama's attempt to save the steady-state theory, like Hoyle and Narlikar's, explored the hypothesis that the slope of -1.8 was the result of a deficit of strong sources (and not an excess of weak sources) produced by some local irregularity in their distribution. But Sciama found irregularities on a scale of 300 Mpc implausible and suggested instead a mixed model in which a substantial fraction of the radio sources were located inside the Milky Way.[45] He showed that if there were roughly as many galactic as extragalactic sources, it was possible to account within the steady-state theory for the observed slope of −1.8. However, as Peter Scott expressed it in a critical examination, this required "a number of special assumptions which, although individually plausible, appears when taken together to present a somewhat artificial situation."[46] Few astronomers aware of Sciama's theory would disagree with this evaluation.

Sciama's model required that a major part of the galactic radio background was produced by a large number of faint galactic sources, which was a view abandoned by most radio astronomers and one difficult to defend physically. Although Sciama found astrophysical mechanisms that would make his scheme work, these implied extreme conditions that lacked observational support and appeared as wishful thinking. Among other things, the model predicted a strong anisotropy in the distribution of radio sources of low flux density. Sciama's theory was no more successful than that of Hoyle and Narlikar in persuading astronomers that the steady-state theory had survived the onslaught of Ryle's number count program. The large majority saw the two theories as artificial and forced attempts to deny the obvious conclusion—that the universe is evolving. As more and more radio sources became identified with extragalactic objects, the model could only be sustained by appealing to unlikely circumstances, and by 1966 it was squarely contradicted by observations. Sciama admitted in 1963 that his theory was "extremely tentative," but maintained that as long as it was not disproved, neither was the steady-state model. In a review article of 1965, he concluded with forced optimism that "the steady-state model remains in the field, bloody but unbowed."[47]

In addition to the two mentioned types of response, a few astronomers questioned if existing measurements were sufficiently certain and unambiguous to allow conclusions of a cosmological nature. G. Burbidge and collaborators argued that this was not the case and that the $\log N$-$\log S$ test did not prove the universe to be evolving. "There is still considerable ambiguity in the interpretation," they wrote in 1971, "so that this cosmological test has so far failed to exclude any model."[48] Most astronomers disagreed, or rather paid no attention to the dissidents.

The Problem of Quasars

When Sciama lost his faith in steady-state cosmology, it was not because of the radio count tests, but the result of the unexpected discovery of a new kind of enigmatic celestial objects referred to as QSOs or quasi-stellar objects. They were also called quasi-stellar sources or—following a suggestion by Hong-Yee Chiu of Columbia University—just quasars.[49]

As is not unusual in discovery histories, the objects were observed before they were identified as something new and, in this sense, discovered. For example, in 1960 Jesse Greenstein took the spectrum of a celestial object, which he listed as a white dwarf some years later. Later, in 1970, he showed that the object on the ten-year-old photo was in fact a quasar. Also in 1960, at the annual meeting of the American Astronomical Society, Sandage reported in an unscheduled paper that he had found a starlike object of the sixteenth magnitude, which he identified with the radio source 3C 48, that is, number 48 in the Cambridge 3C catalog. The optical spectrum was unusual, and since he was unable to find a redshift, the distance of the object was unknown. Sandage believed he had observed the first radio star, "a relatively nearby star with most peculiar properties."[50] In early 1963 two more "starlike [radio] objects" were described, but again without recognizing their true nature. Then, in February-March 1963, quasars were discovered (and not merely observed) in the sense that they were recognized to be not unusual stars, but an entirely new kind of radio objects at cosmological distances.

On 5 February 1963, the Dutch-born Caltech astronomer Maarten Schmidt studied the spectrum of 3C 273, a recently identified "radio star." He was puzzled over not understanding the line spectrum, until he noticed that four of the lines were spaced regularly, much like the visible hydrogen Balmer lines. By taking the ratio of the wavelength of the lines to the nearest Balmer line, he discovered that the ratio was the the same for all four lines, namely, 1.16. Schmidt concluded that the spectrum was redshifted by a factor of 0.16. This was highly remarkable for a "star" of apparent magnitude 13, for with a recessional velocity of almost 48 000 km·s^{-1} it indicated that the object must be very far away and thus emit an enormous energy flux—about one hundred times that of the Milky Way. After he had discussed the discovery with his colleague Greenstein, the two astronomers found that the spectrum of 3C 48 could be understood in a similar way, but now arising from an object of redshift no less than 0.37.

Quasi-stellar objects, i.e., starlike objects with high redshifts, unusual spectra, and varying intensity, had become a reality.[51] During 1963–64 the number of quasars increased and it was estimated that they made up about 30% of all identified radio sources. The redshifts were generally believed to be cosmological, i.e., the result of the expansion of the universe, although the possibilities that they were gravitational or due to local motions were also discussed. Whatever the nature of the mysterious objects, they at once fascinated astronomers and astrophysicists, and quasars became a favorite subject in scientific journals as well as in the news media.

The most puzzling feature about quasars was their enormous energy output, estimated to be a total of about 10^{53} J, or a visual luminosity about 10^{38} J·s^{-1}. To put it differently, quasars shine with a power about one hundred times the total power of a giant galaxy. What was the source of this mysterious energy? The existence of objects with energies of this order of magnitude was discussed well before quasars were discovered, first by Geoffrey Burbidge at the

1958 radio astronomical symposium in Paris.[52] Quasars brought the energy problem into the center of astrophysical theory and many ideas, more or less speculative, were suggested in the early years of quasar research. They also led to another piece of Gamow poetry:

> Twinkle, twinkle, quasi-star
> Biggest puzzle from afar
> How unlike the other ones
> Brighter than a billion suns
> Twinkle, twinkle, quasi-star
> How I wonder what you are.[53]

Of particular importance was a suggestion put forward by Hoyle and Fowler in 1963, shortly before the first quasar was discovered.[54] According to Hoyle and Fowler the energy source was not nuclear reactions, but gravitational collapse of a massive superstar in the center of a galaxy. This proposal, sometimes taken to be a prediction of quasars, was developed in several later papers. It was a main topic at the First Texas Symposium on Relativistic Astrophysics, which took place 16–18 December 1963 in Dallas.[55] Some three hundred physicists and astronomers attended this important meeting, including Hoyle, Gold, McCrea, Salpeter, Fowler, Greenstein, and the Burbidges. In an after-dinner speech, Gold elegantly summarized the situation as it appeared after the discovery of quasars: "It was, I believe, chiefly Hoyle's genius which produced the extremely attractive idea that here we have a case that allowed one to suggest that the relativists with their sophisticated work were not only magnificent cultural ornaments but might actually be useful to science! Everyone is pleased: the relativists who feel they are being appreciated, who are suddenly experts in a field they hardly knew existed; the astrophysicists for having enlarged their domain, their empire, by the annexation of another subject, general relativity. It is all very pleasing, so let us all hope that it is right. What a shame it would be if we had to go and dismiss all the relativists again."[56] Astrophysically important as the study of supermassive objects and energy production in quasars was, it was of no direct significance for cosmological models. We shall therefore not go further into the topic.

The cosmological significance of quasars was not discussed in the earliest papers on the subject, but it was soon realized that the new objects were possibly at variance with the steady-state theory, namely, if they really were so very far away as believed by most astronomers. The existence of objects that exist only very far away and long ago squarely contradicts the perfect cosmological principle. However, in the mid-1960s it was far from proved that quasars are at cosmological distances, which in this context means more than about 300 Mpc away. If some of them are local objects they would pose no difficulty for the steady-state theory, which, in this area, was defended by Hoyle and Sciama in particular. If the redshifts were gravitational there would be no problems either, but in 1964 Greenstein and Schmidt argued convincingly against this possibility.[57] Both Hoyle and Sciama accepted the force of the argument and admitted that the redshifts were cosmological (but see below).

Did this necessarily imply that all quasars were at cosmological distances? Sciama and Hoyle, using different arguments, thought no. In the spring of 1966 Sciama suggested a composite model inspired by his earlier (1963) mixed model for radio sources. According to this view, quasars consisted of two entirely different populations; one group was cosmological, as indicated by their redshifts, whereas the other, including the sources with unknown redshifts, was hypothesized to consist of galactic objects with almost zero redshift. With a model based on this division it was shown that most observations could be reproduced. Sciama considered his composite model the last possibility of saving the steady-state theory, to which he still felt attached. In a paper written with William Saslaw, an American graduate student of his, the conclusion read: "The composite model stands or falls by the presence or absence of detectable proper motions among the known quasi-stellar sources. Moreover, if the model falls, it looks as though the steady-state theory of the universe falls with it."[58]

Without referring explicitly to the steady-state theory, Hoyle, in collaboration with Geoffrey Burbidge and Wallace Sargent, discussed the physical nature of quasars according to both the cosmological and the local hypothesis (the latter meaning distances from 1 to 10 Mpc). Assuming that the energy of quasars had to be explained by a synchrotron mechanism, they found that the cosmological hypothesis led to certain difficulties that could be avoided on the alternative view. Therefore, they concluded, either the synchrotron model was wrong or the quasars were local objects. Hoyle and Burbidge built up what they considered a plausible case for quasars being objects ejected at relativistic speeds from the nuclei of nearby galaxies. Then it was possible to accept the Doppler interpretation of the redshifts without accepting the cosmological hypothesis.[59] The conclusion of Hoyle and his American colleagues in favor of the local hypothesis was criticized by Sciama and his student Martin Rees, who found that there were no compelling arguments against the hypothesis that quasars with known redshifts are at cosmological distances.[60] Among other objections, if the redshifts had a Doppler origin, one would expect to observe blueshifts as well, originating from quasars approaching us. No such blueshifted quasars were observed.

A possible test of the question would be to examine the equivalent of the Hubble diagram in the case of quasars. In June 1966, Hoyle and Burbidge found that a sample of thirty quasars approximately satisfied a linear $\log N$-$\log S$ relation with slope -1.5, i.e., that $NS^{3/2} \approx$ const. If interpreted in the ordinary way—as a volume/distance effect—this would mean that objects with smaller flux density S should be at greater distances; and thus, assuming that the redshift is cosmological, that small S must be correlated with large z. Since no such correlation appeared from the data, Hoyle and Burbidge concluded that "the red-shifts have nothing to do with distances."[61] If the redshifts were entirely due to the cosmic expansion, some kind of Hubble law would be suspected, but the log$z$-magnitude diagram did not reveal any line or smooth curve. The Hoyle-Burbidge conclusion was criticized by Malcolm Longair, Sciama and Rees, and others in a series of papers appearing in *Nature* in the fall of 1966. At that time the microwave background radiation had been discovered, and together with the evidence from radio astronomy it was usually taken as convincing evidence in favor of big-bang cosmology. For this reason, and because the quasars did not exhibit the blueshifted lines to be expected from the local hypothesis, Longair and most other astrophysicists found the local interpretation of Hoyle and Burbidge to be unlikely. Longair argued that the z-S plot could be explained on the cosmological hypothesis, namely, as a result of the increased luminosity of quasars in earlier epochs.[62]

The contributions of Sciama and Rees in the fall of 1966 are particularly interesting because we can here follow Sciama's final abandonment of the steady-state theory. In a letter to *Nature* of 17 September, Sciama and Rees showed that the Hoyle-Burbidge data on the redshifts and flux densities of quasars exhibited a marked statistical disagreement with predictions from steady-state theory (figure 7.2). To Sciama, this amounted to a refutation of the steady-state model: "[It is] the most decisive evidence so far obtained against the steady state model of the universe. . . . We conclude that if the red-shifts of quasars are cosmological in origin, then the present red-shift-flux density relation for quasars rules out the steady state model of the universe."[63] In a contribution of 3 December, Sciama and Rees added as an extra argument that absorption lines had been detected in the spectrum of the quasar 3C 9 and that these were caused by a cloud with redshift 1.6. "Our consequence of this would be that the steady-state model of the universe could almost certainly be ruled out," they wrote.[64] Not unexpectedly, Hoyle denied that the steady-state model was now put in the grave. In a response written jointly with Geoffrey Burbidge, he claimed that their critics overstated the cosmological consequences that could be drawn from the quasar data. Hoyle and Burbidge focused on the conditional clause in the conclusion of Sciama and Rees—"*if* the red-shifts of quasars are cosmological in origin, then . . ."— and it was precisely the "if" they questioned. Not that they had any new arguments against it, but they felt it was illegitimate to take the cosmological hypothesis as an axiom and accused their opponents of being biased against the local hypothesis.[65]

As far as Hoyle was concerned, the question was open, and he could easily come up with arguments in favor of a local position of quasars. For example, in collaboration with Fowler he reconsidered the Greenstein-Schmidt argument against the redshifts being gravitational and found that it was only compelling for a particular kind of quasar model. Assuming another model, Hoyle and Fowler concluded that the redshifts might well be gravitational and then there was no possibility of drawing strict cosmological consequences either for or against the steady-state theory.[66] At this time Hoyle did not defend the original steady-state theory or unconditionally reject an evolving universe. Observational events, such as the discovery of the microwave background

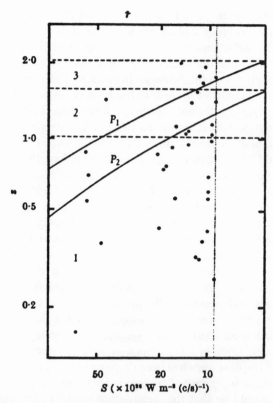

Figure 7.2. The quasar data that persuaded Sciama that the steady-state theory is wrong. The redshifts for thirty-five quasars are plotted against their flux densities. According to the steady-state model there should be an equal number of sources of any given power in the redshift regions 1, 2, and 3. The two curves are lines of constant source power, P_1 and P_2, predicted by the steady-state theory. In region 1 the number of sources with $P > P_1$ and $P > P_2$, respectively, is (0, 2); in the regions 2 and 3 the numbers are (1, 5) and (5, 5). Assuming that the quasar redshifts are cosmological in origin, the distribution thus disagrees with the steady-state prediction. *Source:* Sciama and Rees (1966a), p. 1283. Reprinted with the permission of Macmillan Magazines Limited.

radiation in 1965, had made him more receptive to the big-bang universe, and by 1966 he was for a time inclined to consider this model as likely, or as unlikely, as the steady-state model.

The debate concerning the nature and distances of the quasars continued, but the local hypothesis failed to convince the great majority of astronomers. In 1968 Schmidt confirmed with more data the result of Sciama and Rees, that quasars evolve with the cosmic epoch, and later observations only strengthened the conclusion.[67] However, it was still possible to defend the local hypothesis, and in general the behavior and nature of quasars were so little understood that there was ample room for discussions concerning their cosmological significance. The debate continues to this day, but it suffices to note that by 1970 almost all astronomers accepted the view of Sciama, Rees, and Schmidt that the redshifts of quasars disagree with the steady-state theory.

Sciama's original interest in quasars, and in radio astronomy in general, was closely related to the steady-state model, which he wanted to defend. It was with this purpose that he started examining the number of quasars as a function of their redshifts. At first he was led to believe that the distribution fitted nicely with the steady-state prediction, but his optimism was spoiled by his research student Martin Rees, who did not share his commitment to steady-state cosmology. After having looked at the data, Rees showed that in fact they disagreed with the relation predicted by steady-state theory. There were far too many quasars of large redshift. Sciama found Rees's argument convincing and felt forced to accept the unappealing conclusion. "That was the thing that for me made me give up steady state. . . . it was this study that was decisive, and I had a bad month giving up steady state."[68] Sciama, who for more than a decade had been a leading advocate of and contributor to the steady-state theory, switched rather abruptly from a stationary to an evolving world view. His metamorphosis is interesting also from the point of view of the philosophy of science, contradicting as it does the Planck-Kuhn thesis that supporters of an old paradigm do not convert to a new one, but carry their paradigmatic beliefs with them to their graves.[69] Of course, the quasar data was not the only evidence that convinced Sciama about the incorrectness of the steady-state theory, but when added to the radio source counts and the microwave background radiation, it was the decisive one. Although he was able to switch from one world view to another, the change was only made reluctantly and with regret. Less than one year after his conversion, he wrote: "For me the loss of the steady-state theory has been the cause of great sadness. The steady-state theory has a sweep and beauty that for some unaccountable reason the architect of the universe appears to have overlooked. The universe in fact is a botched job, but I suppose we shall have to make the best of it."[70] Sciama's conversion is illustrated by his book *Modern Astronomy*, which was a greatly revised version of his earlier *Unity of the Universe*. Whereas the latter book dealt extensively with steady-state theory, in his new book, prefaced 1967–69, he only mentioned the theory briefly, as a special instance of relativistic models with negative pressure. Because of the evidence from radio source counts, quasar redshifts, and the cosmic background radiation, "we shall consider it [the steady-state model] no further," he wrote.[71]

Sciama's evaluation of the significance of the quasar redshifts was not shared by most astronomers, who rather saw them as just one more piece of evidence, should one be needed, for the correctness of an evolutionary universe. Most of those who were not already convinced found the radio source counts to be the decisive evidence, or, if they still doubted, they were impressed by the discovery of the cosmic microwave background radiation in 1965, to which we shall shortly turn.

7.2 RELICS FROM THE BIRTH OF THE UNIVERSE

At about the same time as the redshifts of quasars were recognized to contradict the steady-state theory, two other pieces of observational evidence added to the pile of problems that caused the theory's rapid decline and de facto disappearance. Whereas in most earlier tests the confrontation had been between the steady-state theory and the evolving, relativistic universe, the two tests that emerged about 1965 directly involved the big-bang assumption. We shall first look at the origin of helium, and then at the discovery and implications of the microwave cosmic background radiation.

The Helium Problem

As mentioned in section 6.2, there was in the 1950s a competition between the big-bang and steady-state cosmologies over whether the elements that compose our universe were formed cosmologically (as held by the big-bang theory) or are the result of astrophysical processes still going on (as held by the steady-state theory). With the success of the 1957 B^2HF theory, which gave a satisfactory explanation of the formation of heavier elements based on astrophysical processes, it was almost forgotten that at least with regard to one element—helium—the theory failed to reproduce the observational values. And since helium (apart from hydrogen) is by far the most common element in the universe, this was, or ought to have been, a problem. In a talk given in 1959, Alpher, Herman, and Follin pointed out that stellar theories of element formation needed to be complemented with a big bang "since light elements cannot be made in the stars."[72] But the talk was not published, and had it been it would hardly had made much of an impact at a time when cosmological element formation was not in vogue. It was only in the 1960s, when more reliable estimates for the cosmic abundance of helium became known, that the hydrogen-to-helium ratio became established as a cosmological test: A theory which gave a much too low, or much too high, helium abundance would be considered doubtful for that reason.

The early big-bang theory, as developed by Gamow, Alpher, and Herman, gave helium abundances in the range from 12% to 20%, which at the time was seen as satisfactory. The later theory of Alpher, Herman, and Follin indicated a larger amount of helium, between 29% and 36%. During the decade following the 1953 Alpher-Herman-Follin theory, no essential progress took place except that improved observations now indicated a cosmic abundance of helium of about 25–35 %, largely independent of the source. Could this amount of helium be produced in the primordial phase of an evolving universe? Could it be produced astrophysically? In a note of 1955, Bondi, Gold, and Hoyle had drawn attention to the problem, suggesting that hydrogen-to-helium conversion might take place in "black stars"—nonvisible stars located inside dense clouds.[73] Nothing came of this suggestion, and for the next few years there was silence about the origin of helium. In 1961 two American astrophysicists, Donald E. Osterbrock and J. B. Rogerson, concluded that the composition of the sun was roughly the same as that of the planetary nebulae, namely, about 32% helium and 64% hydrogen. Referring to Gamow's theory, they suggested that the amount of helium could be "the original abundance of helium from the time the universe formed, for the build-up of elements to helium can be understood without difficulty on the explosive formation picture."[74] The big-bang theory was not quite forgotten, but the reminder of Osterbrock and Rogerson led to no immediate reconsideration of it.

At the instigation of Zel'dovich, the Soviet physicist Yuri Smirnov reexamined in 1964 the formation of light elements in a hot big-bang "Gamow universe."[75] Apparently unaware of the calculations of Alpher, Herman, and Follin of 1953, he gave an updated version of Hayashi's 1950 theory and concluded that the hot big-bang theory was no good. Smirnov found that, depending on the initial density, it would give either too much deuterium or too much helium. A different and more accurate analysis of the problem was presented by Hoyle and his younger colleague at Cambridge, Roger Tayler, in a paper of the same year.[76] Tayler had done research on stellar evolution under the supervision of Bondi in the early 1950s, and after six years of absence from astrophysics, during which he mostly worked with plasma physics, he became Hoyle's collaborator in Cambridge. The idea to take up the helium problem arose out of a series of lectures on relativity and cosmology that Hoyle gave in the spring of 1964. Hoyle and Tayler recognized that stellar processes gave a much too low amount of helium, and therefore suggested that in order to get the right amount, neutrons and protons had to have been cooked under more dramatic circumstances, at temperatures of at least 10^{10} K. That a galaxy originally formed from pure hydrogen is unable to produce enough helium within its lifetime was confirmed soon thereafter by J. W. Truran, C. J. Hansen, and Alastair G. W. Cameron.[77]

By incorporating the latest developments in elementary-particle physics—including the existence of two kinds of neutrinos (the old electron-neutrinoand the new muon-neutrino)—Hoyle and Taylor found a theoretical helium abundance in fair agreement with the one observed. The two British astrophysicists based their calculations on hot big-bang assumptions, but refrained from taking the agreement as unequivocal support for a singular origin of matter. As they pointed out, the calculations would give a similar result for certain kinds of hypothetical, supermassive objects which would, in effect, simulate small big bangs. Hoyle had for some time cultivated an interest in such objects of masses from 10^3 to 10^6 times the mass of the sun, which in some cases would bounce—perform series of rapid explosions and implo-sions, somewhat like miniature oscillating universes.[78] The helium problem, according to Hoyle and Tayler, led to the conclusion that "Either the Universe has had at least one high-temperature, high-density phase, or massive objects must play (or have played) a larger part in astrophysical evolution than has hitherto been supposed." Whereas Hoyle preferred the latter solution, Tayler tended toward the former.[79]

Three years later, Hoyle returned to the problem, this time together with his long-time collaborator William Fowler and joined by Robert Wagoner, a young physicist who had recently come to Caltech from Stanford Univer-sity.[80] The three astrophysicists gave a detailed theory of nucleosynthesis in which the light elements, in particular, were examined by means of careful numerical calculations. In many respects the theory was a development of the old (and unpublished) Fermi-Turkevich calculations, but it was more exten-sive and used more and better experimental data. For example, Hoyle and his collaborators found that the reaction $^4\text{He} + {}^3\text{He} \rightarrow {}^7\text{Be} + \gamma$, which had been studied in 1949 in order to bridge the mass gap, proceeded at a rate one hun-dred times as fast as assumed by Fermi and Turkevich. No less than 144 dif-ferent nuclear reactions were included by Hoyle, Fowler, and Wagoner. They found that big-bang assumptions were able to reproduce the abundances of deuterium, helium, and lithium reasonably well, but that most of the heavier elements had to be the results of stellar processes. The main results are shown graphically in figure 7.3. With regard to helium, the conclusion was largely the same as in the earlier work, namely, that the element was produced during a short time interval at temperatures exceeding 10^{10} K. This might either mean the primordial phase of an exploding universe or supermassive bodies. In the latter case, a helium percentage as large as forty could be obtained, whereas the singular origin required a roughly uniform helium abundance of about 27%. There is no doubt that Hoyle preferred the first solution. Wagoner recol-lected how "Fred's continuing enthusiasm for supermassive objects . . . had infected Willy [Fowler] and me," and that Hoyle "really wanted those guys [supermassive objects], within the steady state theory, to make the helium, as well as the microwave background radiation."[81] Whatever Hoyle's prefer-ences, he discussed the question in a completely open way. Whether related to a big bang or supermassive objects, the Wagoner-Fowler-Hoyle work was an important contribution to the new physical cosmology. According to McCrea, "It was this paper that caused many physicists to accept hot big-bang cosmol-ogy as a serious quantitative science"[82]—an ironic result of a work done in part by the leading steady-state cosmologist of the time.

The helium problem was the topic of the Eleventh Herstmonceux Confer-ence, which took place 12–13 April 1967, the series of conferences being named after the medieval-looking castle in southeastern England where the Royal Greenwich Observatory was sited. The conference was attended by sixty-eight participants, including Hoyle, Tayler, Bondi, Shakeshaft, and

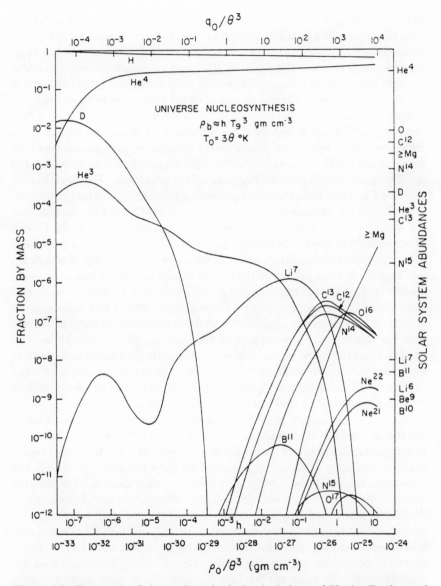

Figure 7.3. The result of the nuclear-physical calculations of Hoyle, Fowler, and Wagoner concerning the formation and abundances of the light elements under the assumption of a universal fireball or a supermassive object. The symbol θ represents the photon temperature in terms of the microwave background temperature 3 K. The symbol ρ_0 denotes the baryon density, h is an adjustable parameter, and $T_9 = 10^9$ K. Element abundances for the solar system are given on the right-hand ordinate. *Source:* Wagoner, Fowler, and Hoyle (1967), p. 21.

Scheuer.[83] Hoyle lectured on his recent work with Wagoner and Fowler and used the opportunity to suggest that if all the observed helium did not have a cosmological origin then also the cosmological origin of the microwave background radiation must be questioned. Although no conclusion was reached at the conference, the general attitude was in favor of helium being produced in the primeval universe. The development of the helium problem is a story of gradually improved calculations in better and better agreement with a big-bang origin. (The same holds for the abundance of deuterium, which in the 1970s became of increasing significance.) Most of the new generation of astrophysicists felt Hoyle's preference for supermassive bodies to be artificial or irrelevant, and simply took the big-bang theory for granted. This was the case with the Canadian-born Princeton physicist James Peebles, who in 1966, inspired by the recent discovery of the cosmic microwave background radiation (of which he was a codiscoverer), made detailed calculations of the helium abundance.[84] His work was entirely based on the big-bang model and he did not even mention either the steady-state theory or the possibility of supermassive objects. Peebles calculated the helium abundance on the assumption of a cosmic background temperature of 3 K and a present mass density of either 7×10^{-31} g·cm^{-3} or 1.8×10^{-29} g·cm^{-3}. In the first case he got 26% helium, in the second 28%, in excellent agreement with the composition of the sun. Peebles took his result as definite support of the big-bang theory, should such be needed. In an interview of 1990, he was asked if he gave much credibility to the steady-state theory in the mid-1960s. "No, I must confess I hadn't looked at it closely," he answered. "I remember being amazed that grown people could get excited about such speculative ideas."[85]

However, to opponents of the big-bang theory, grown or not, it was reassuring that at least it was possible to account for the light elements without assuming a singular origin of the universe. The alternative in the form of supermassive, bouncing objects was preferred by Geoffrey Burbidge also,[86] but it was speculative in the sense that it rested on the existence of bodies for which there was no observational evidence whatever. As the calculations of Wagoner, Fowler, and Hoyle were refined, evidence for a big-bang origin of deuterium and helium seemed to increase. In 1973, Hubert Reeves and collaborators concluded that agreement with observational data for the abundances of deuterium and the two helium isotopes could be obtained on the assumption of a big bang, but that supermassive objects were unlikely to produce the right abundances. "The problem of the origin," they wrote, "involves a certain metaphysical aspect which may be either appealing or revolting. Whatever is the issue, the big bang has one advantage over many other models: calculations can be made and are presently available."[87] And these calculations clearly pointed toward an origin in the big bang. Although the helium problem was never a strict cosmological test in the same way as, for example, the radio source counts, it did much to discredit what was left of the steady-state theory.

Microwaves from the Primeval Universe

The discovery of the cosmic microwave background radiation constituted in effect the final blow to an already dying theory. As early as 1948 Alpher and Herman had predicted the existence of a blackbody-distributed radiation of a present temperature of about 5 K, and interpreted it as a remnant of the decoupling between matter and radiation in the early universe. However, neither their original prediction nor their and Gamow's several repetitions of it attracted interest among cosmologists. In the early 1960s it was effectively forgotten, a most remarkable fact. The serendipitous but (equally remarkably!) Nobel Prize–rewarded discovery of the cosmic background radiation in 1965 has been recounted many times, in more or less detail, and a relatively brief account of the discovery history will suffice for our purpose.[88]

Early attempts to measure the temperature of space were unrelated to what later was realized to be the cosmic background radiation, although in retrospect some of them came close. For example, in 1955 a French radio astronomer, Émile Le Roux, surveyed the sky at wavelength 33 cm and found a temperature of 3 ± 2 K with a high degree of isotropy. Although neither he nor others recognized his result to have any cosmological significance, he suggested in his dissertation of 1956 that the radiation was of extragalactic origin. Le Roux's work was later claimed to be an anticipation of the 1965 discovery of the cosmic microwave background radiation.[89] A rather similar case appeared in the Soviet Union, where a young astrophysicist, T. A. Shmaonov, measured the cosmic background radiation at wavelength 3.2 cm. He reported in a Soviet journal of technical physics in 1957 that the effective temperature of the radiation was 4 ± 3 K and that its intensity was independent of the direction of his antenna.[90] As in the case of Le Roux, no further significance was attached to the measurements. It was only in 1983 that Shmaonov realized what he had measured twenty-six years earlier. Again, in 1962 an American astronomer, William Rose, attempted to measure the cosmic background radiation at the Naval Research Laboratory. He estimated its temperature to be about 3 K. However, the result was not very certain and since Rose was unable to confirm it, he did not publish his result.[91]

Ed Ohm, a physicist at the Bell Laboratories' facility near Crawford, New Jersey, worked within a very different context. He had constructed a radiometer in 1961 to receive microwave signals from the Echo balloon, a reflector of radio and television signals launched by a rocket in 1960. In observations at wavelength 11 cm, he found an excess temperature of 3.3 K in the antenna, but the result attracted little attention among astronomers.[92] It was mentioned by two Soviet astrophysicists, Andrei G. Doroshkevich and Igor Novikov, in a 1963 discussion of the radiation content of the universe in which they related the excess temperature to the background radiation predicted by the Gamow theory. They wrote: "Measurements in the region of frequencies $10^9 - 5 \times 10^{10}$ cps [$\lambda = 6$ to 30 cm] are extremely important for experimental checking of the

Gamow theory. . . . According to the Gamow theory, at the present time it should be possible to observe equilibrium Planck radiation with a temperature of 1–10 °K. . . . Measurements reported in [Ohm (1961)] at frequency $\nu = 2.4 \times 10^9$ cps give a temperature 2.3 ± 0.2 °K, which coincides with theoretically computed atmospheric noise (2.4 °K). Additional measurements in this region (preferably on an artificial earth satellite) will assist in final solution of the problem of the correctness of the Gamow theory."[93]

The paper by Doroshkevich and Novikov was the first one to emphasize the blackbody spectrum of the radiation and also that it should be possible to actually observe it. The two Russian physicists pointed out that the background radiation would lie in a part of the electromagnetic spectrum where the galactic radiation was weak; although the integrated amount of microwave background energy would be comparable with the light energy from the galaxies, the background would not be masked by the latter. However, the work of Doroshkevich and Novikov seems not to have been noticed by Western cosmologists.

Yakov Zel'dovich also referred in 1963 to the background radiation to be expected from Gamow's theory, but he estimated its temperature to about 20 K.[94] Such a high temperature would correspond to an unrealistically high energy density of radiation, from which Zel'dovich concluded that "Gamow's theory must be rejected." It was in the same context that Doroshkevich and Novikov suggested their test as the "final solution" of Gamow's hot big-bang theory. The Soviet astrophysicists apparently misunderstood Ohm's report, which they interpreted as indicating a cosmic background radiation of temperature close to the absolute zero. What Ohm had called the "sky temperature" and found to be 2.3 K at the zenith, Doroshkevich, Novikov, and Zel'dovich seem to have believed contained also the background radiation, which in fact it did not. Thus, Zel'dovich concluded in early 1965 that "experimental data indicate that the Planck spectrum temperature is below 1 K."[95]

At that time Zel'dovich advocated a cold big-bang theory, in which he argued that "the notion of [primordial] matter consisting of protons, electrons and neutrinos is the only one possible."[96] At $t = 0$ the temperature of the primordial substance, composed of equal numbers of protons, electrons, and neutrinos, would be zero. There was no room for neutrons in Zel'dovich's big bang. Under normal high-density circumstances, protons and electrons will combine to form neutrons, but Zel'dovich suggested that the high concentration of neutrinos would prevent this process. Without neutrons there could be no primordial element formation, and according to the cold big-bang scenario the end product of the initial expansion would simply be molecular hydrogen. Although a big-bang theory, Zel'dovich's thus agreed with the steady-state theory that the chemical elements were synthesized in the stars from primordial hydrogen. The short-lived theory was taken up by some Soviet astrophysicists, but attracted no interest among their Western colleagues. According to Tayler, it was "contrived and only to be adopted when other ideas fail."[97] In spite of its short lifetime—it was proposed in 1962 and abandoned at the latest in 1965—the theory merits attention because it was the first revival of the big-bang theory since the Gamow-Alpher-Herman theory. It differed greatly from this theory, but all the same it was a nuclear-physical big-bang theory inspired by the original one. In view of the traditional ideologically based resistance against big-bang theories, it is ironic that this first revival (or revision) took place in the Soviet Union.

Zel'dovich's temporary advocacy of the cold big bang and his dismissal of Gamow's theory may have prevented him from following up his 1963 remarks concerning the cosmic background radiation. Zel'dovich was one among several astrophysisists who might have—and should have?—discovered, or (re)predicted, the cosmic background radiation; but he did not. In the fall of 1963 Doroshkevich and Novikov discussed their idea of the cosmic background radiation with Zel'dovich, who, in spite of his lack of confidence in the hot big-bang model, encouraged his younger colleagues to pursue the matter. Novikov recalled that they discussed the possibility of observation with some Soviet radio astronomers in 1964. Among these was Vladimir Alexandrovich Kotel'nikov, an academician and important figure in Soviet radio astronomy. However, the response was discouraging. They informed Zel'dovich and Novikov that the level of noise in the receiver would be much higher than the predicted signal, and also they did not take seriously the theoreticians' "chatter" about the very early universe. "People smiled," Novikov recalled.[98] Zel'dovich did not consider the background radiation hypothesis very important and seems to have suppressed it. In 1965, after the radiation had been discovered by Penzias and Wilson, he asked why Doroshkevich and Novikov had not thought about it. When Zel'dovich was reminded that this was in fact what they had done in their paper of 1963, he was surprised. He had forgotten about it.

The Soviet astrophysicists were not alone in glimpsing the cosmological relevance of cosmic microwaves, yet failing to clearly realize their significance. Hoyle came close, but not close enough (section 4.5). He was one of the very few astronomers who was aware of and interested in the background radiation. Thus, in the summer of 1956, Hoyle, then at Caltech, was invited to spend some time with Gamow, who had a consultancy job in California. The two cosmologists, personally on friendly terms in spite of their scientific disagreements, were led to discuss the background radiation, which Gamow at that time believed was 10 K or more. Hoyle wanted a temperature near zero in order to accommodate the spectroscopic result of McKellar of 1941; the background, if it existed, could be less than McKellar's 2.3 K, but not more. The discussion between Hoyle and Gamow led to nothing. "We missed the chance of spotting the discovery made nine years later," Hoyle later remarked.[99] Again, at the Paris symposium in 1958 Hoyle mentioned that according to the big-bang models, "the present-day temperature should not be much in excess of 1 K."[100] The remark was not followed up. Hoyle's insistence on relating the temperature of the hypothetical background to McKellar's result proved to be correct, but he seems to have overestimated the significance of the theoretical values. "I was . . . curiously blind to the thought that McKellar had got the right result!" he recalled in 1995.[101]

Whether they were aware of the works of Alpher and Herman or not, in the early 1960s several astrophysicists must have realized that the hot big-bang theory leads to the existence of a background radiation of present temperature about 5 K. In the first draft of their paper on helium production, Hoyle and Tayler gave a rough calculation of the relationship between the present nucleon density and the radiation temperature, which they estimated to be $n \approx 10^{-10} T^3$ cm^{-3}. With $n \approx 10^{-6}$ cm^{-3} this gave $T \approx 20$ K, a value they understandably judged to be "almost certainly too high." The overestimate, based on several uncertainties, was used to cast doubt on the Alpher-Hermann-Gamow theory. However, when the paper appeared in print, there was no reference to the background radiation. Hoyle, who produced the final version of the paper, left out the mention of the radiation, perhaps because he felt the estimate to be too unreliable.[102] Tayler recalled that he gave talks at Cambridge and Manchester in the summer of 1964 in which he mentioned the existence of a low-temperature background radiation, but also that it would be very difficult to detect.[103] It remains a fact that until the fall of 1964 nobody—including Gamow, Alpher, and Herman—combined the belief in the existence of a cosmic background radiation with a realization that it could actually be measured, and in this way serve as an important cosmological test.

In 1963, Arno Penzias and Robert Wilson, like Ohm working at Bell Labs, started preparing the radiometer for use in radio astronomy. Wilson was twenty-seven years old, fresh from a postdoctoral appointment at Caltech, and the three-years-older, German-born Penzias was a graduate of Columbia University with two years of experience at Bell Labs. The Penzias family, which was Jewish, had escaped to England in 1939 and from there gone to the United States. The horn antenna, originally designed for communication (via the Echo balloon) with the Bell Company's Telstar satellite, became available in 1963, when the Bell Company decided to pull out of the infant communications satellite business. Penzias and Wilson decided to use the modified instrument to measure the radio noise from the halo of the Milky Way and also the radio signals from Cassiopeia A, a supernova remnant. However, they immediately recognized that something was quite wrong. Measuring at a wavelength of 7.4 cm, Penzias and Wilson found an antenna temperature of 7.5 K where it should have been only 3.3 K; moreover, this extra intensity (corresponding to an excess temperature), which was first believed to be an effect of noise, turned out to be independent of the direction in which the antenna pointed. The data obtained by the two physicists indicated that what they were measuring could not be of atmospheric, solar, or galactic origin. Since their measurements were very reliable, the result constituted a problem that threatened to make the instrument useless for highly sensitive radio astronomical measurements. This, and not cosmological questions, was what concerned Penzias and Wilson. They therefore used most of a year to understand the reason for the discrepancy, carefully checking all hypotheses and possible sources of error. Privately they were led to the tentative conclusion that the excess radiation was of cosmic origin, whatever that might signify.

Although Penzias did not think of cosmology in connection with his engineer-like work with Wilson, he had earlier made indirect contact with the cosmic background. He originally planned to use radio astronomy in the search for interstellar molecules, and in 1961 he noticed in Herzberg's book on molecular spectroscopy that interstellar cyanogen was excited at 2.3 K. However, when struggling with the antenna temperature problem a couple of years later Penzias did not connect it with his half-forgotten knowledge of cyanogen's excitation temperature.[104]

Theory, Experiment, Discovery

The mystery was solved in the early spring of 1965, primarily through the work of Dicke and coworkers at Princeton University. Robert H. Dicke was born in 1916 in St. Louis, Missouri, and he had taught at Princeton University since 1946. Exceptionally among modern physicists, he had a distinguished career in both experimentally and theoretical physics. Equally exceptional at the time, he transgressed the barrier between quantum theory and general relativity and refused to follow the trend toward superspecialization. Dicke's interest in cosmology derived from his work in gravitation theory and was inspired by the cosmological theories of Dirac and Jordan. His interest in general relativity first developed during a sabbatical year spent at Harvard in 1954. Three years later he participated in the Chapel Hill Conference on the Role of Gravitation in Physics at the University of North Carolina together with Bondi, Wheeler, Peter Bergmann, Bryce DeWitt, and others. On this occasion Dicke suggested a new theory of gravitation, according to which the vacuum was assumed to have a structure, to be treated as a polarizable medium with permittivity ε. Applying this idea to the universe, he was led to the following speculation: "A cataclysmic production of particles could also be obtained within the framework of the theory. The universe might be visualized as initially free of particles ($t < 0$) and containing only gravitational energy. . . . ε would be decreasing with time, varying as $(-t)^{2/3}$. One might postulate in a completely ad hoc fashion the creation of heavy neutral bosons at a rate varying as $\varepsilon^{-n}(d\varepsilon/dt)^2$ with $n > 1/3$. This would result in a cataclysmic production of bosons at $t = 0$. The heavy bosons would then quickly decay into protons and electrons."[105]

As the quotation indicates, Dicke was afraid neither of speculating nor of addressing grand, fundamental questions. Both in style and content his works differed from the positivistic, engineering-like approach which characterized so much of American cosmology and theoretical physics at the time. Dicke was deeply fascinated by the classical discussions of Descartes, Newton, Bishop Bentley, and Ernst Mach, and admitted that work in cosmology was necessarily guided by philosophical considerations. This was an attitude closer to the British tradition than to the one dominating American science.

"Having its roots in philosophic speculations," Dicke wrote in 1963, "cosmology evolved gradually into a physical science, but a science with so little observational basis that philosophical considerations still play a crucial if not dominant role."[106] Among the broader considerations that fascinated Dicke was Dirac's large-number hypothesis, which he developed in his own way into one of the early versions of the anthropic principle.[107]

In 1961 Dicke together with his student Carl Brans developed a new theory of gravitation "which is more satisfactory from the standpoint of Mach's principle than general relativity."[108] According to the new theory, the constant of gravitation was not a fixed constant of nature, but a quantity depending on the structure of the universe through a hypothetical scalar φ field. The reciprocal of the field acts as the constant of gravitation, $G = [\varphi(x,y,z,t)]^{-1}$, and so the constant of gravitation can depend on time, as in the theories of Dirac and Jordan. The dependency depends on the value of a certain parameter (ω) which is not given by theory and which in the limit $\omega \to \infty$ yields the ordinary theory of relativity. The Brans-Dicke theory led to a number of geo- and astrophysical predictions which were much discussed in the 1960s. It also had cosmological consequences, although in general these did not differ much from those of relativistic evolution theory. Brans and Dicke assumed a big-bang universe and showed that for ω not too small their theory gave almost the same solutions as the Friedmann-Lemaître theory. Dicke was well aware of the steady-state theory, but he was convinced that the universe was evolving, and therefore believed that the model of the Cambridge physicists could not describe the real universe.

Although the scalar-tensor theory of Brans and Dicke was a big-bang theory, it belonged to a different tradition from the theories of Lemaître and Gamow. In their work of 1961, Brans and Dicke did not refer to the previous works of Gamow, Alpher, and Herman. It was while investigating the cosmological consequences of the Brans-Dicke theory for the early universe that Dicke, in 1963, was brought to consider the radiation arising in the primordial universe. However, because of the similarity between the Brans-Dicke cosmology and the ordinary relativistic cosmology, his considerations did not depend strongly on the scalar-tensor theory and would also have validity without the particular theory he had established together with Brans.

At that time Dicke was attracted to the idea of an oscillating universe, which he believed avoided the problem of explaining the original creation of matter—or at least pushed back the problem indefinitely. He argued that in order to decompose the heavy elements in a big crunch, the subsequent big bounce would have to be extremely hot and filled with blackbody radiation. Light from the stars in the last phase of the oscillation would have been blueshifted during the collapse and relaxed to a thermal blackbody spectrum if the contraction were deep enough to make space optically thick. Dicke speculated that the heated and thermalized starlight could have decomposed the heavy elements from the last cycle of the oscillation to provide hydrogen for the next cycle. Privately he had deduced by about 1963 that the radiation would cool off as the universe expanded, but retain its spectral composition.

TABLE 7.1
Dicke and Peeble's March 1965 Picture of the Universe

Time	Temperature (K)	Nucleons per cm³	Events
<0.3 s	2×10^{10}	10^{22}	Universe opaque to neutrinos; thermal neutron abundance
200 s	10^9	10^{17}	Neutron capture reactions lead to helium formation
8×10^5y	3×10^3	10	Hydrogen plasma recombines; formation of pregalactic systems
10^{11}y	10	2×10^{-7}	Present universe

Interestingly, in his pioneering work of the late 1940s Gamow was motivated by the same idea that inspired Dicke, the oscillating universe. Without knowing about the Bell measurements, Dicke suggested in the late summer of 1964 to James Peebles, a former student of his, to calculate the relic blackbody radiation for which he, Dicke, had already made a rough calculation that indicated an upper limit of about 40 K. Peebles did not share Dicke's enthusiasm for an oscillating universe, but just found it an interesting physical problem. In February 1965 Peebles presented his work at a colloquium at the Johns Hopkins Applied Physics Laboratory, where he discussed both a cold and a hot early universe.[109] He estimated that in the latter case the present temperature of the background radiation would be about 10 K. At about that time Peebles submitted a paper to the *Physical Review* with his suggestion, but it was rejected on the ground that Gamow, Alpher, and Herman had made a similar suggestion much earlier.[110] This was news to Peebles.

At the same time, February 1965, the two Princeton physicists started a collaboration with Peter Roll and David Wilkinson, who on Dicke's suggestion had constructed a radiometer to measure thermal radiation at a wavelength of 3 cm. In a review paper submitted in early March 1965, Dicke and Peebles gave a qualitative discussion of their work, which they said was motivated by that "central puzzle" and "interesting embarrassment" constituted by the origin of the universe out of a singularity. They suggested the picture of the evolution of the universe which is summarized in table 7.1. With regard to the cosmic background radiation liberated in the past from an opaque, expanding universe, they wrote, mistakenly: "Apparently this point has been noted only recently."[111] Peebles still did not know that his submitted paper had been rejected.

Before the Princeton group had obtained any experimental results, they learned about the experiments at nearby Crawford Hill, which, to them, indicated the existence of the sought-for radiation. A popular journal aptly summarized the events in March 1965 as follows: "Dicke . . . had a theory and nothing to support it. Penzias had noise but no theory. They put the two together and got a perfect fit."[112] The contact was made after Peebles' lecture at Johns Hopkins University, where he mentioned Dicke's idea and the preparations for an experiment at Princeton; the news was passed along to Penzias, who arranged for a meeting with the Princeton group in March 1965. According to Penzias, the route was as follows: "I mentioned our problem to Bernard Burke during a casual telephone conversation on another matter. He replied that a preprint from Princeton had come across his desk shortly before, predicting a ten-degree background at 3 cm. It was written by P. J. E. Peebles and predicted, using certain assumptions, a thermal background of radiation as a residuum of the hot, highly condensed early state of the evolution of the universe." [113] Penzias immediately called Dicke, who sent him a copy of Peebles' reprint.

It was only then that the two Bell physicists realized that they had made an important cosmological discovery. Yet they received the solution with mixed feelings. "Although we were pleased to have *some* sort of answer, both of us at first felt a little distant from cosmology," Wilson later recalled; "I had taken my cosmology from Hoyle at Caltech, and I very much liked the steady-state universe. Philosophically, I still sort of like it. I think Arno and I both felt that it was nice to have one explanation but that there may well have been others."[114] Shortly after Peebles had become aware of the data of Penzias and Wilson, he gave a talk at the American Physical Society in which he reported the Penzias-Wilson result and its hot big-bang interpretation. As Peebles pointed out on this occasion, the measured microwave intensity was clearly distinct from other background radiations such as starlight and radio waves.[115]

The Bell and Princeton physicists published their works as companion papers in the July 1965 issue of the *Astrophysical Journal*.[116] Penzias and Wilson reported their finding of an excess temperature of 3.5 ± 1.0 K at a wavelength of 7.3 cm, without mentioning its implications for cosmology. This was left for the Princeton physicists, who argued that the observed radiation was part of the blackbody radiation remaining from the primordial decoupling of matter and radiation. In this they did not go beyond the explanation suggested much earlier by Alpher and Herman, but at the time neither of the groups seems to have been aware of that work. This is, of course, a rather surprising fact. Neither did they know about, or recall, Gamow's calculations. Given that Peebles' first manuscript was rejected by the *Physical Review* with a reminder of the work of Gamow, Alpher, and Herman, the Princeton physicists must have realized their blunder at about the time their paper appeared in the *Astrophysical Journal*. The paper, dated 7 May, included a reference to Peebles' forthcoming paper ("in press"), which at the time was believed to be going through the refereeing process. The four Princeton physicists referred to the $\alpha\beta\gamma$ paper as well as the Alpher-Herman-Follin paper of 1953, but in neither of these publications was the background radiation mentioned.

Dicke had actually heard a talk Gamow gave at Princeton University on his theory of element formation from a big bang, but he mistakenly believed that Gamow's early universe was cold and so paid no further attention to it. When Penzias realized the omission he sent Gamow a letter and included a copy of his and Wilson's paper on the 3 K radiation. Gamow, understandably annoyed by the neglect of his work, replied by summarizing his early work. "Thus, you see the world did not start with almighty Dicke," he added sarcastically.[117] It was only two years later that the Princeton physicists privately acknowledged, in letters to Gamow, Alpher, and Herman, their failure in not being acquainted with the history of their subject. In a paper of June 1967 Peebles and Wilkinson reviewed the "revolutionary development in cosmology" which had taken place as a result of the work of the Princeton and Bell Labs physicists two years before. They excused their lack of awareness of earlier work in big-bang theory with the somewhat strange argument that the Princeton researchers were "by training physicists rather than cosmologists."[118] A little later Dicke wrote a long letter to Gamow, apparently apologizing for his oversight, and he expressed his wish "to forget past oversights and be friends."[119]

The contributions of Alpher and Herman, who first predicted the background radiation, were even more ignored, which left them with a feeling of bitterness. Together with Gamow, they wrote a semihistorical review in order to set the record straight, but it was only gradually and incompletely that the significance of their contributions was recognized.[120] Alpher and Hermann were eventually honored for their work; they received the Prix George Vanderlinden of the Belgian Royal Society (1975), the John Price Wetherill Gold Medal of the Franklin Institute (1980), the New York Academy of Sciences Award (1981), and the Henry Draper Medal of the National Academy of Sciences (1993). Yet they found it difficult to forget about the delay of recognition and the circumstances surrounding the celebrated 1965 discovery. They came to regard the incident as significant for a general tendency in modern science. As they wrote in 1990:

> Too many authors evidently rely on more recent publications, many of which continue to propagate errors, rather than consulting original source material. One wonders about the forces that shape the activities of some scientific authors. A number of questions have been raised . . . which beg to be looked at with proper objectivity by historians and sociologists of science, particularly in terms of what such matters say about scholarship and integrity in science, which is one of the most important of human endeavors. We do not accept the argument of some that correct attribution does not matter, but that only the furtherance of science matters. This view does not reflect the ideals and realities of the scientific enterprise. A correct history of science as a human endeavor does matter, both for the present and for the future. . . . For many years we have contemplated the possibility that the reason for a lack of acceptance of the worth of our early work lay in our both being employed by large industrial research laboratories at the time when the background radiation was first observed, and for some years thereafter.[121]

Most historians of science will find no difficulty in sympathizing with the views of Alpher and Herman.

In their July 1965 paper, the Princeton physicists discussed the Penzias-Wilson result in a broad cosmological context. The paper reflected Dicke's predilection for a closed oscillating universe—which "relieves us of the

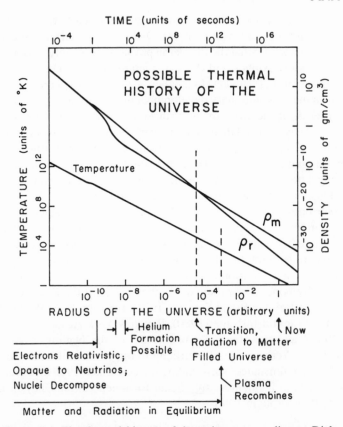

TIME (units of seconds)

RADIUS OF THE UNIVERSE (arbitrary units)

Figure 7.4. The thermal history of the universe according to Dicke, Peebles, Roll, and Wilkinson in the spring of 1965. A present universe with matter density 2×10^{-29} g·cm^{-3} and thermal radiation of temperature 3.5 K is assumed. The temperature scale to the left refers to the temperature of the thermal background radiation. *Source:* Dicke et al. (1965), p. 418.

necessity of understanding the origin of matter at any finite time in the past"—but the authors stressed that their conclusions were valid also for other models involving a hot big bang or "fireball." The essence of their view of the history of the universe was summarized in a figure (figure 7.4), which was an updated version of Gamow's old divine creation curve. However, this connection with the past was also unnoticed. It is ironic that the beginning of the standard big-bang theory was indebted to a heterodox theory of gravitation, the Brans-Dicke theory. Although this theory played no role in the 1965 paper, it remained Dicke's favored theory in a cosmological context also. After the standard relativistic theory had been established by Peebles and others, Dicke returned to the cosmological consequences of the scalar-tensor theory. Because of the scalar field, the expansion rate of the cosmic fireball is larger in the Brans-Dicke theory than in ordinary relativistic cosmology. This affects the production of helium, which, depending on the average density of matter in the universe, will result in either an abundance of 32% or no helium at all.[122]

In their paper of 1965, Dicke and his coauthors stressed that the helium abundance was closely connected with the temperature of the observed cosmic background radiation, and that the combination of the two phenomena "provides some important evidence on possible cosmologies." Drawing on the content of Peebles' unpublished paper, they argued that at temperatures much larger than 10^{10} K the ylem (a word they did not use) would consist of nearly equal amounts of protons and neutrons. With the cooling of the universe the neutrons and protons would combine to form deuterons, which in turn would burn to helium. From this starting point, which was the same as the one adopted by Alpher, Herman, and Follin in 1953, they went on: "Evidently the amount of helium produced depends on the density of matter at the time helium formation became possible. If at this time the nucleon density were great enough, an appreciable amount of helium would have been produced before the density fell too low for reactions to occur. Thus, from an upper limit on the possible helium abundance in the protogalaxy we can place an upper limit on the matter density at the time of helium formation (which occurs at a fairly definite temperature, almost independent of density) and hence, given the density of matter in the present universe, we have a lower limit on the present radiation temperature."[123]

In his ill-fated paper submitted to the *Physical Review*, Peebles calculated $T = 3.5$ K and found that an assumed cosmic helium abundance of 25% necessitated a present matter density in the universe of $\rho \leq 3 \times 10^{-32}$ g·cm^{-3}. Given that the observed value was estimated to be twenty times as high, this was not a very promising result. However, the Princeton group suggested that "the [observational] estimate probably is not reliable enough to rule out this low density." This was not a very convincing suggestion, for it was generally recognized that the real mean density of matter was probably larger than the one inferred from observations. At any rate, in Peebles' improved calculations of 1966 the discrepancy had vanished.

The observation reported by Penzias and Wilson quickly became identified as the event constituting the discovery of the cosmic background radiation. However, by itself the observation of an excess thermal radiation of unexplained origin did not constitute an important discovery; until the intervention of the Princeton theoreticians, Penzias and Wilson had merely found something they could not explain. This anomaly was only *turned into* a discovery by the Princeton physicists' theoretical interpretation and by later experimental work which confirmed and extended the results obtained in 1965. To say that Penzias and Wilson just happened to discover the radiation in 1964–65— sleepwalking like Kepler as portrayed by Arthur Koestler—is to oversimplify the discovery process. An important part of this process, as of other discovery processes, was the scientific community's *decision* that the measurements made up a discovery. This decision depended crucially on the theoretical interpretation by the Princeton physicists and the entire theoretical context in which the measurements were, or came to be, embedded. Had it not been for this interpretation, as well as the interaction of theory and experiment in general, McKellar's and Adams's works of 1940–41 might as well have been celebrated as the discovery of the background radiation.

The discovery of the cosmic background radiation created a sensation in the scientific world, and was widely reported also in newspapers and popular journals. The first account of what was going on did not appear in a scientific journal, but in the *New York Times*, where Walter Sullivan gave a detailed report on 21 May 1965. The story was considered important enough to be placed on the upper part of the front page—a sure indication of success which automatically ensures a great deal of political attention also. "It is clear," wrote Sullivan, "that Dr. Dicke and others would like to see an oscillating universe come out triumphant. The idea of a universe born 'from nothing' raises philosophical as well as scientific problems." He further noticed that the discovery seemed to be incompatible with Hoyle's steady-state theory: "The study of the Bell Laboratories' observations at Princeton leaves open the question of whether there has been one explosion or the universe oscillates. However, both Dr. Sandage and Dr. Dicke clearly doubt the steady state theory in which there is no explosion at all."[124]

The Dicke-Peebles-Roll-Wilkinson big-bang interpretation of Penzias and Wilson's measurements was accepted by the majority of astronomers, who were already convinced that the steady-state alternative was ruled out because of its inability to account for the radio source counts. Still, to count as a full confirmation, the background radiation would have to be detected for other wavelengths too; only if the spectrum was in fact blackbody distributed and isotropic would it confirm the predictions of the Gamow hot big-bang theory. Wilson hesitated for some time to join the big-bang chorus. In an interview of 1982, he recalled: "Some of the steady-state people were pleased by the way we had gone about things. We felt that, at least until they had had a chance to think about our results, we shouldn't go out on a theoretical limb that we couldn't support. For me, the last nail in the coffin of the steady-state theory wasn't driven in for quite a while—not until the blackbody curve was really verified. That's the point when I stopped worrying about it."[125] The majority of astrophysicists did not share Wilson's reservations. Many of them did not wait for new data to draw their conclusion; eager to bury the discredited and disliked steady-state theory, they prematurely (but correctly) concluded that the meager experimental data indicated a blackbody distribution. In fact, this could not be inferred with any certainty until the early 1970s.

The first confirmation came in early 1966, when Roll and Wilkinson reported the measurement of a 3.0 K radiation at a wavelength of 3.2 cm, and during the following years the spectrum was pieced together by many individual measurements (table 7.2). It was of particular importance when George Field and John Hitchcock pointed out in the spring of 1966 that the energy of

TABLE 7.2
Early Measurements of the Cosmic Microwave Background

Year	Scientists	λ (cm)	T (K)	ΔT (K)
1965	A. A. Penzias and R. W. Wilson	7.35	3.5	± 1.0
1966	P. G. Roll and D. T. Wilkinson	3.20	3.0	± 0.5
1966	T. F. Howell and J. R. Shakeshaft	20.70	2.8	± 0.6
1966	G. B. Field and J. L. Hitchcock	0.26	3.2	± 0.5
1966	P. Thaddeus and J. F. Clauser	0.26	2.8	± 0.5
1967	T. F. Howell and J. R. Shakeshaft	49.20	3.7	± 1.2
1967	A. A. Penzias and R. W. Wilson	30.00	2.5	± 0.3
1967	R. A. Stokes, R. B. Partridge, and			
	D. T. Wilkinson	3.20	2.7	± 0.2
1967	D. T. Wilkinson	0.86	2.6	± 0.2
1967	M. Ewing, B. F. Burke, and D. H. Staelin	0.92	3.2	± 0.3

the background radiation corresponded to the excitation energy of CN molecules, as revealed by the old measurements of McKellar and Adams.[126] Whereas McKellar's excitation temperature was 2.3 K, Field and Hitchcock calculated a temperature of 3.2 K ± 0.5 K, in excellent agreement with other determinations of the microwave background. Their result was based on similar data from two clouds in very different parts of the sky, which supported the hypothesis of the universality of the excitation mechanism.

Steady-State Theory and the Cosmic Background

The microwave background radiation obviously posed a serious problem to the steady-state theory, in which there was no mechanism corresponding to the fireball of the big-bang theory. The implication was pointed out by Peebles in a summer school lecture he gave in August 1965. As if anticipating future modifications of the steady-state theory, he mentioned that the flat disagreement was with steady-state cosmology "in its original form."

At that time the steady-state theory seemed to lie in ruins, almost destroyed by quasars, radio source counts, and the helium problem, and with the new radiation heralding its immediate death. On 6 September, Hoyle reviewed the situation at the annual meeting of the British Association for the Advancement of Science, taking place in Cambridge. He admitted that "there seems no way in which such a [3 K, Planck-distributed] background can be explained in terms of current astrophysical processes," and that there was now strong evidence for a universe that had evolved from a superdense state. Hoyle was ready to discard the original steady-state theory, but not quite ready to accept the rival big-bang theory. As he explained—once again—the latter theory seemed to involve a universal singularity, and this he considered objectionable in principle: "I have always had a rooted objection to this conclusion. It seems as objectionable to me as if phenomena should be discovered in the laboratory which not only defied present physical laws but which also defied all possible physical laws. On the other hand, I see no objection to supposing that present laws are incomplete, for they are almost surely incomplete. The issue therefore presents itself as to how the physical laws must be modified in order to prevent a universal singularity, in other words how to prevent a collapse of physics."[127] Hoyle then went on: "It was with this background to the problem that several of us suggested, some twenty years ago, that matter might be created continuously. The idea was to keep the universe in a steady-state with creation of matter compensating the effects of expansion. In such a theory the density in the universe would not be higher in the past than it is at present. From the data that I have presented here it seems likely that the idea will now have to be discarded, at any rate in the form it has become widely known—the steady-state universe." Hoyle had already worked out an alternative to both the big bang and the original steady-state theory which prevented a "collapse of physics" (see section 7.3). After having reconsidered the situation he soon reached the conclusion that the cosmic microwave radiation did not, after all, necessitate a break with steady-state ideas.

In the first couple of years after the discovery by Penzias and Wilson, doubts could still be raised if the radiation was really isotropic and blackbody distributed, but already with the Roll-Wilkinson confirmation of 1966 this seemed somewhat contrived to most experts. If the background radiation were to be explained by theories of the steady-state type, it had to be an explanation that referred to physical mechanisms that operate at present, and which result in an isotropic, blackbody-distributed radiation of the right temperature. One type of explanation built on the concept of thermalization. This idea had its roots in Hoyle's observation that the energy density of all the starlight in our galaxy is of the same order of magnitude as that of the 3 K radiation, about 10^{-20} J·cm^{-3}. Moreover, the cosmic rays were also found to have approximately the same energy density, which also turned up in the energy balance from the conversion of hydrogen to helium for all galaxies. Could this be purely coincidental? Hoyle thought not. If there existed a mechanism through which a major part of the radiation produced by the galaxies became thermalized—that is, converted into long wavelengths of a blackbody-like shape—then the coincidences would become understandable and an explanation for the 3 K radiation would be established. Together with Chandra Wickramasinghe, a twenty-eight-year-old colleague at the new Institute of Theoretical Astronomy in Cambridge, Hoyle suggested that interstellar (and intergalactic) dust particles of suitable composition might well act as thermalizers.[128] The hypothesis of interstellar, thermalizing grains may seem unacceptably ad hoc, but in fact it was not invented by Hoyle and Wickramasinghe as a way to save the steady-state theory. The two physicists had earlier studied the astronomical significance of graphite grains. In 1962 they argued that the existence of such grains would lead to a better explanation of the observed interstellar extinction, and also that they would act as efficient catalysts in the production of interstellar hydrogen and other molecules.[129]

In 1967 there was evidence for tiny interstellar grains consisting of a graphite core surrounded by an ice mantle, and if such grains were assumed to include impurity atoms they would absorb light and reemit it in the far infrared region, and in this way produce a microwave background of temperature about 3 K. The idea was developed by Hoyle, Wickramasinghe, and Narlikar, who considered it a possible rescue of the steady-state theory: "Our own view is that the 'black body' explanation has been overstated," wrote Narlikar and Wickramasinghe in 1968; "it is also premature to conclude from the present observations that the steady state theory of the universe is untenable."[130] Many years later Hoyle claimed that "There has never been a difficulty in the steady-state theory over the energy-density of the cosmic microwave background."[131] Perhaps not, but unfortunately for Hoyle and his few followers, the majority of astrophysicists disagreed. The hypothesis of thermalization by interstellar grains attracted little interest and even less respect. The large majority of astrophysicists, perfectly satisfied with the big-bang explanation, just ignored it. Apart from the rather forced and ad hoc nature of the explanation—no interstellar dust grains of the required composition had been found—it was pointed out by the critics that the interpretation was unlikely to reproduce the observed 3 K radiation data.[132]

Apart from the thermalization hypothesis, which has continued to be developed up to the present, there were in the late 1960s also a few other suggestions to avoid the big-bang interpretation. Couldn't the background be the result of the summed flux of discrete radio sources—and thus be of no obvious cosmological significance? Apparently not, for the flux from the known sources simply did not agree with the 3 K observations. But then one could speculate about radio sources with entirely new properties and fit these so as to account for the microwave radiation. This idea was pursued by Sciama who, in the summer of 1966, suggested the existence of a new kind of radio sources which collectively would produce the observed background radiation. Sciama reasoned that the excitation of CN molecules did not necessarily require a thermal photon bath of temperature about 3 K, but might also be the result of collisions between protons and electrons in intergalactic space. This was Sciama's last defense of the steady-state theory, which he abandoned a couple of months later, under the impact of quasar counts. "If my explanation is correct," he wrote, "the existence of the microwave background would be consistent with the steady-state model of the universe." However, this would require the hypothetical sources to be about three thousand times as numerous as normal galaxies, and Sciama recognized his proposal to be "entirely speculative." A speculation along the same line was proposed by Gold and Franco Pacini in 1968, but neither this nor Sciama's idea was followed up.[133] Most astrophysicists undoubtedly considered such exercises artificial and unnecessary. Geoffrey Burbidge and Arthur Wolfe also investigated the possibility of the background radiation being produced by a cosmic distribution of discrete sources. They found that under certain assumptions it was just possible to construct a discrete model that reproduced the observations in accordance with the steady-state theory.[134] However, even then the radiation would not be precisely blackbody distributed and isotropic. This and other discrete-source hypotheses did not fare any better than the suggestions of Sciama and of Gold and Pacini.

There is no doubt that, in the minds of the overwhelming majority of physicists, the discovery of the 3 K radiation, and its subsequent confirmation and extension to other wavelengths, constituted an impressive argument in favor of the big-bang theory of the universe. It did not amount to a strict proof, but in any practical (i.e., sociological) sense it amounted to a proof that the steady-state theory was just dead. The fact that this theory was still defended by a few astrophysicists did not change the situation from a broad, social point of view. But from other points of view it is interesting to follow, if only incompletely, the attempts to revive the steady-state theory and avoid the big-bang conclusion. Although these attempts were certainly influenced by the difficulties in making the original steady-state model agree with the increasing number of contradictory observations, they were more than just adaptations to the experimental situation. They were also rooted in new theoretical viewpoints that had nothing to do with the accelerated progress in observational cosmology.

7.3 HOYLE'S MANY ALTERNATIVES

From about 1960, steady-state advocates increasingly realized the need for a reexamination of the theory which went beyond special modifications introduced in order to cope with new observational data. In spite of several proposals made during the 1950s, the theory was basically the same as originally suggested in 1948. In order to breathe new life into the steady-state theory, some of its proponents found it necessary to reconsider its foundation, and they were willing to change the theory quite considerably. Some of the proposals differed drastically from the original steady-state theory. Still, they all belonged to the steady-state tradition insofar as they operated with matter creation and avoided a unique origin of the universe. The proliferation of steady-state versions in the 1960s was embarrassing to those who continued to keep a faith in the deductive Bondi-Gold approach based on the perfect cosmological principle. Purists took a pride in this approach being unique— that it resulted in a definite, easily testable model of the universe—and now it appeared that the steady-state assumption, contrary to the original intentions, was developed into a confusing variety of theories. The appealing steady-state methodology was betrayed, transformed into a series of theories protected from refutation—into the kind of theoretical mess they had always criticized in the relativistic evolution theory.

To Bondi and Gold this was a disappointing development, which soon made them consider cosmology a less attractive subject. In part as a consequence of this development (and presumably also under the impact of the observational challenges), the two steady-state pioneers' interest in cosmology waned. They turned, successfully, to other aspects of physics and astronomy, and left the cosmological scene to the indefatigable Hoyle in particular. In general, the early 1960s witnessed an almost complete dissolution of the discipline and coherence that had characterized steady-state theory in its earlier phase and made it a tough competitor to the evolutionary models.

McCrea was one steady-state advocate who felt a need to reexamine the fundamental hypotheses of the theory. It, and cosmology in general, had turned out to lead to little progress, and McCrea believed that the situation called for some conceptually innovative idea. For this reason he made various suggestions of a philosophical nature to introduce new general principles into cosmology, but none of them survived for long. In 1964 he decided that the time was ripe to question the traditional assumption of steady-state cosmology, going back to 1948, that matter was created uniformly through space. He suggested instead what later was called multiplicative creation, i.e., that continual creation of matter takes place near existing matter, which was postulated to have the capacity to form fresh matter.[135] This meant that all matter must be assumed to be located in galaxies, and none or very little in intergalactic space. To avoid the consequences of ever-growing galaxies, he suggested that occasionally galaxies would eject fragments that would become nuclei or embryos for the growth of new galaxies. In this way a steady-state universe could be preserved, and one which, McCrea believed, was conceptually simpler and in better accord with ordinary physics. Inspired by McCrea's suggestion, the American physicist Richard Stothers proposed that matter is created in massive bursts, and that quasars are the manifestations of this instantaneous creation.[136] Contrary to McCrea, Stothers suggested that the new matter was created in the low-density regions of the universe. As he pointed out, his cosmic-explosion proposal was in general agreement with the Hoyle-Tayler hypothesis of mini big bangs as possible seats of helium production.

McCrea's attempt to reformulate steady-state theory was tentative and purely qualitative. It was not developed quantitatively, and to many astronomers it may merely have signified the dissolution of the steady-state position. The idea of multiplicative creation of matter was not dead, though. It turned up again in steady-state theory in the late 1960s, and some years later also in Dirac's development of a cosmology with a varying gravitational contant, but then in rather different contexts. In 1964, McCrea's idea was taken up by two physicists at King's College, I. W. Roxburgh and P. G. Saffman, who used it to reexamine the possibility of galaxy formation in a steady-state universe.[137] As previously mentioned, this problem was unsolved in both big-bang and steady-state theories. In contrast to Harwit's earlier conclusion, Roxburgh and Saffman argued that it was possible to have gravitational condensations of a form agreeing with the steady-state distribution of galaxies. However, to obtain this result they had to assume the existence of initial instabilities. These could be arbitrarily small, but not zero, and the two physicists thus did not solve the classical problem of explaining the formation of galaxies without postulating special initial states.

The attempts to resuscitate steady-state theory were not very successful. They attracted little interest among the majority of physicists and astronomers, who never really considered the new theories as serious candidates for the real universe. The ingenious works of Hoyle, Narlikar, and a few others seemed of little importance compared with the evidence against steady-state theory coming from radio source counts, quasar redshifts, and the cosmic microwave background. Because of this lack of impact, and because much of the work of Hoyle and Narlikar was only indirectly related to cosmology, we shall only give a rather cursory treatment of these theoretical developments.

C-Field Theory

In Hoyle's theory of 1948–49, a noncovariant form was adopted with the argument that no fully covariant theory could account for the preferred direction which was defined at every point of space-time as a result of the continual creation of matter. This was widely seen as an unsatisfactory feature, and in the summer of 1959 Hoyle found a covariant law of creation that permanently maintained an isotropic and homogeneous state of the universe.[138] Extending McCrea's approach of 1951, he wrote the energy-momentum tensor as a sum of four terms, including not only the creation field, but also a nuclear field tensor (in addition to the gravitational and electromagnetic terms). As in McCrea's theory, the form of Einstein's cosmological field equations was not changed. The new element was the creation field, which Hoyle for simplicity took to be a massless scalar field and from which the creation tensor C_{mn} could be derived. "This field presumably arises, if it exists at all, from the microscopic processes of fundamental physics," he wrote. "Although at present nothing is known of what these processes might be, it is reasonable to suppose that the source of the field would be proportional to the mass density ρ." Developing this idea, Hoyle found a set of cosmological equations which, for a steady-state solution, gave the result that $\rho = 3H^2/8\pi G$, i.e., the very same relationship between mass density and the Hubble constant as in the earlier version.

Hoyle's new formulation was mathematically satisfactory in that it presented steady-state theory covariantly, in close harmony with general relativity, but from an observational point of view it contained nothing new. As we have noticed previously, Hoyle's approach to cosmology differed considerably from that of Bondi and Gold, and the new theory underlined Hoyle's willingness to continue along a path that separated steady-state theory from its original foundation. Thus, it opened up the possibility of a secular drift of the Hubble constant, that is, a universe in which this constant changes slowly in time. This was, of course, at variance with the original steady-state idea, in which the Hubble constant was a true constant, but to the pragmatic Hoyle it was a satisfactory feature:

[Apart] from the overriding observational criterion . . . the "best" system of cosmology is the one that admits of the greatest variety of physical possibilities. If this criterion is admitted, a universe with slow drift is to be preferred to an entirely steady-state universe at fixed H [Hubble constant]. In the latter case only one unique set of properties can be worked through, namely those that correspond to the one unique value of H. In the case of secular drift, on the other hand, an infinity of different sets of properties can be examined. . . . Defining H^{-1} as the length of a "generation," with H always taking its instantaneous value, it is possible if the drift were sufficiently slow to proceed backwards along the time-axis through an infinity of generations and yet for the universe never exactly to have repeated itself in any two of these generations.[139]

Clearly, this was a conclusion that could not appeal to Bondi and Gold. What was adrift was not only the value of the Hubble constant, but the steady-state theory itself.

Hoyle's formulation of 1960 was criticized by Bonnor and McVittie, who argued that it was considerably weaker than the formalism of ordinary general relativity, and that Hoyle's field equations did not uniquely determine the motion of matter. According to McCrea and his student R. L. Agacy, this was a misplaced criticism, but they admitted that Hoyle's theory "seems not to offer any fresh physical insight," and that its main qualification was to offer a convenient mathematical formalism for the concept of continual creation of matter.[140] The theory was also examined by the Indian astrophysicists Amalkumar Raychaudhuri and S. Banerji, who found that it allowed for both expanding and contracting space. However, in the latter case the density would continually decrease, leading to an empty world. "One seems to have an explanation as to why the universe is found expanding rather than contracting," they commented.[141]

Hoyle's new formulation soon gave rise to further developments, almost all of them due to Hoyle, either by himself or, more often, in collaboration with Narlikar. At a meeting of the Royal Society on the state of relativity theory in 1962, the two physicists presented a new version, which they developed in various directions throughout the 1960s.[142] The new C-field formulation was based on an action principle in accordance with the general theory of relativity, but a term was added to the action corresponding to the creation of matter. That this was possible was first shown by Maurice Pryce, the British physicist who back in the late 1930s had served as Hoyle's supervisor. Rather than publishing the result himself, Pryce communicated it to Hoyle and Narlikar, who gave him full credit for his contribution. From Pryce's action there followed a set of cosmological equations, including a steady-state solution corresponding to a nonvanishing C field. Hoyle and Narlikar showed that in this state the universe would expand exponentially, and that the density of matter and the creation rate were given by $3H^2/4\pi G$ and $9H^3/4\pi G$, respectively. Thus there was the same relationship between creation rate and density as in the original theory, namely, creation rate $= 3H\rho$. Since the Hubble constant could be expressed in terms of the coupling constant of the C field, important empirical quantities such as the rate of expansion of the universe and the mean density of matter were expressed by the elementary creation process. However, since the value of the coupling constant could not be determined independently, this result, like the C field itself, was a postulate of no real empirical significance. "The essential point," wrote Hoyle and Narlikar, "is that creation of matter keeps the universe going, it keeps ρ at a more or less constant level and the initial non-homogeneity and non-isotropy is expanded away, over specified physical distances." This was in contrast to the relativistic standard cosmology, where homogeneity and isotropy were arbitrarily imposed as boundary conditions in the early universe.

The Hoyle-Narlikar C-field cosmology put the much discussed question of matter creation versus violation of energy conservation in a new perspective. In the new formulation there was strict conservation of energy (and momentum) and this would, Hoyle and Narlikar believed, put an end to the criticism of steady-state theory as being based on an unexplained origin of matter. The physical picture allowing matter creation without violation of energy conservation was this: at the creation point the mass-energy of the created particle was compensated by the negative energy of the C field. Similarly, because the C field has a negative energy it is able to produce a repulsive gravitational effect which drives the expansion of the universe. The introduction of such a mechanism, involving new particles or fields with negative energy, had precedents in McCrea's and Pirani's earlier theories, but it was the first time it was built into a coherent, quantitative theory. As shown by Hoyle and Narlikar, it had as a consequence that supermassive bodies, such as those considered by Hoyle and Fowler, would not collapse into a singularity (black hole), as they would according to general relativity. The negative energy density of the C field would produce a repulsive effect sufficiently strong to overcome the gravitational attraction, with the result that the body would pulsate and never end in a singularity.[143]

From a methodological point of view, the C-field theory can be seen as the completion of the program that Hoyle had had in mind for steady-state theory since its very beginning. This program contrasted with the more deductivist aspirations of Bondi and Gold, and approached the methods of ordinary relativistic cosmology. Hoyle and Narlikar had no use for the perfect cosmological principle, and their universe was one which had *developed* into a steady state; it had not been so always and was not so everywhere. Their new cosmic scenario included some initial state of the universe in which a perturbation in the C field was introduced, and Hoyle and Narlikar showed that this situation would asymptotically produce the present steady-state universe. At the discussion meeting of the Royal Society, this departure from traditional steady-state ideas led Richard Schlegel to ask if the philosophical appeal of the Bondi-Gold theory was not thereby lost, to which Hoyle replied: "It is a question of from what point of view you feel the aesthetic appeal. I feel that more in the equations. I feel that if I can see a set of equations where a small perturbation will make one go into a steady-state solution, I like that better. That is to me more aesthetic." To which Bondi, who presided over the meeting, commented: "We do not all agree."[144]

The *C*-field theory was developed by Hoyle and Narlikar in a large number of wide-ranging papers, only a few of which were primarily of cosmological significance. In collaboration with the Burbidges, Hoyle applied the theory to galaxy formation, with a result that differed considerably from earlier applications of the steady-state theory to this problem.[145] Rather than being formed continually at a uniform rate, Hoyle found that the formation of galaxies was confined to certain episodes, so that galaxies would appear with ages t, $t + T$, $t + 2T$, etc., where t is the time interval that has elapsed since the last episode of galaxy formation and T is the average duration of an unspecified cosmic cycle. By comparing with observations of peculiar galaxies, Hoyle and the Burbidges found evidence for such episodic galaxy formation, disagreeing with both evolutionary cosmology and the traditional steady-state cosmology. Hoyle used the opportunity to spell out again the difference between his new steady-state theory and the old one:

> According to the view of one of the present authors [Hoyle], the "steady" aspect of the steady-state cosmology is a by-product of the theory, not its primary postulate. The primary postulate is that the world lines are half-lines, not complete lines. . . . Although the stationary character of the resulting line element and the constancy of the intergalactic density can be taken as justifying the name "steady-state," the adoption of this name can cause difficulties, particularly in the status of the theory vis-à-vis the perfect cosmological principle. The latter asserts that all average properties of the universe are time-invariant as well as space-invariant. . . . What determines the size of the region throughout one must average? The above considerations suggest that, for some properties, no finite volume of space may be adequate. To the somewhat pallid objection that the theory thereby loses much of its attractive simplicity, we would answer that nothing known to us in physics or astronomy is simple and that we see no reason at all why phenomena on a large scale should be simple.[146]

During the early years of the 1960s Hoyle and Narlikar spent much effort in investigating electromagnetic theory within a direct-particle-interaction framework (see below), and in 1964 they took this approach over into gravitation and cosmology. Their "new theory of gravitation" was equivalent to that of Einstein for all observable phenomena, but it had certain theoretically satisfactory features which went beyond standard gravitation theory. For example, it required a positive gravitational constant, whereas $G > 0$ is introduced empirically in the ordinary theory, that is, to account for the fact that bodies gravitate rather than repel; moreover, in the new theory there was no room for the cosmological term (the Λ constant), which in relativity theory may be there or not. Like the earlier field version, the new theory led asymptotically, for t $\rightarrow \infty$, to a steady state with the same relationship between the mass density and Hubble's constant.[147]

Whereas Hoyle and Narlikar had hitherto followed the standard assumption of a smoothed-out, homogeneous universe, in 1966 they abandoned this assumption and with it the idea of matter being created uniformly through space.[148] As an alternative, they explored the possibility of discrete matter formation around existing concentrations of supermassive bodies—pockets of creation. With this idea, resembling the one that had been advocated qualitatively by McCrea two years earlier, they found the mass of a massive body to grow at a rate proportional to the square of its mass. When Hoyle first presented this idea, in a lecture of 6 September 1965, the immediate reaction of many of his listeners may have been that now Hoyle had gone too far in his speculations. Ryle, not a friend of either Hoyle or the steady-state theory, commented: "I think that after this few students will go into the steady-state in detail."[149]

Hoyle and Narlikar's idea of matter pouring out from localized sources in strong gravitational fields anticipated the idea of white holes, time-reversed versions of black holes, which at that time began to be discussed in relativistic astrophysics, first, perhaps, by Novikov.[150] Hoyle believed that the idea agreed with observations of certain quasars, which behaved in ways seemingly defying the ordinary laws of physics. This led him, in his George Darwin Lecture of 11 October 1968, to speculate about an entirely new physics allowing for such peculiarities. As Hoyle noted, his view of matter pouring into space was essentially the same as Jeans had mentioned back in 1928, in order to explain the structure of spiral galaxies.[151] Looking back on twenty years of steady-state theory, Hoyle expressed his surprise at having been unable to make other astronomers understand—or accept—the theory. Admitting that part of the reason might lie in "a lack of clarity in my exposition," he felt that it was as much due to "the emotional atmosphere which unfortunately has always interfered with a rational discussion of the theory."[152]

Even if ignored by astronomers, initially the new theory of gravitation received some interest from theoretical physicists. After all, it is not every day that a new theory of gravitation is announced. Even before publication, it was covered by *Newsweek* in an article dealing primarily with the impact of quasar research on cosmology. The journal described Hoyle as "a compact 48-year-old Briton with thick eyeglasses and a rolling Yorkshire accent . . . [who] easily occupies the position as No. 1 man in the tight little pecking order of cosmologists."[153] The new theory was found attractive by the theoretical physicist Abdus Salam, with whom Hoyle had discussed it; "Its accuracy smells right," Salam reportedly said. Robert Oppenheimer, on the other hand, dismissed the theory as "fuzzy and inconsistent." It was also dismissed by Zel'dovich, who aptly remarked that "one could agree to such a radical innovation only on esthetic grounds: ' I don't like Friedmann's solution.' "[154]

Young Stephen Hawking, just starting on his brilliant career as a theoretical astrophysicist, was at the time attracted to the theories of Hoyle and Narlikar. As a research student under Sciama, he was well acquainted with the latest developments in steady-state theory. He became particularly interested after conversations with Narlikar, who happened to occupy the office next to Hawking's at Cambridge University's Department of Applied Mathematics and Theoretical Physics. Sciama first set Hawking to work on the arrow of time in cosmology, and suggested that he study Hogarth's paper (see below). However, Hawking did not find the topic satisfactory. Although Hawking found the new Hoyle-Narlikar theory of gravitation more interesting, he also found it problematical because it seemed to exclude Robertson-Walker models and hence to be inapplicable to the real universe. As a way of overcoming this difficulty, he mentioned the possibility of negative masses (in a gravitational and inertial sense), but, as he added, "the introduction of negative masses would probably raise more difficulties than it would solve."[155] Pirani and the American physicist Stanley Deser, of Brandeis University, also subjected the Hoyle-Narlikar theory to critical analysis.[156] They argued that, contrary to the claims of Hoyle and Narlikar, the new theory was inferior to the ordinary theory of relativity. According to Pirani and Deser, the new theory did not yield an unambiguous determination of the sign of the gravitational constant, and it also seemed unable to satisfy the classical tests of the general theory of relativity; even Kepler's laws did not follow from theory, they claimed.

The mid-1960s marked a watershed in cosmology, not only because of the new observational results, but also because of theoretical innovations within the theory of general relativity. First and foremost, the question of a singular origin of the universe was reconsidered with an unexpected and, to many cosmologists, undesirable result. The question was not of much importance to the controversy between big-bang and steady-state cosmologies, where it was rather the postulated initial superdense state of matter which was in focus. Whether this state had an infinite density or merely a density of the order of the atomic nucleus was of less importance. However, indirectly and psychologically the big-bang theory would seem to be weakened if it could be proved that the initial state was really singular; for then there would be no hope of understanding the origin in ordinary physical terms, a situation often pointed out by Hoyle and his allies. As we have seen, most relativist cosmologists avoided the question, assumed that it was unsolvable, or believed that the singularity was an artifact produced by mathematical symmetry assumptions.

The problem was examined in considerable detail by the two Russian physicists Evgeny Lifshitz and Isaac Khalatnikov in the early 1960s. Restricting their investigation to standard general relativity without a cosmological constant, they concluded that the general case of an arbitrary distribution of matter and gravitation does not lead to the appearance of a singularity.[157] To most cosmologists this was a comforting conclusion, but in early 1965 the British mathematician Roger Penrose, using new (topological) mathematical methods, spoiled the comfort. He proved that a gravitationally collapsing star will inevitably end in a space-time singularity, that is, turn into a black hole. Half a year later Hawking showed that a similar result applied cosmologically, and then that the optimistic conclusion of Lifshitz and Khalatnikov was wrong. Further work by Penrose, Hawking, and others resulted in a comprehensive singularity theorem, the essence of which is: yes, a universe governed by the classical (i.e., nonquantum) general theory of relativity must necessarily have started in a space-time singularity.[158]

Bubble Universes

The move away from the standard steady-state theory reached its culmination with yet another version, which Hoyle and Narlikar correctly referred to as "a radical departure from the steady state concept." In order to explain the energy of cosmic rays and quasars, they decided in 1966 on "throwing overboard the usual framework of the steady state theory" with its gentle and uniform creation of matter. The radical departure was obtained by increasing the coupling constant of the C field by a factor of no less than about 10^{20}—and thus also the density of the universe by the same factor! This step led to an entirely different universe, the main feature of which was the idea of separate bubble universes in which the creation process was temporarily cut off and which would therefore expand much more rapidly than their surroundings.[159] The bubbles would eventually be filled with matter and C fields from the denser outside, but other evacuated regions would be formed at the same time and develop as new bubble universes. If the bubbles developed synchronously, Hoyle and Narlikar found that the universe would follow a series of expansions and contractions; not as in oscillating models, but developing around an exponential steady-state expansion (figure 7.5). The astonishing picture of our universe as a pulsating bubble was believed to offer a better framework for explaining high-energy cosmic phenomena and the formation of galaxies.[160] Although many astronomers may have considered the idea more science fiction than science, Hoyle and Narlikar took it very seriously. They expressed great confidence in the theory and affirmed that we indeed live in an evacuated localized bubble. Part of the dying steady-state research program as it was, the idea was largely ignored. However, about fifteen years later, multiple universes produced by inflating domains began to be discussed within the framework of the new inflationary-universe model.[161] Although these ideas developed independently of the Hoyle-Narlikar theory, there is a great deal of similarity between the models. This fact has often been pointed out by Hoyle and Narlikar.[162]

The general conclusion of Hoyle and Narlikar's bubble-universe model, that our observable universe is a nonrepresentative part of the universe as a whole, was not without precedents. For example, in his earlier-mentioned pulsating-universe theory of 1960 the Czechoslovakian physicist Pachner suggested that our universe is only one of many: "We shall suppose the existence of many closed universes so embedded into the 'cosmical space of higher number than four' that their hypersurfaces do not intersect each other. Since there exists no physical interaction under [*sic*] them, they are incapable of

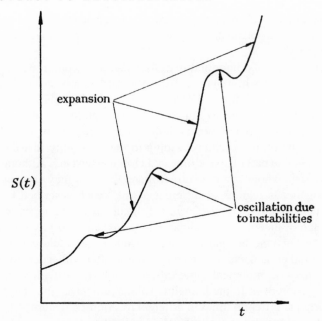

Figure 7.5. Hoyle and Narlikar's modified steady-state bubble universe with periods of expansion and contraction corresponding to different strengths of the creation process. *Source:* Hoyle and Narlikar (1966b), p. 171. Reprinted with the permission of the Royal Society.

being observed, but this does not signify that they do not exist."[163] Another early example of a bubble-universe model was suggested by R. G. Giovanelli, an Australian physicist. Like Hoyle and Narlikar, Giovanelli was led to abandon the homogeneity postulate in order to establish a self-perpetuating universe. Inspired by what he considered the aesthetic attractiveness of steady-state cosmology, he assumed that neither the observed expansion nor the mass density of our local region is representative of the entire universe. As he wrote: "It is conceivable that some regions may be expanding, some contracting (according to local observers), and that this state of affairs may reverse from time to time: in other words, the universe may be in a state of compressional oscillation, with the possibility of travelling compressional waves."[164] On this basis, Giovanelli arrived at his own strange version of the cosmological principle: "The time-averaged properties of any one region may be the same for all parts of the universe, though at any one time the properties of individual regions might differ greatly." Although Giovanelli's proposal was not a steady-state model (it did not include continual creation of matter), it was clearly inspired by the general principles of steady-state cosmology. Like the proposal of Hoyle and Narlikar, and also later ideas of bubble universes, it seemed remote from empirical confrontation. The natural question for such theories is: how does one obtain empirical information, however indirect, from "other universes"? And if the answer is—as it seems it must be—that no information can be obtained, then how can we ever know that there are such other universes?

And yet, strange and unorthodox as the idea may seem, roughly the same kind of universe can be obtained from standard relativistic cosmology. In works of 1933–34 Lemaître and Tolman, both orthodox relativists, examined nonhomogeneous solutions to Einstein's field equations. In his development of Lemaître's work, Tolman observed that in this kind of universe the curvature will depend on the position, leading to the possibility that the universe can be open in one part of space-time and closed elsewhere. Although nonhomogeneous as a whole, the Lemaître-Tolman universe may contain independent homogeneous regions of different density, each described by a Friedmann-Lemaître metric. "Hence," wrote Tolman, "it would appear wise at the present stage of theoretical development, to envisage the possibility that regions of the universe beyond the range of our present telescopes might be contracting rather than expanding and contain matter with a density and stage of evolutionary development quite different from those with which we are familiar."[165] Moreover, in the Lemaître-Tolman universe "the" big bang may have occurred at different times in different locations, a feature also found (in a different version, of course) in the Hoyle-Narlikar theory.

These speculations about separate universes—whether of Hoyle and Narlikar, Giovanelli, Dingle, or Tolman—are variations on a theme first discussed in a scientific context by Boltzmann seventy years before it was brought up in the revised steady-state theory. In 1895 the Austrian physicist noticed that the probability that one part of the universe is in a certain thermodynamic state far from equilibrium is normally very small, but that it will depend on the size of the universe. He elaborated: "If we assume the universe great enough, we can make the probability of one relatively small part being in any given state (however far from the state of thermal equilibrium), as great as we please. We can also make the probability great that, though the whole universe is in thermal equilibrium, our world is in its present state . . . If this assumption were correct, our world would return more and more to thermal equilibrium; but because the whole universe is so great, it might be probable that at some future time some other world might deviate as far from thermal equilibrium as our world does at the present."[166] Although the scientific contexts of later many-universe models differ greatly from that of Boltzmann's speculation, the basic philosophical problems are essentially the same.

Cosmology's Arrow of Time

Yet another approach to steady-state cosmology, relying on and interacting with the approaches mentioned above, was explored by Hoyle and Narlikar in the early 1960s, when they realized that studies of action-at-a-distance electrodynamics might have important cosmological consequences. This avenue had its roots basically in two interrelated sets of problems, the unidirectionality of time and the formulation of electrodynamics in terms of direct interaction between particles. Both of these problems were old, going back well before the steady-state theory was invented.

At the end of the nineteenth century, the problem of the direction of time was taken up by several physicists, first in connection with the statistical thermodynamics developed by Boltzmann and others. The problem is, essentially, that all the basic laws of physics are time symmetric and so do not distinguish between "before" and "after." And yet most natural processes evolve in one, and only one, way, which coincides with (or perhaps defines) our subjective feeling of a natural direction of time, from the past to the future. The problem that occupied some physicists was to find in the basic laws of nature the cause for this unidirectionality, which then would be explained physically. For a period thermodynamics was seen as a candidate, because entropy always increases, but the entropic theory of time turned out to be unsatisfactory. Within his probabilistic theory of entropy, Boltzmann proposed in 1897 to define the direction of time as the direction toward more probable states. If so, it would be different for different parts of the universe at different epochs. Boltzmann believed that such a view was necessary in order to avoid "a unidirectional change of the entire universe from a definite initial state to a final state."[167]

Another candidate for time's arrow, first explored by Boltzmann and Max Planck in the 1890s, was the laws of electrodynamics. However, although at first sight Maxwell's equations may seem to single out a preferred direction of time (such as an electromagnetic wave, which is first emitted and then absorbed), this is not so. To every retarded solution, describing a process going forward in time, there corresponds an advanced solution going the other way; this latter solution is normally discarded as unphysical, but this was done arbitrarily and could not be justified theoretically. Electrodynamics is, like mechanics, time symmetric, and neither relativity nor quantum theory provides the laws of nature with a satisfactory arrow of time. The same was the case with the Friedmann-Lemaître equations of the expanding universe, which were derived from relativity theory. Yet the fact that the universe expands, and does not contract, suggested that cosmology might provide time's arrow. As mentioned in section 2.1, this idea was examined by Bronstein and Landau in 1932–33. The whole question of the physical basis of the direction of time continued to be discussed during the first half of the twentieth century, but gradually it became an occupation of philosophers rather than physicists.[168]

The first formulations of electrodynamics were based on direct particle interaction with instantaneous propagation of signals (action at a distance), but with the success of Maxwell's field formulation during the latter part of the nineteenth century, this concept largely disappeared from physics. However, although Maxwell's theory is a field theory, the corresponding equations can be derived also from an action-at-a-distant standpoint. This was first done by the German physicist and astronomer Karl Schwarzschild in 1903, and in the

1920s the Dutchmen Hugo Tetrode and Adriaan Fokker developed independently a relativistically invariant theory of electrodynamics based on direct interactions. This formulation of electrodynamics makes use of advanced as well as retarded interactions, which implies that there is no electromagnetic radiation in the theory. Since electromagnetic waves do exist, this was a serious disadvantage of the direct-interaction theories. The problem was solved in the 1940s by John Wheeler and his student Richard Feynman, who realized that the problem required that cosmological boundary conditions be taken into account. In important papers of 1945 and 1949, they developed an action-at-a-distance absorber theory of electrodynamics. The essence of the Wheeler-Feynman theory was that the universe (all other particles) must respond to a signal from a certain particle in just such a way that the advanced effects cancel and problems of causality violation are thereby avoided.[169] Wheeler and Feynman were not interested in cosmology, and the universe they used as an absorber was the simplest possible one, an infinite and homogeneous static universe.

The Wheeler-Feynman theory entered cosmology through the work of J. E. Hogarth, a Canadian physicist who prepared his Ph.D. thesis under McCrea at the University of London.[170] In his thesis work of 1953, and more elaborately in an article of 1962, Hogarth examined the Wheeler-Feynman absorber theory in connection with expanding universes. Being a student of McCrea and later, in the early 1960s, attending Bondi's seminar at King's College, it is not surprising that he took a particular interest in the steady-state universe. Although Hogarth only published his analysis in 1962, it was generally known to many people before that time. In part inspired by it, the problem of a cosmological arrow of time—using the expansion of the universe to define the direction of time—and its connection to the electromagnetic arrow of time were taken up by a few cosmologists. It is worth noticing that the interest in the subject seems to have been particularly strong among, and originally limited to, physicists associated with the steady-state program. For example, Gold made it the subject of his address at the 1958 Solvay conference and also discussed it in his Richtmyer Memorial Lecture in 1962, and so did Bondi in his Halley Lecture for 1962.[171] In all three cases the message was that the large-scale motion of the universe is responsible for the arrow of time. According to Gold, the thermodynamic arrow of time was directly determined by the expansion of the universe. Neither Gold nor Bondi discussed particular cosmological models, except that Bondi argued against oscillating models.

The problem under examination, whether a source will emit or absorb radiation, can be formulated as a question of the relative influences from the rest of the universe: in order to obtain agreement with observations, i.e., to obtain retarded signals, influences from the future must be stronger than those from the past. Hogarth examined the electrodynamic arrow of time in three types of universes, the static, the evolutionary Einstein–de Sitter, and the steady-state models (which he called "the stationary density model"). In the first case he found that no arrow of time could be established and he concluded that the Einstein–de Sitter model led to advanced fields; only models with continual creation of matter would lead to observation of retarded fields.[172] This occurs because of the new matter to be created in the future that did not exist in the past. The influence from the future will be enhanced, and with a creation rate large enough it will dominate. Hogarth's calculations showed that the creation rate predicted by steady-state theory was large enough to secure retarded signals.

Hogarth did not relate his result to the cosmological controversy, but other scientists were quick to do just that. After all, if only the steady-state model gives the retarded signals that we experience, isn't this a strong argument in favor of the model? Even before the publication of Hogarth's paper, Sciama presented it as support of a continual creation of matter and hence of steady-state cosmology.[173] With Bondi acting as mediator of Hogarth's results, they were developed by Narlikar and Hoyle in 1962–64. Narlikar examined Hogarth's problem with neutrinos instead of photons, showing that in this case also the steady-state model behaved more favorably than the Einstein–de Sitter model.[174] In a generalization of Hogarth's work, Hoyle and Narlikar used its conclusions as a means of discriminating between evolution and steady-state (C-field) models. In view of the strong empirical evidence in favor of evolutionary relativistic models, their conclusion, based on a series of intricate theoretical arguments, was unusually strong, if not surprising: "The arrow of time cannot be explained in terms of the Einstein–de Sitter cosmology. Nor we believe can it be explained in other evolutionary cosmologies. . . . The implication we believe is that evolutionary cosmologies, at any rate those based on the Robertson-Walker line element, must be discarded. Our opinion is that these considerations carry more weight than the rather uncertain observational attempts that have been made to distinguish between different cosmologies."[175]

Several years later, Narlikar expressed the same conviction, that the understanding of time's arrow in cosmology was "less equivocal than the observational data" and a strong indication of the superiority of the steady-state model over evolutionary cosmologies. "My own interest in the steady state cosmology has sustained over the years," he wrote, "not because of the uncertain observational situation, but because of a remarkable physical property the theory has [viz., its ability to explain the arrow of time]." He considered this "a powerful test of cosmological theories" because "it has the merit of being clear cut and free from . . . observational uncertainties."[176]

If Hoyle and Narlikar believed that the question of the arrow of time would have a major effect on the cosmological controversy, they were disappointed. Their challenge was not answered by big-bang theoreticians, and most advocates of the steady-state theory who dealt with the question (including Gold, Sciama, McCrea, Pirani, and Bondi) showed little interest in using it in the cosmological controversy, which, in any case, they were losing interest in. Working with the conventional Maxwell field theory, Sciama was led to a different conclusion than Hoyle and Narlikar, namely, that the Einstein–de

Sitter model can in some cases produce the desired retarded signals and thus is acceptable from the point of view of temporal directionality.[177] Perhaps the only support to the Hoyle-Narlikar claim came from P. E. Roe, a Cambridge astrophysicist, who in 1969 concluded that the steady-state theory was the only simple cosmological model with a cosmological time agreeing with the electrodynamic one.[178] To almost all other astronomers and physicists "the rather uncertain observational attempts" were far more impressive evidence than the arrow of time. They considered the steady-state theory to be refuted by radio astronomical observations, the cosmic background radiation, and the helium content of the universe, and saw no reason to take up the complex action-at-a-distance theories of Hoyle and Narlikar.

The question of the cosmological context of the direction of time was discussed at a meeting held at Cornell University in the spring of 1963 and organized by Gold. At this meeting, Hogarth, Gold, Bondi, Hoyle, Narlikar, and Sciama discussed the cosmological time's arrow with other theoretical physicists, including Feynman, Wheeler, Penrose, Chandrasekhar, and Léon Rosenfeld.[179] It was symptomatic of the interest in the subject that no big-bang cosmologists participated. Feynman's participation deserves a comment: in the proceedings of the conference there appears a sceptical Mr. X, who does not believe that the electromagnetic arrow of time comes from an action-at-a-distance formulation of electrodynamics, but from ordinary statistical mechanics. It was an open secret that Mr. X was Feynman, and that he did not want to have his name officially associated with the conference. Although Feynman was never seriously interested in cosmology, he did have an interest in general relativity and a good knowledge of the situation in cosmology. This is shown by a lecture series on gravitation physics he gave at Caltech in the academic year 1962–63, one among many signs of the revitalization of this area of science. Feynman was highly critical of the steady-state theory, which he found ingenious but artificial. Referring to Hoyle's 1948 version, he stated that it is "against the rules to explain a result [the universal expansion] by making a convenient change in the theory."[180] In other words, Feynman shared the view of several other physicists and astronomers, that the steady-state theory was ad hoc. This does not mean that Feynman was an uncritical big-bang supporter or that he did not understand the background of the steady-state theory. This background was the fundamental question of how the matter of the universe has originated. As Feynman realized, "other cosmological theories sweep creation of matter under the rug by simply assuming a time at which the matter was already there, . . . [so] the steady-state theory can hardly be accused on being unreasonable on this point."[181]

The efforts of Hoyle and Narlikar to reformulate steady-state theory did not stop with the works mentioned here. They continued to develop their theories and proposed several new modifications during the 1970s. Among these was a version based on Dirac's large-number hypothesis, which incorporated several features of Dirac's cosmology, including multiplicative creation of matter and a gravitational constant decreasing in time (in atomic units).[182] The search for connections between cosmology and local physics had always been a strong motive in Hoyle's research program; it was, for example, clearly stated in his original article of 1948. That such connections were indicated by the apparently coincidental values of combinations of natural constants was an idea which Hoyle had taken over from Eddington and Dirac. As Hoyle increasingly focused on the physics-cosmology frontier, this line of thought came to the forefront of his and Narlikar's thinking. To mention but one example, in 1968 he suggested in a true Eddingtonian spirit that the relationship

$$\frac{c}{H} = \frac{e^2}{GmM} \frac{4\pi e^2}{mc^2} = \text{ca.} 10^{26} \text{m}$$

was satisfied—not only approximately, but *exactly*.[183] However, these later developments of the Hoyle-Narlikar program had almost no influence on the main course of cosmology, which in the 1970s was completely dominated by the reborn big-bang theory.

7.4 THE TERMINATION OF THE CONTROVERSY

In a practical sense, the controversy between steady-state and big-bang cosmologies came to an end during the last part of the 1960s. However, it did not stop completely or abruptly, and there was no event, theoretical or observational, which qualifies as decisive in this respect. Rather, the controversy faded out in the sense that the now standard hot big-bang model became the nearly undisputed new paradigm of cosmology, and the new generation of cosmologists stopped worrying (or even knowing) about the steady-state theory. As we have seen, during the years 1960–66 there was a strong trend in favor of the big-bang model, consisting mainly in new observations, and, since 1964–65, a corresponding renaissance of the physical big-bang program of Gamow and his collaborators. However, none of the observations, either separately or collectively, were able to settle the controversy definitively. In particular, the discovery of the cosmic microwave background, although admittedly of very great importance, was not quite the crucial experiment it has often been claimed to be.

Capitulation?

The discoveries during the first half of the 1960s had an impact on the controversy in two ways, which together amounted to a nearly complete victory for the big-bang view. First, the discoveries seemed to the large majority of astronomers and physicists to be inconsistent with the steady-state view, and so they marginalized the steady-state camp; those who still defended a nonevolutionary universe were increasingly seen as outsiders, defending their cause by means of artificial theoretical arguments and ad hoc hypotheses. Second, the buildup of a strong cause for the big-bang universe resulted in an internal dissolution of the steady-state front, which became complete with the series of theories produced by Hoyle and Narlikar. The spirit of these theories went against the original steady-state theory and was not acceptable to many of those whose sympathy for the steady-state theory was rooted in the methodological virtues of the original (Bondi-Gold) theory. Bondi made his position clear at the 1964 Padova conference, when he pointed out that the Hoyle-Narlikar theory was not a proper steady-state theory at all. According to Bondi, the new theory had neither the testability nor the logical consistency of the original steady-state theory, and for this reason he could see no justification for putting the theory forward.[184]

The accumulated effect of this development was that most of the steady-state advocates gave up the theory, either explicitly or, more often, implicitly. After 1966, it was largely left to Hoyle and Narlikar to defend the steady-state theory, and that in versions which had rather little to do with what traditionally was associated with the theory. The situation about 1970 was that the number of scientists actively advocating some kind of steady-state alternative had shrunk to two—Hoyle and Narlikar—or perhaps one or two more. The extreme smallness of the steady-state population at that time does not automatically mean that the controversy had ended, for in some cases scientific controversies may go on with only a single scientist challenging the rest of the professional community. What matters is the response of the community, the mainstream scientists. If they feel the challengers' views sufficiently interesting or provocative to engage in a discussion with them (or for whatever other reason), then the controversy may go on. But if there is practically no mutual communication between the two parties—if it is restricted to one-way criticism with no responses—it is no longer reasonable to speak of the disagreement as a controversy. By and large, the Hoyle-Narlikar theories seem to belong to the latter category. The alternatives to the big-bang theory remained very much isolated from other work in relativistic cosmology, and so we may conclude that the controversy had indeed terminated at the latest by about 1970. The disagreement, however, was still there.

Hoyle and Narlikar were the most important of the steady-state physicists in the 1960s, but they were not alone in defending the idea of an eternal universe with continual creation of matter. T. Y. Thomas, an American physicist, argued in 1966 that continual creation of matter is possible within the standard framework of the general theory of relativity, and he concluded that "this result strongly supports the Bondi-Gold postulate of the continual creation of mass in space."[185] To mention another example, the Yugoslavian scientist A. O. Zupančič claimed to have explained steady-state matter creation by means of an unorthodox, Eddington-inspired interpretation of the uncertainty relations.[186] However, these and a few other works related to the steady-state theory had no connection to the Hoyle-Narlikar theory and were of no importance to the dying controversy.

When Hoyle delivered his Bakerian Lecture of 1968, he was well aware that the steady-state theory had lost most of its power and former fascination. Reviewing the crises that had faced the theory over the past twenty years, he nonetheless concluded that he and the other steady-state theoreticians had done a very good job indeed. "I think it is fair to say," he said, "that the theory has demonstrated strong survival qualities, which is what one should properly look for in a theory. There is a close parallel between theory and observation on the one hand and mutations and natural selection on the other. Theory supplies the mutations, observations provides the natural selection. Theories are never proved right. The best they can do is to survive."[187] But did the steady-state theory really survive at that time? In the sense that it was still defended by a few scientists, it did. But if we are referring to the community of astronomers and physicists working in cosmology, it was scarcely the case. After all, mere survival is not a sign of scientific health, which not only requires progress in theory and increasing empirical content, but also (and related to these qualities) that the theory is accepted, or, as a minimum, found interesting, by a substantial part of the scientific community. In this sense the health of the steady-state theory was rapidly deteriorating during the 1970s. Yet, gravely ill as the theory was, it remained alive.

There soon developed among mainstream cosmologists a widespread feeling that a truly scientific cosmology only emerged about 1966, with the disappearance of the steady-state rival and the fusion of theory and observation within the big-bang picture. Sciama spoke about the "renaissance of observational cosmology," which he saw as brought about by the calculations of the helium content of the universe and the discovery of the cosmic background radiation in particular.[188] According to R. J. Weymann, the microwave background completely transformed cosmology. "It is clear," he wrote in 1967, "that the detection of this radiation and the many investigations stemming from it are only the introduction to a very exciting new area which may fairly be called the *science*, rather than the *art* of cosmology."[189] The origin myth of scientific cosmology was repeated on 11 December 1978, when Penzias and Wilson were awarded the Nobel Prize "for their discovery of cosmic microwave background radiation." In his presentation speech the Swedish physicist Lamek Hulthén praised Penzias and Wilson for their discovery, "after which cosmology is a science, open to verification by experiment and observation."[190] Whatever the validity of this view—and it is questionable—it is a fact that many physicists and astronomers felt that a new era had begun in cosmology in 1965, and that they identified the new era with the big-bang theory.

With regard to the Nobel Prize, it is, of course, somewhat surprising that Penzias and Wilson received the award for a "discovery" of something they did not understand nor were able to intrepret correctly. But then one should keep in mind that the Nobel institution, in agreement with Alfred Nobel's will of 1896, has always been predisposed toward this kind of purely explorative or phenomenal discovery. Thus it may be useful to compare the Penzias-

Wilson discovery with the discovery of fission in 1938: Otto Hahn and Fritz Strassmann found that neutron-irradiated uranium contained barium, and it was for this observation Hahn was awarded the prestigious prize. Hahn and Strassmann simply did not know what was going on, and it was only after the fission hypothesis of Lise Meitner and her nephew Robert Frisch in late 1938 that they realized they had split the uranium nucleus. The analogy between, on the one hand, Penzias and Wilson, and Hahn and Strassmann, and, on the other, between Meitner and Frisch, and Dicke and Peebles, is clear. Meitner, Frisch, Dicke, and Peebles did not receive the Nobel Prize. As far as the justification of the prize for the microwave background is concerned, it will be the year 2023 before the Nobel archives are opened and then can provide us with the necessary information about the arguments behind the decision.

Most review articles from the late 1960s differed from earlier ones by focusing narrowly on the big-bang theory and leaving out the more critical and philosophically oriented aspects that had characterized the reviews of, say, Bondi and McCrea. A review article by Zel'dovich and Novikov is characteristic in this respect. Written in the fall of 1966, it briefly dismissed steady-state cosmology and other alternatives. According to the two Russians, the Friedman-Lemaître theory was "definitely superior to the steady-state or C-field theories" because it rested on known laws of physics. Referring to the cosmic microwave background observations, they concluded that "the hot [big-bang] model seems inescapable."[191] The term "seems" (rather than "is") should not be taken as a possible concession to the steady-state theory, which, in the view of Zel'dovich and Novikov, was out of the question. It referred to the cold big bang, the hypothesis advocated by Zel'dovich a few years earlier.

Similarly, the textbooks which began to appear in the 1970s identified cosmology—often named *physical cosmology*, in order to dissociate it from older views—with relativistic hot big-bang theory, relegating the steady-state theory (if mentioning it at all) to brief chapters on other cosmological ideas. The working history of the new generation of cosmologists tended to place events prior to about 1965 back into a prescientific, semimythical stage, governed by either sterile mathematics or superstitious ideas.[192] Whatever the ideology of presentations, it is a fact that cosmology experienced a kind of takeoff in this period. This is evidenced by bibliographic data, which indicate a strong quantitative growth after about 1965. The annual number of scientific articles on cosmology increased between 1962 and 1972 from about 50 to 250 (figure 7.6). On the other hand, impressive as this growth is, quantitative data alone do not provide sufficient evidence for a revolution in cosmology in the mid-1960s.[193] Moreover, as a percentage of the total number of abstracted articles, the cosmology articles only made up a modest 0.3%, less than half the percentage of the 1930s. The growth in the cosmology literature may also be compared with the development in the number of active astronomers, which for the years between 1900 and 1940 was about 1800 and during the 1950s increased from about 2000 to 3500.[194] Revolution or not, cosmology was still a very small science in 1970.

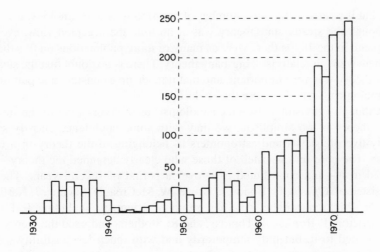

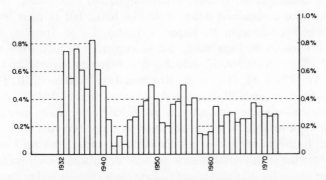

Figure 7.6. The development of the scientific literature on cosmology, based on the annual number of publications listed under "cosmology" and "cosmogony" in *Physics Abstracts*. The upper figure shows the absolute number of publications, the lower one the percentage of the total number of publications abstracted in *Physics Abstracts*. *Source:* Ryan and Shepley (1976), p. 223. Reprinted with the permission of the American Institute of Physics.

It is evidently difficult to estimate the relative strength or popularity of the steady-state theory in the period up to about 1960. Usually scientists do not declare themselves supporters or enemies of a theory, but merely discuss it in a more or less objective way. During the period treated here (ca. 1948–66) about forty physicists and astronomers contributed to various aspects of steady-state theory, most of whom should probably be counted as either supporters or sympathizers, if only for a time. This may seem only a small number, but if related to the total number of contributions to cosmology, it is not. Although the steady-state theory was at no time the accepted view, even in England, in the 1950s there were as many or more publications on this theory as there were on relativistic big-bang theory. There is no doubt that the steady-state theory was very important, and that it made up a considerable part of the cosmological scene for more than a decade.

Although a division between cosmologists as in favor or not of the steady-state theory is too simplistic, we may with some confidence classify some thirty-five physicists and astronomers as belonging to the theory in a wide sense. The core set consisted of those who clearly defended the theory—believed in it—and contributed to it over a period of several years. The set consisted of T. Gold, H. Bondi, F. Hoyle, W. McCrea, D. Sciama, J. Narlikar, and N. Wickramasinghe. Another group of scientists, including F. Pirani, R. Lyttleton, P. Roman, T. Thomas, and R. Stothers, endorsed the theory and contributed to it, but only temporarily and with much less visibility. Other scientists expressed sympathy for the theory in general and examined it in one or two works, but without supporting it in any strong sense. In this group I count D. Littlewood, M. Johnson, J. Dungey, H. Spencer Jones, J. Hogarth, P. Roe, G. Burbidge, W. Fowler, I. Roxburgh, and P. Saffman. Then there were those who entertained steady-state-like ideas, but in their own, often very unorthodox versions: R. Kapp, V. Bailey, C. de Turville, T. Araki, A. Gião, J. Pachner, R. Giovanelli, and A. Zupančič. Finally, some scientists examined aspects of the steady-state theory in either a noncommittal or a critical way: H. Nariai, M. Harwit, A. Raychaudhuri, S. Banerji, R. D. Kemp, W. Davidson, J. Hunter, W. Bonnor, A. Braccesi, and G. Velo.[195] Altogether I have counted forty scientists actively interested in the steady-state theory, of which it may be reasonable to classify thirty as supporters (but not necessarily believers). In this count I have not included scientists who we know had a sympathy for the steady-state theory but who never stated their preference in public. For example, W. Rindler and R. Penrose were attached to the theory at least in a loose sense. However, since they never clearly said so, and since their references to cosmology were rather peripheral, it is not reasonable to include them in the group of scientists associated with the steady-state theory.

The abandonment of the steady-state theory can be illustrated by statistical data from polls conducted among astronomers in 1959 and 1980 (table 7.3).[196] Given that most of the astronomers participating in the polls were Americans, a surprisingly large number favored the steady-state theory in 1959 (but then the statistics are meager). I am not aware of a single American astronomer who clearly defended the steady-state theory in public. The shift toward the big bang from 1959 to 1980 is, of course, marked and expected, but it may be relevant to point out that in 1980 the unpopularity of steady-state theory was not quite matched by a corresponding popularity of the big-bang theory. There is little doubt that had the 1980 poll been restricted to astronomers active in cosmological research, and had it not covered mainly Americans, there would have been a greater disparity between the two cosmologicals models.

379

TABLE 7.3
Popularity of Cosmological Models

	1959		1980	
	Favorable (%)	Unfavorable (%)	Favorable (%)	Unfavorable (%)
Big bang	33	36	69	7
Steady state	24	55	2	91
No. of responses	33		308	

Why and when did the steady-state advocates give up their defense of what came to be seen as a lost cause?[197] Only very few of them actually converted to the big-bang view, i.e., explicitly abandoned the steady-state theory and began working within the new paradigm. The most notable example is Sciama, who converted in the fall of 1966, mainly under the impact of the quasar number counts, which convinced him of the failure of the steady-state program. McCrea also experienced a kind of conversion, but a much more gradual and reserved one, which was restricted to his eventual public recognition of the superiority of the big-bang theory. The discovery of the microwave radiation did not make a great impression on him. He recalled that, "I myself was rather cool at first, because I thought that such a small effect could be almost anything."[198] In 1968, in a review article on the occasion of the fifty-year anniversary of Einstein's path-breaking contribution, he admitted that the simple steady-state theory was wrong and that observation seemed to fit remarkably well with the big-bang cosmology; but he was not yet willing to accept this "other extreme" because of the methodological and conceptual difficulties associated with it. One year later McCrea seemed to have given in, now accepting as a fact "that the universe started with a hot big bang."[199] Other steady-state advocates just quietly left the field of cosmology, without ever publicly admitting that the universe probably did have a violent beginning some ten or twenty billion years ago. Felix Pirani, who in the 1950s contributed to the steady-state theory and counted himself as a supporter, focused in the 1960s on gravitational radiation and other branches of mathematical physics, without engaging in the cosmological debate. His interest in cosmology faded and, he recalls, his earlier enthusiasm for the steady-state theory was almost extinguished by the discovery of the cosmic microwave background.[200]

Gold's and Bondi's farewell to steady-state theory followed a similar, undramatic course. Cosmology had never been their main occupation, and from about 1960 they drifted away from the field, dissatisfied with both the victorious big-bang theory and the turn given to the steady-state theory by Hoyle and Narlikar. With a few exceptions, they no longer contributed to the defense of steady-state theory, but their basic sympathy for the theory remained intact. There is little doubt that, at least through the 1960s, they preferred the steady-state universe in spite of recognizing the successes of the big-bang theory. Bondi defended the steady-state view in general terms in a couple of addresses, though more from a philosophical perspective than as a viable scientific alternative; and Gold sought in 1968, in his work with Pacini, to avoid the big-bang interpretation of the cosmic background radiation. Although they no longer wanted to be associated with the steady-state theory (and especially not the Hoyle-Narlikar versions), they continued to express cautious dissatisfaction with the now dominant big-bang theory. But, characteristically, this criticism mostly turned up in retrospective addresses and popular articles, not in scientific publications, and it was limited to general considerations of a methodological nature. Without saying so directly, Bondi and Gold admitted that the battle was lost; but they did it with regret and without embracing the victor. This defeatist attitude was, of course, very different from that of the fighting Hoyle, who apparently disapproved of the lack of courage shown by his former allies. In 1982, Hoyle wrote:

> There is no sense in which I would think it helpful to return to the concept of a universe that is strictly steady . . . My 1948 form of the steady state theory was rather like a phonograph record stuck in a groove, playing endlessly the same phrase, while the Bondi-Gold form of the theory was like a record that plays only one note, the sort of test record one uses to check a stereo system. . . . Yet in giving universal significance to the Hubble constant, the carrier frequency, the steady state theory of 1948 achieved a breakthrough that seems more and more relevant as the years go by. It was unfortunate that in 1965 those of us who had either worked with the theory or who viewed it with some sympathy allowed ourselves to be bamboozled into impassivity, thus permitting the astronomical world to plunge with avidity into what has come increasingly to look like a blind alley.[201]

To most astronomers, the death of the steady-state theory was a reality by 1970. They would agree with Michael Rowan-Robinson, who in 1972 wrote a brief "Steady State Obituary?," only many of them would probably have dropped the question mark.[202] But whereas the controversy between steady-state and big-bang cosmologies had de facto terminated at that time, this did not imply complete consensus. Even disregarding the efforts of Hoyle and Narlikar, the controversy continued, but now in a different form where the steady-state theory played almost no visible role. It was transformed into an asymmetric controversy over the validity of the big-bang theory, in which the critics offered no single alternative, if an alternative at all. Some of the opponents in this new version of the controversy came from the former steady-state camp (including Hoyle and Narlikar), or had been associated with it (such as Geoffrey Burbidge); but it included also astronomers and physicists with no preference for, or prior association with, the steady-state theory. This controversy, which started about 1970, on the ruins of the steady-state theory, so to speak, has continued until this very day. It is an interesting subject for the sociologist of science (more, perhaps, than for the historian of science), but it is on the periphery of the present study, for which reason I shall mention it, albeit briefly.

Continued Resistance

With the victory of the standard big-bang theory, some scientists felt that it developed into a dogma, which disregarded contradictory evidence and ignored conceptual difficulties. Comparisons with the Aristotelian world picture defended dogmatically by the church against Galileo's sound objections became commonplace. (Generally, and ironically, much of the anti-big-bang rhetoric had a striking similarity with Dingle's attack on the steady-state theory in the early 1950s.) The core of the arguments of the new opposition was observational, a continual concern with phenomena that did not fit with the big-bang theory.

An early and typical example, summarizing much of the earlier criticism, was provided by Geoffrey Burbidge in 1971.[203] Burbidge, one of Hoyle's close collaborators, was highly suspicious of the momentum of the big-bang bandwagon, but in spite of having sympathy for the steady-state theory he had never clearly supported it. Burbidge did not consider himself a cosmologist, and claimed that he was impartial with regard to the rival cosmological views, a claim which is difficult to harmonize with his publications. In accordance with Hoyle and others, he now concluded that none of the arguments traditionally claimed to vindicate the big-bang theory were beyond serious doubt. His conclusion about this unsatisfactory state was not because the steady-state theory was correct (although he considered it a possibility), but because there was very good reason to question the big-bang theory. This more agnostic view was characteristic of the opposition that evolved over the next decades. "Was there really a big bang?" Burbidge asked, and answered: "I believe that the answer clearly must be that we do not know, and that if we are ever to find an answer much more effort must be devoted to cosmological tests, with a much more openminded approach, and that much more original thinking must be done to attempt to explain the large amount of observational material, and not only that material that can be used in a narrow sense to fit preconceived ideas." Burbidge's objections were, in varying modifications, to be repeated and elaborated in many later papers by big-bang opponents.

The position of Fowler paralleled to some extent that of his friend Burbidge. Neither Burbidge nor Fowler considered themselves cosmologists, and they preferred to stay on the periphery of the cosmological debate. Whereas Burbidge later would side with Hoyle and Narlikar in their attack on the big-bang orthodoxy, Fowler was more cautious in being drawn into the controversy. Yet there is little doubt that his sympathy lay with a kind of steady-state theory, or, rather, that he considered the big-bang theory with some suspicion. In a discussion of the Fowler-Hoyle-Wagoner theory at the Weizman Institute of Science in 1967, he introduced his talk by suggesting that the big-bang idea owed its origin to the Genesis of the Old Testament: "The idea of the big bang is a very old one and, indeed, there are some who say that the idea is exactly 5727 years old. As a matter of fact there is some confusion as to whether the event itself took place or whether the idea was born 5727 years ago. However, there is no confusion as to where the idea originated."[204] At this point in his talk a slide was projected showing a satellite photograph of Israel. However, if Fowler was opposed to the big-bang theory in 1967, it neither influenced his scientific work nor caused him to join the opposition which was later organized.

The accusation that the big-bang theory was doctrinaire and quasi-religious was an old theme in Hoyle's fight against that theory. It reappeared regularly in his later writings also. In a textbook of 1980, Hoyle and Narlikar thus wrote about the doctrine of a sudden beginning of the universe: "Many people are happy to accept this position. They accept [the big-bang picture] without looking for any physical explanation of the abrupt beginning of the particles. The abrupt beginning is deliberately regarded as *meta*physical—that is, *outside* physics. The physical laws are therefore considered to break down at $\tau = 0$ *and to do so inherently*. To many people, this thought process seems highly satisfactory because a 'something' outside physics can then be introduced at $\tau = 0$. By a semantic maneuver, the word 'something' is replaced by 'god,' except that the first letter became a capital, God, in order to warn us that we must not carry the enquiry any further."[205] As we have seen, by 1980 there was nothing new in this kind of objection. It was an integral part of the steady-state opposition, and it appealed also to many people not affiliated with the opposition. For example, it seems to have been a main reason for Popper's dismissal of the big-bang theory as metaphysical or nonscientific.

In writings from the 1970s and 1980s, Hoyle claimed that the prediction of the microwave background radiation was the only, and limited, success of the big-bang theory. According to Hoyle, the fashionable theory failed in other predictions, and also when it came to fertility, i.e., in making new connections with observation, it was no good. "Over the past seventeen years," he wrote in 1982, "astronomers and physicists the world over have made numerous investigations, with the outcome essentially nil. It has been a fruitless churning of mathematical symbols, exactly the hallmark of an incorrect theory."[206] If the big-bang theory failed so miserably, why then was it so popular? Hoyle believed that the answer had to be sought in sociology and religious traditions rather than in science. The evolutionary world view and religion are traditionally seen as antagonists, but Hoyle found the struggle to be "over who should be top dog in prosecuting the same basic point of view, and in particular over who should benefit most from State patronage." He maintained that there never was a fundamental disagreement between established science and the church. "Science since Charles Darwin has gradually taken over the material support that formerly was afforded by the State to the Church. It was this that the conflict was really about."[207]

Among the most vociferous opponents of the new big-bang orthodoxy were Hannes Alfvén and Halton Arp, neither of whom can be counted as steady-state scientists. Alfvén in fact rejected the steady-state theory, and Arp did not express any particular interest in it. The Swedish physicist Alfvén was an expert in plasma physics and a Nobel Prize winner of 1970 for his fundamental contributions to one aspect of this field, magnetohydrodynamics. Developing ideas due to his senior colleague, the theoretical physicist Oskar Klein, he arrived in the early 1960s at a cosmological view which differed from both big-bang and steady-state cosmology.[208] Klein and Alfvén rejected the big-bang theory, which they found unscientific and mythical. They based their objections in part on methodological arguments similar to those that had led Bondi, Gold, and Hoyle to propose the steady-state theory. Although Klein expressed some sympathy with the perfect cosmological principle, which he found was "a natural induction from astrophysical experience,"[209] the Klein-Alfvén cosmology rejected the continual creation of matter. The two Swedes stressed that only known laws should enter cosmology—hence no creation of matter, either in a cataclysmic or a continual version. In a lecture delivered in Stockholm in 1968, Alfvén expressed his view as follows: "One is naturally rather reluctant to accept new physical laws until one feels convinced that an observed phenomenon cannot possibly be explained according to laws for which there are satisfactory proofs in other ways. This means that one should prefer a cosmology introducing no new laws of nature to one which is founded on ad hoc assumptions. As a general rule an ad hoc assumption should never be made until it is obvious that other ways of approach are excluded. From this point of view there is no reason to accept the hypothesis of a 'continuous creation.' "[210]

The characteristic feature of the cosmological view of Klein and Alfvén was that it assumed a universe with equal amounts of matter and antimatter separated by cosmic electromagnetic fields. Ordinary matter was called "koinomatter," and for a plasma of both koinomatter and antimatter Alfvén coined the name "ambiplasma." Alfvén suggested that the universe consisted originally of a very dilute and cold mixture of antiprotons and ordinary protons; due to gravitational condensation, annihilation processes would increase and produce a radiation pressure that eventually would halt the collapse and turn it into the observed expansion (figure 7.7). The change from contraction to expansion was found to take place when the ambiplasma had reached a density of about 10^4 particles per m^3. At the turning point the Hubble constant was of course zero, and Alfvén and his collaborators found that it would reach a maximum shortly afterward, and later decrease asymptotically to zero. The "plasma universe" thus included a kind of cosmic explosion, but not from a singularity and not covering the entire universe. According to Alfvén and his few followers, the incomplete collapse and its subsequent expansion were experienced only by a part of the universe, the "metagalaxy" making up our observable universe.

Although Alfvén's cosmological theory received some support from other plasma physicists, it was ignored by most astronomers and cosmologists. It was never developed into a serious alternative to the big-bang theory, but Alfvén's constant criticism of this theory, and his claim to have an alternative, received considerable attention outside mainstream cosmology. Much in line with Dingle's criticism of cosmology in the 1950s, Alfvén claimed that the big-bang dogma of the 1970s represented "an antiscientific attitude and a re-

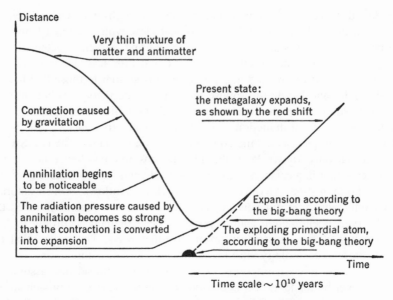

Figure 7.7. Alfvén's plasma universe. *Source:* Alfvén (1966), p. 78.

vival of myth."[211] In Eric Lerner's attempt to revive interest in Alfvén's plasma cosmology, there are many features resembling the earlier, pre-1948 steady-state theory. Thus Lerner emphasizes the reusing and recycling of energy in processes with no time limits. On the methodological level, he (following Alfvén) stresses the need to keep to empirically confirmed processes and to avoid grand theoretical schemes.[212]

Contrary to Alfvén, Halton Arp did not argue for an alternative cosmology, but only that the big-bang model was contradicted by observational facts. Arp was an experienced observational astronomer—he graduated from Caltech in 1953, together with Sandage—and an expert in galaxies and quasars. In a large number of papers, starting in 1966, he reported discordant observations for peculiar galaxies and, in particular, quasars. For example, he found that many quasars were associated with nearby galaxies, and so could not, in spite of their large redshifts, be at cosmological distances, as claimed according to the conventional view. The observational claims of Arp and some other astronomers were generally received with skepticism. Nonetheless, they were taken sufficiently seriously that a public debate on the subject was arranged by the Astronomy Section of the American Association for the Advancement of Science at the end of 1972.[213] The debate, which took place as a duel between Arp and John Bahcall, a Princeton astrophysicist, did not lead to any agreement. As expected, the controversy over discordant redshifts continued. Although Arp did not appeal to the steady-state theory, but stuck to observational issues, the controversy was related to the larger cosmological controversy insofar as G. Burbidge, Hoyle, and Narlikar (and, initially, also Sciama) used the observations as part of their ammunition against the big-bang theory.

When Arp summarized his arguments in 1987, he found that the discordant observations might best be interpreted by the theories proposed by Hoyle and Narlikar, but this was of minor importance to him; what mattered was that the observations contradicted the conventional big-bang theory.[214]

As the hot big-bang theory increasingly dominated cosmology, and as the total number of scientists involved in cosmology increased, opponents of the new paradigm also gathered strength. However, there was little unity among the dissenters except for their shared dislike of the standard theory. It was a countertrend that mainstream cosmologists preferred to pay little attention to, but nonetheless it made enough noise to annoy them. Peebles felt that the charges against the establishment view were exaggerated and unfair, and that its defenders were anything but dogmatic. In his widely used textbook in cosmology, he wrote: "The darker sides to the tradition of dissent in cosmology are the overreactions of those who consider themselves the guardians of the true and canonical faith (as presented in this book), and the tendency for dissent to slip into pathological science. There are documented examples of frivolous criticism of dissenting arguments, which is irrational and destructive." Yet he believed that the view of the establishment was fair on average, and claimed that "most cosmologists would be delighted to abandon the standard model for something new to think about if only the alternatives looked reasonably promising."[215] The dissenters did not, of course, share this view. Hoyle felt that the steady-state theory had been treated unfairly since its very beginning, and that the scientific establishment tried to suppress it by all means. By defending the theory he had, he complained in 1989, exposed himself to "sanctions from referees, editors of journals, peer reviews, and all the rest."[216]

A representative example of the views of the dissenters may be found in a paper Burbidge gave at a conference in honor of Arp's sixtieth birthday in 1987. The conference included contributions from a large number of big-bang opponents, including Hoyle, Narlikar, Arp, Alfvén, and Jean-Pierre Vigier. Burbidge portrayed "the doctrinaire approach to the hot big bang" as building on theoretical prejudices and being closer to religion than to science. In relation to the problems of the big-bang theory in accounting for the formation of galaxies, he claimed that whatever the difficulties, most cosmologists would refuse to leave the big-bang framework. It had become a faith, not a science:

> This is because the *beliefs* developed over the last twenty years in this area have led to a total imbalance in the direction of research. The observational evidence stemming from the Hubble law and the microwave radiation has led to a single-minded approach based on the hot big-bang. Far-out theoretical ideas are only taken seriously, if they are related to the very early universe. There, anything and everything goes. We are told that the unity between particle physics and astrophysics is upon us. Suppose there was no initial dense state? Suppose that matter and radiation were never strongly coupled together? Suppose that the laws of physics have evolved, as has everything else? The current climate of opinion requires that these questions *not* be asked.[217]

Discordant observations were the main message in an article written by Arp, Hoyle, Narlikar, Wickramasinghe, and G. Burbidge in 1990, where they examined the observational material claimed to constitute a proof of big-bang cosmology.[218] The authors concluded, as expected, that the big-bang theory failed miserably when faced with the totality of observations, and that some new theoretical idea (perhaps along the line of the Hoyle-Narlikar theory of pocket creation) had to be considered. Just as predictably, the anti-big-bang manifesto was countered by a pro-big-bang manifesto, concluding that "the standard model passes all believable tests to date."[219] In this last phase (so far) of the controversy, the steady-state alternative was barely visible. It concerned the relationship between the big-bang theory and observations, whereas the former controversy between big-bang and steady-state cosmologies was now considered history.

This is not quite the end of the story, though. Whatever the opinion of the majority of cosmologists, Hoyle and his associates have continued to develop cosmological models in order to offer a viable alternative to the big-bang theory. In 1993 Hoyle, Burbidge, and Narlikar presented what soon became known as the QSSC, the quasi-steady-state cosmological model.[220] This model, which was essentially a development of earlier theories proposed by Hoyle and Narlikar in the 1960s, is claimed to explain the radio source counts, the microwave background radiation (with a temperature of 2.73 K), the largest quasar redshifts, and also the abundances of the light isotopes. Insofar as there is no beginning of the universe and creation of matter takes place, it is a steady-state theory. But it deviates from the classical theory in that creation only takes place in the strong gravitational fields associated with already existing matter, and in that the Hubble constant can vary secularly. The creation of matter in the QSSC is described as "little big bangs," which each involve enormous masses of about 10^{16} times the mass of the sun. Moreover, the Hoyle-Narlikar-Burbidge universe is expanding, but not uniformly; superposed on it there are oscillations of a gigantic time scale. The temporal variation of the scale factor reflects the mixed exponential and oscillatory behavior. It is given by $R(t) = \exp(t/P)[1 + \alpha \cos(2\pi t/Q)]$, where α is a constant and P and Q are two characteristic cosmic time scales. In order to make the theory agree with observations, Hoyle, Burbidge, and Narlikar take $\alpha = 0.75$ and $P = 20Q = 8 \times 10^{11}$ years. The cyclical period is thus $Q =$ forty billion years, and our epoch is at the time 0.85Q, corresponding to an age since the last minimum of 16.5×10^{11} years (see figure 7.8). With these values, the QSSC model results in a Hubble constant of $H = 64.7$ km·s^{-1}·Mpc^{-1}, a deceleration parameter $q_0 = 1.75$, and a present mass density of $\rho_0 = 2.26 (3H^2/8\pi G) =$ ca. 1.8×10^{-29} g·cm^{-3}.

In epochs when the universe was relatively more dense, near the oscillatory minima, matter creation is supposed to have been most intense. Hoyle and his colleagues argue that matter, arising from the C field, appears as primordial Planck particles, i.e., hypothetical particles of mass $(3\hbar c/4\pi G)^{1/2} =$ ca. 10^{-5} g. These are assumed to be highly unstable and to convert into the baryons of

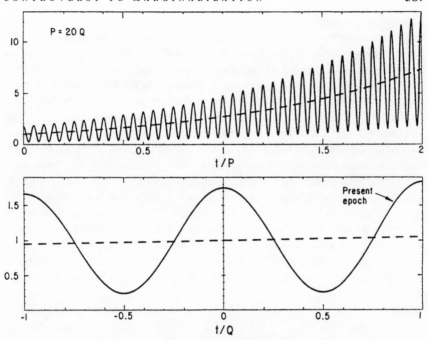

Figure 7.8. The variation of the scale factor with time in the QSSC model. *Source:* Hoyle, Burbidge, and Narlikar (1994b), p. 732. Reprinted with the permission of the European Southern Observatory.

which ordinary matter is composed. In this way Hoyle and his collaborators manage to calculate the abundance of the light isotopes in good agreement with measured data. As to the microwave background, it is generated by stellar hydrogen burning into helium, accumulated through many creation cycles of length Q. The radiation is thermalized by cosmic iron whiskers originating from supernovae exploding in the previous cycle. The spectrum arising from this mechanism is much like the Planck distribution, but not exactly so. It deviates from it at long wavelengths.

As in much previous work of Hoyle and Narlikar, the justification behind the QSSC is the same deep dissatisfaction with the big-bang model which motivated Hoyle, Bondi, and Gold to propose the classical steady-state theory about forty-five years earlier. For example, in 1994 Hoyle, Burbidge, and Narlikar expressed themselves as follows: "We believe that by using the C-field approach our theory allows creation through a field of negative energy described by a physically motivated action principle, whereas in the standard model the creation process is excluded from discussion. . . . The Friedmann model has nothing to offer in the area of cosmogony. Initial condensations have to be *assumed.*"[221]

As will be seen from this brief sketch, most of the features of the QSSC model are repetitions or modifications of views argued already in the 1960s.

But the new theory of Hoyle, Narlikar, and Burbidge has some additional empirical content which distinguishes it observationally from the big-bang theory. Whatever its merits, however, it is unlikely that the community of cosmologists will change its opinion of steady-state theory because of the QSSC model or other new proposals of steady-state models.[222] Most astronomers and physicists will bet without hesitation that theories of the steady-state type will never return. Perhaps their confidence in the standard big-bang type of cosmology will turn out to be unwarranted, but this is irrelevant from a historical point of view. What we can say, and should say, is that no developments in science since 1970 have been of such a nature that they have seriously induced scientists to reconsider the basic elements of big-bang cosmology. The world, so it seems, has not always been, and there is no need for continual creation of matter. That's it. Some ten billion years ago the world came into existence—was created—and since then it has evolved according to the laws of general relativity. To infer from this stabilization of basic ideas that this is also how the universe has actually evolved is, of course, a different story. It is a story which cannot be told with complete confidence at present.

CHAPTER 8

Epilogue: Dynamics of a Controversy

WE HAVE FOLLOWED the controversy between two radically different conceptions of the universe through a period of more than twenty years. Was this an
extraordinary development? Was it an aberration compared with the normal
development of science? The answer is, not really. Controversies occur frequently in the development of science and are much more common than scientists and many historians of science want to admit. Indeed, they are part and
parcel of the very scientific development. The case examined here is unusual
mainly because of its subject, duration, and intensity. Contrary to other twentieth-century controversies, the debate between steady-state and big-bang theories concerned the world picture in the most real sense of the term. Its subject
was not a particular phenomenon or part of the physical universe, but nothing
less than the universe itself. In this sense it is comparable to other controversies over the world picture in the history of science, such as the famous struggle between the Ptolemaic and Copernican world views during the early
phases of the scientific revolution. Whether this comparison is admitted or
not, it may be useful to look at the modern cosmological controversy in the
perspective of controversies in general.

First of all, I have repeatedly referred to the debate between the two cosmological theories as a controversy (and I shall continue to do so), but what,
exactly, is a scientific controversy?[1] Generally speaking, if two groups of scientists disagree, the disagreement may be, or may become, a controversy. This
is a minimum condition, but of course not any disagreement qualifies as a
controversy. It should be of some duration, be expressed in public, and take
place by means of arguments and counterarguments. That is, it should contain
elements of a social and methodological nature. Moreover, a controversy is
more than just a debate or a dispute: the parties must be committed to one of
the opposing views, hold it important enough to defend, and attack the rival
view. In many cases, and certainly in the present one, it is this commitment
which heats the controversy and extends it to cover nonscientific aspects also.
Whether or not public and personal passions enter the controversy, it involves
more than a quarrel between two individual scientists. The relevant *scientific
community* is part of it, and only if the community considers the disagreement
worth taking seriously will it develop into a controversy. Usually, major parts
of the scientific community will be engaged on both sides of the disagreement,
although often with a disproportionate engagement.

It goes without saying that the participation of scientists in a controversy
does not automatically make it scientific. A scientific controversy is concerned with claims of knowledge held to be determinable by legitimate scien-

tific means, such as experiment and theoretical reasoning. If not, it would not be scientific. This does not preclude other factors from entering the controversy (they almost always do), but the parties will agree that these are external and should not count in the normally expected final resolution of the controversy. According to this definition, the cosmological controversy was scientific in spite of involving a certain element of religious, moral, and political principle. Such mixed controversies usually differ from purely scientific ones in relating to some kind of application of the science in question; the application may have moral or social consequences, the desirability of which is debated among both scientists and nonscientists. Now there is no such thing as applied cosmology, but the controversy nonetheless included discussions concerning which cosmological model was the more desirable one from a human and spiritual perspective. However, I have argued that this element was more noisy than significant in the controversy, and that practically all the scientists involved were careful to distinguish between scientific and nonscientific arguments. In particular, the latter kind of argument did not, or at most very rarely, appear in the professional journals, where the controversy was presented as a fairly standard scientific debate. Its subject matter was grander and more awesome, but from a methodological point of view the controversy between the two theories of the universe did not differ fundamentally from other, more mundane scientific disagreements. The broader philosophical and religious considerations definitely influenced the dispute, but mainly in shaping (some of) the contestants' preferences and not in the context of justification.

A controversy can arise if a standard view or paradigmatic belief is questioned by a new theory, experimental claim, or general point of view. The new theory may cause the controversy by disturbing the peace and security of the paradigmatic belief system. But the new idea only becomes controversial—threatening to the standard view—if it receives enough support to make it impossible for the scientific community to ignore it (which is not to say that scientists normally respond to outsider views by ignoring them). There may be all kinds of reasons why scientists become interested in the new theory: the important thing is that some of them do. In the absence of serious interest, even the most controversial theory will fail to spark a controversy. For example, what I have called the Nernst-MacMillan cosmology never succeeded in engaging relativistic cosmologists in a controversy. There were several cosmological controversies in the 1930s, in particular centered around Milne's theory, but the steady-state version of Nernst, MacMillan, and Millikan was by and large ignored by relativists and astronomers, and never gathered support from within these groups. The reason was neither complacent conservatism nor a deliberate attempt to keep intruders out of the field, but the glaring lack of a theoretical framework in the prewar steady-state theories as compared to the framework of relativistic cosmology. If offered a strong theoretical alternative, such as Milne's, many relativist cosmologists were willing to engage in a debate. The revolt against relativistic evolution cosmology, which started in 1948, also quickly found support among a minority of physicists and astronomers. The reason was not that they found the steady-state theory particularly impressive from a theoretical point of view, but rather that they found it conceptually attractive and recognized its considerable effectiveness in solving empirical and conceptual problems.

Controversies in the history of science can be controversies over priority, for example, disagreements about which scientist first made a discovery or suggested a theory. Priority controversies may arise many years after the relevant knowledge claims, at a time when the claims have been accepted as true or otherwise proved valuable. Only in rare cases do disputes over wrong and discarded theories take place. At any rate, in the present case we do not need to consider the issue closely, for priority disputes played almost no role at all. Neither the big-bang nor the steady-state model was discovered in any ordinary sense, and there never was any question that the steady-state idea came from Bondi, Gold, and Hoyle. Although Hoyle soon came to be seen as the number-one steady-state theoretician, the roles of Bondi and Gold in the formulation of the theory were always recognized. On the other hand, when the big-bang model was revived and became the accepted framework of cosmology in the 1960s, the question of its origin also became potentially controversial. Who deserved the honor? As we have seen, there were at least three candidates: Lemaître, Gamow and his collaborators, and Dicke and his collaborators. (Friedmann sometimes enters as a candidate as well, but I have argued that this is a misplaced claim.) A kind of priority dispute arose when Alpher and Herman complained about the lack of acknowledgment of their and Gamow's early work, but the dispute never developed into a proper controversy. The question of who first suggested the big-bang model is of minor importance and had, at any rate, nothing to do with the steady-state controversy. All that one can do is to describe the course of events leading to the modern big-bang concept as precisely as possible. From such a description, I conclude that if a particular scientist should be called the father of the big-bang theory, the paternity belongs to George Gamow.

A controversy takes place between two (or more) parties associated with certain views that are, or seem to be, incompatible. The views may be ordinary theories or broader research programs. In the case of a science which has not yet developed into a mature state with certain generally accepted rules and a corpus of positive knowledge, the competing views may more appropriately be referred to as research programs. This was the case in the cosmological controversy, which arose at a time when cosmology was not yet recognized as a truly scientific discipline, and when there was much uncertainty about its foundation. According to the relativistic research program, the universe was evolving in time, perhaps from a previous state of highly compressed matter, and its evolution was governed by the Friedmann-Lemaître equations. Gamow's big-bang cosmology may be seen as a subclass of this program, but with its emphasis on nuclear-physical processes in the very early universe, it also went beyond the relativistic program as it was usually conceived at the time. In fact, most of the controversy was not really between steady-state theory and big-bang cosmology, but between steady-state and relativistic evolution cosmologies. During the period 1953–63, when the controversy was most intense, the physical big-bang model played a subordinate role. As the steady-state protagonists often pointed out, the two sides entered the debate on very unequal footing, methodologically speaking. Whereas the classical steady-state theory resulted in definite predictions, and hence was eminently falsifiable, the relativistic evolution theories were, as a whole, much more flexible and seemed to evade critical testing. In the long run this turned out to be an advantage for the latter kind of theory, which could adjust to new observations—a luxury the classical steady-state theory did not enjoy.

Competing theories or research programs are not static entities, but often change during time under the impact of criticism and new observations. Sensitivity to criticism is one of the hallmarks of science, and one of the features that distinguishes scientific controversies from other intellectual disputes. One beneficial result of a controversy is that it provokes such change more effectively than most other situations. Sometimes the two theories may change so drastically that a compromise is reached, either as an amalgamation of the modified theories or in the form of a new, more general theory incorporating aspects of the two original ones. In that case a controversy may be resolved without having a winner and a loser. In the cosmological controversy there were a few attempts at compromise, but the differences between the steady-state and the big-bang models were too fundamental for a fruitful compromise theory to be established. After all, how can a theory with a beginning in time of the universe be reconciled with a theory according to which the universe is eternal? Some of Hoyle and Narlikar's versions incorporated big-bang as well as steady-state features, but they were not taken seriously by the big-bang majority. Still, the last word has not been said about the structure and evolution of the universe, and it is at least conceivable that some compromise will be obtained in the future.

It is generally agreed that the steady-state theory did cosmology a great service by questioning basic assumptions in the relativistic evolution program, and thereby forcing cosmologists to reevaluate their science. The new theory, and the controversy that followed, helped cosmologists transform cosmology from a protoscience to a mature science. This critical process took place in the 1950s, when the two sides of the controversy were rather sharply defined. As evidence mounted in favor of the evolution view, beginning about 1960, the steady-state theory came into a state of crisis and responded by developing a series of modifications, some of which were of a rather ad hoc nature. At the same time, the controversy began to focus more on the hypothetical early universe, that is, it became a controversy between the big-bang assumption and the assumption of a temporally infinite universe. Although much of the controversy can be summarized in the question: has the universe existed forever, or only for a finite time?, this was not, in fact, what it was mostly about. Recall that oscillatory models and the one-cycle contraction-expansion models favored by some big-bang physicists have an infinite duration, at least of a sort. The basic question of the controversy was rather: has the universe evolved from a superdense state of matter completely different from its present state? It was this question that was answered affirmatively by most astronomers and physicists in the mid-1960s, thereby bringing an end to the controversy.

Following a proposal of Ernan McMullin, one may distinguish between controversies of fact, of theory, and of principle.[2] In a controversy of fact, scientists disagree about the observational basis of a knowledge claim, that is, whether something reported to exist does actually exist. A controversy of theory involves different theoretical views, whereas a controversy of principle is more extensive in relating also to, for example, basic methodological or ontological principles. These three categories are not mutually exclusive, of course, and in heated controversies they may appear together. This was clearly the case with the cosmological controversy in the 1950s, which included a relatively minor dispute of fact (the Stebbins-Whitford effect), but was predominantly a controversy of theory with a strong element of controversy of principle. The measurement data were rarely criticized by the steady-state protagonists, who, however, often questioned their reliability and certainty, and thus their value as a basis for theoretical interpretation. From the very beginning of the controversy, there was disagreement about fundamental principles of science, such as the extrapolation of laboratory data to the entire universe and the status of the cosmological principles. A cosmological model involving a singularity at the beginning of time is not scientific, the steady-state physicists claimed; whereas the relativists argued that continual creation of matter is both unfounded and unscientific. And so on. As we have seen, for a period the dispute involved basic notions such as the meaning of science and the criteria to be adopted in order to distinguish more-scientific from less-scientific hypotheses. On the other hand, I have also argued that this dispute was never total and that all the scientists involved recognized the most basic rule of scientific judgement, that the ultimate arbiter is observation.

A controversy terminates when a disproportionate part of the scientific community decides that one of the competing views has lost its force and is no longer worth dealing with. "Termination" is here used descriptively, as a neutral term for the fact that the controversy has ceased to exist. This may take place in three different ways. The two sides may come to agree that one of the views (or a third alternative) is the most satisfactory because it offers a better account of observations, is theoretically superior, more coherent, or the like. In this case, where consensus is achieved by epistemically based agreement, the controversy is said to be resolved. Scientific controversies may also end by just being abandoned, withering away because of lack of interest or because the protagonists of one of the competing views vanish from the scene or become so few and marginalized that they no longer count. In still other cases, termination is reached by closure, which means that a measure of external authority intervenes in order to declare the controversy ended. The decision is then influenced by nonepistemic factors, which force the controversy to an end although it may still be unresolved.

As we have seen, the cosmological controversy was not simply resolved, although a few steady-state protagonists, most notably Sciama, came to accept that the universe is indeed expanding from a superdense state of matter. His position was similar to that of the Irish chemist Richard Kirwan, who, in spite of having been a leading advocate of the phlogiston theory, surrendered to Lavoisier's new chemistry. "I lay down my arms and abandon the cause of phlogiston," he wrote in 1792.[3] Not all chemists followed Kirwan, and not all cosmologists followed Sciama. Joseph Priestley, the discoverer of oxygen (which he interpreted as dephlogisticated air), remained faithful to the phlogiston theory to his death, and continued to criticize the oxygen theory. Two hundred years later, Hoyle followed a similar course. Unfortunately I can see no modern cosmologist serving as a proper analogy to Lavoisier.

The cosmological controversy was not abandoned in any strict sense, even though the number of steady-state defenders declined and almost none replaced them. About 1970 the cosmological community had decided that the steady-state theory was a lost cause and relegated it to the history of scientific mistakes, although admittedly a fruitful one. This decision was based on epistemic reasons, and it mattered little that a few scientists remained unconvinced. Hoyle often intimated that it was external authority more than epistemic factors that killed the steady-state theory, i.e., that the controversy was in reality closed by the powerful big-bang establishment. Nonepistemic factors played a role all along, but the controversy did not terminate because one side lost research opportunities, was treated unfairly by referees, or was found politically unacceptable by those in power. The only place where political and ideological intervention was of major importance was in the Soviet Union. And since both steady-state and big-bang cosmologies were found ideologically unacceptable in that country, the result did not affect the controversy to any degree.

The big-bang theory was not socially negotiated or the result of a particular sociocultural climate favorable to that kind of cosmology. Its emergence and development were possibly influenced by such factors, but I find it difficult to document exactly how and to what extent. On the whole, the theory was found superior to other views of the universe on epistemic grounds and not for other reasons. The controversy terminated in a mixture of resolution and abandonment, because an overwhelming part of the cosmological community concluded that there is convincing evidence that the universe has indeed evolved from a big bang in accordance with the laws of general relativity and particle physics. The rapid rise in popularity of the big-bang view was undoubtedly helped by a kind of bandwagon effect, but the cosmologists' conclusion was epistemically, not socially, based.

Commenting on the termination of another famous controversy of the twentieth century, the one over continental drift, Ronald Giere concludes: "In this case, as in most scientific cases, a few scientists failed to realize or refused to admit that the major issues had been resolved. Thus the controversy continued to exist as a social phenomenon, though on a much reduced scale. The 'end' came later. It is a tribute to the relative objectivity of scientific procedures that holdouts are generally few in number and recognized as holdouts."[4] This could as well refer to the cosmological controversy. In fact, these two postwar controversies, the one in cosmology and the other in geology, have much in common. They were both scientific controversies, though the one in the earth sciences was more strictly so: the public discussions and ideological aspects that colored the cosmological controversy were missing. It is to be noticed that in the case of the continental drift controversy (and many other controversies, including the one between oxygen and phlogiston), it was a new and eventually victorious theory which challenged an established view. The fact that the revolutionary drift theory is now accepted as true makes the controversy an instructive success story. In the cosmological case also there was a majority view challenged by a new and revolutionary theory. But in this case the revolution failed. Or one may say that the challenge helped create a new phase in cosmology, a revolutionary one, maybe, but also one solidly rooted in the old tradition. At any rate, the fact that the steady-state theory failed to accomplish its great goals does not make the controversy less interesting or less instructive.

A Cosmological Chronology, 1917–1971

1917   Einstein's cosmological field equations. The spherical Einstein world. De Sitter's alternative of a model devoid of matter. Harkins suggests nuclear synthesis of chemical elements.

1918   MacMillan suggests a kind of steady-state universe.

1920   The "Great Debate" between Shapley and Curtis.

1921   Nernst's cosmic world picture with eternal recycling of matter and energy.

1922   Lanczos's nonstatic world model. Friedmann's equations and his theory of expanding and other models.

1923   Weyl introduces "Weyl's principle," that the galaxies lie on a pencil of geodesics diverging from a common event in the past.

1924   Friedmann's second paper, on models with constant negative curvature.

1925   Lemaître's first derivation of a redshift-distance relation. Hubble's discovery of Cepheids in spiral nebulae.

1926   Millikan and Cameron claim that cosmic-ray studies support an unchanging universe with interstellar element formation.

1927   Lemaître's theory of the expanding universe, with $H = 625$ km·s^{-1}·Mpc^{-1}.

1928   Suzuki considers cosmological helium formation. Jeans's long time scale with galactic ages up to 10^{13} years.

1929   Hubble's law: the apparent velocities of galaxies are proportional to their distances. Robertson's general metric for homogeneous and isotropic universes, whether stationary or not. Atkinson and Houterman apply quantum mechanics to stellar nucleosynthesis.

1930   Eddington rediscovers Lemaître's work. The expanding universe becomes widely accepted. R. J. Trumpler presents strong evidence for the presence of interstellar obscuring matter.

1931   Hubble and Humason confirm the Hubble law and derive $H = 558$ km·s^{-1}·Mpc^{-1}. Einstein abandons the cosmological constant. Tolman applies thermodynamics to the expanding universe. Lemaître's hypothesis of a big-bang universe. The Lemaître model. Eddington suggests that the expanding universe is connected with quantum mechanics.

1932   The Einstein–de Sitter zero-curvature world model. Milne starts his cosmological program.

1933   Robertson's review of relativistic cosmology. Milne formulates the cosmological principle.

1934   Tolman's textbook *Relativity, Thermodynamics and Cosmology*.

1935   Milne's *Relativity, Gravitation and World-Structure*. Bok argues for a cosmic time scale of 3–5 billion years. Gamow suggests neutron capture as mechanism of nucleosynthesis.

1936   Hubble's *Realm of the Nebulae*.

1937   Dirac's large-number hypothesis and cosmological theory with a varying gravitational constant. Goldschmidt publishes data for the abundances of isotopes.

1938   Jordan develops his own version of Dirac's cosmology, including creation of matter. Von Weizsäcker's hypothesis of primordial element formation. Washington conference on stellar energy sources.

1939   Bethe's theory of stellar energy production (CN cycle).

1942   Washington conference on stellar evolution and cosmology. Gamow's first (fission) version of big-bang element formation.

1946   Gamow suggests cold big-bang theory with element formation in early, expanding universe. Lemaître's *L'Hypothèse de l'Atome Primitif.*

1947   Stellar theories of heavy-element formation (Hoyle, van Albada).

1948   Alpher's dissertation on big-bang nucleosynthesis. $\alpha\beta\gamma$ paper and hot big-bang theory (Alpher, Gamow). Theory of early, radiation-filled universe. Alpher and Herman improve Gamow's model and predict cosmic background radiation of $T = 5$ K. The steady-state theories of Hoyle, Bondi, and Gold. Stebbins-Whitford effect indicates intrinsic reddening, in contradiction with steady-state theory.

1949   Mayer and Teller suggest theory of polyneutrons.

1950   Alpher and Herman's article in *Reviews of Modern Physics.* Hayashi suggests improvement of Gamow-Alpher-Herman theory and calculates present hydrogen-helium ratio. Hoyle's *Nature of the Universe.*

1951   McCrea's version of steady-state theory with negative cosmic pressure. Pope Pius XII's address on cosmology and religion.

1952   Gamow's *Creation of the Universe.* Bondi's *Cosmology.* Salpeter suggests triple-alpha burning in stars. Baade's revised distance scale gives Hubble time about 3.6 billion years.

1953   Big-bang theory of Alpher, Follin, and Herman. Dingle attacks steady-state cosmology.

1954   Debate between Whitrow and Bondi concerning the nature of cosmology.

1955   Sciama's steady-state theory of galaxy formation. Ryle interprets 2C radio-astronomical results in favor of evolving universe.

1956   Rindler clarifies the horizon problem in cosmology. Humason, Mayall, and Sandage conclude that deceleration parameter is $q_0 = 2.5 \pm 1$. Bonnor argues that big-bang cosmology cannot explain galaxy formation.

1957   B²HF theory of stellar element formation (Burbidge, Burbidge, Hoyle, Fowler).

1958   Sandage estimates the Hubble time to be about thirteen billion years. Paris symposium on radio astronomy. Hoyle suggests angular diameter–redshift test. Hot steady-state theory (Hoyle, Gold). Solvay Congress on the structure and evolution of the universe.

1959   Lyttleton and Bondi's theory of an electrical universe.

1960   Hoyle's covariant formulation of steady-state theory.

1961   Ryle concludes that the steady-state theory is disproved by the 3C radio astronomical survey. Sandage concludes that $0 \le q_0 \le 2$, contradicting the steady-state theory. Brans-Dicke scalar-tensor theory.

1962   Zel'dovich argues for a cold big bang. Satellite experiment (Kraushaar and Clark) disproves hypothesis of symmetrical creation of matter and antimatter. C-field theory (Hoyle, Narlikar). Hogarth's analysis of the relationship between cosmology and the direction of time.

1963   X-ray astronomical data prove incompatible with hot version of steady-state theory. Discovery of quasi-stellar objects (Schmidt, Greenstein). First Texas Symposium on Relativistic Astrophysics. Doroshkevich and Novikov suggest detection of microwave background as test of Gamow-Alpher-Herman theory.

1964   Bolton *et al.* confirm the 3C result of a log$N$-log$S$ slope incompatible with the steady-state theory. Hoyle and Tayler analyze helium problem and find agreement with hot big-bang theory. Dicke suggests calculation of relic microwave radiation from the big bang.

1965   Penzias and Wilson's discovery of the microwave background radiation. Dicke, Peebles, Roll, and Wilkinson interpret the Penzias-Wilson result as instance of cosmic background radiation of $T = 3.5$ K. First singularity theorems (Penrose, Hawking).

1966   Sciama abandons the steady-state theory because it disagrees with the red-shifts of quasars. Based on the big-bang theory, Peebles calculates the abundance of helium to be 27%. Hoyle and Narlikar suggest discrete matter creation in pulsating quasi-steady-state universe.

1967   Improved theory of nucleosynthesis (Wagoner, Fowler, Hoyle) agrees with hot big bang. Standard big-bang theory is established.

1971   Peebles' *Physical Cosmology* signifies the birth of a new paradigm.

Technical Glossary

absolute luminosity — The total radiant energy output by a star per unit time. A measure of the star's intrinsic brightness.

absolute magnitude — The apparent magnitude that a star would have at a distance of 10 pc without absorption.

accretion — A process by which a star accumulates matter as it moves through a cloud of interstellar gas.

Andromeda — The nearest spiral galaxy, at a distance of about 674 kpc from the Earth. Also known as M31, its catalog number.

anthropic principle — A metascientific principle according to which the structure and development of the universe is (in some way) determined by the existence of human beings.

antiparticle — Most elementary particles have an antiparticle equal in mass and most other properties, but opposite in charge. When particles and antiparticles meet, they annihilate each other, producing high-frequency electromagnetic radiation.

apparent luminosity — The total energy received per unit time and per unit receiving area from a celestial body.

apparent magnitude — Measure of the observed brightness of a celestial object as seen from the earth. It depends on the object's intrinsic brightness, its distance from the earth, and the amount of absorption.

atomic number — The number of protons in an atomic nucleus, equal to the element's position in the periodic table.

beta decay — The radioactive transformation of a neutron—whether located in the atomic nucleus or not—into a proton, an electron, and an antineutrino.

binding energy — The minimum energy needed to split an atomic nucleus into its constituents.

blackbody radiation — Radiation with the same spectral intensity as the radiation emitted from a totally absorbing heated body. Sometimes called cavity or thermal radiation, its spectral distribution follow's Planck's law.

Boltzmann's constant — A fundamental constant of physics that relates the temperature scale to units of energy. The numerical value is $1.38 \times 10^{-16} \mathrm{erg \cdot K^{-1}}$, or $8.62 \times 10^{-5} \mathrm{eV \cdot K^{-1}}$.

Cepheid variables — Very bright, pulsating stars with a well-defined relation between their absolute luminosities and periods of pulsation. Useful distance indicators for galaxies nearer than about 3 Mpc.

cosmic rays — High-energy charged particles, mostly protons and alpha particles, which stream from interstellar space down to earth. The energy is typically 10^9 eV, but may be as large as 10^{20} eV.

cosmogony — The study of the creation or origin of the universe. Sometimes also used for theories of the origin of celestial objects like the earth, moon, and sun. In modern parlance, cosmogony is considered a branch of cosmology.

cross section — Numerical expression for the reaction rate of a particular nuclear process. Often expressed in the unit barn = 10^{-28} m^2.

curvature — The departure of the geometry of the universe from flat, Euclidean geometry. The curvature parameter (k) is +1 for a closed geometry and −1 for an open geometry.

deceleration parameter — A dimensionless number that characterizes the rate at which the expansion of the universe is slowing down as a result of gravity.

Dirac equation — A quantum-mechanical equation of motion that satisfies the laws of the special theory of relativity and is valid for particles with half-integral spin such as electrons. Dirac's equation passes into the ordinary Schrödinger quantum equation in the nonrelativistic limit.

Doppler effect — The change in frequency of any signal received from a source in relative motion in the line of sight. The light of approaching sources is shifted toward the blue, that of receding sources toward the red.

electron volt — Unit of energy (or mass), abbreviated eV, convenient in atomic physics. 1 eV = 1.602×10^{-12} erg = 1.602×10^{-19} J.

energy conservation — Fundamental law of physics, according to which the total energy remains unchanged in any physical process. Also known as the first law of thermodynamics.

entropy — According to the second law of thermodynamics, the entropy of any isolated system can never decrease in time. The entropy is a measure of the degree of disorder (suitably defined) of a physical system.

equilibrium — The unchanging state of a system resulting from the balance between competing forces or processes.

ether — The medium through which light and gravitation were supposed to propagate according to classical, prerelativistic views.

gravitational constant — An expression of the Newtonian gravitational force between two bodies of unit mass separated by a unit distance. Denoted G, it is equal to 6.67×10^{-11}m^3·kg·s^2.

homogeneity — The property that a system (or the universe) has the same physical properties at all points of space.

horizon — The maximum distance that an observer can see in principle.

inflationary universe — A class of theories of the very early universe, according to which the infant universe expanded exponentially for a brief period. After the inflation era, the model coincides with the standard big-bang theory.

isotopy — Two isotopes have the same number of protons, and hence belong to the same chemical element, but differ in their numbers of neutrons.

isotropy — The property that a system (or the universe) appears the same in all directions.

Kelvin — Unit of thermodynamic temperature. The "absolute" Kelvin scale is related to the Celsius scale by T (K) = t (C) + 273.15.

line element — The squared distance between two nearby points of space-time.

mass-energy equivalence — Mass (m) can be transformed to energy (E), and energy to mass, according to the formula $E = mc^2$, where c is the velocity of light, ca. 3×10^8 m·s^{-1}.

neutrino — Neutral and massless elementary particle that accompanies the electron in beta decay. Another type of neutrino is associated with the muon, a heavy relative of the electron.

parsec — Astronomical unit of distance, defined as the distance at which the earth-sun radius subtends an angle of one second of arc. 1 pc (parsec) = 3.26 light years = 3.09×10^{16} m. One million pc, or one megaparsec, is abbreviated Mpc.

photon — Corpuscle of light and other electromagnetic radiation.

Planck's constant — Fundamental constant of nature giving the ratio between the energy and frequency of a light quantum (photon). Always denoted h, the value of the constant is 6.63×10^{-34} J·s.

plasma — A completely ionized gas in which the temperature is too high for atoms to exist and which consists of free electrons and free atomic nuclei.

singularity — A place, either in space or in time (or in both), where some physical quantity—such as mass or electrical density—becomes infinite. Since infinities have no place in physics, singularities are usually believed to be nonreal.

Stefan-Boltzmann law — The relation that the energy density of blackbody radiation is proportional to the fourth power of the temperature of the body.

tensor — A mathematical object defined by certain rules of transformation. Ordinary vectors (such as velocities and momenta) are simple types of tensors.

thermal equilibrium — A physical system in which all parts have exchanged heat. In this state, the entropy of the system remains constant.

uncertainty principle — A fundamental principle of quantum mechanics, according to which certain (conjugate) physical quantities cannot be simultaneously known with complete certainty or precision. The best known pair of such quantities is the position and velocity of a particle.

x rays — Energetic electromagnetic radiation with wavelength in the range ca. 0.03 Å to 3 Å (1 Å = 1 Ångström = 10^{-10} m).

N O T E S

CHAPTER 1
BACKGROUND: FROM EINSTEIN TO HUBBLE

1. Quoted from the extract given in M. K. Munitz, ed. (1957b), pp. 231–49, on p. 240.
2. Ibid., p. 248.
3. S. G. Brush (1987); see also S. Toulmin and J. Goodfield 1982.
4. Quoted in M. J. Crowe (1994), pp. 147–48.
5. W. Thomson (1884), pp. 37–38.
6. A. M. Clerke (1890), p. 368.
7. Nineteenth- and early twentieth-century cosmology is covered in M. A. Hoskin (1982); J. D. North (1994); S. L. Jaki (1972); E. R. Paul (1993); M. J. Crowe (1994).
8. A. Einstein (1917); English translation in Einstein et al. (1923), pp. 175–78, from which the quotations are taken. Einstein's work and its historical context are analyzed in several works. See, e.g., J. D. North (1965); P. Kerzberg (1992). For a concise account of Einstein's road to the theory of general relativity, see J. Mehra (1974).
9. Newton's four letters to Bentley are included in M. K. Munitz, ed. (1957b), pp. 211–19.
10. A. Einstein et al. (1923), p. 180.
11. Einstein to Ehrenfest, 4 February 1917, as quoted in A. Pais (1982), p. 285.
12. Einstein to Besso, December 1916, in P. Speziali, ed. (1979), pp. 58–60.
13. A. Einstein et al. (1923), p. 183.
14. Ibid., p. 188.
15. Mach's principle is notoriously ambiguous and has been formulated in numerous, often very different, versions. For the difference between the versions of Mach and Einstein, see J. B. Barbour (1990).
16. The factor c^2 in front of κ occurs because T_{44} is defined as an energy density. If it is defined as a mass density, the factor is 1.
17. A. Einstein (1936); here quoted from Einstein (1950), p. 81.
18. C. Ray (1990).
19. Einstein to Besso, 20 August 1918, in P. Speziali, ed. (1979), p. 68.
20. A. Einstein (1919); English translation in Einstein et al. (1923), pp. 189–98, on p. 193.
21. P. Speziali ed. (1979), p. 59.
22. C. Kahn and F. Kahn (1975).
23. W. de Sitter (1917). The Italian mathematician Tullio Levi-Città also wrote down equations for all spatially uniform and static universes satisfying Einstein's field equations. Although Levi-Città thus independently duplicated much of de Sitter's work, his influence was much more limited. Contrary to de Sitter, Levi-Città focused on the mathematical aspects of Einstein's theory. T. Levi-Città (1917).
24. W. de Sitter (1917), p. 27.
25. A. Einstein (1918).

26. Einstein to de Sitter, 24 March 1917, as quoted in J. Eisenstaedt (1993).

27. Cf. the subtitle of P. Kerzberg (1992).

28. For example, in Hector MacPherson's survey of the development of cosmology, written in 1929, less than a page was devoted to the theory of "Dr. Albert Einstein of Berlin and his co-worker, Dr. Willem De Sitter of Leyden." H. MacPherson (1929), p. 126.

29. E. Borel (1922), p. 246; English translation as *Space and Time* (New York: Dover Publications, 1960).

30. C. Lanczos (1922).

31. G. Lemaître (1925), p. 188.

32. G. J. Whitrow (1978), p. 582.

33. R. C. Tolman (1929a). There is more on Friedmann's work in section 2.1.

34. V. M. Slipher (1913).

35. V. M. Slipher (1915); see also N. S. Hetherington (1971).

36. H. Weyl (1923a), p. 323; (1930).

37. L. Silberstein (1924).

38. G. Lemaître (1925), p. 191.

39. H. P. Robertson (1928), p. 847.

40. D. E. Osterbrock, R. S. Brashear, and J. A. Gwinn (1990).

41. H. Leavitt (1912); Leavitt's paper appeared under the name of Edward C. Pickering, the director of Harvard College Observatory.

42. J. D. Fernie (1969).

43. E. P. Hubble (1925).

44. For this idea and the controversy around it, see R. Berendzen, R. Hart and D. Seeley (1976). The main papers in the "Great Debate," written by Heber D. Curtis (for the island-universe view) and Harlow Shapley (against it), are reprinted in M. J. Crowe (1994), pp. 273–327.

45. R. Berendzen and M. Hoskin (1971).

46. E. P. Hubble (1926), pp. 368–69.

47. R. Smith (1982), pp. 199–200; see also R. Smith (1979).

48. E. P. Hubble (1929).

49. D. E. Osterbrock (1990), p. 276.

50. M. L. Humason (1929). According to D. E. Osterbrock (1990), Hubble wrote most of Humason's early papers.

51. H. Shapley (1929).

52. E. P. Hubble and M. L. Humason (1931).

53. As indicated by the title of A. S. Sharov and I. D. Novikov (1993).

54. Letter of 23 September 1931, as quoted in R. Smith (1982), p. 192. See also R. Smith (1990).

55. E. P. Hubble (1936a); quotations from pp. 553 and 517. The disturbing values of the radius and density were $R = 145$ Mpc and $\rho \approx 10^{-26}$ g·cm^{-3}. See also E. P. Hubble (1937), p. 65.

56. E. P. Hubble (1953).

57. N. S. Hetherington (1982).

58. H. N. Russell (1929).

59. R. C. Tolman (1929a), p. 304.

60. *Observatory* 53 (1930): 38–39. For de Sitter's view, see also W. de Sitter (1930c).

CHAPTER 2
LEMAÎTRE'S FIREWORKS UNIVERSE

1. H. P. Robertson (1929), p. 822. In this work Robertson derived for the first time the general metric for a homogeneous and isotropic universe, later generally known as the Robertson-Walker metric.

2. Quoted in V. Ya. Frenkel (1988), p. 651.

3. Ibid., p. 661.

4. V. A. Fock (1964), p. 473.

5. A. Friedmann (1922); English translation in J. Bernstein and G. Feinberg, eds. (1986), pp. 59–67, which also includes Einstein's responses. The 1922 work was restricted to spaces with positive curvature, but two years later Friedmann examined also, for the first time, nonstatic models with constant negative curvature (A. Friedmann [1924]).

6. Quoted from E. A. Tropp, V. Ya. Frenkel, and A. D. Chernin (1993), p. 157.

7. Ibid., p. 161.

8. H. Weyl (1924); C. Lanczos (1922).

9. N. S. Hetherington (1973).

10. A. Einstein (1922), (1923).

11. R. Smith (1982), p. 199.

12. E. A. Tropp, V. Ya. Frenkel, and A. D. Chernin (1993), p. 174.

13. V. Fock (1964), p. 474. The "mathematicity" of Friedmann's work is emphasized also in V. P. Vizgin and G. E. Gorelik (1987), where Friedmann is nonetheless hailed as the discoverer of the expanding universe.

14. V. Fréedericksz [Frederiks] and A. Schechter (1928).

15. A. Einstein (1929); here quoted from P. Kerzberg (1992), p. 335.

16. For details about Lemaître's life and work, see O. Godart and M. Heller (1985); A. Deprit (1984); and *Revue des Questions Scientifiques* 155 (1984): 139–224, a special issue devoted to Lemaître. Part of my discussion draws upon H. Kragh (1987).

17. Quoted in O. Godart and M. Heller (1985), p. 162.

18. Quoted in A. V. Douglas (1956), p. 111. The letter, of 24 December 1924, was to the Belgian physicist Théophile de Donder.

19. Lemaître to Shapley, 15 February 1924. By permission of the Harvard University Archives. Harvard College Observatory: Records of Director Harlow Shapley, 1930–1940, box 11, Harvard University Archives.

20. G. Lemaître (1927); J. Bernstein and G. Feinberg, eds. (1986), pp. 92–100.

21. G. Lemaître (1958b); O. Godart and M. Heller (1979b).

22. None of Lemaître's works are listed in the *Physics Citation Index, 1920–20* (Philadelphia: ISI, 1982), which also includes citations from major astronomy journals such as the *Astrophysical Journal* and *Monthly Notices*.

23. G. Lemaître (1929).

24. A. S. Eddington (1933), p. 46.

25. The letter, a draft version of Lemaître to Eddington of early 1930, was located in the Lemaître archive in Louvain by Jean Eisenstaedt, from whom it is quoted. See J. Eisenstaedt (1993), p. 361.

26. G. C. McVittie (1967). In another recollection, McVittie recalled Eddington's words as, "I'm sure Lemaître must have sent me a reprint, he's just sent me another, but I'd forgotten about it." (AIP interview of 21 March 1978, conducted by David

DeVorkin). This is one of the interviews conducted during the 1970s at the Center for History of Physics at the American Institute of Physics (AIP). The Sources for History of Modern Astrophysics project includes over 400 taped interviews transcribed into about 10,000 pages. For a description of the project, see S. R. Weart and D. H. DeVorkin (1981). Eddington's "letter" was a review of L. Silberstein (1930), and appeared as Eddington (1930a).

27. A. S. Eddington (1930b).

28. G. Lemaître, "A homogeneous universe of constant mass and increasing radius," *Monthly Notices* 91 (1931): 483–90. The English translation leaves out minor parts of the text and most of the footnotes. As a historical document the French original should be consulted. Thus, the English version condenses a one-page-long discussion of empirical data to the phrase "from a discussion of available data."

29. De Sitter to Shapley, 17 April 1930, as quoted in Smith (1982), p. 187.

30. *Nature* 128 (1931): 706–9.

31. W. de Sitter (1930a), p. 171.

32. W. de Sitter (1930b).

33. L. Silberstein (1930); A. S. Eddington (1930a).

34. F. Zwicky (1929).

35. J. Q. Stewart (1931).

36. S. Sambursky (1937). A similar proposal was made earlier in J. A. Chalmers and B. Chalmers (1935).

37. W. H. McCrea and G. C. McVittie (1930–31).

38. G. C. McVittie (1930–31).

39. H. P. Robertson (1933); O. Heckmann (1932).

40. E. W. Barnes (1933). The book, prefaced November 1932, was an extensively revised version of Barnes's Gifford Lectures of 1927–29.

41. G. Maneff (1932); R. Zaykoff (1933).

42. Including J. Jeans (1930); W. de Sitter (1932); A. S. Eddington (1933). The development of cosmology during the 1930s is covered in J. Merleau-Ponty (1965); J. North (1965); P. Kerzberg (1992).

43. J. Jeans, "The universe: space finite and expanding," *The Times*, 14, 18, 23, 26, and 27 May 1932.

44. A. Einstein (1931).

45. Einstein to Tolman, 27 June 1931. Einstein Collection, box 6, Boston University. In his letter of reply of 14 September, Tolman expressed satisfaction with Einstein's view that a state of $R \approx 0$ had to be avoided.

46. A. Einstein and W. de Sitter (1932).

47. P. Kerzberg (1992), p. 361.

48. M. Bronstein (1933), p. 74; see also M. Bronstein and L. Landau (1933). On Bronstein, see G. E. Gorelik and V. Ya. Frenkel (1994).

49. The quotations are from J. A. Harrison, ed. (1965), vol. 16, pp. 207, 214. For more about Poe's cosmology, see A. Cappi (1994).

50. F. J. Tipler (1988), p. 46.

51. Quoted in S. G. Brush (1978), p. 73. The original source is Nietzsche's *Der Wille zur Macht*, written about 1886, but only published posthumously in 1901.

52. F. Soddy (1904), pp. 178–88.

53. F. Soddy (1909), pp. 241–42.

54. J. Jeans (1929), p. 316.

55. Ibid., p. 317.

56. R. C. Tolman (1922). Twelve years later Tolman returned to the problem, still unable to provide a solution for the discrepancy except that he now suggested that it was one of several phenomena "which indicate the possibility that the present composition of the matter in this portion of the universe results from a past history that involved exceedingly high temperatures." Tolman (1934a), pp. 140–46.

57. O. Stern (1925), (1926). Stern's treatment was criticized and improved in R. C. Tolman (1926).

58. W. Lenz (1926). Lenz was led to the unattractive conclusion that either the temperature of space was 300 K or the radius of the universe was about one hundred thousand times as large as generally assumed.

59. R. C. Tolman (1928); see also F. Zwicky (1928).

60. S. Suzuki (1928). In another work Suzuki extended his calculations to deal also with the thermal disintegration of transuranic elements (Suzuki [1929]).

61. H. C. Urey and C. A. Bradley (1931).

62. R. C. Tolman (1931a), (1931b).

63. A. S. Eddington (1930b).

64. G. Lemaître (1931a).

65. R. C. Tolman (1930a); A. S. Eddington (1930b).

66. G. Lemaître (1931a).

67. *Nature* 127 (1931): 447–53.

68. H. T. H. Piaggio (1931). Millikan's hypothesis will be dealt with in section 4.1.

69. R. C. Tolman (1931a), p. 1641.

70. A. S. Eddington (1928), p. 85.

71. *Nature* 127 (1931): 706.

72. G. Lemaître (1931b).

73. N. Bohr (1931). For Lemaître's interest in the subject, see G. Lemaître (1933b).

74. G. Lemaître (1931b). Strictly speaking, Lemaître's method of expression was inconsistent. If it is meaningless to speak of time until after the original explosion, how can the world have begun "a little before"?

75. G. Lemaître (1949b), p. 452.

76. G. Lemaître (1931b).

77. O. Godart and M. Heller (1985), p. 73. Lemaître's papers are at the Institute of Astronomy and Geophysics of the Catholic University of Louvain, where they were first examined by Odon Godart, a collaborator of Lemaître since 1934.

78. *Nature* 128 (1931): 700–722.

79. Lemaître to C. R. B. Educational Foundation, Inc., 14 January 1932. By permission of the Harvard University Archives. Harvard College Observatory: Records of Director Harlow Shapley, 1930–1940, box 40, Harvard University Archives.

80. G. Lemaître (1931d).

81. W. de Sitter (1930b).

82. G. Lemaître (1931d).

83. G. Lemaître (1949b), p. 448. Nine years later, in his Solvay address, he came up with an age of the universe between twenty and sixty billion years. Lemaître (1958a).

84. G. Lemaître (1934).

85. G. Lemaître (1933d), p. 1086.

86. See also G. Lemaître (1933c). This feature was one of the factors that caused the cosmological constant and the Lemaître model to be reconsidered in the late 1960s. See, e.g., V. Petrosian and E. E. Salpeter (1970); V. Petrosian (1974).

87. Einstein to Lemaître, 26 September 1947. The letter was a response to a letter from Lemaître of 30 July 1947, in which Lemaître summarized his arguments as given in his recently completed manuscript for the volume to be edited by Paul Schilpp (Lemaître [1949b]). Lemaître made a final attempt to justify his view in a letter to Einstein of 3 October 1947. Einstein Collection, box 6, Boston University.

88. For references and details, see M. De Maria and A. Russo (1989).

89. G. Lemaître and M. S. Vallarta (1933).

90. G. Lemaître (1933a). The paper first appeared in 1932, in *Publication du Laboratoire d'Astronomie et de Géodésie d l'Université de Louvain* 9: 171–205. The kind of anisotropic model suggested by Lemaître became known in later literature as the Tolman-Bondi model, referring to work done by Tolman in 1934 and Bondi in 1947. On Lemaître's proof, see O. Godart (1992).

91. A. Einstein (1945); here quoted from the fifth edition of 1956, p. 126.

92. "Theories of universe shaken," *Science News Letter* 19 (10 January 1931): 23.

93. "Salvation without belief in Jonah's tale," *The Literary Digest* 115 (11 March 1933): 23; building on an interview reported in D. Aikman, "Lemaitre follows two paths to truth," *New York Times Magazine*, 19 February 1933. However, according to *Newsweek*'s report, Einstein praised Lemaître's "beautiful and satisfying interpretation of cosmic rays," and did not mention the creation. "Visiting eminence," *Newsweek* 21 (23 January 1933): 30–31.

94. *Newsweek* 22 (4 December 1933): 50.

95. H. P. Robertson (1932).

96. E. W. Barnes (1933), p. 408.

97. O. Heckmann (1932), p. 106

98. W. de Sitter (1930b), p. 218.

99. R. C. Tolman (1934a), p. 484.

100. Ibid., p. 486. A similar warning appeared in R. C. Tolman (1932).

101. E. P. Hubble (1937), p. 62. The characterization of theoretical cosmology as a "shadowy realm" appears on p. 53. Hubble did not refer to Lemaître by name, but he did so in his "Effects of red shifts" from the year before, on p. 542. Hubble also expressed his deep-rooted empiricism in *The Realm of the Nebulae*. "Not until the empirical results are exhausted, need we pass to the dreamy realms of speculations," he wrote (p. 202).

102. A. S. Eddington (1936); Eddington (1946), published posthumously and arranged by Noel B. Slater.

103. A. S. Eddington (1944), p. 204.

104. G. Lemaître (1946). An English translation appeared in 1950 as *The Primeval Atom: An Essay on Cosmogony* (New York: Van Nostrand). An extended version, including a French translation of Lemaître's 1958 Solvay address and a biographic and bibliographic chapter by Odon Godart, was published in 1972 in Brussels by Éditions Culture et Civilization.

105. G. Lemaître (1961), p. 603.

106. Ibid.

107. G. Lemaître (1949b), p. 453.

108. O. Godart and M. Heller (1985), p. 133.

109. Lemaître's conception of elementary-particle physics was probably shared by several physicists with an interest in fundamental theory. Thus, Stephen Hawking has described his choice of research field in 1962 as follows: "I thought that elementary

particles were less attractive, because, although they were finding lots of new particles, there was no proper theory of elementary particles. All they could do was arrange the particles in families, like in botany. In cosmology, on the other hand, there was a well-defined theory—Einstein's general theory of relativity." Quoted in M. White and J. Gribbin (1988), p. 58.

110. Quoted in E. J. Lerner (1991), p. 214.

111. See also O. Godart and M. Heller (1979a).

112. "Salvation without belief in Jonah's tale," *The Literary Digest* 115 (11 March 1933): 23.

113. The text, found among Lemaître's papers after his death, was later published under the title "L'univers, problème accessible à la science humaine." See O. Godart and M. Heller (1978).

114. G. Lemaître (1958a), p. 7.

115. To mention just one example: "Both the Abbé Lemaître and Sir Edmund Whittaker frankly preferred the Big-Bang picture because it could be reconciled with religious teachings about the Creation more satisfactorily than its rivals." S. Toulmin and J. Goodfield (1982), p. 260.

116. Quoted in H. Vecchierello (1934), p. 23.

117. Part of this section builds on an earlier work of mine, which may be consulted for more detailed references. H. Kragh (1982).

118. E. A. Milne (1933). The first indication of Milne's interest in cosmology appeared in an anonymous review of some of Tolman's latest works. Milne (1930). For background on Milne and his view on science, see A. J. Harder (1974).

119. E. A. Milne (1933), p. 4.

120. E. A. Milne (1935), p. 292.

121. On this question, and on Milne's kinematic relativity theory in general, see the analysis in M. Johnson (1947).

122. E. A. Milne (1949b), p. 428.

123. E. A. Milne (1933), pp. 14–15.

124. E. A. Milne (1935), p. 266. Milne presented his philosophical views clearly and explicitly in Milne (1934). For an analysis of Milne's philosophy of science in the 1930s, see G. Gale and J. Urani (1993).

125. E. A. Milne (1937), p. 998.

126. J. Urani and G. Gale (1993).

127. J. B. S. Haldane (1936).

128. J. B. S. Haldane (1939), pp. 68, 76.

129. J. B. S. Haldane (1945a), p. 132; see also Haldane (1945b). Haldane's interest in speculative cosmology went back to the 1920s; cf., e.g., Haldane (1928).

130. In an untitled companion paper to J. B. S. Haldane (1945b), on pp. 135–36.

131. R. Coutrez (1945).

132. G. Lemaître (1945).

133. E. A. Milne (1949), pp. 10–11. The volume was issued as number 1065 in the series Actualités Scientifiques et Industrielles. More about the association of Milne's system with Christian belief is in section 4.5.

134. Ibid., p. 11.

135. Ibid., p. 65.

136. W. H. McCrea and E. A. Milne (1934).

137. A more detailed account of Dirac's cosmology can be found in H. Kragh (1990), pp. 223–46; see also P. S. Wesson (1978), pp. 6–48.

138. A. S. Eddington (1939), p. 170. Eddington obtained his number as $(137 - 1) \times 2^{256}$. The number 137 is the inverse fine-structure constant, that is, $2\pi e^2/hc$.

139. P. A. M. Dirac (1937).

140. P. A. M. Dirac (1938).

141. Ibid., p. 203.

142. E. Teller (1948).

143. S. Chandrasekhar (1937); D. S. Kothari (1938); F. L. Arnot (1938).

144. H. Dingle (1933), p. 155.

145. H. Dingle (1937b).

146. H. Dingle (1937a), p. 785.

147. E. A. Milne (1937), p. 998. The debate took place under the title "Physical science and philosophy" in the same issue, pp. 1000–1010. It was reopened in *Nature* in 1941, now with Eddington as the target of the criticism of Dingle and Jeans. The 1937 debate is considered in H. Kragh (1982).

148. H. Dingle (1934), p. 819. Although neither Dingle nor Milne would admit it, their positions had a good deal in common.

149. F. L. Arnot (1941). Contrary to Milne and Dirac, Arnot rejected the Doppler interpretation of the redshifts, which he attributed to a variation in the velocity of light. His universe was static in τ time, with radius 2×10^9 light years and density 1.9×10^{-28} g·cm^{-3}.

150. P. Jordan (1952), p. 137.

151. A. E. Haas (1936).

152. P. Jordan (1937), (1939).

153. P. Jordan (1944).

154. P. Jordan (1938).

155. P. Jordan and C. Müller (1947); P. Jordan (1952).

156. A concise survey of Jordan's cosmological field theory in the 1950s is given in D. R. Brill (1962).

157. J. D. North (1965), p. 125.

158. H. N. Russell (1921).

159. E. Rutherford (1929).

160. J. Jeans (1928c), p. 409.

161. W. de Sitter (1932), p. 133.

162. H. Spencer Jones (1934), p. 414.

163. W. de Sitter, *Nature* 138 (1931): 708.

164. *Observatory* 56 (1933): 184 (meeting of the Royal Astronomical Society of 12 May 1933); W. de Sitter (1933).

165. R. Tolman (1934), p. 486.

166. E. P. Hubble (1937), pp. 42–43.

167. E. P. Hubble (1942), pp. 109, 112.

168. E. Öpik (1933), p. 79.

169. *Observatory* 58 (1935): 108–14; B. J. Bok (1936).

170. V. Ambarzumian (1936).

171. C. Payne-Gaposchkin (1944).

172. S. Chandrasekhar (1944); B. Bok (1946).

173. J. H. Oort (1931).

174. E. P. Hubble (1936b), p. 16.

175. R. C. Tolman (1949); G. Gamow (1949), p. 367.

176. H. Hönl (1949).

177. R. C. Tolman (1949), memorandum dated 22 March 1948.

178. G. C. Omer (1949).

179. G. Gamow (1949).

180. G. Whitrow (1949), pp. 123–26.

181. A. Einstein (1945), p. 132.

182. E. A. Milne (1948), p. 224.

183. P. A. M. Dirac (1938), p. 204.

CHAPTER 3
GAMOW'S BIG BANG

1. J. Stark, as quoted in *Nature* 68 (1903): 230. Similar ideas were ventured in Stark (1902), pp. 34–35. At that time many physicists believed in the existence of positive electrons, mirror particles of the ordinary negative electrons, but it soon turned out that electrons are only negative (or so it was believed until 1932, when the positron was discovered in the cosmic radiation).

2. W. D. Harkins (1917), quotations from pp. 876, 878.

3. J. Perrin (1919), p. 94.

4. The fascinating story is detailed in J. D. Burchfield (1975).

5. E. Rutherford and F. Soddy (1903), p. 591.

6. E. Rutherford (1907), p. 165.

7. A. S. Eddington (1917). For context and background, see K. Hufbauer (1981). The annihilation process can be traced even farther back in time, well before the nuclear atom. See J. Jeans (1901), where young Jeans suggested the possibility that "positive and negative electrons would rush together and annihilate one another" (p. 426).

8. A. S. Eddington (1919), p. 376.

9. A. S. Eddington (1920).

10. A. S. Eddington (1926b), p. 301. An almost identical sentence appeared in a paper of 1 May 1926 in which Eddington made it clear that the critics he was addressing were, in particular, Jeans and Nernst. Eddington (1926a).

11. D. Britz (1990); F. Close (1991), pp. 19–21.

12. G. Gamow (1928). A thorough historical analysis is provided in R. H. Stuewer (1986).

13. M. von Laue (1929). Laue suggested that the reverse Gamow theory was an argument in favor of Walther Nernst's ideas of element formation; see section 4.1.

14. I. B. Kriplovich (1992).

15. G. Gamow (1963).

16. R. d'E. Atkinson and F. G. Houtermans (1929). Originally Atkinson and Houtermans had entitled their paper "Wie man ein Helium Kern in einen potential Topf köchen" (How to cook a helium nucleus in a potential pot), but the imaginative title was changed by the editor of *Zeitschrift für Physik*. See G. Gamow (1970), p. 73.

17. The details of the theory were criticized by the British physicist Alan H. Wilson, who found somewhat different values for the rate of proton-nucleus reactions for the light elements. A. H. Wilson (1931).

18. R. Peierls, ed. (1986), vol. 9, p. 88. Bohr's view was first made publically known in Gamow's 1931 textbook on nuclear theory and radioactivity (G. Gamow [1931]). For details on the impact of Bohr's view, see G. E. Gorelik and V. Ya. Frenkel (1994), pp. 63–82.

19. Pauli to Bohr, 17 July 1929, reprinted in J. Kalckar, ed. (1985), vol. 6, p. 447.

20. R. C. Tolman (1934a), pp. 382, 486; see also R. C. Tolman (1934c).

21. L. Landau (1932), p. 287. Landau's work is often considered the first prediction of the formation of black holes.

22. M. Bronstein (1933), p. 76. See also section 2.1.

23. C. H. Payne (1925), p. 197.

24. B. Strömgren (1932). Eddington independently reached the same conclusion by a different approach (A. S. Eddington [1932]).

25. R. d'E. Atkinson (1931), p. 250; see also G. Steensholt (1932), in which the calculations of Atkinson and Wilson were further refined.

26. On early thermonuclear reactions and their relevance for astrophysics, see J. Hendry (1987).

27. T. E. Sterne (1933).

28. R. d'E. Atkinson (1936).

29. G. Gamow (1970), p. 15.

30. G. Gamow and D. Iwanenko (1926).

31. Interview with C. F. von Weizsäcker, conducted by T. S. Kuhn, 9 July 1963. Archive for History of Quantum Physics, Niels Bohr Archive, Copenhagen (AHQP).

32. Quoted in R. H. Stuewer (1986), p. 147.

33. Quoted in V. Ya. Frenkel (1994), p. 783.

34. "Gamow interviews Gamow," *Stanford Daily*, 25 June 1936. Gamow sent a copy of the interview to Bohr, explaining that Miller's was a famous drinking place (AHQP).

35. G. Gamow and L. Landau (1933).

36. G. Gamow (1935).

37. G. Gamow (1938).

38. H. J. Walke (1935), p. 350.

39. W. Baade and F. Zwicky (1934). The possible existence of stellar cores consisting mainly of neutrons was first suggested in T. E. Sterne (1933).

40. L. Landau (1938); see also J. R. Oppenheimer and R. Serber (1938).

41. G. Gamow (1935), p. 413. The idea was repeated in Gamow (1937), (preface dated May 1936), p. 238.

42. "The problem of stellar energy," *Nature* 141 (1938): 982.

43. G. Gamow (1970), p. 136. On Bethe, see J. Bernstein (1979).

44. S. A. Blumberg and L. G. Panos (1990), p. 93.

45. H. A. Bethe and C. L. Critchfield (1938).

46. G. Gamow (1940), p. 99 in the third edition of 1952. Train rides seem to have stimulated Bethe's creativity. Ten years later, this time on a train between New York and Schenectady, Bethe made the first calculation of the Lamb shift and thereby an important contribution to the new quantum electrodynamics.

47. Preface to 1940 edition, p. vi.

48. H. A. Bethe (1939); received 7 September 1938. For an excellent, nontechnical survey, see Bethe (1942). See also Bethe's Nobel lecture of 1967, reprinted in Bethe (1968).

49. K. Hufbauer (1991), p. 37.

50. W. A. Fowler (1984), which was the 1983 Nobel lecture. There is more about Fowler and the Kellogg Laboratory in section 5.4.

51. H. N. Russell (1939), p. 18.

52. G. Gamow (1939a), p. 622.

53. F. Hoyle and R. A. Lyttleton (1939a).

54. *Science* 87 (1938): 487–90. The program for 2–3 May 1938 included the following speakers: Carl D. Anderson, "Some aspects of the cosmic-ray problem" and "The basic constituents of matter"; Manuel S. Vallarta, "The influence of the magnetic field of the earth on cosmic-ray particles"; J. F. Carlson, "The theory of cosmic-ray particles"; Arthur H. Compton, "Whence cosmic rays?" and "Recent research on cosmic rays"; Harlow Shapley, "The distribution of matter in the metagalaxy"; Gregory Breit, "The nature of the forces between primordial particles"; Georges Lemaître, "The significance of the clusters of nebulae"; William D. Harkins, "The heat of the stars and the building of the atoms in the universe"; Eugene Guth, "The relativistic theory of primordial particles"; and Arthur E. Haas, "Cosmic constants."

55. UDIS 46/7, University of Notre Dame Archives. I am grateful to Ernan McMullin for having helped me with material concerning the Notre Dame symposium.

56. H. A. Bethe (1939), p. 436.

57. C. F. von Weizsäcker (1938). All quotations are from the translation in K. R. Lang and O. Gingerich, eds. (1979), pp. 309–19. In part I of his work, appearing in *Physikalische Zeitschrift* 38 (1937): 176–91, Weizsäcker dealt with the proton-proton reaction and argued for a proton-proton cycle, which, however, differed in its details from the one proposed by Bethe and Critchfield. For Weizsäcker's early interest in astrophysics, see also C. F. von Weizsäcker (1937).

58. L. Farkas and P. Harteck (1931).

59. C. F. von Weizsäcker (1939).

60. Weizsäcker, letter to the author, 15 December 1994.

61. E.g., H. Kienle (1943); A. Unsöld (1948). Unsöld's paper was dated 3 October 1944, but due to the war and the difficult conditions after 1945 it only appeared in the spring of 1948. For German contributions to cosmology during the war, see also C. F. von Weizsäcker (1948), which was part of the series of FIAT reviews (Field Information Agency, Technical) arranged by the British, French, and American military authorities.

62. O. Heckmann (1942), p. 100.

63. G. Gamow (1938), p. 114.

64. G. Gamow (1940), p. 201. Preface dated 1 January 1940.

65. G. Gamow and E. Teller (1939).

66. V. M. Goldschmidt (1937a); see also Goldschmidt (1937b). On Goldschmidt's career and work, see H. E. Suess (1988).

67. S. Chandrasekhar and L. R. Henrich (1942).

68. Ibid., p. 298.

69. G. Gamow and J. A. Fleming (1942). A shorter report appeared in *Carnegie Institution of Washington, Year Book* no. 41 (1941–42): 64–65.

70. The reality of the expansion of the universe was still questioned by some scientists. Llewellyn Thomas, a British-American physicist, argued at the conference that the redshift-distance relation did not indicate universal expansion but was the result of the interaction between traveling photons and free electrons in intergalactic space.

71. G. Gamow (1942).

72. BSC 19.4 (Bohr Scientific Correspondence), Niels Bohr Archive, Copenhagen. Blegdamsvejen is the street where Bohr's institute was (and still is) located. Gamow was not much interested in politics, but he feared and hated communism. According to Teller, he was violently anticommunist. See S. A. Blumberg and L. G. Panos (1990), p. 44.

73. G. Gamow (1946a).

74. Ibid. The entire quote was italicized by Gamow.

75. G. Gamow (1939b).

76. G. Gamow (1942), (1946a). Italics added.

77. G. Wataghin (1944); G. Wataghin and C. Lattes (1946); O. Klein, G. Beskow, and L. Treffenberg (1946). For comprehensive reviews of the field, see D. ter Haar (1950); R. A. Alpher and R. C. Herman (1950).

78. O. Klein, G. Beskow, and L. Treffenberg (1946), p. 7.

79. G. B. van Albada (1947); F. Hoyle (1947).

80. F. C. Frank (1948).

81. D. ter Haar (1950), p. 142.

82. J. Podolanski and D. ter Haar (1954).

83. Gamow to Einstein, 24 September 1946 (Einstein Collection, box 4, Boston University); G. Gamow (1946b). As in other extracts of Gamow's letters, I have kept to his original spelling.

84. Einstein to Gamow, September 1946 (Einstein Collection, box 4, Boston University), where Einstein also objected that "it is not possible to infer from that part of the universe, which is accessible to us, to the mean material density of the entire world and then to the spatial finiteness and total expansion of the world."

85. K. Gödel (1949).

86. I rely in part on the recollections of Alpher and Herman as presented in R. A. Alpher and R. C. Herman (1975), (1988), (1990), (1993).

87. I thank Alpher for the information (letter to author, 9 November 1994). The mentioned works are R. C. Tolman (1934a); H. Weyl, *Space-Time-Matter* (New York: Dover Publications, 1950), the English translation of Weyl (1923a); W. Pauli (1921); and G. Gamow (1937). Bethe's "bible" consisted of three extensive articles in *Reviews of Modern Physics*, in part coauthored by R. F. Bacher and M. S. Livingston. The papers are reprinted in H. A. Bethe (1986).

88. Abstract of D. J. Hughes (1946).

89. "Early stages of the universe," *Science News Letters* 53 (24 April 1948): 259.

90. Alpher and Herman, letter to author, 20 July 1994. For the association with the nuclear bomb, see the cartoon in the *Washington Post*, 1948, as reproduced in R. A. Alpher and R. C. Herman (1988), p. 28.

91. R. A. Alpher (1983).

92. R. A. Alpher, H. A. Bethe, and G. Gamow (1948).

93. M. A. Dennis (1994), which includes an account of the origin and early history of the Applied Physics Laboratory.

94. G. Gamow (1966), p. 445.

95. G. Gamow (1949), p. 369. Gamow undoubtedly enjoyed being able to smuggle the fake reference into a serious volume celebrating Einstein's seventieth birthday.

96. R. A. Alpher and R. C. Herman (1988), p. 28. Back in 1931, young Bethe and his colleagues Guido Beck and Wolfgang Riezler had outdone Gamow with a joke in *Die Naturwissenschaften*. For the joke, and Gamow's attempt to counter it, see M. Delbrück (1972).

97. G. Gamow, *The Creation of the Universe*, 2nd ed. (London: Macmillan, 1961), p. 64.

98. G. Gamow (1948b).

99. G. Gamow (1954a), p. 63.

100. R. A. Alpher (1948), p. 1581.

101. The widely different values reported in the paper were the result of a trivial error of calculation where a factor of 10^4 had been used instead of the correct 10^{-4}.

102. G. Gamow (1970), pp. 149–50.

103. Einstein to Gamow, 4 August 1948 (Gamow file, Library of Congress), reproduced in facsimile in R. A. Alpher and R. C. Herman (1972), p. 310. On the bottom of Einstein's letter, Gamow added the following comment: "Of cauvse [*sic*], the old man agrees with almost anything nowadays." The paper Einstein commented on was "The evolution of the universe," which Gamow had sent to Einstein on 9 July 1948, asking for his opinion. Einstein Collection, box 4, Boston University.

104. The archbishop of Laodicea to Gamow, 15 March 1952; The secretary of state, the Vatican City, to Gamow, 16 July 1952. Gamow file, Library of Congress. The paper was G. Gamow (1951).

105. The constants have their usual meaning: e = electronic charge, $\hbar$ = Planck's constant divided by 2π, m = proton mass, c = velocity of light, and G = constant of gravitation; ε is the binding energy of the deuteron. For Gamow's fascination with combinations of natural constants, see R. A. Alpher (1973); and H. Kragh (1991).

106. "Creation of galaxies," *Science News Letter* 53 (31 July 1948): 71. Gamow found his *Nature* calculations interesting enough to include them as an appendix in the third, revised edition of his textbook of nuclear theory (G. Gamow and C. L. Critchfield [1949], pp. 334–37).

107. The Gamow file in the Library of Congress contains much correspondence between Alpher and Gamow from this period, some of it relating to the errors in the *Nature* paper.

108. R. A. Alpher and R. C. Herman (1948b).

109. R. A. Alpher and R. C. Herman (1949).

110. E. M. Lifshitz (1946). Alpher was thoroughly familiar with the problem, which was the subject of his ill-fated dissertation of 1945–46; it was Lifshiftz's article that made him turn to primordial nucleosynthesis.

111. R. A. Alpher and R. C. Herman (1949), p. 1094.

112. Ibid., p. 1091; A. Einstein (1945), p. 126.

113. Gamow, interview by Charles Weiner, AIP, 25 April 1968.

114. R. A. Alpher, R. C. Herman, and G. Gamow (1949).

115. Gamow file, Library of Congress.

116. For this reason the Gamow-Teller-Metropolis-Ulam work remained unpublished. Gamow summarized it in 1949 ("On relativistic cosmogony"), at a time when he still had confidence in the program.

117. Alpher, letter to author, 20 July 1994.

118. M. G. Mayer and E. Teller (1949).

119. M. G. Mayer and E. Teller (1950).

120. Ibid., p. 84.

121. Ibid., p. 88.

122. W. Band (1950).

123. R. E. Peierls, K. S. Singwi, and D. Wroe (1952). A historical review of poly-neutron theories is given in R. J. Tayler (1983).

124. J. S. Smart (1949).

125. R. A. Alpher and R. C. Herman (1951). In an appendix, Theodore H. Berlin of Johns Hopkins University analyzed the rate equations mathematically and derived an exact solution. He proved that such a solution could only be obtained if the expansion of the universe is ignored (T. H. Berlin [1951]).

126. C. Hayashi (1950).

127. R. A. Alpher, J. W. Follin, and R. C. Herman (1953). A preliminary account was given at a meeting of the American Physical Society in Washington on 1 May 1953; see "Initial conditions in the expanding universe and element synthesis," *Physical Review* 91 (1953): 479. See also the review article, Alpher and Herman (1953).

128. As recalled by Chen Ning Yang, who as a graduate student at the University of Chicago's Institute of Nuclear Studies attended the evening lectures. See E. Fermi, *Collected Papers*, vol. 2 (Rome: University of Chicago Press, 1965), p. 673.

129. See M. G. Mayer and E. Teller (1950), p. 69.

130. E. Fermi (1949), pp. 707–20. I am grateful to Domenico Meli for having provided me with an English translation.

131. Gamow to Alpher, 12 January 1949. Gamow file, Library of Congress.

132. The first published account of the calculations appeared in G. Gamow (1949), where the paper was referred to as "as yet unpublished." In their review from the spring of 1950, Alpher and Herman discussed in more detail the Fermi-Turkevich work (on pp. 193–96, 201–2).

133. Gamow to Alpher and Herman, 6 June 1949 (Gamow file, Library of Congress). In the letter the carbon isotope is written as $_5C^{10}$, which is obviously a slip of the pen. The abbreviation m.u. refers to an atomic mass unit and is equal to 932 MeV.

134. For Wigner's proposal, which remained unpublished, see G. Gamow (1949), p. 371; and (1952b), p. 70.

135. E. Fermi (1949), p. 720.

136. On Carson and the self-heating idea, see G. Gamow (1952b), p. 72. The neutron-aggregate idea was mentioned in R. A. Alpher and R. C. Herman (1950), p. 201, and the photoeffect idea in Gamow (1950).

137. G. Gamow (1949), p. 371. In 1953, he similarly referred to the mass-5 problem as "not *yet* removed" (Gamow, p. 12). Emphasis added.

138. R. A. Alpher and R. C. Herman (1953), p. 34.

139. A. S. Eddington (1926), p. 371.

140. E. Regener (1933).

141. R. A. Alpher and R. C. Herman (1949), (1950).

142. Gamow to Alpher, undated (Gamow file, Library of Congress). Cf. the recollections of Alpher and Herman: "In 1948 and 1949 he [Gamow] argued with us personally and in correspondence that even if the concept of a remnant blackbody background radiation was real, it was not useful because of the presence of starlight at the Earth at about the same energy density" (R. A. Alpher and R. C. Herman [1988], p. 31.

143. Arthur Chernin has argued that Gamow's 1953 calculation is in fact justified and contains no errors, although it rests on a certain assumption not explicitly stated. This leaves Gamow's argument correct, but no less obscure. In his (nationalistically colored) endeavors to honor the Russian-born Gamow, Chernin claims that Gamow predicted the cosmic background radiation in his paper of 1946. This is, of course, incorrect. A. D. Chernin (1994).

144. G. Gamow (1956b), p. 1731; (1953).

145. In 1975 Alpher and Herman wrote that "At the time we made the black-body radiation prediction we did not pursue the question of its detectability" (p. 336). However, on one occasion Alpher and Herman did explore the possibility of detecting the radiation. They consulted radar experts at Johns Hopkins University, the Naval Research Laboratory, and the National Bureau of Standards, who told them that detection would not be possible with the technology then available. See S. Weinberg (1977), p. 130.

146. A. McKellar (1940).

147. W. S. Adams (1941).

148. G. Herzberg (1950), p. 496.

149. R. Dicke, et al. (1946).

150. Quoted in S. S. Schweber (1986), p. 66, where the American pragmatism is discussed. See also Schweber (1989).

151. Gamow file, Library of Congress. I am not sure exactly what conference it was or when it took place.

152. R. A. Alpher and R. C. Herman (1988), p. 26. The disciplinary orientation of the big-bang program is illustrated by the extensive bibliography in Alpher and Herman's review article of 1950. It contains 108 references to physics literature (journals, reports or books) of which 64 are papers in the *Physical Review* or *Reviews of Modern Physics*; 54 of the references are to science publications of a general kind (such as *Nature*); and only 27 are to publications in the astronomical or astrophysical literature.

153. Summarized in E. M. Burbidge and G. R. Burbidge (1953). At the same conference, Mayer reported on her and Teller's polyneutron theory.

154. O. Gingerich (1994).

155. "The great event," *Time* 57 (30 April 1951): 80–81.

156. R. A. Alpher and R. C. Herman (1990), p. 144. A search through the literature for 1954–63 reveals only one research publication which clearly belongs to the Gamow big-bang tradition, a paper by Hayashi and Nishida of 1957 (for this, see section 5.4). Alpher, Herman, and Follin gave papers at American Physical Society meetings in 1959 (on "Initial conditions in the expanding universe and element synthesis," and "Light element formation during early stages of the expanding universe") and in 1960 (on "Formation of D, Li, Be, and B in an expanding universe"). They did not publish their work.

157. G. Gamow (1954a), p. 62.

158. R. D. Richtmyer, review of Gamow (1970) in *Science* 171 (1971): 997–98. According to Gamow, "the last good article on theoretical physics" was Dirac's paper on antiparticles, written in 1931. See R. Hobart Ellis (1968).

159. Quoted in O. Gingerich (1994), p. 36.

160. C. Weiner (1973).

161. Interview in A. Lightman and R. Brawer (1990), p. 208.

162. Discussion remark in P. Ledoux, ed. (1954).

CHAPTER 4
THE STEADY-STATE ALTERNATIVE

1. R. Schlegel (1958). Section 4.1 is essentially an expanded version of H. Kragh (1995).

2. W. D. MacMillan (1918), p. 49. The "singular points" referred to stars.

3. W. D. MacMillan (1925), p. 99.

4. In a review of A. Veronnet (1926), appearing in *Astrophysical Journal* 66 (1927): 139–43.

5. W. D. MacMillan (1920), p. 73; see also the popular exposition in MacMillan (1929).

6. W. D. MacMillan (1923), p. 105.

7. Ibid., p. 106.

8. W. D. MacMillan (1932). For the tired-light hypothesis, see also section 2.4. As another possibility, MacMillan suggested that the energy "evaporated" from the

photons would exist as a kind of low-frequency cosmic background radiation. Had MacMillan lived to 1965—he died in 1948—he might have welcomed the discovery of the cosmic microwave radiation as a confirmation of his old idea.

9. W. M. Rankine (1881).

10. A. Holmes (1913), pp. 120–21.

11. Quoted from P. A. Y. Gunter (1971), p. 539. Gunter claims that Bergson's was a precursor of the steady-state universe, but in fact Bergson's speculations were purely philosophical and did not relate to either physics or astronomy.

12. G. T. Toulmin (1789), as quoted in S. Toulmin (1962). For further historical background, see chaps. 2, 3 in C. C. Gillispie (1951).

13. R. A. Millikan (1930), p. 640.

14. W. Kohlhörster (1924), p. 66.

15. R. A. Millikan (1925).

16. R. A. Millikan and G. H. Cameron (1926).

17. R. A. Millikan and G. H. Cameron (1928), p. 539.

18. R. Kargon (1983); see also R. C. Tobey (1971), pp. 137–54. Millikan gave a popular version of his theory in an address before the Society of Chemical Engineers on 4 September 1928. R. A. Millikan (1928).

19. R. A. Millikan and G. H. Cameron (1928), p. 556.

20. Ibid., p. 554. Millikan insisted on calling protons positive electrons. The real positive electron or positron, the antiparticle of the ordinary electron, was only discovered in 1932, but even then Millikan continued to use positive electron for the proton.

21. M. De Maria and A. Russo (1990).

22. J. Jeans (1928a), p. 470; see also Jeans (1926b), (1926a).

23. R. A. Millikan (1930).

24. W. D. MacMillan (1926).

25. "Cosmic optimism," editorial, *New York Times*, 31 December 1930, as quoted in M. De Maria and A. Russo (1990), p. 406. According to De Maria and Russo, Millikan's view expressed the "typically American optimistic and progressive ideology of science."

26. G. I. Pokrowski (1929).

27. S. B. Stone (1930).

28. E. C. Stoner (1929).

29. R. A. Millikan (1935), pp. 454–56.

30. R. Kargon (1982), pp. 144–47.

31. R. A. Millikan and G. H. Cameron (1928), p. 556.

32. R. A. Millikan (1931), p. 170.

33. W. R. Inge (1933), p. 50. For another British defense of a perpetually creative universe à la Millikan, see H. T. H. Piaggio (1931).

34. W. R. Inge (1933), pp. 64–65.

35. R. Kapp (1940), pp. 247–55.

36. For a representative example of cosmochemical speculations, see N. Lockyer (1900). For Crookes, see below.

37. S. Arrhenius (1909), p. 184. The book, first published in 1907, became very popular and by 1909 had been translated into German, English, French, Russian, Finnish, and Hungarian. The friendship between Nernst and Arrhenius went back to the mid-1880s, but from about 1900 their relations were strained and soon turned into open enmity.

38. W. Nernst (1912).

39. W. Nernst (1921), p. 2. In spite of the fact that cosmophysics occupied a large part of Nernst's scientific life for two decades, this aspect of his work is not well known. For example, in his biography of Nernst, Kurt Mendelssohn only refers briefly and inadequately to his work in astrophysics and his arguments for a kind of steady-state universe. K. Mendelssohn (1973), pp. 115–16.

40. W. Nernst (1935), p. 528.

41. W. Nernst (1916).

42. W. Nernst (1921), p. 37; see also P. Günther (1924), pp. 454–457.

43. E. Wiechert (1921). Nernst acknowledged the affinity between his and Wiechert's ideas.

44. W. Nernst (1922), which was Nernst's inaugural lecture of 15 October 1921 as president of the University of Berlin. For Nernst's address as an example of the Weimar zeitgeist, see P. Forman (1972), pp. 84–87.

45. W. Nernst (1928). Nernst first suggested the principle in his work of 1921.

46. In his review of A. Veronnet (1926) (see note 4 above).

47. See section 3.1. See also J. Bromberg (1976).

48. W. Nernst (1935a), p. 520. Enrico Fermi, Otto Hahn, and Lise Meitner believed about 1934 that they had detected elements with atomic numbers between 92 and 96 in experiments with neutrons bombarding uranium. The reports turned out to be wrong.

49. W. Nernst (1928), p. 141.

50. J. Becquerel (1908), p. 1308.

51. G. LeBon (1907), pp. 92–93.

52. W. Nernst (1907), p. 392. Nernst may have taken the conjecture, including the name neutron, from W. Sutherland (1899). The idea of a corpuscular ether consisting of neutrons lost its appeal in the 1920s with the demise of the ether and with Rutherford's introduction of hypothetical neutrons as proton-electron composites. Such neutrons were frequently discussed about 1930, in a few cases within an ether framework. See, for example, V. Posejpal (1930). For earlier attempts to elevate the ether to a physically—and chemically—active medium, see H. Kragh (1989).

53. W. Nernst (1937), p. 660.

54. W. Nernst (1938).

55. W. Nernst (1935b).

56. M. von Laue (1929), p. 733. The suggestion was taken up by the Hungarian physicist Johann Kudar, who worked in Berlin (J. Kudar [1929]).

57. W. Nernst, "Zur Energiebilanz des Weltalls," abstracted in *Vierteljahrschrift der Astronomischen Gesellschaft* 72 (1937): 311.

58. A. S. Eddington (1927–29), p. 111. In *The Internal Constitution of Stars*, Eddington addressed "a criticism urged by Nernst, Jeans and others" concerning his calculations of stellar energy output (on p. 3).

59. A. S. Eddington (1927–29), p. 117.

60. A. S. Eddington (1928), p. 86.

61. J. Jeans (1930), pp. 74–76.

62. J. Jeans (1929), p. 311. In a reply to the Czechoslovakian chemist Bohuslav Brauner, who had suggested a rejuvenating universe à la MacMillan, Jeans pointed out that this kind of "widely desired cyclic universe" would already be in a state of maximum entropy. "With universes, as with humanity, the only possible life is progress to the grave," he concluded. Jeans (1928d).

63. According to C. F. von Weizsäcker (1964), pp. 151–52.

64. O. L. Reiser (1952).

65. The vacuum of quantum field theory, being endowed with fluctuations out of which particles may be formed, is very different from the classical nothingness. It is not matter, not even energy, but in its fluctuations it possesses the potentialities of matter creation. One cannot help noticing the similarity between the classical ether of Nernst, Wiechert, and Lodge and the modern quantum vacuum. From this point of view, Nernst's theory may well be said to include also the notion of creation out of nothing.

66. J. Dalton (1808), p. 212.

67. W. Crookes (1887), p. 572.

68. O. Lodge (1920), (1923), (1924). As mentioned, radiation pressure also played an important role in the cosmological views of Arrhenius.

69. O. Lodge (1904), p. 228.

70. O. Lodge (1926), p. 88. For Lodge's world view and cosmological speculations, see P. Rowlands (1990), pp. 270–98.

71. O. Lodge (1926), p. 96.

72. J. Jeans (1928c), p. 360.

73. K. Pearson (1888–89); Pearson (1900), pp. 265–67; J. Jeans (1904).

74. J. Jeans (1928c), p. 421.

75. A. E. Milne (1946); (1948), pp. 156–69.

76. A. E. Milne (1948), p. 167.

77. R. C. Tolman (1929b), p. 266; see also J. D. North (1965), pp. 198–201.

78. Y. Mimura (1935), followed by numerous other papers in the same journal by Mimura, T. Sibata, H. Takeno, and K. Itimaru. The Hiroshima program continued until August 1945, when the atomic bomb destroyed the city; it was not resumed after the war. For a summary of the program, see H. Takeno (1960–61).

79. E. Schrödinger (1937).

80. E. Schrödinger (1939), p. 901.

81. A. Rüger (1988); see also J. Audretsch (1987).

82. E. Schrödinger (1940).

83. H. Urbantke (1992). In E. Schrödinger (1956), wave mechanics was discussed in expanding universes. But Schrödinger did not mention the concept of gravitational matter creation, which at that time would have been most relevant in connection with the steady-state theory. Apparently without knowing of Schrödinger's work, Bryce DeWitt argued in 1953 that a varying curved metric can produce particles through its interaction with the vacuum fluctuations of a scalar field. See B. DeWitt (1953).

84. Fred Hoyle, interview 15 August 1989, as quoted in A. Lightman and R. Brawer (1990), p. 53; see also F. Hoyle (1994).

85. F. Hoyle (1985), p. 91.

86. A. Lightman and R. Brawer (1990), p. 56; see also F. Hoyle (1992).

87. F. Hoyle (1938); H. A. Bethe, F. Hoyle, and R. Peierls (1939).

88. Pauli to Peierls, 25 May 1938, and Pauli to Uhlenbeck, 9 July 1938, in K. von Meyenn, ed. (1985), which contains references to Hoyle's unpublished work on pp. 561–608. The theory of Pauli and Fierz appeared in M. Fierz (1939); and M. Fierz and W. Pauli (1939).

89. F. Hoyle (1939).

90. F. Hoyle (1985), p. 145. According to another source, Dirac warned him in about 1939 that theoretical physics was running out of steam and indicated that there was not much to do even for good people. F. Hoyle (1986).

91. F. Hoyle (1982a), which includes a reconstruction of Hoyle's original thought experiment.

92. E. Schrödinger (1935).

93. F. Hoyle (1982a), p. 19. Hoyle's recollection may to some degree rationalize his motives for leaving quantum theory. In his autobiographies of 1985 and 1994, he does not mention the crisis at all. Also, if Hoyle reached the decision to quit quantum theory in 1938 it makes it difficult to understand why he contributed to the very same field as late as the summer of 1939. There were, undoubtedly, additional motives of a less ideal kind, such as the fear of not being good enough, a fear which the unfortunate incident with the redrawn paper of 1938 must have strengthened.

94. F. Hoyle and R. A. Lyttleton (1939b).

95. F. Hoyle and R. A. Lyttleton (1939a), which was a reply to G. Gamow (1939a); see also R. A. Lyttleton and F. Hoyle (1940).

96. H. Bondi, interview by David DeVorkin, AIP, 20 March 1978.

97. H. Bondi (1990a), p. 18.

98. Letters of recommendation, copies included in letter from Helene Bondi to Einstein, 7 January 1941. Einstein Collection, box 24, Boston University.

99. Einstein requested HIAS, a Jewish immigrant aid society, to take action and offered his further assistance in the case (Einstein to J. Hershfield, 4 February 1941. Einstein Collection, box 24, Boston University.)

100. T. Gold, interview by Spencer Weart, AIP, 1 April 1978, which is the source of much of the biographical information.

101. T. Gold (1982), p. 62.

102. F. Hoyle (1965a), p. 92.

103. H. Bondi (1942).

104. F. Hoyle (1965a), p. 92.

105. H. Bondi (1990a), p. 41.

106. T. Gold (1982), p. 63. In another version of the story, the figure is 10^{48} and the particular problem a calculation of the shock wave originating when a star passes a gas cloud. F. Hoyle (1965a), p. 93.

107. H. Bondi, interview by David DeVorkin, AIP, 20 March 1978.

108. F. Hoyle (1986), p. 448; see also Hoyle (1994), pp. 219–30.

109. R. J. Pumphrey and T. Gold (1947); T. Gold (1948). Until the appearance of the steady-state theory, Gold's entire production consisted of two notes in *Nature*, coauthored with Pumphrey, and two consecutive papers in the *Proceedings*.

110. T. Gold (1989), p. 105. It was only much later, in the 1980s, that Gold's theory of hearing as a phenomenon involving an active receiver gained support.

111. H. Bondi and F. Hoyle (1944); Bondi (1990), p. 50.

112. F. Hoyle (1946), (1947).

113. R. J. Tayler, ed. (1987), p. 145.

114. A. Lightman and R. Brawer (1990), p. 55.

115. O. Heckmann (1942).

116. H. Bondi (1947).

117. H. Bondi (1948a). Bondi expressed his thanks to McVittie and Gold for helpful suggestions. Since he did not mention Hoyle, we may assume that Hoyle's role was merely to encourage Bondi in taking on the job.

118. G. Temple (1939), p. 466.

119. H. Bondi (1948a), p. 107.

120. H. Bondi (1990b).

121. F. Hoyle (1982b), p. 51. This essay was first published separately as F. Hoyle (1980). In 1980 Hoyle dated the visit to the cinema to have taken place in 1946; in 1990 he believed it was late 1946 or early 1947. See F. Hoyle (1990).

122. H. Bondi (1982); see also Bondi (1993b); H. Bondi, T. Gold, and F. Hoyle (1995).

123. F. Hoyle (1965a), p. 93.

124. "From the psychoanalytical standpoint, the lost pencil sharpener [if it was a pencil-sharpener!], for instance, suggests anxieties of lost potency; an ensuing feminine disarrangement (of the room) ensues. The theory of continuous creation might then have emerged as an answer to anxieties of diminishing procreative power; hence the importance given to the recollection of this incident." L. Feuer (1977), p. 404.

125. H. Bondi (1990b), p. 191. Bondi only refers to the Lemaître-Eddington model, but the same kind of objection holds for Lemaître's big-bang model.

126. H. Bondi (1948a), p. 111.

127. H. Bondi and T. Gold (1948), p. 264.

128. F. Hoyle (1948), p. 374.

129. According to Hoyle, he knew of references to the continuous creation of matter "that go back more than twenty years." F. Hoyle (1950a), p. 122.

130. F. Hoyle (1982b), p. 51; (1947).

131. "According to Hoyle," *Time* 56 (20 November 1950): 84–91. According to the account given by the magazine, apparently based on an interview with Hoyle, Lyttleton was a leading figure in the new cosmology. In a letter to the author of 19 February 1995, Hoyle confirms that Lyttleton had no role in the development of the steady-state theory.

132. F. Hoyle (1982b), p. 50.

133. W. H. McCrea (1950).

134. T. Gold, interview by Spencer Weart, AIP, 1 April 1978.

135. H. Bondi (1952a), preface dated 27 October 1950.

136. H. Bondi (1948a), p. 114. For a thorough, philosophically oriented examination of the works of Bondi, Gold, and Hoyle, and the relationship between the steady-state theory and Milne's cosmology, see J. Merleau-Ponty (1965), pp. 214–55. A modern analysis of steady-state philosophy about 1948 is given in Y. Balashov (1994).

137. F. Hoyle (1949d). According to the preface, dated 27 June 1948, the main part of the book was written during the summer of 1947.

138. The reply, of 10 May 1948, is excerpted in F. Hoyle (1982b), p. 53.

139. Lecture no. 557, Kapitza Club minute book, AHQP 38.3. The Kapitza Club—so named after the Russian physicist Peter Kapitza who founded the club in 1922—was an informal discussion club for physicists at Cambridge University of which both Hoyle and Bondi were members (but not Gold). One year earlier, Gold had given another talk to the club, on "Physics of the ear" (no. 540), and on 3 February 1948 Bondi had lectured on "Cosmology" (no. 551).

140. F. Hoyle (1948), p. 372. All the following quotations are from this paper.

141. F. Hoyle (1989a), p. 101.

142. H. Bondi and T. Gold (1948), p. 254. If not otherwise mentioned, all following quotations are from this paper.

143. T. Gold, interview by Spencer Weart, AIP, 1 April 1978.

144. F. Hoyle (1950a), p. 125. In the British edition of his book, Hoyle did not refer to skyscrapers, an American phenomenon, but to St. Paul's Cathedral.

145. F. Hoyle (1955a), p. 290.

146. A somewhat similar feature appeared in a new formulation of electrodynamics proposed by Dirac in 1951, where he reintroduced a kind of ether. Dirac's electrodynamic ether suggested to Bondi and Gold that there might be a connection to steady-state cosmology, but Dirac denied that there was any connection between the two kinds of preferred motions. See H. Bondi and T. Gold (1952), and Dirac's reply in the same journal.

147. H. Weyl (1923b). A detailed historical analysis is provided in P. Kerzberg (1986).

148. T. Gold (1949).

149. H. Bondi (1952a), p. 144.

150. W. H. McCrea (1950a).

151. B. R. Martin (1976), p. 52, based on an interview with Bondi on 5 July 1976. I am grateful to Dr. Martin for having supplied me with a copy of his thesis work.

152. *Observatory* 69 (1949): 49.

153. W. H. McCrea (1950a).

154. F. Hoyle (1949a). The lectures took place on 26 October, and 2 and 9 November.

155. F. Hoyle and R. A. Lyttleton (1940).

156. Bondi recalled that he did not know of Alpher and Herman's prediction (see H. Bondi [1990b], p. 196). Hoyle, however, was aware of it. He knew about the 5 K prediction of 1948 and also studied carefully Alpher and Herman's article in *Reviews of Modern Physics* from 1950. F. Hoyle, letter to author, 19 February 1995.

157. F. Hoyle (1949b).

158. *Observatory* 68 (1948): 214–16.

159. According to Gold's recollections, Born gave "an impassioned speech" against the steady-state theory, and the general response at the meeting was intense and largely negative (T. Gold, interview by Spencer Weart, AIP, 1 April 1978). Born was not much occupied with cosmology, but there is little doubt that he disliked the steady-state theory because of its rationalistic features. Born had earlier criticized Eddington and Milne for cultivating theory at the expense of experiment. See M. Born (1956), based on an address given in 1943.

160. *Observatory* 69 (1949): 47–50.

161. Ibid., p. 49.

162. E. A. Milne (1952).

163. Ibid., p. 77.

164. F. Hoyle (1949c).

165. F. Hoyle (1950a). All following references are to the American edition rather than to the first British edition published by Blackwell in 1950 with a preface dated 3 April 1950. A revised edition appeared in 1960, and reprints by Pelican Books in 1963, 1965, and 1968.

166. Ibid., p. 124.

167. Ibid., p. 139

168. F. Hoyle (1965a), p. 98.

169. F. Hoyle (1982b), p. 54.

170. H. Dingle, review of F. Hoyle (1950a), in *Nature* 166 (1950): 82–83.

171. H. Dingle (1952), p. 164. The essay on "Modern theories of the origin of the universe" (pp. 151–69) was dated November 1949.

172. G. P. Thomson (1951).

173. F. K. Edmondson, *Sky and Telescope* 10 (1951): 273–74. K. F. Mather, a geologist, also noticed Hoyle's tendency to manipulation, and warned readers that the book should be read with caution. See *Science* 113 (1951): 427.

174. R. E. Williamson (1951).

175. In his book *Worlds in Collision* (London: Victor Gollancz, 1950) and in articles in *Reader's Digest*, Immanuel Velikovsky challenged established astronomy and geology by claiming that the history of the earth was completely different from the one ordinarily accepted. Velikovsky was quickly denounced by professional scientists, but his claims, and the way in which they were handled by leading astronomers, caused a controversy which was much discussed in the early 1950s. Incidentally, Hoyle met Velikovsky shortly after, at a meeting in Princeton in 1953. See F. Hoyle (1994), p. 285.

176. H. Bondi (1955a), pp. 158, 161; see also section 5.2.

177. F. Hoyle and R. A. Lyttleton (1948), p. 90.

178. Ibid. For a further discussion of the topic, see section 5.2.

179. D. O'Connell (1952–53), p. 136.

180. P. Jordan (1949), with Born's introduction on p. 637. Jordan also discussed his theory in relation to the steady-state cosmology in Jordan (1951).

181. T. Gold (1949).

182. E.g., A. J. Rutgers (1950).

183. R. O. Kapp (1949), (1950).

184. F. Hoyle, *Nature* 165 (1950): 68–69. Kapp admitted in a rejoinder that Hoyle's reasoning was "beyond my mathematics," but maintained that his view had certain advantages over Hoyle's. See Kapp, *Nature* 165 (1950): 687–88.

185. R. O. Kapp (1953); (1955), with a discussion between Kapp and Bondi on pp. 239–43.

186. Ibid., p. 180.

187. R. O. Kapp (1960). The book received a caustic review by McVittie in *Science* 133 (1961): 1231–36.

188. P. Couderc (1952), (updated translation of French original of 1950), p. 220. See also G. J. Whitrow (1954a), an essay review of Couderc's book based on a BBC talk of 10 February 1953.

189. P. Couderc (1952), pp. 213–14.

190. O. Heckmann (1951). The Stebbins-Whitford effect will be examined in section 5.4.

191. P. Michelmore (1962), p. 253. The remark was made in a conversation with Manfred Clynes, a young pianist and acquaintance of Einstein, who apparently reported it to Michelmore.

192. O. Godart and M. Heller (1985), p. 139. Lemaître knew Hoyle personally, especially from a two-week drive they made together through Italy and Austria in 1957. In spite of their very different cosmological (and religious) views they got along very well. See F. Hoyle (1993).

193. E. Finlay-Freundlich (1951).

194. D. ter Haar (1950).

195. Alpher and Herman, letter to author, 20 July 1994.

196. H. Bondi (1952a); G. Gamow (1951). In the second edition of *Cosmology*, prefaced March 1959, Bondi did refer to Gamow, Alpher, and Herman, but only to point out that their theory failed in accounting for the formation of the elements (p. 58).

197. F. Hoyle (1950b); G. Gamow and C. L. Critchfield (1949). The latter was prefaced September 1947, but included a couple of appendixes written later. One of these, on "The origin of elements" (pp. 334–37), summarized the most recent results of the Gamow-Alpher-Herman theory as of the fall of 1948.

198. Alpher and Herman, letter to author, 23 August 1995.

199. W. H. McCrea (1939).

200. Gamow to Alpher, 23 April 1961, apparently in reply to a question from Alpher. Gamow file, Library of Congress.

201. W. H. McCrea (1950a), p. 10.

202. W. H. McCrea (1950b).

203. G. Gamow (1965), pp. 60–61. Reproduced with the permission of Cambridge University Press.

CHAPTER 5
CREATION AND CONTROVERSY

1. D. E. Littlewood (1955).

2. M. Johnson (1951a).

3. J. W. Dungey (1955).

4. R. Stoops, ed. (1958), p. 78.

5. Discussion section in O. Heckmann (1962), p. 439.

6. H. Nariai (1952). Another non-British steady-state candidate is A. Gião, to be mentioned below.

7. H. Nariai (1961–62), p. 40.

8. H. Nariai (1955).

9. W. H. McCrea (1951).

10. The expression is due to Leopold Infeld, the Polish mathematical physicist and collaborator of Einstein. L. Infeld (1957), p. 398.

11. E. T. Whittaker (1935).

12. As McCrea noted, these results were well known from relativistic cosmology. See R. C. Tolman (1934a), pp. 381–83.

13. W. H. McCrea (1953), p. 350.

14. W. H. McCrea (1963a), p. 195.

15. G. C. McVittie (1987), p. 68.

16. W. H. McCrea (1990).

17. G. C. McVittie, interview by David DeVorkin, AIP, 21 March 1978. It is difficult to reconcile this sentiment with McVittie's work of 1952.

18. G. C. McVittie (1952); see also G. C. McVittie (1953).

19. W. Bonnor (1955), p. 23.

20. A. Gião (1964). This was part of the proceedings of a lecture course held in Lisbon from 9 to 20 September 1963 and sponsored by the Gulbenkian Foundation and the NATO Science Committee. Indicating the unsettled situation in cosmology, the course mostly dealt with unorthodox cosmological theories, and with the exception of McVittie's contribution none of the lectures were concerned with empirical aspects.

21. F. Hoyle (1994), p. 404.

22. C. de Turville (1960).

23. T. Araki (1953).

24. K. Pearson (1900), pp. 265–67.

25. F. A. E. Pirani (1955). The gravitinos appearing in later supergravity theories have only their name in common with Pirani's hypothetical particles.

26. F. A. E. Pirani, letter to author, 25 August 1994.

27. V. A. Bailey (1959).

28. First, perhaps, in J. S. Dowker (1964).

29. F. Hoyle (1958a), p. 57.

30. P. Roman (1960), p. 17.

31. W. Davidson (1959).

32. W. Bonnor (1957a).

33. W. Bonnor (1960).

34. Ibid., p. 480.

35. A biographical sketch and a bibliography are included in M. A. H. MacCallum, ed. (1985), pp. xiii–xx.

36. R. A. Lyttleton and H. Bondi (1959).

37. See E. Whittaker (1960), vol. 1, pp. 149–51.

38. H. Bondi and T. Gold (1948), p. 267.

39. L. G. Chambers (1961a), (1961b), (1963).

40. However, in a letter to the author of January 1995, Bondi wrote that he never saw the electrical universe as a version of the steady-state theory, although it originated from a similar iconoclastic mental attitude.

41. W. F. G. Swann (1961b).

42. H. Bondi et al., eds. (1960), p. 40; see also R. A. Lyttleton (1960).

43. R. A. Lyttleton and H. Bondi (1960), p. 443.

44. R. A. Lyttleton (1956).

45. F. Hoyle (1960a).

46. R. A. Lyttleton (1960), p. 33.

47. R. A. Millikan (1917), pp. 80–83.

48. A. Piccard and E. Kessler (1925).

49. A. M. Hillas and T. E. Cranshaw (1959). Hillas and Cranshaw knew about the Bondi-Lyttleton hypothesis from a clip in the *Manchester Guardian*, 13 May 1959.

50. Reply to Hillas and Cranshaw, in *Nature* 184 (1959): 974.

51. A. M. Hillas and T. E. Cranshaw (1960).

52. J. G. King (1960).

53. J. C. Zorn, G. E. Chamberlain, and V. W. Hughes (1963); V. W. Hughes (1964).

54. Ibid., p. 277.

55. D. W. Sciama, interview by Spencer Weart, AIP, 14 April 1978. G. F. R. Ellis, A. Lanza, and J. Miller, eds. (1993) contains a brief biographical section and a complete bibliography.

56. Interview with D. W. Sciama, 25 January 1989, as quoted in A. Lightman and R. Brawer (1990), p. 141. Similarly in the AIP interview of 1978 (note 55 above): "They were famous as a group and were very irreverent, overthrowing the establishment and all that kind of things. They were the glamour in-group in this sense."

57. D. W. Sciama (1953). Sciama acknowledged Bondi and Gold for "their constant advice and encouragement." For an analysis of Sciama's theory of inertia, see J. Merleau-Ponty (1965), pp. 241–45.

58. D. W. Sciama (1955a), p. 42. See also the elaborated discussion in Sciama (1959), pp. 145–50, 177–80. There is a clear methodological similarity between the self-propagating universe of the steady-state theory and the so-called bootstrap theory in elementary-particle physics which was much discussed in the 1960s. According to

the bootstrap theory, as worked out by Geoffrey Chew in particular, it should be possible to understand microphysics as following uniquely from requirements of self-consistency. For a detailed analysis, see J. T. Cushing (1990).

59. D. W. Sciama (1960–61), p. 7.

60. W. H. McCrea (1973), p. 96.

61. G. C. McVittie (1951), p. 75. McVittie's criticism was directed not only against the steady-state theory, but also against Milne's theory. He insisted that, from a methodological point of view, the two theories belonged to the same group.

62. M. Johnson (1951b).

63. E. Öpik (1954c), pp. 91, 106. The paper was an expanded version of Öpik (1954b).

64. H. Spencer Jones (1951–54); reprinted in a slightly different version in *Science News* 32 (1954): 19–32.

65. E.g., front page of *New York Times*, 24 May 1952, "British astronomer supports theory that creation is continuing." See also editorial of 1 June 1952 under the heading "Continuous creation."

66. O. Klein (1954), p. 43.

67. H. Dingle (1953), and a shorter version in *Observatory* 73 (1953): 42–48.

68. H. Dingle (1940).

69. H. Dingle and Viscount Samuel (1961), pp. 71–79. Dingle's objections were eventually published in Dingle (1960), (1958), (1957). For objections on behalf of orthodox physics, see W. H. McCrea (1957); H. Bondi (1957a). The controversy over special relativity is detailed in H. Chang (1993).

70. H. Dingle and Viscount Samuel (1961), p. 74.

71. H. Dingle (1953), pp. 403–4.

72. H. Dingle (1956), reprinted in G. Piel et al., eds. (1957), pp. 131–38.

73. *The Times*, 14 February 1953; *New York Times*, 14 February 1953; "British astronomer assails new theory," *Science* 120 (1954): 513–21. An abridged version of the address also appeared in German, in *Naturwissenschaftliche Rundschau* 8 (1953): 176–80.

74. T. Gold, interview by Spencer Weart, AIP, 1978. Wolfgang Rindler, who as a young mathematical physicist felt attracted to the steady-state theory, similarly recalled that Dingle's attack annoyed him and that he dismissed him as a crank. Letter to author, 8 September 1995.

75. A. Grünbaum, *Scientific American* 189 (December 1953): 6–7; P. Morrison, ibid. (September 1953): 14.

76. Cited in A. Grünbaum (1989), p. 32. An almost identical version was published in J. Leslie, ed. (1990), pp. 92–112.

77. M. Bunge (1979), p. 24; the first edition appeared as *Causality: The Place of the Causal Principle in Modern Science* (Cambridge, Mass.: Harvard University Press, 1959). See also J. D. North (1965), pp. 399–405.

78. F. Zwicky (1966), p. 217.

79. A. Grünbaum (1952).

80. A. Grünbaum (1989), (1993). According to Grünbaum, steady-state creation of matter is not particularly objectionable philosophically and cannot be used to discard the model, which has been rejected on empirical and not philosophical grounds. See Grünbaum (1991).

81. M. K. Munitz (1954). A slightly changed version was published three years later in Munitz (1957a), pp. 156–72.

82. M. K. Munitz (1952); (1957a), p. 89.

83. See, e.g., B. Wynne (1979).

84. W. B. Bonnor, review of Munitz (1957a), in *Observatory* 77 (1957): 248.

85. F. Hoyle (1952), p. 51.

86. This and the following quotations are from M. K. Munitz (1952).

87. O. L. Reiser (1952). Reiser's universe was eternal and infinite in time, and creation of matter was balanced by the dissolution of an equivalent amount of matter into an undifferentiated "ocean of cosmic energy." As Reiser noted, his cyclic-creative universe had been anticipated by MacMillan in the 1920s.

88. M. K. Munitz (1957a), p. 162. For the conceptual problems in cosmological creation processes, see also J. D. North (1965), pp. 399–406. The distinction between creation and origination has continued to occupy philosophers. Some of them argue, in accordance with Munitz, that the former concept is meaningless without an agency or external cause of some kind. For a modern philosophical analysis of the problem, see A. Grünbaum (1989).

89. Bonnor, review of M. K. Munitz (1957a), see note 84 above.

90. F. Hoyle (1952), p. 51.

91. M. Bunge (1962).

92. M. Scriven (1954). Part of the prize was awarded to J. T. Davies of King's College, London, and honorable mentions were given to the essays of Gerald Whitrow and Ernst Öpik. The essays of these three scholars, together with those of the American physicist Richard Schlegel and the German philosopher B. Abramenko, appeared under the common title "The age of the universe" in the same issue of the journal.

93. B. Ellis (1955).

94. G. J. Whitrow and H. Bondi (1954), based in part on a review of Bondi's *Cosmology* which Whitrow gave on the BBC on 23 September 1952.

95. G. J. Whitrow (1949), p. 40.

96. G. J. Whitrow (1962).

97. G. J. Whitrow (1954a).

98. G. J. Whitrow (1953).

99. P. Kerzberg (1992), pp. 189–90; A. S. Eddington (1933), p. 73.

100. F. Hoyle (1948), p. 380; and also his reply to Kapp in *Nature* 165 (1950): 68. See also Hoyle (1950a), p. 101.

101. H. Bondi and T. Gold (1954), with Whitrow's rejoinder on pp. 37–38. A further exchange of opinions between Hoyle and Whitrow, appearing under the same title, followed (*Observatory* 74 [1954]: 253–54).

102. F. A. E. Pirani (1954), followed by Whitrow's comment.

103. T. Gold (1955), which was a reply to an anonymous author on "The age of the universe" in (*Nature* 164 [1955]: 68–69), and was followed by this author's reply as well as Hoyle's contribution (ibid., p. 808).

104. W. Rindler (1956). For a detailed historical survey and a complete bibliography, see F. Tipler, C. Clarke and G. F. R. Ellis (1980).

105. W. Rindler, letter to author, 8 September 1995.

106. G. J. Whitrow (1959), pp. 138–41. The existence of causally unconnected bubble universes is also included in some relativistic models (such as the Lemaître-Eddington model), but appears most radically in the steady-state model.

107. G. C. McVittie (1961), p. 172. As mentioned in section 4.4, a similar point was raised by Reginald Kapp in 1950.

108. R. Schlegel (1962); see also J. D. North (1965), pp. 379–83, 422–23.

109. H. D. Ursell (1962), followed by a brief exchange of opinions between Schlegel and Ursell. See also R. D. Kemp (1964).

110. H. Bondi (1956).

111. G. J. Whitrow and H. Bondi (1954), p. 279.

112. H. Bondi (1952b).

113. This is according to Hoyle, who was a council member at the time and recalled the incident in F. Hoyle (1982b), pp. 20–21.

114. H. Bondi (1955a), p. 158. Hoyle also realized the problematic nature of observations. E.g., "When one speaks of an observation, it is very rare that the genuine raw observation is intended. Usually we mean an interpretive fusion of raw observation with established theory." F. Hoyle (1962), p. 141. For a larger perspective on Bondi's article and the question of the relative reliability of theory and observation in astronomy, see N. S. Hetherington (1988).

115. Quoted in D. Overbye (1991), p. 40.

116. F. Hoyle and R. A. Lyttleton (1948), p. 90. See also section 4.4.

117. E. H. Hutten (1962). For Ryle's disproof, see section 6.3.

118. T. Gold (1956), p. 1722.

119. W. H. McCrea (1960–61).

120. W. H. McCrea (1960), p. 13.

121. Ibid., p. 19.

122. W. H. McCrea (1963a), p. 199. Address delivered 8 February 1963. See also McCrea (1962).

123. A. W. K. Metzner and P. Morrison (1959).

124. W. Davidson (1960a), with a reply from McCrea; see also Davidson (1962–63). Other objections, from a logical and philosophical perspective, were stated by two Hungarian astronomers, B. Balázs and G. Paál (1961). An American philosopher, Carlton Berenda, interpreted McCrea's principle as a Kantian antimony: The universe must have begun in the past, and the universe could not have begun in the past, therefore no knowledge of cosmic events remote in space and time is possible. See C. W. Berenda (1964).

125. W. H. McCrea (1960–61), p. 1035.

126. W. H. McCrea (1970).

127. R. Harré (1962–63).

128. Ibid., p. 119. Harré's argument, too lengthy to give here, rested on the difference between the principle of sufficient reason and the principle of universal causality.

129. N. R. Hanson (1963).

130. A. C. B. Lovell (1959), p. 109.

131. A. Grünbaum (1989), p. 21.

132. S. Toulmin and J. Goodfield (1982), p. 258. The preface of the original edition, published in 1965, is dated 1964. See also S. Toulmin (1962), where a similar distrust of cosmological theory is aired.

133. S. Toulmin and J. Goodfield (1982), p. 262.

134. Bondi suspects, but is not certain, that he first read Popper's work in German, perhaps in the late 1940s. Letter to author, January 1995.

135. H. Bondi (1994).

136. H. Bondi (1958–59), p. 71.

137. H. Bondi (1966a), p. 396.

138. H. Bondi and C. W. Kilmister (1959–60), p. 56.

139. E.g., H. Bondi (1960), pp. 18–20; (1973), p. 11; (1990b), p. 194; (1992), a celebration of Popper on his ninetieth birthday.

140. K. R. Popper (1940).

141. My account of Popper's view is based on his letter to me of 10 June 1994. According to this letter, in the early versions of what became *Logik der Forschung*, there was a section in which Popper applied his theory of science to cosmology. Unfortunately, the section was omitted from the book.

142. H. Bondi et al., eds. (1960).

143. W. Bonnor (1964), p. 3.

144. Ibid., p. 119.

145. W. Bonnor (1957b), pp. 162–63.

146. Bonnor first considered oscillating world models in 1954, when he suggested a way in which a contracting closed model might be changed into an expanding one without passing through a singular state. W. Bonnor (1954).

147. J. Pachner (1965).

148. J. Pachner (1960), further developed in Pachner (1961).

149. W. Bonnor (1964), p. 176.

150. G. C. McVittie interview by David DeVorkin, AIP, 21 March 1978.

151. C. F. von Weiszäcker (1964), p. 149.

152. G. Gamow, interview by Charles Weiner, AIP, 25 April 1968.

153. H. Bondi (1966a), p. 396. A similar point was made in A. Bondi (1967), p. 82: "We have got to take the motion of the universe, and not its law of motion. It is boring to describe separately the motion of the apple and of the moon and so on. But if there is nothing but one apple falling, then you would be silly if you did anything but describe that motion."

154. H. Bondi et al., eds. (1960), pp. 44–45.

155. W. Bonnor (1964), p. 158.

156. G. C. McVittie (1951), p. 71.

157. G. C. McVittie (1961b), p. 1231.

158. G. C. McVittie (1965), pp. 8–9, which is also the source of the other quotations.

159. G. C. McVittie (1966).

160. B. Russell (1931), pp. 110, 108. As far as Eddington is concerned, Russell's remark was more smart than true. Eddington in fact warned against drawing religious implications from physics. For example: "I repudiate the idea of proving the distinctive beliefs of religion either from the data of physical science or by the methods of physical science." A. S. Eddington (1928), p. 333.

161. E. Whittaker (1946), pp. 116–17.

162. Ibid., p. 121. In 1943, Whittaker referred to the beginning of the world as "an operation of the Divine Will to constitute Nature from nothingness." E. T. Whittaker (1943), p. 63.

163. E. A. Milne (1952), p. 58.

164. Ibid., p. 23.

165. E. L. Mascall (1957), pp. 117–25.

166. Some authors have suggested that dislike of religion was part of the motivation of the steady-state theory. This is possible, but there is no historical evidence which supports the speculation. S. L. Jaki (1978), p. 269; E. McMullin (1981), p. 34. In a letter to the author of 19 February 1995, Hoyle states that the religious connotations of the

big-bang theory were opposed by Bondi, Gold, and himself, but that they were not of primary importance.

167. They continued to do so. For a recent example of Bondi's dislike of religion, see H. Bondi (1993a).

168. F. Hoyle (1955a), pp. 310, 312. For other expressions of Hoyle's antireligious views, see Hoyle (1956b), pp. 139, 157; (1977), pp. 5–9. Hoyle's emotional preference for a steady-state universe has been seen as "someting that gives him a quasi-religious satisfaction." See E. L. Mascall (1957), p. 159.

169. F. Hoyle (1977), p. 7.

170. F. Hoyle (1950a), p. 9.

171. M. Davidson (1955).

172. A. Lightman and R. Brawer (1990), p. 121.

173. W. A. Inge (1933), p. 64; see also section 4.1.

174. A. N. Whitehead (1929), pp. 521, 523; A. C. B. Lovell (1990), pp. 352–55.

175. D. W. Sciama, interview by Spencer Weart, AIP, 14 April 1978.

176. W. H. McCrea (1977), p. 72.

177. T. Gold, interview by Spencer Weart, AIP, 1978. I have found nothing in the writings of McCrea that supports the contention made by Gold.

178. Quoted in J. Farley (1974), p. 81.

179. Ibid., p. 120.

180. Quoted in S. Toulmin (1962), p. 147.

181. G. Gamow (1952b), note added August 1952.

182. G. Gamow (1952a). Quotation from the pope's address also appeared in a recognized Soviet physics journal, where the author noticed that "it is understandable that this theory [Gamow's] has been pleasing to people of a religious state of mind." D. A. Frank-Kamenetskij (1959), p. 606.

183. Address "Un Ora," Italian original in *Acta Apostolicae Sedis—Commentarium Officiale* 44 (1952): 31–43. Quotations from the translation in P. J. McLaughlin (1957), pp. 137–47. The address is also reprinted in excerpts in *Bulletin of the Atomic Scientists* 8 (1952): 142–46, 165. Predictably, the Pope's speech received much attention outside clerical and scientific circles also. See, for example, "Behind every door: God," *Time* 58 (3 December 1951): 75–77.

184. The asterisk refers to a reference in the pope's address to A. Unsöld (1948). The other major reference was to Whittaker's *Space and Spirit*, from which the pope approvingly quoted a longer passage.

185. E. L. Mascall (1957), p. 155; and similarly in E. McMullin (1981), p. 39.

186. J. Turek (1986). On Lemaître's view, see section 2.2. Ernan McMullin, who attended a graduate seminar with Lemaître in 1951, recalled "Lemaître storming into the class on his return from the Academy meeting in Rome, his usual jocularity entirely missing. He was emphatic in his insistence that the Big Bang model was still very tentative, and further that one could not exclude the possibility of a previous cosmic stage of contraction." E. McMullin (1981), p. 53.

187. Discourse given in French on 7 September 1952, translated into English in P. J. McLaughlin (1957), pp. 185–94. In spite of its caution, the address was widely seen as one more attempt to argue the existence of God from the results of science. "Pope says science proves God exists," *New York Times*, 8 September 1952, p. 23.

188. W. Bonnor (1964), p. 117. In a letter to the author of 9 October 1994, Bonnor confirms that his dissatisfaction with the big-bang theory was rooted in his atheism.

189. R. J. Russell, W. R. Stoeger, and G. V. Coyne, eds. (1990), pp. M9, M12. Holmes Rolston expressed the dilemma of the church in the following elegant way: "They [Christian scholars] know that the religion that is married to science today is a widow tomorrow, while the religion that is divorced from science leaves no offspring tomorrow." Ibid., p. 87.

190. E. Nicolaïdis (1990); L. R. Graham (1972), pp. 139–94; see also G. A. Wetter (1958), pp. 60–66.

191. Quoted in E. A. Tropp, V. Ya. Frenkel, and A. D. Chernin (1993), p. 223.

192. Quoted in O. Struve and V. Zebergs (1962), p. 34.

193. V. E. Llov, as quoted in M. W. Mikulak (1955), p. 170.

194. I. Prokofieva (1950).

195. Introductory remark to O. I. Schmidt (1951), p. 60. I have used the translation in *Bulletin of the Atomic Scientists* 8 (1952): 146. See also the sharp attack in P. Labérenne (1952).

196. V. Ambarzumian (1959). On Ambarzumian, see also R. A. McCutcheon (1990).

197. Quoted in L. R. Graham (1987), p. 401.

198. M. W. Mikulak (1958). Some reviews and theoretical works on relativistic models of the expanding universe did appear, in both the 1930s (M. Bronstein) and the 1950s (L. Landau, E. M. Lifschitz, V. A. Ambarzumian, V. A. Fock). Mikulak's studies are somewhat one sided and decidedly anti-Soviet, probably reflecting the cold war period. A fuller and more balanced view is given in L. R. Graham (1972).

199. "Russian astronomers hold theory of cosmos origin surpasses West," *New York Times*, 14 July 1949, front page.

200. Quoted from a book review by Otto Struve in *Astrophysical Journal* 110 (1949): 315–18. The book was B. A. Vorontzoff-Velyaminov, *Gaseous Nebulae and New Stars* (in Russian) (Moscow: U.S.S.R. Academy of Sciences, 1948).

201. O. Struve and V. Zebergs (1962), p. 32.

202. F. Hoyle (1989a), p. 101.

203. Quoted from L. R. Graham (1972), p. 171. M. W. Mikulak (1958), p. 47, refers to other Soviet rejections of the steady-state theory.

204. Meeting report in *Soviet Astronomy-AJ* 1 (1957): 306–7.

205. I. S. Shklovsky and S. B. Pikel'ner (1961).

206. I. D. Novikov, letter to author, 3 February 1995.

207. Ya. B. Zel'dovich (1963a) (Russian original dated December 1962).

208. Ibid., p. 948.

209. Ya. B. Zel'dovich (1963b), p. 102 (Russian original dated August 1962).

210. Ya. B. Zel'dovich and I. D. Novikov (1983), p. xxi (Russian original dated 1975).

211. Ya. B. Zel'dovich (1965), p. 313.

212. Ibid., p. 370.

213. I. D. Novikov and Ya. B. Zel'dovich (1967), p. 633.

214. M. F. Shirokov and I. Z. Fisher (1963) (Russian original dated September 1962).

215. E. A. Tropp, V. Ya. Frenkel, and A. D. Chernin (1993), p. 225.

216. A. D. Sakharov (1967), p. 33.

217. M. A. Markov (1967) (Russian original dated September 1966).

218. F. Hoyle (1956b), p. 139.

CHAPTER 6
THE UNIVERSE OBSERVED

1. W. Bonnor (1964), pp. 163–64.

2. A. Behr (1951). Before knowing of Baade's revision, Gamow used Behr's recent study to argue that neither the introduction of the cosmological constant nor the steady-state theory was needed in order to make the time scale problem disappear. G. Gamow (1952b), p. 41.

3. W. Baade (1952); reprinted in K. R. Lang and O. Gingerich, eds. (1979), pp. 750–52. For an early, popular account, see G. W. Gray (1953).

4. A. D. Thackeray and A. J. Wesselink (1953). The two astronomers at the Radcliffe Observatory near Pretoria, South Africa, determined the distance to the Magellanic Clouds to be about 44 kpc, which was almost double the distance previously estimated by Shapley.

5. W. H. McCrea (1953), p. 361.

6. M. L. Humason, N. U. Mayall, and A. R. Sandage (1956).

7. A. R. Sandage (1958), p. 525.

8. See the interview with Sandage in A. Lightman and R. Brawer (1990), pp. 67–84. A description of Sandage's career also appears in D. Overbye (1991), pp. 11–66.

9. F. Hoyle (1955a), p. 291. At the time, the best determinations of the Hubble constant did not really justify Hoyle's optimism. The value $T = 5.4 \times 10^9$ years gives a reduction factor of 9, which is not enough to solve the problem.

10. See the account in S. G. Brush (1982), where references to the primary sources can be found.

11. F. Hoyle (1961), p. 6. λ is the cosmological constant. The same point was made in the Varenna lecture of 1961, where Hoyle concluded that the time scale issue "constitutes a major difficulty for normal $\lambda = 0$ cosmologies." Hoyle (1962), p. 155.

12. A. Sandage (1961), p. 389.

13. M. J. Pierce et al. (1994). For newspaper coverage, see, e.g., *New York Times*, 1 November 1994. The value of 7.3×10^9 years presupposes a density parameter (ρ/ρ_c) equal to 1; with $\rho/\rho_c = 0$ the result is 11.2×10^9 years.

14. G. Gamow (1954b). Gamow's argument was contradicted by Geoffrey Burbidge, who found it oversimplified and suggested that it was not possible to relate evolutionary schemes of galaxies to a particular type of cosmological model. G. R. Burbidge (1958). Much later, Hoyle argued that "for anthropic reasons, we may quite well live in an infrequent system." F. Hoyle (1989b), p. 85.

15. I. King (1961), p. 130.

16. F. Hoyle (1962), p. 167; Hoyle and J. V. Narlikar (1962b).

17. O. Struve and V. Zebergs (1962), pp. 80–86; J. Stebbins (1952).

18. J. Stebbins and A. E. Whitford (1948).

19. J. Stebbins (1950).

20. H. Bondi (1956), p. 153. Similarly in F. Hoyle (1955a), p. 299: "This conclusion, if we accept it, destroys the steady-state theory at once."

21. G. Gamow (1952b), p. 41; O. Heckmann (1951), p. 90. A similar claim was made in G. C. McVittie (1951). Gamow first interpreted the observations of Stebbins and "Whiteford" as a galactic age effect in Gamow (1949), p. 369. He repeated the claim in 1954, when he concluded that the effect "seems to provide a direct evidence for the evolution of the universe at large, in direct contradiction to the steady-state theory." Gamow (1954b).

22. G. de Vaucouleurs (1948). In a meeting of the Royal Astronomical Society in July 1951, McVittie also concluded that there was no firm evidence for a reddening-distance relationship. *Observatory* 71 (1951): 174–76.

23. H. Bondi, T. Gold, and D. Sciama (1954).

24. A. E. Whitford (1954).

25. A. D. Code (1959).

26. *Astronomical Journal* 61 (1956): 352–53.

27. *Sky and Telescope* 16 (March 1957): 222.

28. B. R. Martin (1976), p. 73; Sciama, interview by Spencer Weart, AIP, 14 April 1977.

29. For example, by Whitrow in G. O. Jones, J. Rotblat, and G. J. Whitrow (1956), p. 223. The Stebbins-Whitford effect was only definitely dismissed in 1968, when Jeremy Oke and Sandage concluded that there was no observational evidence for an age effect increasing with distance. See J. B. Oke and A. Sandage (1968). For Whitford's recollections of the events, see A. E. Whitford (1986), p. 18. According to Whitford, the reason for the "hidden" withdrawal of the Stebbins-Whitford effect was practical. He was on his way to the Lick Observatory and therefore had no time to make a proper publication. Interview by David DeVorkin, AIP, 15 July 1977.

30. K. Just (1960), p. 46; see also Just (1959). Gamow (1964) agreed that Just's result, if confirmed, would be another piece of evidence against the steady-state theory.

31. M. Kohler (1933).

32. For more details, see, e.g., G. C. McVittie (1959a).

33. H. P. Robertson (1955); F. Hoyle and A. Sandage (1956).

34. W. Mattig (1958).

35. W. A. Baum (1953). The erroneous reference to Baum's paper of 1953 appears in many texts of cosmology, both old and new. For an example, see G. F. R. Ellis (1988), p. 391.

36. M. L. Humason, N. U. Mayall, and A. R. Sandage (1956), p. 162.

37. A. R. Sandage (1957), p. 96, originally published in *Scientific American* 192 (September 1956): 170–82.

38. F. Hoyle (1956c).

39. A. R. Sandage, interview by Spencer Weart, AIP, 22 May 1978.

40. W. A. Baum (1957).

41. G. C. McVittie (1959b), p. 533.

42. A. R. Sandage (1961), p. 366.

43. J. Peach (1973).

44. F. Hoyle (1959a). See also the extended analysis in Hoyle (1962); W. Davidson (1960b). According to J. D. North (1965), p. 252, Davidson was the first to derive the relevant formulas, in his unpublished Ph.D. thesis of 1958.

45. W. Davidson (1961); see also A. C. B. Lovell (1990), pp. 289–90.

46. H. P. Palmer (1963), p. 233.

47. K. I. Kellermann (1993).

48. A good, semihistorical survey, concentrating on the 1950s and 1960s, is given in J. R. Gribbin (1976).

49. J. Jeans (1902), (1928).

50. G. Gamow (1949), (1952a), (1953), Gamow (1954c).

51. C. F. von Weizsäcker (1943); G. Gamow and J. A. Hynek (1945).

52. C. F. von Weizsäcker (1951a).

53. Gamow to C. Møller, 19 November 1952. Møller collection, Niels Bohr Archive, Copenhagen. The Danish physicist Christian Møller presented Gamow's paper, which was published in 1953 as "Expanding universes." Gamow ended his letter: "I am missing the good old Københown [*sic*], but I am afraid to come so close to the Iron Curtain."

54. Gamow to Bohr, 16 September 1953. BSC 28.3, AHQP.

55. G. Gamow (1952a).

56. W. Bonnor (1956), (1957b).

57. C. F. von Weizsäcker (1951b). Among the other participants were B. Strömgren, H. Alfvén, G. C. McVittie, J. von Neumann, F. Zwicky, W. Heisenberg, and L. Spitzer.

58. D. W. Sciama (1955b). The theory of the twenty-eight-year-old Sciama was described to the Royal Astronomical Society at its meeting of 8 October 1954 by Gold. See *Observatory* 74 (1954): 234–36. See also Sciama (1959), pp. 155–60.

59. F. Hoyle (1953).

60. D. W. Sciama (1964b).

61. F. Hoyle (1958a); T. Gold and F. Hoyle (1959); see also "Propose heavens are hot," *Science News Letter* 76 (11 July 1959): 82. Part of Hoyle's Solvay report relied on his collaboration with Gold.

62. M. Harwit (1961), (1961–62). The first paper was refereed by Bondi, who became interested in the matter and suggested that Harwit look at steady-state universes that were denser than the Hoyle version would permit. This was the origin of the second of Harwit's papers. Harwit, letter to author, 11 October 1994.

63. E. M. Burbidge, G. R. Burbidge, and F. Hoyle (1963), p. 874.

64. G. Gamow (1954a), p. 62. See also Gamow (1964), p. 40.

65. H. Bondi (1952a), p. 167. Characteristically, Bondi did not mention Gamow's big-bang theory, but referred to the big-bang view under the name of Lemaître.

66. T. Gold (1954).

67. G. Gamow (1950), p. 40.

68. E. J. Öpik (1951); E. E. Salpeter (1952); see also Öpik (1954a).

69. H. A. Bethe (1939), p. 444.

70. H. Bondi and E. E. Salpeter (1952).

71. This is Salpeter's characterization of his approach in an interview by Spencer Weart, AIP, 30 March 1978.

72. W. A. Fowler, interview by Charles Weiner, AIP, 5–6 February 1973. The "funny little man" recalls the immediate response rather differently; see F. Hoyle (1994), p. 264.

73. F. Hoyle (1954). A preliminary announcement, together with the experimental verification, was given at the meeting of the American Physical Society in Albuquerque, 2–5 September 1953 (F. Hoyle et al. [1953]). The full confirmation of the resonance was reported in D. N. F. Dunbar et al. (1953). See also F. Hoyle (1965), pp. 143–50; (1986). Hoyle's prediction has been cited as an important case of the anthropic principle, namely, a prediction based on the existence of carbon-based life in the universe. See J. D. Barrow and F. J. Tipler (1986), pp. 252–53.

74. C. Hayashi and M. Nishida (1956).

75. F. Hoyle (1954), p. 122.

76. W. A. Fowler, G. R. Burbidge, and E. M. Burbidge (1955).

77. G. R. Burbidge et al. (1956). Californium, element number 98 in the periodic

system, was discovered, or manufactured, in 1950. We recall that the hypothesis of transuranic elements as the sources of stellar energy was made as early as the 1920s, when it was suggested by Jeans, Nernst, Kohlhörster, and others (see section 4.1).

78. F. Hoyle et al. (1956); E. M. Burbidge et al. (1957). For a retrospective comment, see Hoyle (1981).

79. A. G. W. Cameron (1957a). A more technical report, with a limited circulation, was Cameron (1957b).

80. E. M. Burbidge et al. (1957), p. 550.

81. H. E. Suess and H. C. Urey (1956).

82. E. M. Burbidge, et al. (1957), p. 608.

83. Ibid., p. 640.

84. F. Hoyle (1958b), discussion section on p. 293.

85. Ibid., p. 296.

86. G. Gamow (1950), p. 56. The Dutch astronomer G. B. van Albada suggested in the late 1940s that the elements were synthesized in red giant stars.

87. G. Gamow (1970), p. 127.

88. R. V. Wagoner (1990).

89. H. Bondi (1966a), p. 400.

90. G. Gamow (1970), pp. 124–127.

91. Radio astronomy is one of the few branches of modern astronomy that has received thorough historical attention. For the early development of radio astronomy, see J. S. Hey (1973); D. O. Edge and M. J. Mulkay (1976); W. T. Sullivan, ed. (1982), (1984); F. Graham Smith and B. Lovell (1983). Further references to the literature can be found in these sources and the sources mentioned below.

92. H. C. Van de Hulst (1945). The paper was originally published in Dutch, in *Nederlandsch Tijdschrift voor Natuurkunde* 11 (1945): 210–24.

93. F. Graham-Smith (1986).

94. R. L. F. Boyd, ed. (1951). Part of the mimeographed report, including Gold's presentation and the comments by Hoyle, McVittie, and Ryle, is reproduced in K. R. Lang and O. Gingerich (1979), pp. 782–85, from which the quotations are taken. See also D. O. Edge and M. J. Mulkay (1976), pp. 97–98.

95. *Observatory* 71 (1951): 213–14. Gold mentioned on this occasion that he first made his proposal "three years ago," that is, in 1948, which must have been in an unpublished discussion.

96. F. Hoyle (1982b), p. 54.

97. *Observatory* 75 (1955): 107.

98. Interview by Woodruff T. Sullivan, 1976, as quoted in Sullivan (1990), p. 316.

99. B. Mills (1952).

100. Quoted in W. T. Sullivan (1990), p. 321.

101. Ibid.

102. J. R. Shakeshaft (1954).

103. B. R. Martin (1978). For an alternative explanation in terms of previous technical commitments, see D. O. Edge and M. J. Mulkay (1976), pp. 138–43.

104. E. P. Hubble (1936a); see also E. P. Hubble (1936b), pp. 182–97.

105. M. Ryle and P. A. G. Scheuer (1955), received 8 March.

106. M. Ryle (1955), pp. 146–47. The incompatibility between the 2C observations and the steady-state theory was also emphasized by Shakeshaft at a meeting of the Royal Astronomical Society on 13 May (ibid., p. 106).

107. Quoted from an interview in D. O. Edge and M. J. Mulkay (1976), p. 162.

108. McCrea to Ryle, 4 March 1955, quoted in W. T. Sullivan (1990), p. 326.

109. Gold, interview by Spencer Weart, AIP, 1 April 1978.

110. See D. O. Edge and M. J. Mulkay (1976), p. 157.

111. *Newsweek* 63 (25 May 1964): 67.

112. F. Hoyle (1990), p. 228.

113. B. Y. Mills and O. B. Slee (1957), p. 194.

114. In a review paper presented to the 1958 General Assembly of the International Astronomical Union, as quoted in D. O. Edge and M. J. Mulkay (1976), p. 169.

115. Ibid., p. 175.

116. C. Hazard and D. Walsh (1959).

117. R. H. Brown (1959); B. Y. Mills (1959).

118. A. C. B. Lovell (1958), p. 201.

119. I. S. Shklovsky (1960), preface dated September 1958.

120. G. C. McVittie (1962a), p. 265.

121. M. Ryle (1959), p. 526; see also Ryle (1958).

CHAPTER 7
FROM CONTROVERSY TO MARGINALIZATION

1. J. Eisenstaedt (1986).

2. A. Schild (1960), p. 778.

3. J. A. Wheeler (1962), p. 40.

4. E. P. Hubble (1936b), p. 6.

5. M. S. Longair (1993), p. 160.

6. *New York Times*, 1 February 1962.

7. W. L. Kraushaar and G. W. Clark (1962b), (1962a). The final report of the experiment, confirming the earlier conclusion, appeared in W. Kraushaar et al. (1965).

8. G. C. McVittie (1962c), p. 2878.

9. G. R. Burbidge and F. Hoyle (1956); see also Burbidge and Hoyle (1958). Other early discussions of antimatter in a cosmological context include M. Goldhaber (1956); R. A. Alpher and R. C. Herman (1958); W. F. G. Swann (1961a); C. J. Kevane (1961).

10. G. Steigman (1969).

11. F. Hoyle (1969a).

12. For details on the development of x-ray astronomy, see R. F. Hirsh (1983); W. Tucker and R. Giacconi (1985).

13. F. Hoyle (1963).

14. R. J. Gould and G. R. Burbidge (1963).

15. G. B. Field and R. C. Henry (1964).

16. S. L. Jaki (1978), p. 271.

17. R. J. Gould and D. W. Sciama (1964).

18. Quoted in W. Tucker and R. Giacconi (1985), p. 49.

19. D. O. Edge, P. Scheuer, and J. Shakeshaft (1958).

20. R. Minkowski (1959), p. 538.

21. D. O. Edge and M. J. Mulkay (1976), p. 192.

22. M. Ryle and R. W. Clarke (1961); the brief presentation by Ryle in *Observatory* 81 (1961): 57–60; P. F. Scott and M. Ryle (1961); M. Ryle (1961); see also A. Hewish (1961–62), where Ryle's conclusions were confirmed.

23. M. Ryle and R. W. Clarke (1961), p. 361.

24. W. H. McCrea (1984a), p. 381; M. Harwit, letter to author, 4 October 1994.

25. D. O. Edge and M. J. Mulkay (1976), p. 194; J. Bolton, F. F. Gardner, and M. B. Mackey (1964); M. Ryle (1968).

26. V. V. Narlikar (1932).

27. J. V. Narlikar, letter to author, 14 October 1994.

28. E. Schücking and O. Heckmann (1958).

29. O. Heckmann (1961), p. 603.

30. J. V. Narlikar (1962a).

31. *Observatory* 81 (1961): 60.

32. M. Harwit, letter to author, 9 October 1994. According to Harwit's recollection, "The Cambridge intellectual climate in astrophysics was scandalous at the time. Redmond, at the Observatory, Ryle at the Radio Observatory, and Hoyle at the Department of Applied Mathematics and Theoretical Physics, had a constant feud going on. . . . [Apart from Shakeshaft] the others pretty much stayed within their own groups at the time."

33. H. Bondi, *Cosmology*, 2nd ed. (Cambridge: Cambridge University Press, 1960), p. 167.

34. " 'Big Bang' theory of cosmos backed," *New York Times*, 11 February 1961.

35. Gold to Gamow, 4 April 1961. Gamow file, Library of Congress.

36. M. Ryle (1962).

37. "Rival cosmologies," *New York Times*, 27 August 1961.

38. H. Bondi (1964), p. 163.

39. W. H. McCrea (1963b), p. 220. Emphasis added.

40. W. Priester (1958). Like most other astronomers outside the English-speaking world, Priester ignored the steady-state model in his analysis of radio source counts.

41. W. Davidson (1961–62); Davidson and M. Davies (1964).

42. W. H. McCrea (1963a), p. 196.

43. F. Hoyle and J. V. Narlikar (1961–62), (1962a); presented at meeting of the Royal Astronomical Society on 14 April (*Observatory* 81 [1961]: 86–89).

44. M. Ryle (1963), p. 229.

45. D. W. Sciama (1963a), (1964a).

46. P. F. Scott (1963).

47. D. W. Sciama (1965), p. 15.

48. K. Brecher, G. Burbidge, and P. A. Strittmatter (1971), p. 108.

49. The name was introduced in H.-Y. Chiu (1964).

50. According to the meeting report in *Sky and Telescope* 21 (March 1961): 148.

51. M. Schmidt (1963); J. L. Greenstein and T. A. Matthews (1963). For discovery histories of quasars, see M. Schmidt (1990); D. O. Edge and M. J. Mulkay (1976), pp. 202–12; M. Harwit (1981), pp. 137–40. A good contemporary review of the field is E. M. Burbidge (1967).

52. G. R. Burbidge (1959).

53. *Newsweek* 63 (25 May 1964): 63. Reproduced together with other quasar poems in I. Robinson, A. Schild, and E. L. Schucking, eds. (1965), pp. 471–72.

54. F. Hoyle and W. A. Fowler (1962–63), (1963).

55. I. Robinson, A. Schild, and E. L. Schucking, eds. (1965).

56. Ibid., p. 470.

57. J. L. Greenstein and M. Schmidt (1964).

58. D. W. Sciama and W. C. Saslaw (1966).

59. F. Hoyle, G. R. Burbidge, and W. L. W. Sargent (1966); Hoyle and Burbidge (1966c).

60. M. J. Rees and D. W. Sciama (1966a).

61. F. Hoyle and G. R. Burbidge (1966b).

62. M. S. Longair (1966).

63. D. W. Sciama and M. J. Rees (1966a).

64. D. W. Sciama and M. J. Rees (1966b).

65. F. Hoyle and G. R. Burbidge (1966a).

66. F. Hoyle and W. A. Fowler (1967).

67. M. Schmidt (1968).

68. Interview with Sciama quoted in A. Lightman and R. Brawer (1990), p. 144; see also L. John, ed. (1973), p. 59

69. On this principle, see D. L. Hull, P. D. Tessner, and A. M. Diamond (1978).

70. D. W. Sciama (1967).

71. D. W. Sciama (1971), p. 117.

72. Paper on "Light element formation during early stages of the expanding universe" given to American Physical Society meeting in Los Alamos; abstract in *Bulletin of the American Physical Society* 4 (1959): 476. In 1958 G. Burbidge pointed out the problems in assuming all the helium to have been produced by hydrogen burning in stars. However, he did not see the problem as evidence for an evolving universe. Instead, he suggested a mechanism according to which the galaxies in their early epoch released far more energy than they do today. G. R. Burbidge (1958).

73. H. Bondi, T. Gold, and F. Hoyle (1955). According to Bondi, he was aware that the helium abundance might well be a fossil of the big bang, and he stressed the need for its observational determination. However, he does not seem to have published his ideas on this question. See H. Bondi (1988).

74. D. E. Osterbrock and J. B. Rogerson (1961). Their reference to the big-bang theory was to G. Gamow (1949).

75. Yu. N. Smirnov (1965) (Russian original published in 1964).

76. F. Hoyle and R. J. Tayler (1964); see also R. J. Tayler (1990). For a useful survey of nucleosynthesis in the mid-1960s, see R. J. Tayler (1966).

77. J. W. Truran, C. J. Hansen, and A. G. W. Cameron (1965).

78. For Hoyle's continuing fascination with this kind of hypothetical object, see F. Hoyle (1969b).

79. Tayler did not feel attracted by the steady-state theory and was doubtful about helium production in supermassive bodies. He tended toward the big bang, but preferred to leave the matter open. Letter to author, 2 September 1994.

80. R. V. Wagoner, W. A. Fowler, and F. Hoyle (1967). Preliminary results were reported in Hoyle's Halley Lecture of 3 May 1966, "Recent developments in nucleosynthesis," and in Wagoner, Fowler, and Hoyle (1966). See also Wagoner (1990).

81. Quoted in B. Bertotti et al., eds. (1990), p. 174; and in A. Lightman and R. Brawer (1990), p. 177.

82. W. H. McCrea (1984b).

83. *Observatory* 87 (1967): 193–213.

84. P. J. E. Peebles (1966b). The II in the title referred to an earlier, more qualitative report, in which Pebbles got 27–30 % helium. See Peebles (1966a).

85. A. Lightman and R. Brawer (1990), p. 219.

440

86. E.g., G. R. Burbidge (1969).

87. H. Reeves et al. (1973).

88. Discovery accounts include A. A. Penzias (1972); R. W. Wilson (1979); D. T. Wilkinson and P. J. E. Peebles (1983). The last account was based on an unpublished manuscript written in 1968; a slightly updated version with the same title is presented in N. Mandolesi and N. Vittorio, eds. (1990), pp. 17–31. For historical reviews, see T. Ferris (1977); S. Weinberg (1977); J. Bernstein (1984); S. G. Brush (1992); H. Kragh (1993).

89. J.-F. Denisse, É. Le Roux, and J. C. Steinberg (1957): A. Le Floch and F. Bretenaker (1991); see also S. G. Brush (1993).

90. The case is quoted in A. S. Sharov and I. D. Novikov (1993), p. 148. See also I. Novikov (1990), pp. 130–31.

91. See S. G. Brush (1993), p. 578.

92. E. A. Ohm (1961).

93. A. G. Doroshkevich and I. D. Novikov (1964) (Russian original submitted 11 October 1963). Doroshkevich and Novikov referred to a publication of Gamow that did not, in fact, relate to the background radiation, namely, G. Gamow (1949). They did not refer to the works of Alpher and Herman, but according to Novikov they knew about the predictions of Gamow, Alpher, and Herman. Conversation with author, 27 January 1995.

94. Ya. B. Zel'dovich (1964) (Russian original dated July 1963).

95. Ya. B. Zel'dovich (1965), p. 321,

96. Ya. B. Zel'dovich (1963b), p. 1102.

97. R. J. Tayler (1966), p. 509.

98. I. Novikov, conversation with author, 27 January 1995; letter to author, 3 February 1995.

99. F. Hoyle (1981).

100. F. Hoyle (1959a), p. 530.

101. Letter to author, 19 February 1995.

102. Personal communication from R. J. Tayler (letter of 2 September 1994), who kindly provided me with a copy of the relevant pages of the first draft of the article.

103. R. J. Tayler (1990), p. 372.

104. Penzias, quoted in his Nobel autobiography in S. Lundqvist, ed. (1992), p. 440. On Herzberg's mention of the excitation temperature of interstellar cyanogen, see section 3.3.

105. R. H. Dicke (1957), p. 375.

106. R. H. Dicke (1963), p. 500.

107. E.g., R. H. Dicke (1959); see also J. D. Barrow and F. J. Tipler (1986), pp. 245–48.

108. C. Brans and R. H. Dicke (1961), p. 928.

109. P. J. E. Peebles (1993), pp. 146–47.

110. D. Overbye (1991), p. 132; Peebles, letter to author, 24 June 1995.

111. R. H. Dicke and P. J. E. Peebles (1965), p. 448. In a note added in proof, Dicke and Peebles referred to the still unpublished results of Penzias and Wilson.

112. C. P. Gilmore (1965), p. 202.

113. A. A. Penzias (1972), p. 34.

114. J. Bernstein (1984), p. 204.

115. P. J. E. Peebles (1993), p. 148, where the spectrum presented in March 1965 is reproduced.

116. A. A. Penzias and R. W. Wilson (1965); R. H. Dicke et al. (1965).

117. Gamow to Penzias, 29 September 1965 (mistakenly dated 1963), reproduced in facsimile in A. A. Penzias (1972), p. 35. In an interview given by Gamow shortly before his death, he more than suggested that the work of Dicke and his group was not independent of the much earlier work of Alpher, Herman, and himself. Interview by C. Weiner, AIP, 25 April 1968.

118. P. J. E. Peebles and D. T. Wilkinson (1967), p. 37. See also Peebles (1971), which includes a fairly detailed historical summary of the contributions of Gamow, Alpher, and Herman (pp. 125–28, 240–41). According to Dicke, the Princeton physicists' suggestion of a hot big bang was independent not only of Gamow, but also of the Hoyle-Tayler paper, of which they were not aware. R. H. Dicke (1970), p. 65.

119. Gamow to Dicke, 13 December 1967; Dicke to Gamow, 27 December 1967; Peebles to Alpher, 20 June 1967; Gamow to Peebles and Wilkinson, 20 June 1967. Gamow Collection, Library of Congress.

120. R. A. Alpher, G. Gamow, and R. C. Herman (1967). See also the later papers by Alpher and Herman, mentioned in section 3.3.

121. R. A. Alpher and R. C. Herman (1990), pp. 150–51.

122. R. H. Dicke (1968).

123. R. H. Dicke et al. (1965), p. 417.

124. W. Sullivan, "Signals imply a 'big bang' universe," *New York Times*, 21 May 1965. The following Sunday the newspaper included a follow-up story on both quasars and the new radiation ("New light thrown on the birth of the universe," 23 May 1965).

125. J. Bernstein (1984), p. 205

126. G. B. Field and J. L. Hitchcock (1966). Similar results were obtained independently in P. Thaddeus and J. F. Clauser (1966).

127. F. Hoyle (1965c). Hoyle first referred to the 3 K radiation in an article submitted in late June 1965, before the paper of Penzias and Wilson had appeared; see F. Hoyle (1965d). The results of Penzias and Wilson, and the interpretation in favor of a big bang, were known some time in advance of the publication, either through informal, professional channels or through mention in newspapers and popular journals.

128. F. Hoyle and N. C. Wickramasinghe (1967); J. V. Narlikar and N. C. Wickramasinghe (1967).

129. F. Hoyle and N. C. Wickramasinghe (1962).

130. J. V. Narlikar and N. C. Wickramasinghe (1968).

131. F. Hoyle (1990), p. 224. Hoyle and Wickramasinghe have continued to develop their theory, of which they gave a comprehensive review in 1991. The conclusion was the same, namely, that the microwave radiation is not of big-bang origin, but has been thermalized by cosmic grains at an epoch subsequent to the formation of galaxies. F. Hoyle and N. C. Wickramasinghe (1991).

132. Ya. B. Zel'dovich and I. D. Novikov (1967); J. R. Shakeshaft and A. S. Webster (1968).

133. D. W. Sciama (1966); T. Gold and F. Pacini (1968).

134. A. M. Wolfe and G. R. Burbidge (1969). Cyril Hazard and Edwin Salpeter concluded the same year that "No population of sources can fit the observational data on the microwave background radiation within the framework of the steady-state theory and yet have a density less than that of galaxies." Hazard and Salpeter (1969).

135. W. H. McCrea (1964).

136. R. Stothers (1965), (1966).

137. I. W. Roxburgh and P. G. Saffman (1965); see also J. H. Hunter (1966).

138. F. Hoyle (1960b).

139. Ibid., p. 262.

140. W. B. Bonnor and G. C. McVittie (1961); R. L. Agacy and W. H. McCrea (1961–62).

141. A. K. Raychaudhuri and S. Banerji (1964).

142. F. Hoyle and J. V. Narlikar (1962c).

143. F. Hoyle and J. V. Narlikar (1964b), (1964c).

144. F. Hoyle and J. V. Narlikar (1962c).

145. E. M. Burbidge, G. R. Burbidge, and F. Hoyle (1963).

146. Ibid., p. 878.

147. F. Hoyle and J. V. Narlikar (1964a).

148. F. Hoyle and J. V. Narlikar (1966a).

149. According to *Newsweek* 66 (25 October 1965): 77 ("Not according to Hoyle").

150. I. D. Novikov (1965) (Russian original dated November–December 1964). Novikov hypothesized the existence of white holes in an interpretation of quasars as local areas of the universe delayed in expansion. He did not use the name white hole. Novikov's idea was probably inspired by the work of Hoyle and Fowler, but he emphasized that "the arguments advanced share nothing in common with Hoyle's concept of the continuous creation of matter" (p. 862).

151. F. Hoyle (1969b); J. Jeans (1928c), p. 360. Curiously, Hoyle also quoted Jeans's speculation in his 1948 paper on the steady-state theory. A somewhat similar idea was included in Jordan's cosmology of the late 1930s, where entire stars were postulated to be created.

152. F. Hoyle (1969b), p. 18.

153. "Life and death of the universe," *Newsweek* 63 (25 May 1964): 63–67.

154. Ya. B. Zel'dovich (1965), p. 371.

155. S. W. Hawking (1965a). As shown by Bondi in 1957, negative mass (not to be confused with antimatter) can consistently be accounted for within the framework of conventional relativity theory. H. Bondi (1957b).

156. S. Deser and F. A. E. Pirani (1965). McCrea also argued that Hoyle and Narlikar's claim to have determined the sign of the gravitational constant was unfounded (W. H. McCrea [1965]). Hoyle and Narlikar responded to their critics in F. Hoyle and J. V. Narlikar (1966d).

157. E. M. Lifshitz and I. M. Khalatnikov (1963).

158. R. Penrose (1965); S. W. Hawking (1965b); Hawking and Penrose (1970). On the confrontation between the works of Lifschitz and Khalatnikov and that of Penrose, see K. S. Thorne (1994), pp. 459–69. For a general, philosophical perspective on the significance of the singularity theorems, see B. Kanitscheider (1984), pp. 234–66.

159. F. Hoyle and J. V. Narlikar (1966b); Hoyle (1965b), pp. 117–31.

160. F. Hoyle and J. V. Narlikar (1966c).

161. For a brief survey, see G. Gale (1990). Tommy Gold also discussed the possibility of multiple universes before inflation cosmology, but using general arguments that did not rely on Hoyle and Narlikar's model. See T. Gold (1973).

162. E.g., F. Hoyle (1989b). According to one version of the inflation model, "we are now winding up with a model of a stationary Universe, in which the notion of the Big Bang loses its dominant position, being removed to the indefinite past." A. Linde and A. Mezhlumian (1993), p. 32. See also A. Linde, D. Linde, and

A. Mezhlumian (1994). I am grateful to C. Klixbüll Jørgensen for having called my attention to these papers.

163. J. Pachner (1960), p. 673.

164. R. G. Giovanelli (1964), p. 462. Speculations along similar lines had been suggested by Herbert Dingle in 1936, when he hypothesized that the universe as a whole might be static and the observed recession of galaxies merely a local phenomenon. H. Dingle (1936).

165. R. C. Tolman (1934b); see also A. Krasinski (1990).

166. Quoted in S. G. Brush (1976), p. 623.

167. Ibid., p. 636.

168. On physical theories of time, see, e.g., G. J. Whitrow (1980); P. C. W. Davies (1974).

169. J. A. Wheeler and R. P. Feynman (1945), (1949). For the historical background of direct-interaction theories, see J. M. Sánchez-Ron (1983).

170. A historical account of Hogarth's work and its impact on cosmology is given in J. M. Sánchez-Ron (1990).

171. T. Gold (1958), (1962); H. Bondi (1962).

172. J. E. Hogarth (1962).

173. D. W. Sciama (1960–61).

174. J. V. Narlikar (1962b). Hogarth, Sciama, and Narlikar lectured on their works at a conference on relativity held in Warsaw in July 1962. See L. Infeld, ed. (1964), pp. 331–37.

175. F. Hoyle and J. V. Narlikar (1964d); see also Hoyle and Narlikar (1974).

176. J. V. Narlikar (1973).

177. D. W. Sciama (1963b); see also S. Banerji (1966).

178. P. E. Roe (1969).

179. T. Gold, ed. (1967).

180. R. P. Feynman, F. B. Moringo, and W. G. Wagner (1995), p. 183.

181. Ibid., p. 167.

182. F. Hoyle and J. V. Narlikar (1972).

183. F. Hoyle (1969b).

184. H. Bondi (1966b).

185. T. Y. Thomas (1966).

186. A. O. Zupančič (1965).

187. F. Hoyle (1968), p. 12.

188. D. W. Sciama (1969).

189. R. J. Weymann (1967), p. 8.

190. S. Lundqvist, ed. (1992), p. 419.

191. I. D. Novikov and Ya. B. Zeldo'vich (1967), pp. 631, 636.

192. Much of this quasi history is alive in modern, popular expositions of cosmology. In one such work the reader is told that "Before 1965, cosmology was a quiet backwater of science, almost a little ghetto where a few mathematicians could play with their models without annoying anybody else." M. White and J. Gribbin (1988), p. 112.

193. M. P. Ryan and L. C. Shepley (1976), based on *Physics Abstracts*. The most detailed bibliography is B. Kuchowicz (1967), which contains references to 2101 works on nuclear astrophysics and related topics (including cosmology) up to 1964. For the years from 1963 to 1967 Kuchowicz (1968–71) collected 5690 references.

194. O. Struve and V. Zebergs (1962), p. 26.

195. A. Braccesi (1963); G. Velo (1963). Contributions made by the other scientists have already been mentioned.

196. S. G. Brush (1993), p. 587; C. M. Copp (1982). In another paper, Copp investigated the social pattern of cosmology in the 1980s and concluded that cosmology had by then reached a state of normal science in Kuhn's sense. Copp (1985).

197. See also S. G. Brush (1993), pp. 585–87. I believe Brush exaggerates the crucial nature of the discovery of the microwave background in the capitulation of the steady-state theory. Also, I do not agree with his list of steady-state supporters, which includes many scientists who did not, in fact, support the theory. Scientists who only expressed a technical interest in, had a brief flirtation with, or wrote on aspects related to the steady-state theory should not be counted as supporters. (If so, Hoyle would be a major big-bang supporter.) Thus I find it unjustified to enlist W. Davidson, L. G. Chambers, S. Hawking, A. Dauvillier, R. Kemp, A. Banerji, and A. Raychaudhuri among the the steady-state supporters.

198. W. H. McCrea, interview by Robert Smith, AIP, 22 September 1978.

199. W. H. McCrea (1970), which was a lecture delivered in the fall of 1969.

200. Pirani, letter to author, 25 August 1994. Pirani has remained critical of the standard big-bang cosmology, which he sees more as ideology than science. See F. A. E. Pirani (1992).

201. F. Hoyle (1982b), p. 49.

202. M. Rowan-Robinson (1972).

203. G. R. Burbidge (1971).

204. W. A. Fowler (1967), p. 203.

205. F. Hoyle and J. V. Narlikar (1980), p. 428.

206. F. Hoyle (1982c), p. 11.

207. Ibid. See also the critical comment in R. A. Alpher and R. C. Herman (1983).

208. H. Alfvén and O. Klein (1963); Alfvén (1966), (1968), (1971). For Klein's original cosmological hypothesis, dating from 1953, see section 5.2.

209. O. Klein (1966).

210. H. Alfvén (1968), p. 9.

211. H. Alfvén (1977).

212. E. J. Lerner (1991).

213. G. B. Field, H. Arp, and J. N. Bahcall (1973), which includes the presentations of the two discussants as well as reprints of thirty-two relevants papers.

214. H. Arp (1987), pp. 178–83.

215. P. J. E. Peebles (1993), p. 197.

216. F. Hoyle (1989b), p. 81.

217. G. R. Burbidge (1988), p. 225.

218. H. C. Arp et al. (1990).

219. P. J. E. Peebles et al. (1991). The discussion was continued in *Nature* 357 (1992): 287–88.

220. F. Hoyle, G. R. Burbidge, and J. V. Narlikar (1993); see also Hoyle, Burbidge, and Narlikar (1994a), (1994b).

221. Ibid., p. 738.

222. A closed steady-state model was proposed by Peter Phillips, an American physicist, in 1994. Because of the fixed space, matter is here assumed to be continuously annihilated at the same rate as it is created. P. R. Phillips (1994).

CHAPTER 8
EPILOGUE: DYNAMICS OF A CONTROVERSY

1. The following relies on E. McMullin (1987). Among McMullin's many examples, the controversy caused by the steady-state theory also figures (pp. 71–72).

2. Ibid.

3. Ibid., p. 68

4. R. Giere (1987), p. 133. On the controversy over continental drift, see H. Frankel (1987).

BIBLIOGRAPHY

SMALL anonymous articles, reviews, and clips appearing in newspapers or magazines are not included. Neither is unpublished material. References to these can be found in the notes.

Abbreviations of periodicals cited more than once are listed below:

AHES	Archive for History of Exact Sciences
AIHP	Annales de l'Institut Henri Poincaré
AJ	Astronomical Journal
AJP	American Journal of Physics
AN	Astronomische Nachrichten
AnS	Annals of Science
AP	Annalen der Physik
APJ	Astrophysical Journal
AQ	Astronomical Quarterly
ARAA	Annual Review of Astronomy and Astrophysics
AS	American Scientist
ASPL	Astronomical Society of the Pacific, Leaflet
ASSB	Annales de Société Scientifique de Bruxelles
AuJP	Australian Journal of Physics
BAN	Bulletin of the Astronomical Institutes of the Netherlands
BJPS	British Journal for the Philosophy of Science
CA	Comments on Astrophysics
CASP	Comments on Astrophysics and Space Physics
CASS	Comments on Astrophysics and Space Science
CR	Comptes Rendus de l'Académie des Sciences (Paris)
CT	Ciel et Terre
HPA	Helvetica Physica Acta
HSPS	Historical Studies in the Physical Sciences (since 1988: Historical Studies in the Physical and Biological Sciences)
IAJ	Irish Astronomical Journal
JACS	Journal of the American Chemical Society
JETP	Soviet Physics—JETP
JFI	Journal of the Franklin Institute
JHA	Journal for the History of Astronomy
JRASC	Journal of the Royal Astronomical Society of Canada
JWAS	Journal of the Washington Academy of Sciences
LNC	La Nuova Critica
MLPS	Memoirs and Proceedings of the Manchester Literary and Philosophical Society
MNRAS	Monthly Notices of the Royal Astronomical Society
NC	Nuovo Cimento
NW	Die Naturwissenschaften
ONRAS	Occasional Notes of the Royal Astronomical Society

PA	*Popular Astronomy*
PAPS	*Proceedings of the American Philosophical Society*
PASJ	*Publications of the Astronomical Society of Japan*
PASP	*Publications of the Astronomical Society of the Pacifics*
PCPS	*Proceedings of the Cambridge Philosophical Society*
PJ	*Philosophisches Jahrbuch*
PM	*Philosophical Magazine*
PNAS	*Proceedings of the National Academy of Sciences of the U.S.A.*
PPMSJ	*Proceedings of the Physico-Mathematical Society of Japan*
PPS	*Proceedings of the Physical Society, London*
PR	*Physical Review*
PRIA	*Proceedings of the Royal Irish Academy Section A*
PRL	*Physical Review Letters*
PRS	*Proceedings of the Royal Society of London, Series A*
PS	*Philosophy of Science*
PT	*Physics Today*
PTP	*Progress of Theoretical Physics*
PU	*Physics—Uspekhi*
PZ	*Physikalische Zeitschrift*
PZS	*Physikalische Zeitschrift der Sowjetunion*
QJRAS	*Quarterly Journal of the Royal Astronomical Society*
RMP	*Reviews of Modern Physics*
RPP	*Reports on Progress in Physics*
RQS	*Revue des Questions Scientifiques*
SA	*Scientific American*
SAAJ	*Soviet Astronomy—AJ*
SHPS	*Studies in the History and Philosophy of Science*
SPAW	*Sitzungsberichte der Preussischen Akademie der Wissenschaften*
SPU	*Soviet Physics—Uspekhi*
VA	*Vistas in Astronomy*
ZAP	*Zeitschrift für Astrophysik*
ZP	*Zeitschrift für Physik*
ZPC	*Zeitschrift für Physikalische Chemie*

Abramenko, B. (1954), "The age of the universe," *BJPS* 5: 238–52.

Adams, W. S. (1941), "Some results with the Coudé spectrograph of the Mount Wilson Observatory," *APJ* 93: 11–23.

Agacy, R. L., and W. H. McCrea (1961), "A transformation of the de Sitter metric and the law of creation of matter," *MNRAS* 123: 383–90.

Alfvén, H. (1966), *Worlds-Antiworlds: Antimatter in Cosmology.* San Francisco: W. H. Freeman.

———. (1967), "Antimatter and cosmology," *SA* 216 (April): 106–14.

———. (1968), "Antimatter and the development of the metagalaxy," *Uppsatser* (Föreningarna för Matematisk-Naturvetenskaplig Undervisning), no. 3: 1–42.

———. (1971), "Plasma physics applied to cosmology," *PT* 24 (February): 24–33.

———. (1977), "Cosmology: Myth or science?" 1–14 in W. Yourgrau and A. D. Breck, eds. (1977).

Alfvén, H., and O. Klein (1963), "Matter-antimatter annihilation and cosmology," *Arkiv för Fysik* 23: 187–94.

Alpher, R. A. (1948), "A neutron-capture theory of the formation and relative abundance of the elements," *PR* 74: 1577–89.

———. (1973), "Large numbers, cosmology, and Gamow," *AS* 61: 52–58.

———. (1983), "Theology of the big bang," *Religious Humanism* 17 (January): 1–13.

Alpher, R. A., H. Bethe, and G. Gamow (1948), "The origin of chemical elements," *PR* 73: 803–4.

Alpher, R. A., J. W. Follin, and R. C. Herman (1953), "Physical conditions in the initial stages of the expanding universe," *PR* 92: 1347–61.

Alpher, R. A., and G. Gamow (1968), "A possible relation between cosmological quantities and the characteristics of elementary particles," *PNAS* 61: 363–66.

Alpher, R. A., and R. C. Herman (1948a), "On the relative abundance of the elements," *PR* 74: 1737–42.

———. (1948b), "Evolution of the universe," *Nature* 162: 774–75.

———. (1949), "Remarks on the evolution of the expanding universe," *PR* 75: 1089–99.

———. (1950), "Theory of the origin and relative abundance distribution of the elements," *RMP* 22: 153–212.

———. (1951), "Neutron-capture theory of element formation in an expanding universe," *PR* 84: 60–66.

———. (1953), "The origin and abundance distribution of the elements," *Annual Review of Nuclear Science* 2: 1–40.

———. (1958), "On nucleon-antinucleon symmetry in cosmology," *Science* 128: 904.

———. (1972), "Memories of Gamow." 304–13 in F. Reines, ed. (1972).

———. (1975), "Big bang cosmology and the cosmic black-body radiation," *PAPS* 119: 325–48.

———. (1983), "In the beginning," *The Sciences* 23 (April): 2.

———. (1988), "Reflections on early work on 'big bang' cosmology," *PT* 41 (August): 24–34.

———. (1990), "Early work on 'big-bang' cosmology and the cosmic blackbody radiation." 129–58 in B. Bertotti et al., eds. (1990).

———. (1993), "Origins of primordial nucleosynthesis and prediction of cosmic background radiation." 453–75 in N. S. Hetherington, ed. (1993).

Alpher, R. A., R. C. Herman, and G. Gamow (1948), "Thermonuclear reactions in the expanding universe," *PR* 74: 1198–99.

———. (1949), "On the origin of the elements," *PR* 75: 332.

———. (1967), "Thermal cosmic radiation and the formation of protogalaxies," *PNAS* 58: 2179–86.

Ambarzumian, V. (1936), "Double stars and the cosmogonic time-scale," *Nature* 137: 537.

———. (1959), "La méthode en cosmologie," *Le Cosmos: Recherches Internationale*, nos. 14–15: 22–41.

Araki, T. (1953), "Über das Erschaffen der Materie und das sich ausdehnende Universum," *PASJ* 5: 44–47.

Arnot, F. L. (1938), "Cosmological theory," *Nature* 141: 1142–43.

———. (1941), *Time and the Universe: A New Basis for Cosmology*. Sydney: Australasian Medical Publishing Co.

Arp, H. C. (1987), *Quasars, Redshifts, and Controversies*. Berkeley: Interstellar Media.

Arp, H. C., F. Hoyle, G. Burbidge, J. V. Narlikar, and N. C. Wickramasinghe (1990), "The extragalactic universe: an alternative view," *Nature* 346: 807–12.

Arrhenius, G., and H. Levi (1988), "The era of cosmochemistry and geochemistry 1922–1935." 11–36 in G. Marx, ed. (1988), *George de Hevesy Festschrift*. Budapest: Akadémia Kiadó.

Arrhenius, S. (1909), *Världernas Utveckling*, 6th ed. Stockholm: Hugo Geber.

Atkinson, R. d'E. (1931), "Atomic synthesis and stellar energy I, II," *APJ* 73: 250–95, 308–47.

———. (1936), "Atomic synthesis and stellar energy III," *APJ* 84: 73–84.

Atkinson, R. d'E., and F. G. Houtermans (1929), "Zur Frage der Aufbaumöglichkeiten in Sternen," *ZP* 54: 656–65.

Audretsch, J. (1987), "Wellenmechanik und Raumzeit-Struktur: Erwin Schrödinger und die allgemeine Relativitätstheorie," *Physikalische Blätter* 43: 333–37.

Baade, W. (1952), "Extragalactic nebulae. Report to IAU Commission 28," *Transactions of the International Union of Astronomy* 8: 397–99.

Baade, W., and F. Zwicky (1934), "On super-novae," *PNAS* 20: 254–59.

Bagge, E. (1950), "Eine Deutung der Expansions des Kosmos," *ZP* 128: 239–54.

Bailey, V. A. (1959), "The steady-state universe and the deduction of continual creation of matter," *Nature* 184: 537.

Balashov, Y. (1994), "Uniformitarianism in cosmology: background and philosophical implications of the steady-state theory," *SHPS* 25: 933–58.

Balázs, B., and G. Paál (1961), "The interpretation of cosmology," *Nature* 189: 992–93.

Band, W. (1950), "On the origin of the lighter elements," *PR* 80: 813–18.

Banerji, S. (1966), "The arrow of time in homogeneous cosmological models," *PPS* 89: 393–98.

Barbour, J. B. (1990), "The role played by Mach's principle in the genesis of relativistic cosmology." 47–66 in B. Bertotti et al., eds. (1990).

Barnes, E. W. (1933), *Scientific Theory and Religion: The World Described by Science and its Spiritual Interpretation*. Cambridge: Cambridge University Press.

Barrow, J. D., and F. J. Tipler (1986), *The Anthropic Cosmological Principle*. Oxford: Oxford University Press.

Baum, W. A. (1953), "The cosmological distance scale," *AJ* 58: 211.

———. (1957), "Photoelectric determinations of redshifts beyond 0.2*c*," *AJ* 62: 6–7.

Becquerel, J. (1908), "Sur la nature des charges d'électricité positive et sur l'existence des électrons positifs," *CR* 146: 1308–11.

Behr, A. (1951), "Zur Entfehrnungsskala der extragalaktischen Nebel," *AN* 279: 97–104.

Berenda, C. W. (1951), "Notes on Lemaître's cosmology," *The Journal of Philosophy* 48: 338–41.

———. (1964), "On the cosmological indeterminacy principle of McCrae," *PS* 31: 265–70.

Berendzen, R., R. Hart, and D. Seeley (1976), *Man Discovers the Galaxies*. New York: Science History Publications.

Berendzen, R., and M. Hoskin (1971), "Hubble's announcement of Cepheids in spiral nebulae," *ASPL* no. 504.

Berger, A., ed. (1984), *The Big Bang and Georges Lemaître*. Dordrecht: Reidel.

Berlin, T. H. (1951), "The neutron-capture equations of element formation in a static universe: an exact solution," *PR* 84: 66–68.

Bernstein, J. (1979), *Hans Bethe: Prophet of Energy*. New York: Basic Books.

———. (1984), *Three Degrees Above Zero*. New York: Scribner's.

Bernstein, J., and G. Feinberg, eds. (1986), *Cosmological Constants: Papers in Modern Cosmology*. New York: Columbia University Press.

Berny, A. (1913), "Über kosmische Entwicklung," *Das Weltall* 13: 317–24.

Bertotti, B., et al., eds. (1990), *Modern Cosmology in Retrospect*. Cambridge: Cambridge University Press.

Bethe, H. A. (1939), "Energy production in stars," *PR* 55: 434–56.

———. (1942), "Energy production in stars," *AS* 30: 243–64.

———. (1968), "Energy production in stars," *Science* 161: 541–47.

———. (1986), *Basic Bethe: Seminal Articles on Nuclear Physics, 1936–1937*. Vol. 6 of The History of Modern Physics 1800–1950. New York: American Institute of Physics.

Bethe, H. A., and C. L. Critchfield (1938), "The formation of deuterons by proton combination," *PR* 54: 248–54.

Bethe, H. A., F. Hoyle, and R. Peierls (1939), "Interpretation of beta-disintegration data," *Nature* 143: 200–201.

Blumberg, S. A., and L. G. Panos (1990), *Edward Teller: Giant of the Golden Age of Physics*. New York: Charles Scribner's Sons.

Bohr, N. (1931), "The use of the concepts of space and time in atomic theory," *Nature* 127: 43.

Bok, B. J. (1936), "Galactic dynamics and cosmic time scale," *Observatory* 59: 76–85.

———. (1946), "The time-scale of the universe," *MNRAS* 106: 61–75.

Bolton, J., F. F. Gardner, and M. B. Mackey (1964), "The Parkes Catalogue of radio sources, declination zone −20 to −60," *AuJP* 17: 340–72.

Bondi, H. (1942), "On the generation of waves on shallow water by wind," *PRS* 181: 67–71.

———. (1947), "Spherically symmetrical models in general relativity," *MNRAS* 107: 410–25.

———. (1948a), "Review of cosmology," *MNRAS* 108: 104–20.

———. (1948b), "Observation and theory in cosmology," *Observatory* 68: 111–12.

———. (1952a), *Cosmology*. Cambridge: Cambridge University Press.

———. (1952b), "Fact and inference in experiment and observation," *Observatory* 72: 89–93.

———. (1955a), "Fact and inference in theory and in observation," *VA* 1: 155–62.

———. (1955b), "Theories of cosmology," *The Advancement of Science* 12: 33–38.

———. (1956), "The steady-state theory of cosmology and relativity." 152–54 in A. Mercier and M. Kervaire, eds. (1956), *Fünfzig Jahre Relativitätstheorie*. Basel: Birkhäuser.

———. (1957a), "The space traveller's youth," *Discovery* 18: 505–10.

———. (1957b), "Negative mass in general relativity," *RMP* 29: 423–28.

———. (1958–59), "Science and the structure of the universe," *MLPS* 101: 58–71.

———. (1960), *The Universe at Large*. London: Heinemann.

———. (1962), "Physics and cosmology," *Observatory* 82: 133–43.

———. (1964), "Foundations of general relativity and cosmology." 138–68 in A. Gião, ed. (1964b).

———. (1966a), "Some philosophical problems in cosmology." 393–400 in C. A. Mace, ed. (1966), *British Philosophy in the Mid-Century*. London: Allen and Unwin.

———. (1966b), "Steady-state cosmology." 31–36 in L. Rosino et al. (1966).

———. (1967), *Assumption and Myth in Physical Theory*. Cambridge: Cambridge University Press.

———. (1973), "Setting the scene." 11–22 in L. John, ed. (1973).

Bondi, H., (1982), "Steady state origins: Comments I." 58–61 in Y. Terzian and E. M. Bilson, eds. (1982).

———. (1988), "Steady-state cosmology," *QJRAS* 29: 65–67.

———. (1990a), *Science, Churchill & Me: The autobiography of Hermann Bondi.* Oxford: Pergamon Press.

———. (1990b), "The cosmological scene 1945–1952." 189–96 in B. Bertotti et al., eds. (1990).

———. (1992), "The philosopher for science," *Nature* 358: 363.

———. (1993a), "Religious divisions," *Nature* 365: 484.

———. (1993b), "Origins of steady state theory." 475–78 in N. S. Hetherington, ed. (1993).

———. (1994), "Karl Popper (1902–1994)," *Nature* 371: 478.

Bondi, H., and T. Gold (1948), "The steady-state theory of the expanding universe," *MNRAS* 108: 252–70.

———. (1952), "Is there an aether?" *Nature* 169: 146.

———. (1954), "The steady-state theory of the homogeneous expanding universe," *Observatory* 74: 36–37.

Bondi, H., T. Gold, and F. Hoyle (1955), "Black giant stars," *Observatory* 75: 80–81.

———. (1995), "Origins of steady-state theory," *Nature* 373: 10.

Bondi, H., T. Gold, and D. Sciama (1954), "A note on the reported color-index effect on distant galaxies," *APJ* 120: 597–99.

Bondi, H., and F. Hoyle (1944), "On the mechanism of accretion by stars," *MNRAS* 104: 273–82.

Bondi, H., and C. W. Kilmister (1959–60), "The impact of Logik der Forschung," *BJPS* 10: 55–57.

Bondi, H., and E. E. Salpeter (1952), "Thermonuclear reactions and astrophysics," *Nature* 169: 304–5.

Bondi, H., et al., eds. (1960), *Rival Theories of Cosmology: A Symposium and Discussion of Modern Theories of the Structure of the Universe.* London: Oxford University Press.

Bonnor, W. B. (1954), "The stability of cosmological models," *ZAP* 35: 10–20.

———. (1955), "Fifty years of relativity," *Science News* 37: 1–24.

———. (1956), "The formation of the nebulae," *ZAP* 39: 143–59.

———. (1957a), "Jeans' formula for gravitational instability," *MNRAS* 117: 104–17.

———. (1957b), "La formation des nébuleuses en cosmologie relativiste," *AIHP* 15: 158–72.

———. (1960), "The relativistic model of the steady-state universe," *MNRAS* 121: 475–81.

———. (1964), *The Mystery of the Expanding Universe.* New York: Macmillan.

Bonnor, W., and G. C. McVittie (1961), "Hoyle's covariant formulation of the law of creation of matter," *MNRAS* 122: 381–87.

Borel, E. (1922), *L'Espace et le Temps.* Paris: F. Alcan.

Born, M. (1956), *Experiment and Theory in Physics.* New York: Dover.

Braccesi, A. (1963), "On the relation between the production of energy and the composition of matter in a steady state universe," *NC* 29: 543–48.

Bracewell, R. N., ed. (1959), *Paris Symposium on Radio Astronomy.* Stanford, Calif.: Stanford University Press.

Brans, C., and R. H. Dicke (1961), "Mach's principle and a relativistic theory of gravitation," *PR* 124: 925–35.

Brecher, K., G. Burbidge, and P. A. Strittmatter (1971), "Counts of sources and theories," *CASP* 3: 99–109.

Brill, D. R. (1962), "Review of Jordan's extended theory of gravitation." 50–68 in C. Møller, ed. (1962).

Britz, D. (1990), "Cold fusion: An historical parallel," *Centaurus* 33: 368–72.

Bromberg, J. (1976), "The concept of particle creation before and after quantum mechanics," *HSPS* 7: 161–83.

Bronstein, M. (1933), "On the expanding universe," *PZS* 3: 73–82.

Bronstein, M., and L. Landau (1933), "Über den zweiten Wärmesatz und die Zusammenhangsverhältnisse der Welt im Großen," *PZS* 4: 114–18.

Brown, R. H. (1959), "The distribution and identification of the sources." 471–74 in R. N. Bracewell, ed. (1959).

Brush, S. G. (1976), *The Kind of Motion We Call Heat*. Amsterdam: North-Holland.

———. (1978), *The Temperature of History: Phases of Science and Culture in the Nineteenth Century*. New York: Burt Franklin & Co.

———. (1982), "Finding the age of the earth—by physics or by faith?" *Journal of Geological Education* 30: 34–58.

———. (1987), "The nebular hypothesis and the evolutionary world view," *History of Science* 25: 245–78.

———. (1990), "Theories of the origin of the solar system 1956–1985," *RMP* 62: 43–112.

———. (1992), "How cosmology became a science," *SA* 267 (August): 34–40.

———. (1993), "Prediction and theory evaluation: Cosmic microwaves and the revival of the big bang," *Perspectives on Science* 1: 565–602.

Bunge, M. (1962), "Cosmology and magic," *The Monist* 47: 116–41.

———. (1979), *Causality and Modern Science* (New York: Dover).

Burbidge, E. M. (1967), "Quasi-stellar objects," *ARAA* 5: 399–452.

Burbidge, E. M., and G. R. Burbidge (1953), "The abundance of the elements," *Observatory* 73: 69–74.

Burbidge, E. M., G. R. Burbidge, W. A. Fowler, and F. Hoyle (1957), "Synthesis of the elements in stars," *RMP* 29: 547–650.

Burbidge, E. M., G. R. Burbidge, and F. Hoyle (1963), "Condensations in the intergalactic medium," *APJ* 138: 873–88.

Burbidge, G. R. (1958), "Nuclear energy generation and dissipation in galaxies," *PASP* 70: 83–89.

———. (1959), "The theoretical explanation of radio emission." 541–51 in R. N. Bracewell, ed. (1959).

———. (1961), "Galactic explosions as sources of radio emission," *Nature* 190: 1053–56.

———. (1969), "Cosmic helium," *CASP* 1: 101–5.

———. (1971), "Was there really a big bang?" *Nature* 233: 36–40.

———. (1988), "Problems of cosmogony and cosmology." 223–38 in F. Bertola, J. W. Sulentic and B. F. Madore, eds. (1988), *New Ideas in Astronomy*. Cambridge: Cambridge University Press.

Burbidge, G. R., and F. Hoyle (1956), "Matter and anti-matter," *NC* 4: 558–64.

———. (1958), "Anti-matter," *SA* 198 (April): 34–39.

Burbidge, G. R., F. Hoyle, E. M. Burbidge, R. F. Christy, and W. A. Fowler (1956), "Californium-254 and supernovae," *PR* 103: 1145–49.

Burchfield, J. D. (1975), *Lord Kelvin and the Age of the Earth*. Chicago: University of Chicago Press.

Cameron, A. G. W. (1957a), "Nuclear reactions in stars and nucleogenesis," *PASP* 69: 201–22.

———. (1957b), "Stellar Evolution, Nuclear Astrophysics, and Nucleogenesis." Chalk River Report CRL-41.

Cappi, A. (1994), "Edgar Allan Poe's physical cosmology," *QJRAS* 35: 177–92.

Chalmers, J. A., and B. Chalmers (1935), "The expanding universe—an alternative view," *PM* 19: 436–46.

Chambers, L. G. (1961a), "The steady state universe with charge excess," *Nature* 191: 262–63.

———. (1961b), "An unsteady universe with excess of charge," *Nature* 191: 1082–83.

———. (1963), "A Lorentz-invariant universe with charge excess," *Nature* 198: 378–79.

Chandrasekhar, S. (1937), "The cosmological constants," *Nature* 139: 757–58.

———. (1944), "Galactic evidence for the time-scale of the universe," *Science* 99: 133–36.

Chandrasekhar, S., and L. R. Henrich (1942), "An attempt to interpret the relative abundances of the elements and their isotopes," *APJ* 95: 288–98.

Chang, H. (1993), "A misunderstood rebellion: the twin-paradox and Herbert Dingle's vision of science," *SHPS* 24: 741–90.

Chernin, A. D. (1994), "How Gamow calculated the temperature of the background radiation or a few words about the fine art of theoretical physics," *PU* 37: 813–20.

Chiu, H.-Y. (1964), "Gravitational collapse," *PT* 17 (March): 21–34.

Clerke, A. M. (1890), *The System of the Stars*. London: Longmans, Green and Co.

Close, F. (1991), *Too Hot to Handle: The Race for Cold Fusion*. Princeton: Princeton University Press.

Code, A. D. (1959), "Energy distribution curves of galaxies," *PASP* 71: 118–25.

Copp, C. M. (1982), "Relativistic cosmology I: Paradigm commitment and rationality," *AQ* 4: 103–16.

———. (1983), "Relativistic cosmology II: Social structure, skepticism, and cynicism," *AQ* 4: 179–88.

———. (1985), "Professional specialization, perceived anomalies, and rival cosmologies," *Knowledge: Creation, Diffusion, Utilization* 7: 63–95.

Couderc, P. (1952), *The Expansion of the Universe*. London: Faber and Faber.

Coutrez, R. (1945), "Une nouvelle hypothèse cosmogonique établie par Haldane sur les bases de la cosmologie de Milne," *CT* 61: 208–13.

Crookes, W. (1887), "Presidential address to section B," *Report of the British Association for the Advancement of Science*, 1886: 558–76.

Crowe, M. J. (1994), *Modern Theories of the Universe: From Herschel to Hubble*. New York: Dover.

Cushing, J. T. (1990), *Theory Construction and Selection in Modern Physics: The S-Matrix*. Cambridge: Cambridge University Press.

Dalton, J. (1808), *A New System of Chemical Philosophy*. Pt. I. Manchester: R. Bickerstaff.

Dauvillier, A. (1963), "Les hypothèse cosmogoniques et la théorie des cycles cosmiques," *Scientia* 98: 121–26.

Davidson, M. (1955), "Modern cosmology and the theologians," *VA* 1: 166–72.

Davidson, W. (1959), "Steady-state cosmology treated according to general relativity," *MNRAS* 119: 309–24.

———. (1960a), "Interpretation of cosmology," *Nature* 187: 583.

———. (1960b), "Angular measurements in observational cosmology," *MNRAS* 120: 271–85.

———. (1961), "Cosmological significance of angular measurements of distant radio sources," *Nature* 189: 991–92.

———. (1961–62), "The cosmological implications of the recent counts of radio sources," *MNRAS* 123: 425–35.

———. (1962–63), "Philosophical aspects of cosmology," *BJPS* 13: 120–29.

———. (1967), "Fitting cosmological models to the radio source counts," *Nature* 216: 1076–79.

———. (1990), "George McVittie's work in relativity," *VA* 33: 65–69.

Davidson, W., and M. Davies (1963–64), "Interpretation of the counts of radio sources in terms of a 4-parameter family of evolutionary eniverses," *MNRAS* 127: 241–55.

Davidson, W., and J. V. Narlikar (1966), "Cosmological models and their observational validation," *RPP* 29: 539–62.

Davies, J. T. (1954), "The age of the universe," *BJPS* 5: 191–202.

Davies, P. C. W. (1974), *The Physics of Time Asymmetry*. London: Surrey University Press.

Delbrück, M. (1972), "Out of this world." 280–88 in F. Reines, ed. (1972).

De Maria, M., and A. Russo (1989), "'Cosmic ray romancing': the discovery of the latitude effect and the Compton-Millikan controversy," *HSPS* 19: 211–66.

———. (1990), "Cosmic rays and cosmological speculations in the 1920s: the debate between Jeans and Millikan." 401–09 in B. Bertotti et al., eds. (1990).

Denisse, J.-F., É. Le Roux, and J. C. Steinberg (1957), "Nouvelles observations du rayonnement du ciel sur la longueur d'onde 33 cm," *CR* 244: 3030–33.

Dennis, M. A. (1994), "'Our First Line of Defense': two university laboratories in the postwar American state," *Isis* 85: 427–55.

Deprit, A. (1984), "Monsignor Georges Lemaîtres." 363–92 in A. Berger, ed. (1984).

Deser, S., and F. A. E. Pirani (1965), "Critique of a new theory of gravitation," *PRS* 288: 133–45.

De Sitter, W. (1917), "On Einstein's theory of gravitation and its astronomical consequences. Third paper," *MNRAS* 78: 3–28.

———. (1930a), "On the magnitudes, diameters and distances of the extragalactic nebulae, and their apparent radial velocities," *BAN* 5: 157–71.

———. (1930b), "The expanding universe. Discussion of Lemaître's solution of the equations of the inertial field," *BAN* 5: 211–18.

———. (1930c), "On the distances and radial velocities of extra-galactic nebulae, and the explanation of the latter by the relativity theory of inertia," *PNAS* 16: 474–88.

———. (1932), *Kosmos*. Cambridge, Mass.: Harvard University Press.

———. (1933), "On the expanding universe and the time-scale," *MNRAS* 93: 628–34.

DeWitt, B. (1953), "Pair production by a curved metric," *PR* 90: 357.

Dicke, R. H. (1957), "Gravitation without a principle of equivalence," *RMP* 29: 363–76.

———. (1959), "Gravitation—an enigma," *AS* 47: 25–40.

———. (1963), "Cosmology, Mach's principle and relativity," *AJP* 31: 500–509.

———. (1968), "Scalar-tensor gravitation and the cosmic fireball," *APJ* 152: 1–24.

———. (1970), *Gravitation and the Universe*. Philadelphia: American Philosophical Society.

Dicke, R. H., R. Beringer, R. L. Kyhl, and A. B. Vane (1946), "Atmospheric absorption measurements with a microwave radiometer," *PR* 70: 340–48.

Dicke, R. H., and P. J. E. Peebles (1965), "Gravitation and space science," *Space Science Reviews* 4: 419–60.

Dicke, R. H., P. J. E. Peebles, P. G. Roll, and D. T. Wilkinson (1965), "Cosmic black-body radiation," *APJ* 142: 414–19.

Dingle, H. (1933) "On isotropic models of the universe, with special reference to the stability of the homogeneous and static states," *MNRAS* 94: 134–58.

———. (1934), "Physics and the public mind," *Nature* 133: 818–20.

———. (1936), "Physical universe," *JWAS* 26: 183–95.

———. (1937a), "Modern Aristotelianism," *Nature* 139: 784–86.

———. (1937b), *Through Science to Philosophy*. London: Williams & Norgate.

———. (1938), "Science and the unobservable," *Nature* 141: 21–28.

———. (1940), *The Special Theory of Relativity*. London: Methuen.

———. (1952), *The Scientific Adventure*. London: Pitman and Sons.

———. (1953), "Science and modern cosmology," *MNRAS* 113: 393–407.

———. (1956), "Cosmology and science," *SA* 192 (September): 224–36.

———. (1957), "The resolution of the clock paradox," *AuJP* 10: 418–23.

———. (1958), "Clock paradox of relativity," *Science* 127: 158–60.

———. (1960), "Relativity and electromagnetism: An epistemological appraisal," *PS* 27: 233–53.

Dingle, H., and Viscount Samuel (1961), *A Threefold Cord: Philosophy, Science, Religion*. London: Allen and Unwin.

Dirac, P. A. M. (1937), "The cosmological constants," *Nature* 139: 323.

———. (1938), "A new basis for cosmology," *PRS* 165: 199–208.

———. (1973), "Long range forces and broken symmetries," *PRS* 333: 403–18.

Doroshkevich, A. G., and I. D. Novikov (1964), "Mean density of radiation in the metagalaxy and certain problems in relativistic cosmology," *Soviet Physics—Doklady* 9: 111–13.

Douglas, A. V. (1956), *The Life of Arthur Stanley Eddington*. London: Thomas Nelson and Sons.

Dowker, J. S. (1964), "Cosmology and weak interactions," *NC* 32: 1816–18.

Dungey, J. W. (1955), "Deductions from the perfect cosmological principle," *PCPS* 51: 532–35.

Earman, J., M. Janssen, and J. D. Norton, eds. (1993), *The Attraction of Gravitation: New Studies in the History of General Relativity*. Boston: Birkhäuser.

Eddington, A. S. (1917), "Further notes on the radiative equilibrium of the stars," *MNRAS* 77: 59–61.

———. (1919), "The sources of stellar energy," *Observatory* 42: 371–76.

———. (1920), "The internal constitution of the stars," *Nature* 106: 14–20.

———. (1926a), "The sources of stellar energy," *Nature* 117 (suppl.): 25–32.

———. (1926b), *The Internal Constitution of the Stars*. Cambridge: Cambridge University Press.

———. (1927–29), "Sub-atomic energy," *MLPS* 72–73: 101–17.

———. (1928), *The Nature of the Physical World*. Cambridge: Cambridge University Press.

———. (1930a), "Space and its properties," *Nature* 125: 849–50.

———. (1930b), "On the instability of the Einstein world," *MNRAS* 90: 668–78.

———. (1931), "The end of the world: from the standpoint of mathematical physics," *Nature* 127: 447–53.

———. (1932), "The hydrogen content of the stars," *MNRAS* 92: 471–81.

————. (1933), *The Expanding Universe*. Cambridge: Cambridge University Press.

————. (1936), *Relativity Theory of Protons and Electrons*. Cambridge: Cambridge University Press.

————. (1939), *The Philosophy of Physical Science*. Cambridge: Cambridge University Press.

————. (1944), "The recession-constant of the galaxies," *MNRAS* 104: 200–204.

————. (1946), *Fundamental Theory*. Cambridge: Cambridge University Press.

Edge, D. O., and M. J. Mulkay (1976), *Astronomy Transformed: The Emergence of Radio Astronomy in Britain*. New York: Wiley and Sons.

Edge, D. O., P. Scheuer, and J. Shakeshaft (1958), "Evidence on the spatial distribution of radio sources from a survey at a frequency of 159 Mc/s," *MNRAS* 118: 183–96.

Einstein, A. (1917), "Kosmologische Betrachtungen zur allgemeinen Relativitätstheorie," *Sitzungsberichte der königliche Preussische Akademie der Wissenschaften zu Berlin*: 142–52.

————. (1918), "Kritischen zu einer von Herrn de Sitter gegebenen Lösung der Gravitationsgleichungen," *SPAW*: 270–72.

————. (1919), "Spielen Gravitationsfelder im Aufbau der materiellen Elementarteilchen eine wesentliche Rolle?" *SPAW*: 349–56.

————. (1922), "Bemerkung zu der Arbeit von A. Friedmann," *ZP* 11: 326.

————. (1923), "Notiz zu der Arbeit von A. Friedmann," *ZP* 16: 228.

————. (1929), "Space-time." 105–8 in *Encyclopedia Britannica* (London: Encyclopedia Britannica, Inc.), 14th ed. Vol. 21.

————. (1931), "Zum kosmologischen Problem der allgemeinen Relativitätstheorie," *SPAW*: 235–37.

————. (1936), "Physics and Reality," *JFI* 221: 349–82.

————. (1945), *The Meaning of Relativity*. Princeton: Princeton University Press.

————. (1950), *Out of My Later Years*. London: Thames and Hudson.

Einstein, A., and W. de Sitter (1931), "On the relation between the expansion and the mean density of the universe," *PNAS* 18: 213–14.

Einstein, A., et al. (1923), *The Principle of Relativity*. New York: Dover.

Eisenstaedt, J. (1986), "La relativité génerale à l'étiage: 1925–1955," *AHES* 35: 15–85.

————. (1993), "Lemaître and the Schwarzschild solution." 353–89 in J. Earman, M. Janssen, and J. D. Norton, eds. (1993).

Eisenstaedt, J., and A. J. Kox, eds. (1992), *Studies in the History of General Relativity*. Boston: Birkhäuser.

Ellis, B. (1955), "Has the universe a beginning in time?" *Australasian Journal of Philosophy* 33: 32–37.

Ellis, G. F. R. (1984), "Alternatives to the big bang," *ARAA* 22: 157–84.

————. (1988), "The expanding universe: a history of cosmology from 1917 to 1960." 367–431 in D. Howard and J. Stachel, eds. (1988), *Einstein and the History of General Relativity*. Boston: Birkhäuser.

Ellis, G. F. R., A. Lanza, and J. Miller, eds. (1993), *The Renaissance of General Relativity and Cosmology: A Survey to Celebrate the 65th Birthday of Dennis Sciama*. Cambridge: Cambridge University Press.

Ellis, R. H. (1968), "The swashbuckling physicist—a talk with George Gamow," *PT* 21 (February): 101–3.

Engelhardt, H. T., and A. L. Caplan, eds. (1987), *Scientific Controversies: Case Studies in the Resolution and Closure of Disputes in Science and Technology*. Cambridge: Cambridge University Press.

Farkas, L., and P. Harteck (1931), "Thermodynamische Bemerkungen zur Entstehung der Elemente," *NW* 19: 705–6.

Farley, J. (1974), *The Spontaneous Generation Controversy From Descartes to Oparin*. Baltimore: Johns Hopkins University Press.

Fermi, E. (1949), "Teorie sulle origine degli elementi." 707–20 in *Collected Papers*. Vol. 2. Rome: University of Chicago Press.

Fernie, J. D. (1969), "The period-luminosity relation: A historical review," *PASP* 81: 707–31.

Ferris, T. (1977), *The Red Limit: The Search for the Edge of the Universe*. New York: William Morrow and Co.

Feuer, L. (1977), "Teleological principles of science," *Inquiry* 21: 377–407.

Feynman, R. P., F. B. Moringo, and W. G. Wagner (1995), *Feynman Lectures on Gravitation*, ed. B. Hatfield. Reading, Mass.: Addison-Wesley.

Field, G. B., H. Arp, and J. N. Bahcall (1973), *The Redshift Controversy*. Reading, Mass.: Benjamin.

Field, G. B., and R. C. Henry (1964), "Free-free emission by intergalactic hydrogen," *APJ* 140: 1002–12.

Field, G. B., and J. L. Hitchcock (1966), "Cosmic black-body radiation at $\lambda = 2.6$ mm," *PRL* 16: 817–18.

Fierz, M. (1939), "Über die relativistische Theorie kräftefreie Teilchen mit beliebigen Spin," *HPA* 12: 3–37.

Fierz, M., and W. Pauli (1939), "On relativistic wave equations for particles of arbitrary spin in an electromagnetic field," *PRS* 173: 211–32.

Finlay-Freundlich, E. (1951), *Cosmology*. Chicago: University of Chicago Press.

Fock, V. A. (1964), "The researches of A. A. Fridman on the Einstein theory of gravitation," *SPU* 6: 473–74.

Forman, P. (1971), "Weimar culture, causality, and quantum theory, 1918–1927: Adaption by German physicists and mathematicians to a hostile intellectual environment," *HSPS* 3: 1–116.

Fowler, W. A. (1967), "Nucleosynthesis in big and little bangs." 203–25 in G. Alexander, ed. (1967), *High Energy Physics and Nuclear Structure*. Amsterdam: North-Holland.

———. (1984), "The quest for the origin of the elements," *Science* 226: 922–35.

———. (1992),"From steam to stars to the early universe," *ARAA* 30: 1–9.

Fowler, W. A., G. R. Burbidge, and E. M. Burbidge (1955), "Stellar evolution and the synthesis of the elements," *APJ* 122: 271–85.

Frank, F. C. (1948), "An isotopic abundance rule and its bearing on the origin of the nuclei," *PPS* 60: 211.

Frankel, H. (1987), "The continental drift debate." 203–48 in H. T. Engelhardt and A. L. Caplan, eds. (1987).

Frank-Kamenetskij, D. A. (1959), "The origin of the chemical elements," *SPU* 26: 600–619.

Frederiks, V., and A. Schechter (1928), "Notiz zur Frage nach der Berechnung der Aberration und der Parallaxe in Einsteins, de Sitters und Friedmanns Welten in der allgemeinen Relativitätstheorie," *ZP* 51: 584–92.

Frenkel, V. Ya. (1988), "Aleksandr Aleksandrovich Fridman (Friedmann): a biographical essay," *SPU* 31: 645–65.

———. (1994), "George Gamow: world line 1904–1933," *PU* 37: 767–89.

Friedmann, A. (1922), "Über die Krümmung des Raumes," *ZP* 10: 377–86.

————. (1924), "Über die Möglichkeit einer Welt mit konstanter negativer Krümmung des Raumes," *ZP* 21: 326–32.

Gale, G. (1990), "Cosmological fecundity: theories of multiple universes." 189–206 in J. Leslie, ed. (1990).

Gale, G., and J. Urani (1993), "Philosophical midwifery and the birthpangs of modern cosmology," *AJP* 61: 66–73.

Gamow, G. (1928), "Zur Quantentheorie des Atomkernes," *ZP* 51: 203–20.

————. (1931), *Constitution of Atomic Nuclei and Radioactivity.* Oxford: Clarendon Press.

————. (1935), "Nuclear transformations and the origin of the chemical elements," *Ohio Journal of Science* 35: 406–13.

————. (1937), *Structure of Atomic Nuclei and Nuclear Transformations.* Oxford: Clarendon Press.

————. (1938), "Kernumwandlungen als Energiequellen der Sterne," *ZAP* 16: 113–60.

————. (1939a), "Nuclear reactions in stellar evolution," *Nature* 144: 575–77, 620–22.

————. (1939b), "Physical possibilities of stellar evolution," *PR* 55: 718–25.

————. (1940), *The Birth and Death of the Sun.* New York: Viking Press.

————. (1942), "Concerning the origin of chemical elements," *JWAS* 32: 353–55.

————. (1946a), "Expanding universe and the origin of elements," *PR* 70: 572–73.

————. (1946b), "Rotating universe?" *Nature* 158: 549.

————. (1947), *Atomic Energy in Cosmic and Human Life.* Cambridge: Cambridge University Press.

————. (1948a), "The origin of elements and the separation of galaxies," *PR* 74: 505–6.

————. (1948b), "The evolution of the universe," *Nature* 162: 680–82.

————. (1949), "On relativistic cosmogony," *RMP* 21: 367–73.

————. (1950), "Half an hour of creation," *PT* 3 (May): 16–21.

————. (1951), "The origin and evolution of the universe," *AS* 39: 393–407.

————. (1952a), "The role of turbulence in the evolution of the universe," *PR* 86: 251.

————. (1952b), *The Creation of the Universe.* New York: Viking Press.

————. (1953), "Expanding universes and the origin of galaxies," *Kongelige Danske Videnskabernes Selskab, Matematisk-Fysiske Meddelelser* 27: 1–16.

————. (1954a), "Modern cosmology," *SA* 190 (March): 55–63.

————. (1954b), "On the steady-state theory of the universe," *AJ* 59: 200.

————. (1954c), "On the formation of protogalaxies in the turbulent primordial gas," *PNAS* 40: 480–84.

————. (1956a), "The evolutionary universe," *SA* 192 (September): 136–54.

————. (1956b), "The physics of the expanding universe," *VA* 2: 1726–32.

————. (1963), "My early memories of Fritz Houtermans." vii–viii in J. Geis and E. D. Goldberg, eds. (1963), *Earth Science and Meteoritics.* Amsterdam: North-Holland.

————. (1964), "The universe and its origin," 11–56 in H. Messel and S. T. Butler, eds., *The Universe and its Origin* (London: Macmillan.

————. (1965), *Mr Tompkins in Paperback.* Cambridge: Cambridge University Press.

————. (1966), "Cosmological theories of the origin of chemical elements." 443–48 in R. E. Marshak, ed. (1966), *Perspectives in Modern Physics: Essays in Honor of Hans A. Bethe.* New York: Interscience.

————. (1967), "History of the universe," *Science* 158: 766–69.

————. (1970), *My World Line: An Informal Autobiography.* New York: Viking Press.

Gamow, G., and C. L. Critchfield (1949), *Theory of Atomic Nucleus and Nuclear Energy-Sources.* Oxford: Clarendon Press.

Gamow, G., and J. A. Fleming (1942), "Report on the eighth annual Washington Conference of Theoretical Physics, April 23–25, 1942," *Science* 95: 579–81.

Gamow, G., and J. A. Hynek (1945), "A new theory by C. F. von Weizsäcker of the origin of the planetary system," *APJ* 101: 249–54.

Gamow, G., and D. Iwanenko (1926), "Zur Wellentheorie der Materie," *ZP* 39: 865–68.

Gamow, G., and L. Landau (1933), "Internal temperature of stars," *Nature* 132: 567.

Gamow, G., and E. Teller (1939), "On the origin of great nebulae," *PR* 55: 654–57.

Gião, A. (1964a), "On the theory of the cosmological models with special reference to a generalized steady-state model." 5–100 in A. Gião, ed. (1964b).

———. ed. (1964b), *Cosmological Models*. Lisbon: Instituto Gulbenkian de Ciência.

Giere, R. N. (1987), "Controversies involving science and technology: a theoretical perspective." 125–50 in H. T. Engelhardt and A. L. Caplan, eds. (1987).

Gillispie, C. C. (1951), *Genesis and Geology*. Cambridge, Mass.: Harvard University Press.

Gilmore, C. P. (1965), "They're solving the world's greatest mystery," *Popular Science* 187 (November): 102–5, 200–203.

Gingerich, O. (1994), "The summer of 1953: a watershed for astrophysics," *PT* 47 (December): 34–41.

Giovanelli, R. G. (1963–64), "A fluctuation theory of cosmology," *MNRAS* 127: 461–69.

Godart, O. (1992), "Contributions of Lemaître to general relativity (1922–1934)." 437–52 in J. Eisenstaedt and A. J. Kox, eds. (1992).

Godart, O., and M. Heller (1978), "Un travail inconnu de Georges Lemaître," *Revue d'Histoire des Sciences* 31: 345–56.

———. (1979a), "Les relations entre la science et la foi chez Georges Lemaître," *Pontifica Academia delle Scienze, Commentarii* 3: 1–12.

———. (1979b), "Einstein-Lemaître: rencontres d'idées," *RQS* 150: 23–43.

———. (1985), *Cosmology of Lemaître*. Tucson: Pachart Publishing House.

Gödel, K. (1949), "An example of a new type of cosmological solutions of Einstein's field equations of gravitation," *RMP* 21: 447–50.

Gold, T. (1948), "Hearing, I: The cochlea as a frequency analyzer," *PRS* B135: 462–91.

———. (1949), "Creation of matter in the universe," *Nature* 164: 1006.

———. (1951), ["The origin of cosmic radio noise"] in "Proceedings of Conference on Dynamics of Ionized Media." University College, London.

———. (1954), "Relation between modern cosmologies and nuclear astrophysical processes." 68–70 in P. Ledoux, ed. (1954).

———. (1955), "The 'horizon' of the steady-state universe," *Nature* 175: 382.

———. (1956), "Cosmology," *VA* 2: 1721–26.

———. (1958), "The arrow of time." 81–95 in R. Stoops, ed. (1958).

———. (1962), "The arrow of time," *AJP* 30: 403–10.

———. ed. (1967), *The Nature of Time* (Ithaca, N.Y.: Cornell University Press).

———. (1973), "Multiple universes," *Nature* 242: 24–25.

———. (1982), "Steady state origins: comments II." 62–65 in Y. Terzian and E. M. Bilson, eds. (1982).

———. (1989), "New ideas in science," *Journal of Scientific Exploration* 3: 103–12.

Gold, T., and F. Hoyle (1959), "Cosmic rays and radio waves as manifestations of a hot universe." 583–88 in R. N. Bracewell, ed. (1959).

Gold, T., and F. Pacini (1968), "Can the observed microwave background be due to a superposition of sources?" *APJ* 152: L115–L118.

Goldhaber, M. (1956), "Speculations on cosmogony," *Science* 124: 218–19.

Goldschmidt, V. M. (1937a), "Geochemische Verteilungsgesetze der Elemente. IX. Die Mengenverhältnisse der Elemente und der Atom-Arten," *Norske Videnskabs-Akademien i Oslo*, no. 4.

———. (1937b), "The principles of distribution of chemical elements in minerals and rocks," *Journal of the Chemical Society*: 655–73.

Gorelik, G. E., and V. Ya. Frenkel (1994), *Matvei Petrovich Bronstein and Soviet Theoretical Pysics in the Thirties*. Basel: Birkhäuser.

Gould, R. J., and G. R. Burbidge (1963), "X-rays from the galactic center, external galaxies, and the intergalactic medium," *APJ* 138: 969–77.

Gould, R. J., and D. W. Sciama (1964), "Cosmic X- and infrared rays as tools for exploring the large-scale structure of the universe," *APJ* 140: 1634–36.

Graham, L. R. (1972), *Science and Philosophy in the Soviet Union*. New York: Knopf.

———. (1987), *Science, Philosophy and Human Behavior in the Soviet Union*. New York: Columbia University Press.

Graham-Smith, F. (1986), "Martin Ryle," *Biographical Memoirs of Fellows of the Royal Society* 32: 497–524.

Graham-Smith, F., and B. Lovell (1983), "On the discovery of extragalactic radio sources," *JHA* 14: 155–65.

Gray, G. W. (1953), "A larger and older universe," *SA* 188 (June): 56–66.

Greenstein, J. L., and T. A. Matthews (1963), "Red-shift of the unusual radio source: 3C 48," *Nature* 197: 1041–42.

Greenstein, J. L., and M. Schmidt (1964), "The quasi-stellar radio sources 3C 48 and 3C 273," *APJ* 140: 1–34.

Gregory, C. (1945), "On a supplement to the field equations with an application to cosmology," *PR* 67: 179–84.

Gribbin, J. R. (1976), *Galaxy Formation: A Personal View*. New York: John Wiley and Sons.

Grünbaum, A. (1952), "Some highlights of modern cosmology and cosmogony," *The Review of Metaphysics* 5: 481–98.

———. (1989), "The pseudo-problem of creation in physical cosmology," *Epistemologia* 12: 3–32.

———. (1991), "Creation as a pseudo-explanation in current physical cosmology," *Erkenntnis* 35: 233–54.

———. (1993), "Creation in cosmology." 126–36 in N. S. Hetherington, ed. (1993).

Gunter, P. A. Y. (1971), "Bergson's theory of matter and modern cosmology," *Journal for the History of Ideas* 32: 525–42.

Günther, P. (1924), "Die kosmologischen Betrachtungen von Nernst," *Zeitschrift für angewandte Chemie* 37: 454–57.

Haar, D. ter (1950), "Cosmogonical problems and stellar energy," *RMP* 22: 119–52.

Haas, A. E. (1936), "An attempt to a purely theoretical derivation of the mass of the universe," *PR* 49: 411–12.

Haldane, J. B. S. (1928), "The universe and irreversibility," *Nature* 122: 808–9.

———. (1936), "Is space-time simply connected?" *Observatory* 59: 228–29.

———. (1939), *The Marxist Philosophy and the Sciences*. New York: Random House.

———. (1945a), "A new theory of the past," *AS* 33: 129–45.

Haldane, J.B.S., (1945b), "A quantum theory of the origin of the solar system," *Nature* 155: 133–35.

Halpern, O. (1933), "Scattering processes produced by electrons in negative energy states," *PR* 44: 856–57.

Hanson, N. R. (1963), "Some philosophical aspects of contemporary cosmologies." 465–82 in B. Baumrin, ed. (1963), *Philosophy of Science. The Delaware Seminar.* Vol. 2. New York: Interscience.

Harder, A. J. (1974), "E. A. Milne, scientific revolutions and the growth of knowledge," *AnS* 31: 351–63.

Harkins, W. D. (1917), "The evolution of the elements and the stability of complex atoms," *JACS* 39: 856–79.

Harré, W. H. (1962–63), "Philosophical aspects of cosmology," *BJPS* 13: 104–19.

Harrison, J. A., ed. (1965), *The Complete Works of Edgar Allan Poe.* New York: AMS Press.

Harwit, M. (1961), "Can gravitational forces alone account for galaxy formation in a steady state universe? I, " *MNRAS* 122: 47–50.

———. (1961–62), "Can gravitational forces alone account for galaxy formation in a steady state universe? II, Bondi-Gold universes," *MNRAS* 123: 257–63.

———. (1981), *Cosmic Discovery: The Search, Scope, and Heritage of Astronomy.* New York: Basic Books.

Hawking, S. W. (1965a), "On the Hoyle-Narlikar theory of gravitation," *PRS* 286: 313–19.

———. (1965b), "Occurrence of singularities in open universes," *PRL* 15: 689–90.

Hawking, S. W., and R. Penrose (1970), "The singularities of gravitational collapse and cosmology," *PRS* 314: 529–48.

Hayashi, C. (1950), "Proton-neutron concentration ratio in the expanding universe at the stages preceding the formation of the elements," *PTP* 5: 224–35.

Hayashi, C., and M. Nishida (1956), "Formation of light nuclei in the expanding universe," *PTP* 16: 613–24.

Hazard, C., and E. Salpeter (1969), "Discrete sources and the microwave background in steady-state cosmologies," *APJ* Letters 157: L87–L90.

Hazard, C., and D. Walsh (1959), "An experimental investigation of the effects of confusion in a survey of localized radio sources," *MNRAS* 119: 648–56.

Heckmann, O. (1932), "Die Ausdehnung der Welt in ihrer Abhängigkeit von der Zeit," *Nachrichten von der Gesellschaft der Wissenschaften zu Göttingen, Math.-Phys. Klasse*: 97–106.

———. (1942), *Theorien der Kosmologie.* Berlin: Springer-Verlag.

———. (1951), "Theorie und Erfahrung in der Kosmologie," *NW* 38: 84–91.

———. (1961), "On the possible influence of a general rotation on the expansion of the universe," *AJ* 66: 599–603.

———. (1962), "General review of cosmological theories." 429–40 in G. C. McVittie, ed. (1962b).

Hendry, J. (1987), "The scientific origins of controlled fusion technology," *AnS* 44: 143–68.

Herzberg, G. (1950), *Molecular Spectra and Molecular Structure*, 2nd ed. Vol. 1. New York: Van Nostrand.

Hetherington, N. S. (1971), "The measurement of radial velocities of spiral nebulae," *Isis* 62: 309–13.

————. (1973), "The delayed response to suggestions of an expanding universe," *Journal of the British Astronomical Association* 84: 22–28.

————. (1982), "Philosophical values and observation in Edwin Hubble's choice of a model of the universe," *HSPS* 13: 41–68.

————. (1988), *Science and Objectivity: Episodes in the History of Astronomy*. Ames, Iowa: Iowa State University Press.

————. ed. (1993), *Encyclopedia of Cosmology: Historical, Philosophical, and Scientific Foundations of Modern Cosmology*. New York: Garland Publishing.

Hewish, A. (1961–62), "Extrapolation of the number-flux density relation of radio stars by Scheuer's statistical method," *MNRAS* 123: 167–81.

Hey, J. S. (1973), *The Evolution of Radio Astronomy*. London: Paul Elek.

Hillas, A. M., and T. E. Cranshaw (1959), "A comparison of the charges of the electron, proton and neutron," *Nature* 184: 892–93.

————. (1960), "Comparison of the charges of the electron, proton and neutron," *Nature* 186: 459–60.

Hirsh, R. F. (1983), *Glimpsing an Invisible Universe: The Emergence of X-Ray Astronomy*. Cambridge: Cambridge University Press.

Hogarth, J. E. (1962), "Cosmological considerations of the absorber theory of radiation," *PRS* 267: 365–82.

Holmes, A. (1913), *The Age of the Earth*. New York: Harper.

Hönl, H. (1949), "Zwei Bemerkungen zum kosmologischen problem," *AP* 6: 169–76.

Hoskin, M. A. (1982), *Stellar Astronomy: Historical Studies*. New York: Science History Publications.

Hoyle, F. (1938), "β-transitions in a Coulomb field," *PRS* 166: 249–69.

————. (1939), "Quantum electrodynamics, parts I and II," *PCPS* 35: 419–62.

————. (1946), "The synthesis of the elements from hydrogen," *MNRAS* 106: 343–83.

————. (1947), "On the formation of heavy elements in stars," *PPS* 59: 972–78.

————. (1948), "A new model for the expanding universe," *MNRAS* 108: 372–82.

————. (1949a), "Stellar evolution and the expanding universe," *Nature* 163: 197–98.

————. (1949b), "On the cosmological problem," *MNRAS* 109: 365–71.

————. (1949c), "Continuous creation," *The Listener* 41: 567–68.

————. (1949d), *Some Recent Researches in Solar Physics*. Cambridge: Cambridge University Press.

————. (1950a), *The Nature of the Universe*. Oxford: Blackwell.

————. (1950b), "Nuclear energy," *Observatory* 70: 194–95.

————. (1952), "Concepts of the universe," *New York Times Magazine* (1 June): 11–12, 50–51.

————. (1953), "On the fragmentation of gas clouds into galaxies and stars," *SPJ* 118: 513–28.

————. (1954), "On nuclear reactions occurring in very hot stars. I. The synthesis of elements from carbon to nickel," *APJ*, Suppl. 1: 121–46.

————. (1955a), *Frontiers of Astronomy*. New York: Mentor Books.

————. (1955b), "The horizon of the steady-state universe," *Nature* 175: 808.

————. (1956a), "Observational tests in cosmology." 150–51 in A. Mercier and M. Kervaire, eds. (1956).

————. (1956b), *Man and Materialism*. New York: Harper and Brothers.

————. (1956c), "The steady-state universe," *SA* 192 (September): 157–66.

————. (1958a), "The steady state theory." 53–80 in R. Stoops, ed. (1959).

Hoyle, F., (1958b), "Origin of the elements in stars." 281–90 in R. Stoops, ed. (1959).

———. (1959a), "The relation of radio astronomy to cosmology." 529–32 in R. N. Bracewell, ed. (1959).

———. (1959b), "Concluding lecture." 598–601 in R. N. Bracewell, ed. (1959).

———. (1960a), "On the possible consequences of a variability of the elementary charge," *PRS* 257: 431–44.

———. (1960b), "A covariant formulation of the law of creation of matter," *MNRAS* 120: 256–62.

———. (1961), "Observational tests in cosmology," *PPS* 77: 1–16.

———. (1962), "Cosmological tests of gravitational theories." 141–73 in C. Møller, ed. (1962).

———. (1963), "X-rays from outside the solar system," *APJ* 137: 993–95.

———. (1965a), *Encounter with the Future*. New York: Trident Press.

———. (1965b), *Galaxies, Nuclei, and Quasars*. New York: Harper and Row.

———. (1965c), "Recent developments in cosmology," *Nature* 208: 111–14.

———. (1965d), "Origin of cosmic X rays," *PRL* 15: 131–32.

———. (1966), "Recent developments in nucleosynthesis," *Observatory* 86: 217–23.

———. (1968), "Review of recent developments in cosmology," *PRS* 308: 1–17.

———. (1969a), "Speculation on the nature of the nuclei of galaxies," *Nature* 224: 477.

———. (1969b), "Highly condensed objects," *QJRAS* 10: 10–20.

———. (1973), "The origin of the universe," *QJRAS* 14: 278–87.

———. (1975), "On the origin of the microwave background," *APJ* 196: 661–70.

———. (1977), *Ten Faces of the Universe*. London: Heinemann.

———. (1980), *Steady-State Cosmology Re-Visited*. Cardiff: University College Cardiff Press.

———. (1981), "The big bang in astronomy," *New Scientist* 92: 521–27.

———. (1982a), "The universe: past and present reflections," *ARAA* 20: 1–35.

———. (1982b), "Steady state cosmology revisited." 17–57 in Y. Terzian and G. M. Bilson, eds. (1982).

———. (1982c), "The world according to Hoyle," *The Sciences* 22 (August): 9–13.

———. (1985), *The Small World of Fred Hoyle*. London: Michael Joseph.

———. (1986), "Personal comments on the history of nuclear astrophysics," *QJRAS* 27: 445–53.

———. (1988), "The relation of the microwave background to the remarkable properties of slender metallic needles." 236–40 in B.R. Iyer et al., eds. (1988), *Highlights in Gravitation and Cosmology*. Cambridge: Cambridge University Press.

———. (1989a), "Frontiers in cosmology." 97–107 in S. K. Biswas, D. C. V. Mallik, and C. V. Vishveshwara, eds. (1989), *Cosmic Perspectives: Essays Dedicated to the Memory of M. K. V. Bappu*. Cambridge: Cambridge University Press.

———. (1989b), "The steady-state theory revived?" *CA* 13: 81–86.

———. (1990), "An assessment of the evidence against the steady-state theory." 221–31 in B. Bertotti et al., eds. (1990).

———. (1992), "The achievement of Dirac," *Notes and Records of the Royal Society of London* 46: 183–87.

———. (1993), "Final remarks." 693–98 in G. Chincarini, ed. (1993), *Observational Cosmology*. San Francisco: Astronomical Society of the Pacific.

———. (1994), *Home Is Where the Wind Blows: Chapters from a Cosmologist's Life*. Mill Valley, Calif.: University Science Books.

Hoyle, F., and G. R. Burbidge (1966a), "Relation between the red-shifts of quasi-stellar objects and their radio magnitudes," *Nature* 212: 1334.

———. (1966b), "Relation between the red-shifts of quasi-stellar objects and their radio and optical magnitudes," *Nature* 210: 1346–47.

———. (1966c), "On the nature of the quasi-stellar objects," *APJ* 144: 534–52.

Hoyle, F., G. R. Burbidge, and J. V. Narlikar (1993), "A quasi-steady state cosmological model with creation of matter," *APJ* 410: 437–57.

———. (1994a), "Astrophysical deductions from the quasi-steady state cosmology," *MNRAS* 267: 1007–19.

———. (1994b), "Further astrophysical quantities expected in a quasi-steady state universe," *Astronomy and Astrophysics* 289: 729–39.

Hoyle, F., G. R. Burbidge, and W. L. W. Sargent "On the nature of quasi-stellar objects," *Nature* 209: 751–53.

Hoyle, F., D. N. F. Dunbar, W. A. Wenzel, and W. Whaling (1953), "A state in C^{12} predicted from astrophysical evidence," *PR* 92: 1095.

Hoyle, F., and W. A. Fowler (1962–63), "On the nature of strong radio sources," *MNRAS* 125: 169–76.

———. (1963), "Nature of strong radio sources," *Nature* 197: 533–35.

———. (1967), "Gravitational red-shifts in quasi-stellar objects," *Nature* 213: 373–74.

Hoyle, F., W. A. Fowler, G. R. Burbidge, and E. M. Burbidge (1956), "Origin of the elements in stars," *Science* 124: 611–14.

Hoyle, F., and R. A. Lyttleton (1939a), "The evolution of the stars," *Nature* 144: 1019–20.

———. (1939b), "Evolution of stars," *PCPS* 35: 592–609.

———. (1940), "Physical aspects of accretion by stars," *PCPS* 36: 424–27.

———. (1948), "The internal constitution of the stars," *ONRAS* no. 12: 89–108.

Hoyle, F., and J. V. Narlikar (1961–62), "On the counting of radio sources in the steady-state cosmology, I," *MNRAS* 123: 133–66.

———. (1962a), "On the counting of radio sources in the steady-state cosmology, II," *MNRAS* 125: 13–20.

———. (1962b), "The steady-state model and the ages of galaxies," *Observatory* 82: 13–14.

———. (1962c), "Mach's principle and the creation of matter," *PRS* 270: 334–41.

———. (1964a), "A new theory of gravitation," *PRS* 282: 191–207.

———. (1964b), "On the avoidance of singularities in *C*-field cosmology," *PRS* 278: 464–78.

———. (1964c), "The *C*-field as a direct particle field," *PRS* 282: 178–83.

———. (1964d), "Time symmetric electrodynamics and the arrow of time in cosmology," *PRS* 277: 1–23.

———. (1966a), "On the effects of the non-conservation of baryons in cosmology," *PRS* 290: 143–60.

———. (1966b), "A radical departure from the 'steady-state' concept in cosmology," *PRS* 290: 162–76.

———. (1966c), "On the formation of elliptical galaxies," *PRS* 290: 177–85.

———. (1966d), "A conformal theory of gravitation," *PRS* 294: 138–48.

———. (1971), "On the nature of mass," *Nature* 233: 41–44.

———. (1972), "Cosmological models in a conformally invariant gravitational theory," *MNRAS* 155: 305–21, 323–35.

Hoyle, F., and J. V. Narlikar (1974), *Action at a Distance in Physics and Cosmology*. San Francisco: W. H. Freeman.

———. (1980), *The Physics-Astronomy Frontier*. San Francisco: W. H. Freeman.

Hoyle, F., and A. Sandage (1956), "The second-order term in the redshift-magnitude relation," *PASP* 68: 301–7.

Hoyle, F., and R. J. Tayler (1964), "The mystery of the cosmic helium abundance," *Nature* 203: 1108–10.

Hoyle, F., and N. C. Wickramasinghe (1962), "On graphite particles as interstellar grains," *MNRAS* 124: 417–33.

———. (1967), "Impurities in interstellar grains," *Nature* 214: 969–71.

———. (1991), *The Theory of Cosmic Grains*. Dordrecht: Kluwer.

Hubble, E. P. (1925), "Cepheids in spiral nebulae," *Observatory* 48: 139–42.

———. (1926), "Extra-galactic nebulae," *APJ* 64: 321–69.

———. (1929), "A relation between distance and radial velocity among extra-galactic nebulae," *PNAS* 15: 168–73.

———. (1936a), "Effect of red shifts on the distribution of nebulae," *APJ* 84: 517–54.

———. (1936b), *The Realm of the Nebulae*. New Haven: Yale University Press.

———. (1937), *The Observational Approach to Cosmology*. Oxford: Clarendon Press.

———. (1942), "The problem of the expanding universe," *AS* 30: 99–115.

———. (1953), "The law of redshifts," *MNRAS* 113: 43–80.

Hubble, E., and M. L. Humason (1931), "The velocity-distance relation among extragalactic nebulae," *APJ* 74: 43–80.

Hufbauer, K. (1981), "Astronomers take up the stellar-energy problem, 1917–1920," *HSPS* 11: 277–303.

———. (1991), *Exploring the Sun: Solar Science since Galileo*. Baltimore: Johns Hopkins University Press.

Hughes, D. J. (1946), "Radiative capture cross sections for fast neutrons," *PR* 70: 106–7.

Hughes, V. W. (1964), "The Lyttleton-Bondi universe and charge equality." 259–78 in H.-Y. Chiu and W. F. Hoffmann, eds. (1964), *Gravitation and Relativity*. New York: Benjamin.

Hull, D. L., P. D. Tessner, and A. M. Diamond (1978), "Planck's principle," *Science* 202: 717–23.

Humason, M. L. (1929), "The large radial velocity of N.G.C. 7619," *PNAS* 15: 167–68.

Humason, M. L., N. U. Mayall, and A. R. Sandage (1956), "Redshifts and magnitudes of extragalactic nebulae," *AJ* 61: 97–162.

Hunter, J. H. (1966), "Galaxy formation in a steady state universe," *MNRAS* 133: 181–96.

Hutten, E. H. (1962), "Methodological remarks concerning cosmology," *The Monist* 47: 104–15.

Infeld, L. (1957), "Equations of motion in general relativity and the action principle," *RMP* 29: 398–411.

———. ed. (1964), *Relativistic Theories of Gravitation*. Oxford: Pergamon Press.

Inge, W. R. (1933), *God and the Astronomers*. London: Longmans, Green and Co.

Jaki, S. L. (1972), *The Milky Way: An Elusive Road to Science*. New York: Science History Publications.

———. (1978), *The Road of Science and the Ways to God*. Edinburgh: Scottish Academic Press.

Jeans, J. (1901), "The mechanism of radiation," *PM* 2: 421–55.

———. (1902), "The stability of a spiral nebula," *Philosophical Transactions* A 199: 117–27.

———. (1926a), "The evolution of the stars," *Nature* 117: 18–21.

———. (1926b), "Recent developments of cosmical physics," *Nature* 118: 29–40.

———. (1928a), "The wider aspects of cosmogony," *Nature* 121: 463–70.

———. (1928b), "The physics of the universe," *Nature* 122: 689–700.

———. (1928c), *Astronomy and Cosmogony*. Cambridge: Cambridge University Press.

———. (1928d), "What becomes of stellar radiation?" *Nature* 121: 674.

———. (1929), *The Universe Around Us*. New York: Macmillan.

———. (1930), *The Mysterious Universe*. Cambridge: Cambridge University Press.

John, L., ed. (1973), *Cosmology Now*. London: BBC.

Johnson, M. (1947), *Time, Knowledge, and the Nebulae*. New York: Dover.

———. (1951a), "Alternative time-dependence of cosmological quantities or scales," *Observatory* 70: 31–32.

———. (1951b), "The meanings of time and space in philosophies of science," *AS* 39: 412–21.

Jones, G. O., J. Rotblat, and G. J. Whitrow (1956), *Atoms and the Universe*. New York: Scribner's.

Jordan, P. (1937), "Die physikalische Weltkonstanten," *NW* 25: 513–17.

———. (1938), "Zur empirischen Kosmologie," *NW* 26: 417–21.

———. (1939), "Bemerkungen zur Kosmologie," *AP* 32: 64–70.

———. (1944), "Über die Entstehung der Sterne, I," *NW* 45: 183–90.

———. (1949), "Formation of stars and development of the universe," *Nature* 164: 637–40.

———. (1951), "Neuere Gesichtspunkte der kosmologischen Theorienbildung," *PJ* 61: 8–12.

———. (1952), *Schwerkraft und Weltall*. Braunschweig: Vieweg.

Jordan, P., and C. Müller (1947), "Über die Feldgleichungen der Gravitation bei variabler 'Gravitationskonstante'," *Zeitschrift für Naturforschung* 2A: 1–2.

Just, K. (1959), "A test for evolution in cosmology," *APJ* 129: 268–70.

———. (1960), "Eine Widerlegung der 'Steady-State'-Kosmologie." 43–47 in *Colloque sur la Théorie de la Relativité*. Louvain: Libraire Universitaire.

Kahn, C., and F. Kahn (1975), "Letters from Einstein to de Sitter on the nature of the universe," *Nature* 257: 451–54.

Kalckar, J., ed. (1985), *Niels Bohr. Collected Works*. Vol. 6. Amsterdam: North-Holland.

Kanitscheider, B. (1984), *Kosmologie: Geschichte und Systematik in Philosophischer Perspektive*. Stuttgart: Reclam.

Kapp, R. O. (1940), *Science versus Materialism*. London: Methuen.

———. (1949), "Effects of the origin of matter on cosmology," *Observatory* 69: 149–50.

———. (1950), "Development of the universe," *Nature* 165: 68, 687–88.

———. (1953), "Theories about the origin and disappearance of matter," *Observatory* 73: 113–16.

———. (1955), "Hypotheses about the origin and disappearance of matter," *BJPS* 6: 177–85.

———. (1960), *Towards a Unified Cosmology*. London: Hutchison.

Kargon, R. (1982), *The Rise of Robert Millikan: Portrait of a Life in American Science*. Ithaca, N.Y.: Cornell University Press.

Kargon, R. (1983), "The evolution of matter: nuclear physics, cosmic rays, and Robert Millikan's research program." 69–89 in W. R. Shea, ed. (1982), *Otto Hahn and the Rise of Nuclear Physics*. Dordrecht: Reidel.

Kellermann, K. I. (1993), "The cosmological deceleration parameter estimated from the angular-size/redshift relation for compact radio sources," *Nature* 361: 134–36.

Kemp, R. R. (1964), "On the number of particles in a steady-state universe," *PCPS* 60: 176–77.

Kerzberg, P. (1986), "Le principe de Weyl et l'invention d'une cosmologie non-statique," *AHES* 35: 1–89.

———. (1992), *The Invented Universe: The Einstein–de Sitter Controversy (1916–17) and the Rise of Relativistic Cosmology*. Oxford: Clarendon Press.

Kevane, C. J. (1961), "On antimatter and cosmology," *Science* 133: 580–81.

Kienle, H. (1943), "Das Alter der Sterne und die Expansion der Welt," *NW* 130: 149–50.

King, I. (1961), "Cosmological models and the ages of galaxies," *Observatory* 81: 128–31.

King, J. G. (1960), "Search for a small charge carried by molecules," *PRL* 5: 562–65.

Klein, O. (1954), "Some cosmological considerations in connection with the problem of the origin of the elements." 42–51 in P. Ledoux, ed. (1954).

———. (1966), "Instead of cosmology," *Nature* 211: 137–40.

Klein, O., G. Beskow, and L. Treffenberg (1946), "On the origin of the abundance distribution of chemical elements," *Arkiv för Matematik, Astronomi och Fysik* 33 B: 1–7.

Kohler, M. (1933), "Beiträge zum kosmologischen Problem und zur Lichtausbreitung in Schwerefeldern," *AP* 16: 129–61.

Kohlhörster, W. (1924), *Die Durchdringenden Strahlung in der Atmosphäre*. Hamburg: Henri Grand.

Kothari, D. S. (1938), "Cosmological and atomic constants," *Nature* 142: 354–55.

Kragh, H. (1982), "Cosmo-physics in the thirties: towards a history of Dirac cosmology," *HSPS* 13: 69–108.

———. (1987), "The beginning of the world: Georges Lemaître and the expanding universe," *Centaurus* 32: 114–39.

———. (1989), "The aether in late nineteenth century chemistry," *Ambix* 36: 49–65.

———. (1990), *Dirac: A Scientific Biography*. Cambridge: Cambridge University Press.

———. (1991), "Cosmonumerology and empiricism: The Dirac-Gamow dialogue," *AQ* 8: 109–26.

———. (1993), "Cosmic microwave background radiation." 100–105 in N. S. Hetherington, ed. (1993).

———. (1995), "Cosmology between the wars: the Nernst-MacMillan alternative," *JHA* 26: 93–115.

Krasinski, A. (1990), "Early inhomogeneous cosmological models in Einstein's theory." 115–27 in B. Bertotti et al., eds. (1990).

Kraushaar, W. L., and G. W. Clark (1962a), "Gamma ray astronomy," *SA* 206 (May): 52–61.

———. (1962b), "Search for primary cosmic gamma rays with the satellite Explorer XI," *PRL* 8: 106–9.

Kraushaar, W. L., et al. (1965), "Explorer XI experiment on cosmic gamma rays," *APJ* 141: 845–63.

Kriplovich, I. B. (1992), "The eventful life of Fritz Houtermans," *PT* 45 (July): 29–37.

Kuchowicz, B. (1967), *Nuclear Astrophysics. A Bibliographical Survey.* New York: Gordon and Breach.

————. (1968–71), *Nuclear and Relativistic Astrophysics and Nuclidic Cosmochemistry: 1963–1967.* 4 vols. Warsaw: Nuclear Energy Information Center.

Kudar, J. (1929), "Wellenmechanische Begründung der Nernstschen Hypothese von der Wiederentstehung radioaktiver Elemente," *ZP* 53: 166–67.

Labérenne, P. (1952), "Pie XII et la science," *La Pensée*, no. 41: 107–20.

Lanczos, C. (1922), "Bemerkung zur de Sitterchen Welt," *PZ* 2: 539–43.

Landau, L. (1932), "On the theory of stars," *PZS* 1: 285–88.

————. (1938), "Origin of stellar energy," *Nature* 141: 333–34.

Lang, K. R., and O. Gingerich, eds. (1979), *A Source Book in Astronomy and Astrophysics 1900–1975.* Cambridge, Mass.: Harvard University Press.

Laue, M. von (1929), "Notiz zur Quantentheorie des Atomkerns," *ZP* 52: 726–34.

Leavitt, H. (1912), "Periods of twenty-five variable stars in the small Magellanic cloud," *Harvard College Observatory Circular*, no. 173: 1–3.

LeBon, G. (1907), *The Evolution of Matter.* New York: Charles Scribner's Sons.

Ledoux, P., ed. (1954), *Les Processus Nucléaires dans les Astres.* Louvain. Special issue of *Mémoires de la Société Royale des Sciences de Liège* 14.

Le Floch, A., and F. Bretenaker (1991), "Early cosmic background," *Nature* 352: 198.

Lemaître, G. (1925), "Note on de Sitter's universe," *Journal of Mathematical Physics* 4: 188–92.

————. (1927), "Un univers homogéne de masse constante et de rayon croissant rendant compte de la vitesse radiale des nébuleuses extra-galactiques," *ASSB* 47: 49–56.

————. (1929), "La grandeur de l'espace," *RQS* 15: 189–216.

————. (1931a), "The expanding universe," *MNRAS* 91: 490–501.

————. (1931b), "The beginning of the world from the point of view of quantum theory," *Nature* 127: 706.

————. (1931c), "Sur l'interpretation de l'equation de Dirac," *ASSB* 51: 83–93.

————. (1931d), "L'expansion de l'espace," *RQS* 17: 391–40.

————. (1933a), "L'Universe en expansion," *ASSB* 53: 51–85.

————. (1933b), "The uncertainty of the electromagnetic field of a particle," *PR* 43: 148.

————. (1933c), "Condensations sphériques dans l'univers en expansion," *CR* 196: 903–4.

————. (1933d), "La formation des nébuleuses dans l'univers en expansion," *CR* 196: 1085–87.

————. (1934), "Evolution of the expanding universe," *PNAS* 20: 12–17.

————. (1945), "A propos de la théorie de Milne-Haldane," *CT* 61: 213–18.

————. (1946), *L'Hypothèse de l'Atome Primitif: Essai de Cosmogonie.* Neuchatel: Éditions du Griffon.

————. (1949a), "Cosmological applications of relativity," *RMP* 21: 357–66.

————. (1949b), "The cosmological constant." 437–56 in P. A. Schilpp, ed. (1949).

————. (1958a), "The primeval atom hypothesis and the problem of the clusters of galaxies." 1–32 in R. Stoops, ed. (1958).

————. (1958b), "Rencontres avec A. Einstein," *RQS* 129: 129–32.

————. (1961), "Exchange of galaxies between clusters and field," *AJ* 66: 603–6.

————. (1978), ["L'univers, problème accessible a la science humaine"]. 349–59 in O. Godart and M. Heller, eds. (1978).

Lemaître, G., and M. S. Vallarta (1933), "On Compton's latitude effect of cosmic radiation," *PR* 43: 87–91.

Lenz, W. (1926), "Das Gleichgewicht von Materie und Strahlung in Einsteins geschlossener Welt," *PZ* 27: 642–45.

Lerner, E. J. (1991), *The Big Bang Never Happened*. New York: Times Books.

Leslie, J., ed. (1990), *Physical Cosmology and Philosophy*. New York: Macmillan.

Levi-Città, T. (1917), "Realtà fisica di alcuni spazî normali del Bianchi," *Rendiconti, Reale Accademia dei Lincei* 26: 519–31.

Lifshitz, E. M. (1946), "On the gravitational stability of the expanding universe," *Journal of Physics USSR* 10: 116–29.

Lifshitz, E. M., and I. M. Khalatnikov (1963), "Investigations in relativistic cosmology," *Advances in Physics* 12: 185–249.

Lightman, A., and R. Brawer (1990), *Origins: The Lives and Worlds of Modern Cosmology*. Cambridge, Mass.: Harvard University Press.

Linde, A., and A. Mezhlumian (1993), "Stationary universe," *Physics Letters* B 307: 25–33.

Linde, A., D. Linde, and A. Mezhlumian (1994), "From the big bang theory to the theory of a stationary universe," *PR* D 49: 1783–826.

Littlewood, D. E. (1955), "The cosmological principle," *PCPS* 51: 678–83.

Lockyer, N. (1900), *Inorganic Evolution as Studied by Spectrum Analysis*. London: Macmillan.

Lodge, O. (1904), "Modern views on matter," *Annual Report of the Smithsonian Institution* 1903: 215–28.

———. (1920), "Restoration of energy," *Nature* 106: 341.

———. (1923), "Gravitation and light-pressure in spiral nebulae," *Nature* 111: 702.

———. (1924), "Stationary clouds in interstellar space," *Nature* 113: 307.

———. (1926), *Evolution and Creation*. London: Hodder and Stoughton.

Longair, M. S. (1966), "Evidence on the evolutionary character of the universe derived from recent red-shift measurements," *Nature* 211: 949–50.

———. (1993), "Modern cosmology—a critical assessment," *QJRAS* 34: 157–200.

Lovell, A. C. B. (1958), "Radio-astronomical observations which may give information on the structure of the universe." 185–207 in R. Stoops, ed. (1958).

———. (1959), *The Individual and the Universe*. Oxford: Oxford University Press.

———. (1990), *Astronomer by Chance*. New York: Basic Books.

Lovell, B., and Smith, F. G. (1983), "On the discovery of extragalactic radio sources," *JHA* 14: 155–65.

Lundmark, L. (1927), "Arrhenius och kosmologien," *Astronomisk Tidskrift*: 177–81.

Lundqvist, S., ed. (1992), *Nobel Lectures. Physics 1971–1980*. Singapore: World Scientific.

Lyttleton, R. A. (1956), *The Modern Universe*. New York: Harper.

———. (1960), "An electric universe?" 22–33 in H. Bondi et al., eds. (1960).

Lyttleton, R. A., and H. Bondi (1959), "On the physical consequences of a general excess of charge," *PRS* 252: 313–33.

———. (1960), "Note of the preceding paper," *PRS* 257: 442–44.

Lyttleton, R. A., and F. Hoyle (1940), "Evolution of stars," *Observatory* 63: 39–43.

MacCallum, M., ed. (1985), *Galaxies, Axisymmetric Systems and Relativity: Essays Presented to W. B. Bonnor on his 65th Birthday*. Cambridge: Cambridge University Press.

MacMillan, W. D. (1918), "On stellar evolution," *APJ* 48: 35–49.

———. (1920), "The structure of the universe," *Science* 52: 67–74.

———. (1923), "Cosmic evolution," *Scientia* 33: 3–12.

———. (1925), "Some mathematical aspects of cosmology," *Science* 62: 63–72, 96–99, 121–27.

———. (1926), "The new cosmology," *SA* 134: 310–11.

———. (1929), "The heavens and earth." 11–37 in B. Brownell, ed. (1929), *The World Mechanism*. New York: D. van Nostrand.

———. (1932), "Velocities of the spiral nebulae," *Nature* 129: 93.

MacPherson, H. (1929), *Modern Cosmologies: A Historical Sketch of Researches and Theories Concerning the Structure of the Universe*. London: Oxford University Press.

Mandolesi, N., and N. Vittorio, eds. (1990), *The Cosmic Microwave Background: 25 Years Later*. Dordrecht: Kluwer.

Maneff, G. (1932), "Über die Welt in Ausdehnung," *ZAP* 4: 241–46.

Markov, M. A. (1967), "Elementary particles of maximally large masses (quarks and maximons)," *JETP* 24: 584–92.

Martin, B. R. (1976), "The Origins, Development and Capitulation of Steady-State Cosmology: A Sociological Study of Authority and Conflict in Science." M.Sc. thesis, University of Manchester.

———. (1978), "Radio astronomy revisited: a reassessment of the role of competition and conflict in the development of radio astronomy," *The Sociological Review* 26: 27–55.

Mascall, E. L. (1957), *Christian Theology and Natural Science: Some Questions in their Relations*. London: Longmans, Green and Co.

Mattig, W. (1958), "Über den Zusammenhang zwischen der Anzahl der extragalaktischen Objekte und der scheinbaren Helligkeit," *AN* 284: 109–11.

Mayer, M. G., and E. Teller, E. (1949), "On the origin of elements," *PR* 76: 1226–31.

———. (1950), "On the abundance and origin of elements." 59–88 in R. Stoops, ed. (1950), *Les Particules Élémentaires. Proceedings of the Eighth Solvay Congress 1948*. Brussels: Coudenberg.

McCrea, W. H. (1939), "The evolution of theories of space-time and mechanics," *PS* 6: 137–62.

———. (1950a), "The steady-state theory of the expanding universe," *Endeavour* 9: 3–10.

———. (1950b), "Quantum mechanics and astrophysics," *Nature* 166: 884–86.

———. (1951), "Relativity theory and the creation of matter," *PRS* 206: 562–75.

———. (1952), "The primeval atom," *Observatory* 72: 41–43.

———. (1953), "Cosmology," *RPP* 16: 321–63.

———. (1957), "Relativistic ageing," *Nature* 179: 909–10.

———. (1958), "Cosmology," *Endeavour* 17: 5–11.

———. (1960–61), "The interpretation of cosmology," *Nature* 186: 1035.

———. (1960), "The interpretation of cosmology," *LNC* 11: 11–20.

———. (1962), "Information and prediction in cosmology," *The Monist* 47: 94–103.

———. (1963a), "Cosmology—a brief review," *QJRAS* 4: 185–202.

———. (1963b), "Cosmological theories, a survey." 206–21 in H. P. Palmer, R. V. Davies, and M. L. Large, eds. (1963).

———. (1964), "Continual creation," *MNRAS* 128: 335–43.

———. (1965), "The sign of the constant of gravitation," *Nature* 206: 553–55.

———. (1968), "Cosmology after half a century," *Science* 160: 1295–99.

———. (1970a), "A philosophy for big-bang cosmology," *Nature* 228: 21–24.

———. (1970), "Cosmology today," *AS* 58: 521–27.

———. (1973), "The problem of the galaxies." 85–96 in L. John, ed. (1973).

McCrea, W. H. (1977), "Models, laws, and the universe." 59–73 in W. Yourgrau and A. D. Breck, eds. (1977).

———. (1984a), "The influence of radio astronomy on cosmology." 365–84 in W. T. Sullivan, ed. (1984).

———. (1984b), "Nobel prize to astrophysicists," *Physics Bulletin* 35: 16–17.

———. (1990), "George Cunliffe McVittie (1904–88), OBE, FRSE. Pupil of Whittaker and Eddington: pioneer of modern cosmology," *VA* 33: 43–58.

McCrea, W. H., and G. C. Mcvittie (1930–31), "On the contraction of the universe," *MNRAS* 91: 128–33.

McCrea, W. H., and E. A. Milne (1934), "Newtonian universes and the curvature of space," *Quarterly Journal of Mathematics* 5: 73–80.

McCutcheon, R. A. (1990), "The early career of Viktor Amazaspovich Ambartsumian: an interview," *AQ* 7: 143–78.

McKellar, A. (1940), "Evidence for the molecular origin of some hitherto unidentified interstellar lines," *PASP* 52: 187–92.

McLaughlin, P. J. (1957), *The Church and Modern Science*. New York: Philosophical Library.

McMullin, E. (1981), "How should cosmology relate to theology?" 17–57 in A. R. Peacocke, ed. (1981), *The Sciences and Theology in the Twentieth Century*. Stockfield: Oriel Press.

———. (1987), "Scientific controversy and its termination." 49–92 in H. T. Engelhardt and A. L. Caplan, eds. (1987).

McVittie, G. C. (1930–31), "The problem of *n* bodies and the expansion of the universe," *MNRAS* 91: 274–83.

———. (1951), "The cosmological problem," *Science News* 21: 61–75.

———. (1952), "A model universe admitting the interchangeability of stress and matter," *PRS* 211: 295–301.

———. (1953), "The age of the universe in the cosmology of general relativity," *AJ* 58: 129–34.

———. (1959a), "Distance and time in cosmology: the observational data." 445–519 in S. Flügge, ed. (1959), *Handbuch der Physik*. Vol. 53. Berlin: Springer-Verlag.

———. (1959b), "Remarks on cosmology." 533–35 in R. N. Bracewell, ed. (1959).

———. (1961a), *Fact and Theory in Cosmology*. London: Eyre and Spottiswoode.

———. (1961b), "Rationalism versus empiricism in cosmology," *Science* 133: 1231–36.

———. (1962a), "Cosmology and the interpretation of astronomical data." 253–69 in M. A. Lichnerowicz and M. A. Tonnelat, eds. (1962), *Les Théories Relativistes de la Gravitation*. Paris: CNRS.

———. ed. (1962b), *Problems of Extra-Galactic Research*. New York: Macmillan.

———. (1962c), "Photon density and the gamma-ray flux at a point in an expanding universe," *PR* 128: 2871–78.

———. (1965), *General Relativity and Cosmology*, 2nd ed. London: Chapman and Hall.

———. (1966), "The analysis of observations in cosmology." 13–21 in L. Rosino et al. (1966).

———. (1967), "Georges Lemaître," *QJRAS* 8: 294–97.

———. (1987), "An Anglo-Scottish university education." 66–70 in R. Williamson, ed. (1987), *The Making of Physicists*. Bristol: Adam Hilger.

Mehra, J. (1974), *Einstein, Hilbert, and the Theory of Gravitation*. Dordrecht: Reidel.

Mendelssohn, K. (1973), *The World of Walther Nernst: The Rise and Fall of German Science*. New York: Macmillan.

Mercier, A., and M. Kervaire, eds. (1956), *Fünfzig Jahre Relativitätstheorie*. Basel: Birkhäuser.

Merleau-Ponty, J. (1965), *Cosmologie du XXe Siècle: Étude Épistemologique et Historique des Théories de la Cosmologie Contemporaine*. Paris: Gallimard.

Metzner, A. W. K., and P. Morrison, P. (1959), "The flow of information in cosmological models," *MNRAS* 119: 657–64.

Meyenn, K., ed. (1985), *Wolfgang Pauli. Wissenschaftlicher Briefwechsel*. Vol. 2. Berlin: Springer-Verlag.

Michelmore, P. (1962), *Einstein: Profile of the Man*. New York: Dodd.

Mikulak, M. W. (1955), "Soviet cosmology and communist ideology," *The Scientific Monthly*, no. 4: 167–72.

———. (1958), "Soviet philosophic-cosmological thought," *PS* 25: 35–50.

Millikan, R. A. (1917), *The Electron*. Chicago: University of Chicago Press.

———. (1925), "High frequency rays of cosmic origin," *Science* 57: 445–48, 461.

———. (1928), "Available energy," *Science* 68: 279–84.

———. (1930), "Remarks on the history of cosmic radiation," *Science* 71: 640–41.

———. (1931), "Present status of theory and experiment as to atomic disintegration and atomic synthesis," *Nature* 127: 167–70.

———. (1935), *Electrons (+ and −), Protons, Photons, Neutrons, and Cosmic Rays*. Cambridge: Cambridge University Press.

Millikan, R. A., and G. H. Cameron (1926), "High frequency rays of cosmic origin III. Measurements in snow-fed lakes at high altitudes," *PR* 28: 851–68.

———. (1928), "The origin of the cosmic rays," *PR* 32: 533–57.

Mills, B. Y. (1952), "The distribution of the discrete sources of cosmic radio radiation," *Australian Journal of Scientific Research* A5: 266–87.

———. (1959), "A survey of radio sources at 3.5 m wavelength." 498–506 in R. N. Bracewell, ed. (1959).

Mills, B. Y., and O. B. Slee (1957), "A preliminary survey of radio sources in a limited region of the sky at a wavelength of 3.5 meters," *AuJP* 10: 162–94.

Milne, E. A. (1930), "World geometry in its time relations," *Nature* 126: 742–43.

———. (1933), "World-structure and the expansion of the universe," *ZAP* 6: 1–35.

———. (1934), "Some points in the philosophy of physics: time, evolution and creation," *Philosophy* 9: 19–38.

———. (1935), *Relativity, Gravitation and World-Structure*. Oxford: Clarendon Press.

———. (1937), "On the origin of laws of nature," *Nature* 139: 997–99.

———. (1946), "On the spiral character of the external galaxies," *MNRAS* 106: 180–99.

———. (1947), "Last testament of a physicist," *Nature* 159: 486–87.

———. (1948), *Kinematic Relativity*. Oxford: Clarendon Press.

———. (1949a), "Kinematic relativity." 7–32 in *Théories Nouvelles de Relativité*. Paris: Hermann et Cie.

———. (1949b), "Gravitation without general relativity." 409–36 in P. A. Schilpp, ed. (1949).

———. (1952), *Modern Cosmology and the Christian Idea of God*. Oxford: Clarendon Press.

Mimura, Y. (1935), "Relativistic quantum mechanics and wave geometry," *Journal of Science of the Hiroshima University* 5: 99–106.

Minkowski, R. (1959), "Concluding lecture." 536–38 in R. N. Bracewell, ed. (1959).

Møller, C., ed. (1962), *Evidence for Gravitational Theories*. New York: Academic Press.

Munitz, M. K. (1952), "Scientific method in cosmology," *PS* 19: 108–30.

———. (1954), "Creation and the 'new' cosmology," *BJPS* 5: 32–46.

———. (1957a), *Space, Time and Creation: Philosophical Aspects of Scientific Cosmology.* Glencoe, Ill.: Free Press.

———. ed. (1957b), *Theories of the Universe: From Babylonian Myth to Modern Science.* Glencoe, Ill.: Free Press.

Nariai, H. (1952), "On Hoyle's continuous creation theory," *PASJ* 3: 139–43.

———. (1955), "On a new approach to the construction of a smoothed-out model of the universe," *PASJ* 7: 144–60.

———. (1960–61), "On the smoothed-out universe," *LNC* 11: 33–46.

Narlikar, J. V. (1962a), "Rotating universes." 222–27 in C. Møller, ed. (1962).

———. (1962b), "Neutrinos and the arrow of time in cosmology," *PRS* 270: 553–61.

———. (1973), "Steady state defended." 69–84 in L. John, ed. (1973).

———. (1991), "What if the big bang didn't happen?" *New Scientist* 129: 48–51.

Narlikar, J. V., and N. C. Wickramasinghe (1967), "Microwave background in a steady-state universe," *Nature* 216: 43–44.

———. (1968), "Interpretation of cosmic microwave background," *Nature* 217: 1235–36.

Narlikar, V. V. (1932), "The highest atomic number," *Nature* 129: 402.

Nernst, W. (1907), *Theoretische Chemie.* 5th ed. Stuttgart: F. Enke.

———. (1912), "Zur neueren Entwicklung der Thermodynamik," *Verhandlungen der Gesellschaft deutscher Naturforscher under Ärtze* 1: 100–116.

———. (1916), "Über einen Versuch, von quantentheoretischen Betrachtungen zur Annahme stetiger Energieänderungen zurückzukehren," *Verhandlungen der deutsche physikalische Gesellschaft* 18: 83–116.

———. (1921), *Das Weltgebäude im Lichte der Neueren Forschung.* Berlin: Springer.

———. (1922), "Zum Gültigkeitsbereich der Naturgesetze," *NW* 10: 489–95.

———. (1928), "Physico-chemical considerations in astrophysics," *JFI* 206: 135–42.

———. (1935a), "Physikalische Betrachtungen zur Entwicklungstheorie der Sterne," *ZP* 97: 511–34.

———. (1935b), "Einige weitere Anwendungen der Physik auf die Sternentwicklung," *SPAW* 28: 473–79.

———. (1937), "Weitere Prüfung der Annahme eines stationären Zustandes im Weltall," *ZP* 106: 633–61.

———. (1938), "Die Strahlungstemperatur des Universums," *AP* 32: 44–48.

Neuhäusler, A. (1951), "Georges Lemaîtres Kosmogonie," *PJ* 61: 1–7.

Nicolaïdis, E. (1990), "Astronomy and politics in Russia in the early Stalinist period (1928–1932)," *JHA* 21: 345–51.

North, J. D. (1965), *The Measure of the Universe: A History of Modern Cosmology.* London: Oxford University Press.

———. (1994), *Astronomy and Cosmology.* London: Fontana Press.

Novikov, I. D. (1965), "Delayed explosion of a part of the Friedman universe and quasars," *SAAP* 8: 857–63.

———. (1990), *Black Holes and the Universe.* Cambridge: Cambridge University Press.

Novikov, I. D., and Ya. B. Zel'dovich (1967), "Cosmology," *ARAA* 5: 627–48.

O'Connell, D. (1952–53), "According to Hoyle," *IAJ* 2: 127–38.

Ohm, E. A. (1961), "Receiving system," *Bell System Technical Journal* 40: 1065–94.

Oke, J. B., and A. R. Sandage (1968), "Energy distributions, K corrections, and the Stebbins-Whitford effect for giant elliptical galaxies," *APJ* 154: 21–32.

Omer, G. C. (1949), "A nonhomogeneous cosmological model," *APJ* 109: 164–76.

Oort, J. H. (1931), "Some problems concerning the distribution of luminosities and peculiar velocities of extragalactic nebulae," *BAN* 6: 155–60.

Öpik, E. J. (1933), "Meteorites and the age of the universe," *PA* 41: 71–79.

———. (1938), "Stellar structure, source of energy, and evolution,". 342–48 in K. R. Lang and O. Gingerich, eds. (1979); Originally in *Publications de l'Observatoire Astronomique de l'Université de Tartu* 30, no. 3: 1–115.

———. (1951), "Stellar models with variable composition. II. Sequences of models with energy generation proportional to the fifteenth power of temperature," *PRIA* 54: 49–77.

———. (1954a), "The chemical composition of white dwarfs." 131–38 in P. Ledoux, ed. (1954).

———. (1954b), "The age of the universe," *BJPS* 5: 203–14.

———. (1954c), "The time-scale of our universe," *IAJ* 3: 89–108.

Oppenheimer, J. R., and R. Serber (1938), "On the stability of stellar neutron cores," *PR* 54: 540.

Osterbrock, D. E. (1990), "The observational approach to cosmology: U.S. observatories pre-World War II." 247–90 in B. Bertotti et al., eds. (1990).

Osterbrock, D. E., R. S. Brashear, and J. A. Gwinn (1990), "Self-made cosmologist: the education of Edwin Hubble." 1–8 in R. G. Kron, ed. (1990), *Evolution of the Universe of Galaxies: Edwin Hubble Centennial Symposium*. San Francisco: Astronomical Society of the Pacific.

Osterbrock, D. E., and J. B. Rogerson (1961), "The helium and heavy-element content of gaseous nebulae and the sun," *PASP* 73: 129–34.

Overbye, D. (1991), *Lonely Hearts of the Cosmos: The Scientific Quest for the Secret of the Universe*. New York: HarperCollins.

Pachner, J. (1960), "Dynamics of the universe," *Acta Physica Polonica* 19: 662–73.

———. (1961), "Zur relativistischen Kosmologie," *AP* 8: 60–75.

———. (1965), "An oscillating isotropic universe without singularity," *MNRAS* 131: 173–76.

Pais, A. (1982), *'Subtle is the Lord . . .': The Science and Life of Albert Einstein*. Oxford: Oxford University Press.

Palmer, H. P. (1963), "The angular diameters and luminosity function of extragalactic radio sources." 230–35 in H. P. Palmer, R. V. Davies, and M. I. Large, eds. (1963).

Palmer, H. P., R. V. Davies, and M. I. Large, eds. (1963), *Radio Astronomy Today*. Manchester: Manchester University Press.

Paul, E. R. (1993), *The Milky Way and Statistical Cosmology 1890–1924*. Cambridge: Cambridge University Press.

Pauli, W. (1921), "Relativitätstheorie." In *Encyclopädie der Mathematischen Wissenschaften*. Vol. V19. Leipzig: B. G. Teubner.

Payne, C. H. (1925), "Astrophysical data bearing on the relative abundance of the elements," *PNAS* 11: 192–98.

Payne-Gaposchkin, C. H. (1944), "The cosmic time scale," *AS* 32: 222–26.

Peach, J. (1973), "Choosing a universe." 23–36 in L. John, ed. (1973).

Pearson, K. (1888–89), "On a certain atomic hypothesis," *Proceedings of the London Mathematical Society* 20: 38–63.

———. (1900), *The Grammar of Science*, 2nd ed. London: Adams and Charles Black.

Peebles, P. J. E. (1966a), "Primeval helium abundance and the primeval fireball," *PRL* 16: 410–13.

Peebles, P. J. E. (1966b), "Primordial helium abundance and the primordial fireball, II," *APJ* 146: 542–52.

――――. (1967), "Microwave radiation from the big bang." 274–84 in J. Ehlers, ed. (1967), *Relativity Theory and Astrophysics*. Vol. 1. Providence, R.I.: American Mathematical Society.

――――. (1971), *Physical Cosmology*. Princeton: Princeton University Press.

――――. (1972), "Light out of darkness vs. order out of chaos," *CASS* 4: 53–58.

――――. (1993), *Principles of Physical Cosmology*. Princeton: Princeton University Press.

Peebles, P. J. E., D. N. Schram, E. L. Turner, and R. G. Kron (1991), "The case for the relativistic hot big bang cosmology," *Nature* 352: 769–76.

Peebles, P. J. E., and D. T. Wilkinson (1967), "The primeval fireball," *SA* 216 (June): 28–37.

Peierls, R. E., ed. (1986), *Niels Bohr. Collected Works*. Vol. 9. Amsterdam: North-Holland.

Peierls, R. E., K. S. Singwi, and D. Wroe (1952), "The polyneutron theory of the origin of the elements," *PR* 87: 46–50.

Penrose, R. (1965), "Gravitational collapse and space-time singularities," *PRL* 14: 57–59.

Penzias, A. A (1972), "Cosmology and microwave astronomy." 29–47 in F. Reines, ed. (1972).

――――. (1979), "The origin of the elements," *Science* 205: 549–54.

Penzias, A. A., and R. W. Wilson (1965), "A measurement of excess antenna temperature at 4080 Mc/s," *APJ* 142: 419–20.

Perrin, J. (1919), "Matière et lumière: essai de synthèse de la mécanique chimique," *Annales de Physique* 11: 5–108.

Petrosian, V. (1974), "Confrontation of Lemaître models and the cosmological constant with observations." 31–46 in M. S. Longair, ed. (1974), *Confrontation of Cosmological Theories With Observational Data*. Dordrecht: Reidel.

Petrosian, V., and E. E. Salpeter (1970), "Lemaître models and the cosmological constant," *CASP* 2: 109–15.

Phillips, P. R. (1994), "Solutions of the field equations for a steady-state cosmology in a closed space," *MNRAS* 269: 771–78.

Piaggio, H. T. H. (1931), "Science and prediction," *Nature* 127: 454.

Piccard, A., and E. Kessler (1925), "Détermination du rapport des charges électrostatique du proton et de l'électron," *Archive des Sciences Physiques et Naturelles* 7: 340–42.

Piel, G., et al., eds. (1957), *The Universe*. New York: Simon and Schuster.

Pierce, M. J., et al. (1994), "The Hubble constant and Virgo cluster distance from observations of Cepheid variables," *Nature* 371: 385–89.

Pirani, F. A. E. (1954), "The steady-state theory of the homeogeneous expanding universe," *Observatory* 74: 172–73.

――――. (1955), "On the energy-momentum tensor and the creation of matter in relativistic cosmology," *PRS* 228: 455–62.

――――. (1992), "The crisis in cosmology," *New Left Review* no. 191: 69–89.

Podolanski, J., and D. ter Haar (1954), "The origin of the elements: a general survey." 19–41 in P. Ledoux, ed. (1954).

Pokrowski, G. I. (1929), "Über die Synthese von Elementen," *ZP* 54: 123–32.

[Pope] Pius XII (1952), "Science and the catholic church," *Bulletin of Atomic Scientists* 8: 142–46, 165.

Popper, K. R. (1940), "Interpretations of nebular red-shifts," *Nature* 145: 69–70.

Posejpal, V. (1930), "Détermination directe du volume de l'électron," *CR* 191: 1000–1002.

Priester, W. (1958), "Zur Statistik der Radioquellen in der relativistischen Kosmologie," *ZAP* 46: 179–202.

Prokofieva, I. (1950), "Conférence sur les questions idéologiques de l'astronomie," *La Pensée*, no. 28: 10–20.

Pumphrey, R. J., and T. Gold (1947), "Transient reception and the degree of resonance of the human ear," *Nature* 160: 124–25.

Rankine, W. M. (1881), *Miscellaneous Scientific Papers*. London: Charles Griffin and Co.

Ray, C. (1990), "The cosmological constant: Einstein's greatest mistake?" *SHPS* 21: 589–604.

Raychaudhuri, A. K., and S. Banerji (1964), "Cosmological evolution with creation of matter," *ZAP* 58: 187–91.

Rees, M. J., and D. W. Sciama (1966a), "Inverse Compton effect in quasars," *Nature* 211: 805–7.

———. (1966b), "The kinetic temperature and ionization level of intergalactic hydrogen in a steady-state universe," *APJ* 145: 6–20.

Reeves, H., J. Audouze, W. A. Fowler, and D. N. Schramm (1973), "On the origin of light elements," *APJ* 179: 909–30.

Reeves, H., W. A. Fowler, and F. Hoyle (1970), "Galactic cosmic origin of Li, Be and B in stars," *Nature* 226: 727–29.

Regener, E. (1933), "Der Energiestrom der Ultrastrahlung," *ZP* 80: 666–69.

Reines, F., ed. (1972), *Cosmology, Fusion and other Matters. George Gamow Memorial Volume*. London: Adam Hilger.

Reiser, O. L. (1952), "The evolution of cosmologies," *PS* 19: 93–107.

Reynolds, J. H. (1932), "Physical and observational evidence for the expanding universe," *Nature* 130: 458–62.

Rindler, W. (1956), "Visual horizons in world models," *MNRAS* 116: 662–77.

Robertson, H. P. (1928), "On relativistic cosmology," *PM* 5: 835–48.

———. (1929), "On the foundations of relativistic cosmology," *PNAS* 15: 822–29.

———. (1932), "The expanding universe," *Science* 76: 221–26.

———. (1933), "Relativistic cosmology," *RMP* 5: 62–90.

———. (1949), "On the present state of relativistic cosmology," *PAPS* 93: 527–31.

———. (1955), "The theoretical aspects of the nebular redshift," *PASP* 67: 82–98.

Robinson, I., A. Schild, and E. L. Schucking, eds. (1965), *Quasi-Stellar Sources and Gravitational Collapse*. Chicago: University of Chicago Press.

Roe, P. E. (1969), "Time symmetric electrodynamics in Friedmann universes," *MNRAS* 144: 219–30.

Roman, P. (1960), "On the steady-state theory of cosmology," *NC* 18: 9–20.

Rosino, L., et al. (1966), *Atti del Convegno Sulla Cosmologia*. Florence: G. Barbéra.

Rowan-Robinson, M. (1972), "Steady state obituary," *Nature* 240: 439.

Rowlands, P. (1990), *Oliver Lodge and the Liverpool Physical Society*. Liverpool: Liverpool University Press.

Roxburgh, I. W., and P. G. Saffman (1965), "The growth of condensations in a Newtonian model of the steady-state universe," *MNRAS* 129: 181–89.

Rüger, A. (1988), "Atomism from cosmology: Erwin Schrödinger's work on wave mechanics and space-time structure," *HSPS* 18: 377–401.

Russell, B. (1931), *The Scientific Outlook*. New York: Norton.

Russell, H.N. (1921), "A superior limit to the age of the earth's crust," *PRS* 99: 84–86.

———. (1929), "The highest known velocity," *SA* 140: 504–5.

———. (1939), "What keeps the stars shining?" *SA* 161 (July): 18–19.

Russell, R. J., W. R. Stoeger, and G. V. Coyne, eds. (1990), *John Paul II on Science and Religion: Reflections on the New View from Rome*. Vatican City State: Vatican Observatory Publications.

Rutgers, A. J. (1950), "Wordt materie voortdurend geschapen?" *Nederlandsch Tijdschrift voor Natuurkunde* 16: 161–70.

Rutherford, E. (1907), "Some cosmical aspects of radioactivity," *JRASC* 2: 145–65.

———. (1929), "Origin of actinium and age of the earth," *Nature* 123: 313–14.

Rutherford, E., and F. Soddy (1903), "Radioactive change," *PM* 6: 576–91.

Ryan, M. P., and L. C. Shepley (1976), "Resource letter RC-1: Cosmology," *AJP* 44: 223–30.

Ryle, M. (1955), "Radio stars and their cosmological significance," *Observatory* 75: 127–47.

———. (1958), "The nature of cosmic radio sources," *PRS* 248: 289–307.

———. (1959), "The nature of the radio sources." 523–27 in R. N. Bracewell, ed. (1959).

———. (1961), "Radio astronomy and cosmology," *Nature* 190: 852–54.

———. (1962), "The radio luminosity function and the number versus flux-density relationship for the discrete sources." 326–46 in G. C. McVittie, ed. (1962b).

———. (1963), "Radio astronomical tests of cosmological models." 228–30 in H. P. Palmer, R. V. Davies, and M. I. Large, eds. (1963).

———. (1968), "The counts of radio sources," *ARAA* 6: 249–66.

Ryle, M., and R. W. Clarke (1961), "An examination of the steady-state model in the light of some recent observations of radio sources," *MNRAS* 122: 349–62.

Ryle, M., and P. A. G. Scheuer (1955), "The spatial distribution and the nature of radio stars," *PRS* 230: 448–62.

Sakharov, A. D. (1967), "Violation of *CP* invariance, *C* asymmetry, and baryon asymmetry in the universe," *JETP Letter* 5: 32–35.

Salpeter, E. E. (1952), "Nuclear reactions in stars without hydrogen," *APJ* 115: 326–28.

———. (1955), "Nuclear reactions in stars. II. Protons on light nuclei," *PR* 97: 1237–44.

———. (1957), "Nuclear reactions in stars. Buildup for helium," *PR* 107: 516–27.

Sambursky, S. (1937), "Static universe and nebular red shift," *PR* 52: 335–38.

Sánchez-Ron, J.-M. (1983), "The problem of interaction: on the history of the action-at-a-distance concept in physics," *Fundamenta Scientiae* 4: 55–76.

———. (1990), "Steady-state cosmology, the arrow of time, and Hoyle and Narlikar's theories." 233–43 in B. Bertotti et al., eds. (1990).

Sandage, A. R. (1957), "The red shift." 89–98 in N. G. Piel et al., eds. (1957).

———. (1958), "Current problems in the extragalactic distance scale," *APJ* 127: 513–26.

———. (1961), "The ability of the 200-inch telescope to discriminate between selected world models," *APJ* 133: 355–92.

———. (1970), "Cosmology: a search for two numbers," *PT* 23 (February): 34–41.

Scheuer, P. A. G. (1990), "Radio source counts." 331–45 in B. Bertotti et al., eds. (1990).

Schild, A. (1960), "Equivalence principle and red-shift measurements," *AJP* 28: 778–80.

Schilpp, P. A., ed. (1949), *Albert Einstein: Philosopher-Scientist*. Evanston, Ill.: Library of Living Philosophers.

Schlegel, R. (1954), "The age of the universe," *BJPS* 5: 226–36.

———. (1958), "Steady-state theory at Chicago," *AJP* 26: 601–4.

———. (1961), *Time and the Physical World*. East Lansing: Michigan State University Press.

———. (1962), "Transfinite numbers and cosmology," *Nature* 193: 665–66.

Schmidt, M. (1963), "3C 273: a star-like object with a large red-shift," *Nature* 197: 1040.

———. (1968), "Space distribution and luminosity functions of quasi-stellar radio sources," *APJ* 151: 393–409.

———. (1972), "Radio source counts and redshifts in steady state cosmology," *Nature* 240: 399–400.

———. (1990), "The discovery of quasars." 347–54 in B. Bertotti et al., eds. (1990).

Schmidt, O. I. (1951), "Le problème de l'origine de la terre et des planètes," *La Pensée*, no. 39: 60–68.

Schrödinger, E. (1935), "Die gegenwärtige Situation in der Quantenmechanik," *NW* 23: 807–12, 823–28, 844–49.

———. (1937), "Sur la théorie du monde d'Eddington," *NC* 15: 246–54.

———. (1939), "The proper vibrations of the expanding universe," *Physica* 6: 899–912.

———. (1940), "Maxwell's and Dirac's equations in the expanding universe," *PRIA* 46: 25–47.

———. (1957), *Expanding Universes*. Cambridge: Cambridge University Press.

Schücking, E., and E. Heckmann (1958), "World models." 149–58 in R. Stoops, ed. (1958).

Schweber, S. S. (1986), "The empiricist temper regnant: theoretical physics in the United States 1920–1950," *HSPS* 17: 55–98.

———. (1989), "Some reflections on the history of particle physics in the 1950s." 668–93 in L. M. Brown, M. Dresden, and L. Hoddeson, eds. (1989), *Pions to Quarks: Particle Physics in the 1950s*. Cambridge: Cambridge University Press.

Sciama, D. W. (1953), "On the origin of inertia," *MNRAS* 113: 34–42.

———. (1955a), "Evolutionary processes in cosmology," *The Advancement of Science* 12: 38–42.

———. (1955b), "On the formation of galaxies in a steady state universe," *MNRAS* 115: 3–14.

———. (1959), *The Unity of the Universe*. London: Faber and Faber.

———. (1960), "Observational aspects of cosmology," *VA* 3: 311–28.

———. (1960–61), "New developments in cosmology," *LNC* 11: 3–10.

———. (1963a), "On the interpretation of radio source counts," *MNRAS* 126: 195–201.

———. (1963b), "Retarded potentials and the exansion of the universe," *PRS* 273: 484–95.

———. (1964a), "On a possible class of galactic radio sources," *MNRAS* 128: 49–61.

———. (1964b), "On the formation of galaxies and their magnetic fields in a steady state universe," *QJRAS* 5: 196–213.

———. (1965), "Radio astronomy and cosmology," *Science Progress* (U.K.) 53: 1–16.

———. (1966), "On the origin of the microwave background radiation," *Nature* 211: 277–79.

———. (1967), "Cosmology before and after quasars," *SA* 211 (September): 31–33.

Sciama, D. W. (1969), "The recent renaissance of observational cosmology," *Rivista NC* 1, numero speciale: 371–90.

———. (1971), *Modern Astronomy*. Cambridge: Cambridge University Press.

Sciama, D. W., and M. J. Rees (1966a), "Cosmological significance of the relation between red-shift and flux density for quasars," *Nature* 211: 1283.

———. (1966b), "Absorption spectrum of 3C 9," *Nature* 212: 1001–2.

Sciama, D. W., and W. C. Saslaw (1966), "Distribution of quasi-stellar radio sources in steady state cosmology," *Nature* 210: 348–52.

Scott, P. F. (1963), "The experimental evidence for Sciama's explanation of the log$N$/log$S$ curve," *MNRAS* 127: 37–43.

Scott, P. F., and M. Ryle (1961), "The number-flux density relation for radio sources away from the galactic plane," *MNRAS* 122: 389–97.

Scriven, M. (1954), "The age of the universe," *BJPS* 5: 181–90.

Sexl, R. U., and H. Urbantke (1967), "Cosmic particle creation processes," *Acta Physica Austriaca* 26: 339–56.

Shakeshaft, J. R. (1954), "The isotropic component of cosmic radio-frequency radiation," *PM* 45: 1136–44.

Shakeshaft, J. R., and A. S. Webster (1968), "Microwave background in a steady state universe," *Nature* 217: 339–40.

Shapley, H. (1929), "Note on the velocities and magnitudes of external galaxies," *PNAS* 15: 565–70.

Sharov, A. S., and I. D. Novikov (1993), *Edwin Hubble, the Discoverer of the Big Bang Universe*. Cambridge: Cambridge University Press.

Shirokov, M. F., and I. Z. Fisher (1963), "Isotropic space with discrete gravitational-field sources," *SAAJ* 6: 699–705.

Shklovsky, I. S. (1960), *Cosmic Radio Waves*. Cambridge, Mass.: Harvard University Press.

Shklovsky, I. S., and S. B. Pikel'ner (1961), "Concerning the article by F. Hoyle on 'radio-source problems,'" *SAAJ* 5: 146–47.

Silberstein, L. (1924), "The curvature of de Sitter's space-time derived from globular clusters," *MNRAS* 84: 363–66.

———. (1930), *The Size of the Universe*. Oxford: Oxford University Press.

Slipher, V. M. (1913), "The radial velocity of the Andromeda nebula," *Lowell Observatory Bulletin*, no. 58.

———. (1915), "Spectrographic observations of nebulae," *PA* 23: 21–24.

Smart, J. S. (1949), "The effects of nuclear stability on the formation of the chemical elements," *PR* 75: 1379–84.

Smirnov, Yu. N. (1965), "Hydrogen and He-4 formation in the prestellar Gamow universe," *SAAJ* 8: 864–73.

Smith, R. (1979), " The origins of the velocity-distance relation," *JHA* 10, 133–65.

———. (1982), *The Expanding Universe: Astronomy's 'Great Debate' 1900–1931*. Cambridge: Cambridge University Press.

———. (1990), "Edwin P. Hubble and the transformation of cosmology," *PT* 43 (April): 52–58.

Soddy, F. (1904), *Radio-Activity: An Elementary Treatise from the Standpoint of the Disintegration Theory*. London: The Electrician.

———. (1909), *The Interpretation of Radium*. London: John Murray.

Spencer Jones, H. (1934), *General Astronomy*. Cambridge: Cambridge University Press.

———. (1951–54), "Continuous creation," *Proceedings of the Royal Institution of Great Britain* 35: 36–47.

———. (1952), *Life on Other Worlds*. London: English Universities Press.

Speziali, P., ed. (1979), *Albert Einstein-Michele Besso: Correspondence 1903–1955*. Paris: Hermann.

Stark, J. (1902), *Die Elektrizität in Gasen*. Leipzig: Barth.

Stebbins, J. (1950), "The electrical photometry of stars and nebulae," *Observatory* 70: 203–8.

———. (1952), "Measuring starlight by photocell," *SA* 186 (March): 56–59.

Stebbins, J., and A. E. Whitford (1948), "Six-color photometry of stars. VI. The colors of extragalactic nebulae," *APJ* 108: 413–28.

Steensholt, G. (1932), "On the transmutation of elements in stars," *ZAP* 5: 140–52.

Steigman, G. (1969), "Antimatter and cosmology," *Nature* 224: 477–81.

Stern, O. (1925), "Über das Gleichgewicht zwischen Materie und Strahlung," *Zeitschrift für Electrochemie und angewandte physikalische Chemie* 31: 448–49.

———. (1926), "Über die Umwandlung von Atomen in Strahlung," *ZPC* 120: 60–62.

Sterne, T. E. (1933), "The equilibrium theory of the abundance of the elements: a statistical investigation of assemblies in equilibrium in which transmutations occur," *MNRAS* 93: 736–67.

Stewart, J. Q. (1931), "Nebular redshift and universal constants," *PR* 38: 2071.

Stone, S. B. (1930), "The origin of the chemical elements," *JPC* 34: 821–41.

Stoner, E. C. (1929), "Cosmic rays and a cyclic universe," *Proceedings of the Leeds Philosophical and Literary Society* 1: 349–55.

Stoops, R., ed. (1958), *La Structure et l'Évolution de l'Univers*. Brussels: Coudenberg.

Stothers, R. (1965), "Cosmic explosions, *Nature* 206: 82.

———. (1966), "Quasars as the origin of primordial matter in a steady-state universe," *MNRAS* 132: 217–23.

Strömgren, B. (1932), "The opacity of stellar matter and the hydrogen content of the stars," *ZAP* 4: 118–52.

Struve, O., and V. Zebergs (1962), *Astronomy of the 20th Century*. New York: Macmillan.

Stuewer, R. H. (1986), "Gamow's theory of alpha-decay." 147–86 in E. Ullmann-Margalit, ed. (1986), *The Kaleidoscope of Science*. Dordrecht: Reidel.

Suess, H. E. (1988), "V. M. Goldschmidt and the origin of the elements," *Applied Geochemistry* 3: 385–91.

Suess, H. E., and H. C. Urey (1956), "Abundances of the elements," *RMP* 28: 53–74.

Sullivan, W. T., ed. (1982), *Classics in Radio Astronomy*. Dordrecht: Reidel.

———. ed. (1984), *The Early Years of Radio Astronomy: Reflections Fifty Years after Jansky's Discovery*. Cambridge: Cambridge University Press.

———. (1990), "The entry of radio astronomy into cosmology: radio stars and Martin Ryle's 2C survey." 309–30 in B. Bertotti et al., eds. (1990).

Sutherland, W. (1899), "Cathode, Lenard and Röntgen rays," *PM* 47: 269–84.

Suzuki, S. (1928), "On the thermal equilibrium of dissociation of atom-nuclei," *PPMSJ* 10: 166–69.

———. (1929), "A further study on thermo-dissociation of atom-nuclei with a remark on Jeans' new theory of stellar evolution," *PPMSJ* 11: 119–38.

———. (1930), "Die obere Grenze der Energiedichte und einige thermodynamischen Schlussfolgerungen," *PZ* 31: 618–23.

Swann, W. F. G. (1961a), "Matter, anti-matter, and gravitation," *APJ* 133: 73–77.

Swann, W. F. G. (1961b), "An analysis of R. A. Lyttleton and H. Bondi's theory of 'the physical consequences of a general excess of charge'," *APJ* 133: 738–54.

Takeno, H. (1960–61), "Cosmology in terms of wave geometry," *LNC* 11: 21–32.

Tayler, R. J. (1966), "The origin of the elements," *RPP* 29: 490–538.

———. (1983), "The neutron in cosmology," *QJRAS* 24: 1–9.

———. (1990), "Neutrinos, helium and the early universe: a personal view," *QJRAS* 31: 371–75.

———. ed. (1987), *History of the Royal Astronomical Society. Vol. 2: 1920–1980.* Oxford: Blackwell Scientific.

Teller, E. (1948), "On the change of physical constants," *PR* 73: 801–2.

Temple, G. (1939), "Relativistic cosmology," *PPS* 51: 465–78.

Terzian, Y., and E. M. Bilson, eds. (1982), *Cosmology and Astrophysics: Essays in Honor of Thomas Gold.* Ithaca, N.Y.: Cornell University Press.

Thackeray, A. D., and A. J. Wesselink (1953), "Distances of the Magellanic Clouds," *Nature* 171: 693.

Thaddeus, P., and J. F. Clauser (1966), "Cosmic microwave radiation at 2.63 mm from observations of interstellar CN," *PRL* 16: 819–22.

Thomas, T. Y. (1966), "On the creation of mass in an expanding universe," *PNAS* 56: 1349–53.

Thomson, G. P. (1951), "Continuous creation and the edge of space," *New Republic* 124: 21–22.

Thomson, W. (1884), "On mechanical antecedents of motion, heat and light." 34–40 in *Mathematical and Physical Papers.* Vol. 2. Cambridge: Cambridge University Press.

Thorne, K. S. (1994), *Black Holes and Time Warps: Einstein's Outrageous Legacy.* New York: Norton.

Tipler, F. J. (1988), "Olbers's paradox, the beginning of creation, and Johann Mädler," *JHA* 19: 45–48.

Tipler, F. J., C. Clarke, and G. F. R. Ellis (1980), "Singularities and horizons, a review article." 97–206 in A. Held, ed. (1980), *General Relativity and Gravitation.* Vol. 2. New York: Plenum Press.

Tobey, R. C. (1971), *The American Ideology of National Science, 1919–1930.* Pittsburgh: University of Pittsburgh Press.

Tolman, R. C. (1922), "Thermodynamic treatment of the possible formation of helium from hydrogen," *JACS* 44: 1902–8.

———. (1926), "On the equilibrium between radiation and matter," *PNAS* 12: 670–74.

———. (1928), "On the energy and entropy of Einstein's closed universe," *PNAS* 14: 348–53.

———. (1929a), "On the possible line elements for the universe," *PNAS* 15: 297–304.

———. (1929b), "On the astronomical implications of the de Sitter line element for the universe," *APJ* 69: 245–74.

———. (1930a), "The effect of the annihilation of matter on the wave-length of light from the nebulae," *PNAS* 16: 320–27.

———. (1930b), "Discussion of various treatments which have been given to the non-static line element for the universe," *PNAS* 16: 582–94.

———. (1931a), "On the problem of the entropy of the universe as a whole," *PR* 37: 1639–60.

———. (1931b), "Nonstatic model of the universe with reversible annihilation of matter," *PR* 38: 797–814.

———. (1932), "Models of the physical universe," *Science* 75: 367–73.

———. (1934a), *Relativity, Thermodynamics, and Cosmology*. Oxford: Oxford University Press.

———. (1934b), "Effect on inhomogeneity in cosmological models," *PNAS* 20: 169–76.

———. (1934c), "Remarks on the possible failure of energy conservation," *PNAS* 20: 379–83.

———. (1949), "The age of the universe," *RMP* 21: 374–78.

Tolman, R. C., and M. Ward (1932), "On the behavior of non-static models of the universe when the cosmological term is omitted," *PR* 39: 835–43.

Toulmin, S. (1962), "Historical inference in science: geology as a model for cosmology," *The Monist* 47: 142–58.

Toulmin, S., and J. Goodfield (1982), *The Discovery of Time*. Chicago: University of Chicago Press.

Tropp, E. A., V. Ya. Frenkel, and A. D. Chernin (1993), *Alexander A. Friedmann: The Man who made the Universe Expand*. Cambridge: Cambridge University Press.

Truran, J. W., C. J. Hansen, and A. G. W. Cameron (1965), "The helium content of the galaxy," *Canadian Journal of Physics* 43: 1616–35.

Tucker, W., and R. Giacconi (1985), *The X-Ray Universe*. Cambridge, Mass.: Harvard University Press.

Turek, J. (1986), "Georges Lemaître and the Pontifical Academy of Sciences," *Vatican Observatory Publications* 2: 166–75.

Turville, C. de (1960), "A note on continuous creation," *APJ* 131: 741–42.

Unsöld, A. (1948), "Kernphysik und Kosmologie," *ZAP* 24B: 278–305.

Urani, J., and G. Gale (1993), "E. A. Milne and the origins of modern cosmology: an essential presence." 390–419 in J. Earman, M. Janssen, and J. D. Norton, eds. (1993).

Urbantke, H. (1992), "Schrödinger and cosmology." 453–59 in J. Eisenstaedt and A. J. Kox, eds. (1992).

Urey, H. C., and C. A. Bradley (1931), "On the relative abundances of isotopes," *PR* 38: 718–24.

Ursell, H. D. (1962), "Transfinite numbers and cosmology," *Nature* 196: 1015–16.

van Albada, G. B. (1947), "On the origin of the heavy elements," *APJ* 105: 393–405.

Van de Hulst, H. C. (1945), "Origin of the radio waves from space." 302–15 in W. T. Sullivan, ed. (1982).

Vaucouleurs, G. de (1948), "Sur la relation entre l'excès de couleur et le déplacement vers le rouge pour les nébuleuses extragalactiques," *CR* 227: 466–68.

Vecchierello, H. (1934), *Einstein and Relativity; Le Maitre and the Expanding Universe*. Paterson, N.J.: St. Anthony Guild Press.

Velo, G. (1963), "A selection rule for the production of pions in the steady-state theory," *NC* 29: 780–81.

Veronnet, A. (1926), *Constitution et Évolution de l'Univers*. Paris: Gaston Doin et Cie.

Vizgin, V. P., and G. E. Gorelik (1987), "The reception of the theory of relativity in Russia and the USSR." 265–326 in T. F. Glick, ed. (1987), *The Comparative Reception of Relativity*. Dordrecht: Reidel.

Wagoner, R. V. (1990), "Deciphering the nuclear ashes of the early universe: a personal perspective." 159–85 in B. Bertotti et al., eds. (1990).

Wagoner, R. V., W. A. Fowler, and F. Hoyle (1966), "Nucleosynthesis in the early stages of an expanding universe," *Science* 152: 677.

———. (1967), "On the synthesis of elements at very high temperatures," *APJ* 148: 3–49.

Walke, H. J. (1935), "Nuclear synthesis and stellar radiation," *PM* 19: 341–67.

Wataghin, G. (1944), "Statistical mechanics at extremely high temperatures," *PR* 66: 149–54.

Wataghin, G., and C. Lattes (1946), "On the abundance of nuclei in the universe," *PR* 69: 237.

Weart, S. R., and D. H. DeVorkin (1981), "Interviews as sources for history of modern astrophysics," *Isis* 72: 471–77.

Weinberg, S. (1977), *The First Three Minutes: A Modern View of the Origin of the Universe*. New York: Basic Books.

———. (1989), "The cosmological constant problem," *RMP* 61: 1–23.

Weiner, C. (1973), "Physics Today and the spirit of the forties," *PT* 25 (May): 23–28.

Weizsäcker, C. F. von (1937), *Die Atomkerne: Grundlagen und Anwendungen ihrer Theorie*. Leipzig: Akademische Verlagsgesellschaft.

———. (1938), "Über Elementumwandlungen im Innern der Sterne, II," *PZ* 39: 633–46.

———. (1939), "Der zweite Hauptsatz und der Unterschied von Vergangenheit und Zukunft," *AP* 36: 275–83.

———. (1943), "Über die Entstehung des Planetensystems," *ZAP* 22: 319–55.

———. (1948), "Kosmogonie." 413–26 in P. Ten Bruggencate et al., eds. (1948), *Astronomy, Astrophysics and Cosmogony* (Wiesbaden: Office of Military Government for Germany FIAT).

———. (1951a), "The evolution of galaxies and stars," *APJ* 114: 165–86.

———. (1951b), "Turbulence in interstellar matter." 158–64, 200–203 in *Problems of Cosmical Aerodynamics* (Dayton, Ohio: Central Air Documents Office).

———. (1964), *The Relevance of Science: Creation and Cosmogony*. New York: Harper and Row.

Wesson, P. S. (1978), *Cosmology and Geophysics*. Bristol: Adam Hilger.

Wetter, G. A. (1958), *Philosophie und Naturwissenschaft in der Sowjetunion*. Hamburg: Rowohlt.

Weyl, H. (1923a), *Raum, Zeit, Materie*, 5th ed. Berlin: Springer-Verlag.

———. (1923b), "Zur allgemeinen Relativitätstheorie," *PZ* 24: 230–32.

———. (1924), "Observations on the note of Dr L. Silberstein," *PM* 48: 348–49.

———. (1930), "Redshift and cosmology," *PM* 9: 936–43.

Weymann, R. J. (1967), "The cosmic microwave radiation," *ASPL*, no. 461.

Wheeler, J. A. (1962), "The universe in the light of general relativity," *The Monist* 47: 40–76.

Wheeler, J. A., and R. P. Feynman (1945), "Interaction with the absorber as the mechanism of radiation," *RMP* 17: 157–81.

———. (1949), "Classical electrodynamics in terms of direct interparticle action," *RMP* 21: 425–33.

White, M., and J. Gribbin (1988), *Stephen Hawking: A Life in Science*. London: Penguin Books.

Whitehead, A. N. (1929), *Process and Reality: An Essay in Cosmology*. New York: Macmillan.

Whitford, A. E. (1954), "Observational status of the color-excess effect in distant galaxies," *APJ* 120: 599–602.

———. (1986), "A half-century of astronomy," *ARAA* 24: 1–22.

Whitrow, G. J. (1949), *The Structure of the Universe: An Introduction to Cosmology*. London: Hutchison.

———. (1953), "A query concerning the steady-state theory of the homogeneous expanding universe," *Observatory* 73: 205–6.

———. (1954a), "The orthodox theory of the expanding universe," *ONRAS*, no. 17: 81–86.

———. (1954b), "The age of the universe," *BJPS* 5: 215–25.

———. (1959), *The Structure and Evolution of the Universe*. New York: Harper.

———. (1962), "Is the physical universe a self-contained system?" *The Monist* 47: 77–93.

———. (1978), "Theoretical cosmology in the twentieth century." 576–93 in E. G. Forbes, ed. (1978), *Human Implications of Scientific Advance. Proceedings of the XVth International Congress of the History of Science*. Edinburgh: Edinburgh University Press.

———. (1980), *The Natural Philosophy of Time*. Oxford: Clarendon Press.

Whitrow, G. J., and H. Bondi (1954), "Is physical cosmology a science?" *BJPS* 4: 271–83.

Whittaker, E. T. (1935), "Gauss' theorem and the concept of mass in general relativity," *PRS* 149: 384–95.

———. (1943), *The Beginning and End of the World*. London: Oxford University Press.

———. (1946), *Space and Spirit: Theories of the Universe and the Arguments for the Existence of God*. London: Nelson and Sons.

———. (1960), *A History of the Theories of Aether and Electricity*. New York: Harper.

Wiechert, E. (1921), "Der Äther im Weltbild der Physik," *Nachrichten von der königlichen Gesellschaft der Wissenschaften zu Göttingen, Math.-Phys. Klasse*. no. 1: 29–70.

Wilkinson, D. T., and P. J. E. Peebles (1983), "Discovery of the 3K radiation." 175–84 in K. Kellermann and B. Sheets, eds. (1983), *Serendipitous Discoveries in Radio Astronomy*. Green Bank, W.Va.: National Radio Astronomy Observatory.

Williamson, R. E. (1951), "Fred Hoyle's universe," *JRASC* 45: 185–89.

Wilson, A. H. (1931), "The transmutation of elements in stars," *MNRAS* 91: 283–90.

Wilson, R. W. (1979), "The cosmic microwave background radiation," *Science* 205: 866–74.

Wolfe, A. M., and G. R. Burbidge (1969), "Discrete source models to explain the microwave background radiation," *APJ* 156: 345–71.

Wynne, B. (1979), "Physics and psychics: science, symbolic action, and social control in late Victorian England." 167–89 in B. Barnes and S. Shapin, eds. (1979), *Natural Order: Historical Studies of Scientific Culture*. Beverly Hills, Calif.: Sage.

Yourgrau, W., and A. D. Breck, eds. (1977), *Cosmology, History, and Theology*. New York: Plenum Press.

Zaykoff, R. (1933), "Zur relativistichen Kosmogonie," *ZAP* 6: 128–37, 193–97.

Zel'dovich, Ya. B. (1963a), "Problems of present-day physics and astronomy," *SPU* 5: 931–50.

———. (1963b), "Prestellar state of matter," *JETP* 16: 1102–3.

———. (1964), "The theory of the expanding universe as originated by A. A. Friedman," *SPU* 6: 475–94.

———. (1965), "Survey of modern cosmology," *Advances in Astronomy and Astrophysics* 3: 241–379.

Zel'dovich, Ya. B., and I. D. Novikov (1967), "The uniqueness of the interpretation of isotropic cosmic radiation with T = 3° K," *SAAJ* 11: 526–27.

———. (1983), *Relativistic Astrophysics. II: The Structure and Evolution of the Universe.* Chicago: University of Chicago Press.

Zorn, J. C., G. E. Chamberlain, and V. W. Hughes (1963), "Experimental limits for the electron-proton charge difference and for the charge of the neutron," *PR* 129: 2566–76.

Zupančič, A. O. (1965), "Creation rate of matter and the Heisenberg uncertainty principle," *Nature* 207: 279.

Zwicky, F. (1928), "On the thermodynamic equilibrium in the universe," *PNAS* 14: 592–97.

———. (1929), "On the red shift of spectral lines through interstellar space," *PNAS* 15: 773–79.

———. (1953), "Neue Methoden der kosmologischen Forschung," *HPA* 26: 241–54.

———. (1966), *Entdecken, Erfinden, Forschen im Morphologischen Weltbild.* Munich: Droemersche Verlagsanstalt Th. Knaur.